The Urban Book Series

Aims and Scope

The Urban Book Series is a resource for urban studies and geography research worldwide. It provides a unique and innovative resource for the latest developments in the field, nurturing a comprehensive and encompassing publication venue for urban studies, urban geography, planning and regional development.

The series publishes peer-reviewed volumes related to urbanization, sustainability, urban environments, sustainable urbanism, governance, globalization, urban and sustainable development, spatial and area studies, urban management, urban infrastructure, urban dynamics, green cities and urban landscapes. It also invites research which documents urbanization processes and urban dynamics on a national, regional and local level, welcoming case studies, as well as comparative and applied research.

The series will appeal to urbanists, geographers, planners, engineers, architects, policy makers, and to all of those interested in a wide-ranging overview of contemporary urban studies and innovations in the field. It accepts monographs, edited volumes and textbooks.

More information about this series at http://www.springer.com/series/14773

Yuji Murayama · Courage Kamusoko
Akio Yamashita · Ronald C. Estoque
Editors

Urban Development in Asia and Africa

Geospatial Analysis of Metropolises

Springer

Editors
Yuji Murayama
Faculty of Life and Environmental Sciences
University of Tsukuba
Tsukuba, Ibaraki
Japan

Akio Yamashita
Faculty of Life and Environmental Sciences
University of Tsukuba
Tsukuba, Ibaraki
Japan

Courage Kamusoko
Asia Air Survey Co., Ltd.
Kawasaki, Kanagawa
Japan

Ronald C. Estoque
Faculty of Life and Environmental Sciences
University of Tsukuba
Tsukuba, Ibaraki
Japan

ISSN 2365-757X ISSN 2365-7588 (electronic)
The Urban Book Series
ISBN 978-981-10-3240-0 ISBN 978-981-10-3241-7 (eBook)
DOI 10.1007/978-981-10-3241-7

Library of Congress Control Number: 2016961661

Printed on acid-free paper

This Springer imprint is published by Springer Nature
The registered company is Springer Nature Singapore Pte Ltd.
The registered company address is: 152 Beach Road, #21-01/04 Gateway East, Singapore 189721, Singapore

Preface

This book examines the urban growth trends and patterns of various metropolitan regions in Asia and Africa from a geographical perspective. State-of-the-art geospatial tools and techniques from the geographic information systems (GIS) and science, remote sensing, and machine learning disciplines were used for the land change analysis. In addition to the empirical results, the methodological approaches employed and discussed in this book showcase the potential of geospatial analysis (e.g., land change modeling) for improving our understanding of the trends and patterns of urban growth in Asia and Africa. Furthermore, given the complexity of the urban growth process across the world, issues raised in this book will contribute to the improvement of future geospatial analysis of urban growth in the developing regions. This book is written for researchers, academics, practitioners, and graduate students. The inclusion of the origin and brief history of each of the selected metropolitan regions, including the analysis of their urban primacy, spatiotemporal patterns of urban land use/cover changes, driving forces of urban development, and implications for future sustainable development, makes the book an important reference for various related studies.

Most of the contributors to this book are affiliated with the Division of Spatial Information Science, University of Tsukuba, Japan. The division, which was established in 2000 to include geographical information science within the doctoral program in geoenvironmental sciences, provides an enabling research environment where faculty members, staff, and students work together to advance knowledge in GIS and remote sensing techniques in different areas of interest.

Our sincere thanks go to the staff members of the Division of Spatial Information Science, University of Tsukuba, especially to Mr. Hao Hou, Mr. Hao Gong, Mr. Matamyo Simwanda, Mr. Shyamantha Subasinghe, and Mr. Xinmin Zhang.

Finally, we would like to thank the Japan Society for the Promotion of Science which financially supported our research work through Grant-in-Aid for Scientific Research B (No. 26284129, 2014–16, Representative: Yuji Murayama) and Grant-in-Aid for Research Activity Start-Up (No. 15H06067, 2015–16, Representative: Ronald C. Estoque).

Tsukuba, Japan
September 2016

Yuji Murayama
Courage Kamusoko
Akio Yamashita
Ronald C. Estoque

Contents

Editors and Contributors

About the Editors

Yuji Murayama is a professor at the Division of Spatial Information Science, Faculty of Life and Environmental Sciences, University of Tsukuba, Japan. His expertise and fields of interest include GIS, spatial analysis, urban geography, and transportation geography. Publications: Murayama Y (ed) (2012) Progress in geospatial analysis. Springer, Tokyo, 291 pp. Murayama Y, Thapa RB (eds) (2011) Spatial analysis and modeling in geographical transformation process: GIS-based application. Springer, Dordrecht, 300 pp. Kamusoko C, Mundia CN, Murayama Y (eds) (2011) Recent advances in remote sensing and GIS in Sub-Sahara Africa. Nova Publishers, New York, 211 pp. Murayama Y, Du G (eds) (2005) Cities in global perspective: diversity and transition. College of Tourism, Rikkyo University with IGU Commission, Tokyo, 626 pp. Murayama Y (2000) Japanese urban system. Kluwer, Dordrecht, 271 pp.

Courage Kamusoko is a researcher at Asia Air Survey, Japan. His expertise includes land use/cover change modeling, and the design and implementation of geospatial database management systems. His primary research involves analyses of remotely sensed images, land use/cover modeling, and machine learning. In addition to his focus on geospatial research and consultancy, he has dedicated time to teaching practical machine learning for geospatial analysis and modeling. Publication: Kamusoko C, Mundia CN, Murayama Y (eds) (2011) Recent advances in remote sensing and GIS in Sub-Sahara Africa. Nova Publishers, New York, 211 pp.

Akio Yamashita is an assistant professor at the Division of Regional Geography, Faculty of Life and Environmental Sciences, University of Tsukuba, Japan. His expertise includes geography, GIS, and study of urban environmental issues. Papers published: Yamashita A (2014) Aspects of water environmental issues in Jakarta due to its rapid urbanization. Tsukuba Geoenvironmental Sci 10:43–50. Yamashita A (2011) Comparative analysis on land use distributions and their changes in Asian mega cities. In: Taniguchi M (ed) Groundwater and subsurface environments: human impacts in Asian coastal cities. Springer, pp 61–81.

Ronald C. Estoque is a researcher at the Faculty of Life and Environmental Sciences, University of Tsukuba, Japan. His research interests include the applications of geospatial technologies such as remote sensing and GIS, as well as social-ecological approaches, for landscape sustainability studies. One of his most recent research articles, entitled "Quantifying landscape pattern and ecosystem service value changes in four rapidly urbanizing hill stations of Southeast Asia," is published in *Landscape Ecology* (2016), 31, 1481–1507. His other major research articles are published in other international peer-reviewed journals, such as *Cities*, *ISPRS International Journal of Geo-Information*, *Applied Geography*, *Landscape and Urban Planning*, *AMBIO*, *GIScience & Remote Sensing*, *Geocarto International*, *Ecological Indicators*, *Land Use Policy*, *and Science of the Total Environment*.

Contributors

Niloofar Haji Mirza Aghasi Graduate School of Life and Environmental Sciences, University of Tsukuba, Tsukuba, Japan

Chiaki M. Akiyama National Institute for Environmental Studies, Tsukuba, Japan

Enos Chikati Department of Environmental Sciences, University of South Africa, Pretoria, South Africa

Ronald C. Estoque Faculty of Life and Environmental Sciences, University of Tsukuba, Tsukuba, Japan

Courage Kamusoko Asia Air Survey Co., Ltd., Kawasaki, Japan

Syeda Khaleda Department of Disaster Management, Government of Bangladesh, Dhaka, Bangladesh

Duong Dang Khoi Hanoi University of Natural Resources and Environment, Hanoi, Vietnam

Qazi Azizul Mowla Department of Architecture, Bangladesh University of Engineering and Technology, Dhaka, Bangladesh

Charles N. Mundia Institute of Geomatics, GIS and Remote Sensing, Dedan Kimathi University of Technology, Nyeri, Kenya

Kondwani Godwin Munthali Computer Science Department, Chancellor College, University of Malawi, Zomba, Malawi

Yuji Murayama Faculty of Life and Environmental Sciences, University of Tsukuba, Tsukuba, Japan

Tabukeli M. Ruhiiga Department of Geography and Environmental Sciences, North West University, Potchefstroom, South Africa

Rajesh Bahadur Thapa Geospatial Solutions, International Centre for Integrated Mountain Development, Khumaltar, Lalitpur, Nepal; Earth Observation Research Center, Japan Aerospace Exploration Agency (JAXA), Tsukuba, Ibaraki, Japan

Akio Yamashita Faculty of Life and Environmental Sciences, University of Tsukuba, Tsukuba, Japan

Part I
Introduction

Chapter 1
Importance of Remote Sensing and Land Change Modeling for Urbanization Studies

Courage Kamusoko

Abstract Remote sensing analysis and land change modeling provide valuable insights into urban land use/cover changes and growth processes at multiple spatial and temporal scales. This chapter briefly outlines the importance of remote sensing, and land change modeling for urbanization studies in selected countries in Africa and Asia. The methodological approaches discussed in this book showcase the potential of remote sensing and land change modeling analysis in order to improve understanding of urban growth in Africa and Asia. Given the complexity of urban growth processes globally, issues raised in this book will contribute to the improvement of future land use/cover change analysis and modeling, particularly in the developing country context. The geospatial analysis approach based on remote sensing and land change modeling provides a synoptic view of urbanization in Africa and Asia.

1.1 Introduction

According to the United Nations (2015), approximately 54% of the world's population currently lives in urban areas. It is estimated that continuing urbanization will add 2.5 billion people to the world's urban population by 2050, of which 90% of the increase will be concentrated in Asia and Africa (Masser 2001; United Nations 2006, 2012, 2015). While 40 and 48% of the population in Africa and Asia reside in urban areas, urban population is expected to increase to 56 and 64% in these regions by 2050 (United Nations 2015). In addition, the fastest-growing urban agglomerations, which are medium-sized cities and cities with less than 1 million inhabitants will be located in Asia and Africa (United Nations 2015).

The rapid increase in urban population and urbanization poses a number of challenges to planners and policy makers (Yuan et al. 2005; Pacione 2007).

C. Kamusoko (✉)
Asia Air Survey Co., Ltd, Kawasaki, Japan
e-mail: kamas72@gmail.com

Y. Murayama et al. (eds.), *Urban Development in Asia and Africa*,
The Urban Book Series, DOI 10.1007/978-981-10-3241-7_1

For example, most urban areas in developing Africa and Asia are confronted with problems such as failure to provide services to the growing urban population, increasing rural–urban migration, proliferation of informal settlements and epidemics, as well as environmental degradation (Rakodi 1995; Brown 2001). While rapid urbanization is expected to exacerbate these problems, experiences from the developed countries show that urbanization has the potential to boost national economic growth (Collier 2016). Therefore, increasing urbanization presents Africa and Asia with an opportunity to improve the quality of life and social well-being. In order to ensure that urbanization produces optimal economic growth levels, African and Asian countries need to formulate smart and sustainable urban development strategies that can guide socioeconomic development (Collier 2016). This requires accurate, consistent, and timely geospatial information on urbanization trends in order to assess current and future urban growth (Herold et al. 2002). Equally important, geospatial information will be useful for setting policies that promote inclusive and equitable urban, environmental, and socioeconomic development (United Nations 2015).

1.2 Application of Remote Sensing for Urban Land Use/Cover Mapping

The past decades have witnessed the rise of sustainable urban development and smart growth initiatives, particularly in developed countries (Herold et al. 2003, 2005). However, the implementation of sustainable Urban development and smart growth initiatives have been lagging behind in most parts of Africa and Asia due to a number of factors. Chief among these factors is the lack of clear and practical urban planning due to the dearth of geospatial information (International Federation of Surveyors 2010). According to the International Federation of Surveyors (2010), 70% of urban growth in developing countries is not planned. Efforts to produce or update existing urban geospatial information for planning purposes have been hampered by high cost of acquiring geospatial data, especially from conventional land use surveys or aerial photography (Conitz 2000). Nonetheless, increases in the use of remote sensing technology (e.g., high and medium-resolution satellite data) for mapping urban land use/cover have been noted in the past decades (Ward et al. 2000; Guindon et al. 2004; Yuan et al. 2005; Lu and Weng 2007). This is because high and medium-resolution satellite remotely sensed data such as Ikonos, Quickbird, WorldView, Landsat satellite series, Systeme Pour l'Observation de Ia Terre (SPOT), Terra-1 ASTER, RapidEye, and Advanced Land Observing Satellite (ALOS) have relatively good global coverage. It is recognized that some of the high and medium-resolution satellite data have coarse spatial resolution, which limits detailed urban planning. However, the high and medium-resolution satellite data is useful for identifying and mapping land use/cover patterns in urban landscapes at a regional scale.

High and medium-resolution earth observation satellite data have been successfully used to map and monitor urban growth in developed countries (Lo and Choi 2004; Yuan et al. 2005; Bagan and Yamagata 2012). Furthermore, a variety of classification techniques have also been developed and used to classify urban land use/cover. These classification techniques include: the incorporation of structural and textural information (Gong and Howarth 1990; Moller-Jensen 1990); combining satellite images with ancillary data (Harris and Ventura 1995); vegetation—impervious surface—soil models (Ridd 1995); expert systems (Stefanov et al. 2001); hybrid methods that incorporate soft and hard classifications (Lo and Choi 2004); the use of built-up indices (Zha et al. 2003; Xu 2007; Estoque and Murayama 2015); neural networks (Seto and Liu 2003); segmentation and object-based classifications (Guindon et al. 2004); linear spectral mixture analysis (Phinn et al. 2002; Wu and Murray 2003); support vector machines (Ghosh et al. 2014) and random forests (Cao et al. 2009). While significant urban land use/cover classification and urban growth monitoring in the developed countries have been noted, urban landscapes are still poorly quantified in Africa and Asia despite their rapid urbanization. This is attributed to the fact that urban land use/cover classification still poses a number of challenges—such as spectral confusion and mixed pixels—given the heterogeneous nature of the urban landscapes coupled with the relatively small spatial size of surficial materials (Foody 2000; Masser 2001; Stefanov et al. 2001; Alpin 2003; Xian and Crane 2005). For example, in most African urban landscapes, spectral confusion is a problem because gravel (dirt) roads in informal settlements or slums areas have similar spectral responses to those of bare vacant plots and agriculture fields leading to inaccurate land use/cover mapping. In this book, a random forests classification approach is used to classify built-up and non-built-up areas based on Landsat imagery.

1.3 Developments in Urban Land Change Modeling

The improvements in remote sensing technology during the past 40 years, combined with developments in Geographic Information Systems (GIS) have provided an opportunity to advance urban land change modeling (The State of Land Change Modeling 2014). While many land change models (LCMs) have been developed to examine and simulate urban growth, this book focuses on machine learning and cellular automata-based LCMs. Therefore, this chapter will briefly review some of the major milestones in the development of urban LCMs.

The application of urban models in the context of planning can be traced back to von Thünen's agricultural location model, Weber's industrial location models, and Christaller's central place theory (Liu 2009). Later on, static urban growth and land use pattern models such as Burgess's concentric zone model, Hoyt's sector model, and Harry and Ullmans's multiple nuclei model were developed (Liu 2009). However, interest in modeling in general, and urban growth modeling in particular grew during the "quantitative revolution" of the 1950s and 1960s (Liu 2009).

During this period urban-scale models were generated by heuristic techniques for forecasting (Briassoulis 2000). For example, the Chicago Area Transportation Study (CATS) model used the development capacity concept based on historical information on population densities and vacant land to forecast land use (Hamburg and Creightan 1959). With the emergence of computer simulation techniques in the early 1960s, urban-scale LCMs such as the California Urban Futures Model (CUFM), and the integrated land use/transportation models were further developed (Briassoulis 2000).

The California Urban Futures Model (CUFM) was developed by Landis (1994, 1995). It provided a spatially explicit integrated model of the housing market, which was used to analyze various policies as well as to incorporate the environmental variability of a study area. While the CUFM assumed profit maximizing land developers, it lacked robust theoretical foundations in economics and land development (Briassoulis 2000). Second, the model did not take into account the interaction of land use with the transportation network. Third, the model did not include feedbacks from development or excess demand on housing prices. Fourth, the model did not deal with the allocation of other uses such as industrial and commercial. In essence, the CUFM ignored the important driving influence of the location decisions (Briassoulis 2000).

The integrated land use/transportation models were developed in the 1970s and 1980s. Their purpose was to model land use/cover and transportation interactions. More importantly, they were used to analyze the spatial land use impacts of changes in the transportation system (Briassoulis 2000). For example, the Integrated Transportation and Land Use Package (ITLUP) was implemented to link urban land use and transportation models (Putman 1983). More models such as the transportation and land use (TRANUS) model (de la Barra et al. 1984, 1989), the Integrated Land Use, Transportation, Environment (ILUTE) modelling systems were developed. Although the integration between the urban land use systems and transportation have been achieved, criticism on urban growth modeling intensified in the 1970s and 1980s because the models emphasized modeling techniques and lacked theoretical underpinnings (Liu 2009).

However, innovations in GIS and the availability of GIS data (e.g., remotely sensed-derived land use/cover maps) inspired a new wave of developments in urban growth modeling. This was also supported by rapid advancement in computer technology coupled with the decrease in the cost of computer hardware. In addition, developments in spatial, natural, and social sciences concerning bottom-up, dynamic and flexible self-organizing modeling systems complemented by theories that emphasize locally made decision to give rise to global patterns led to application of cellular automata (CA) models for urban growth modeling (Tobler 1979; Wolfram 1984; Couclelis 1985; Engelen 1988; Batty 1998; Wu and Webster 1998). Cellular automata models were originally conceived by Ulam and Von Neumann in the 1940s to provide a formal framework for investigating the behavior of complex and self-reproducible systems (White and Engelen 1993). Generally, CA models are dynamic and discrete space and time systems. Time progresses in discrete steps and all cells change their state simultaneously as a function of their own state, together with the state of the cells in their neighborhood according to a specified set of transition rules (White and Engelen 1993).

The application of CA models for urban growth modeling offered a flexible platform to integrate GIS data at multiple temporal and spatial scales. This is because CA approaches could easily represent complex patterns using simple rules (White and Engelen 1997). According to Torrens (2003), CA models were suitable to simulate urban systems since land use can be presented as a cell. More importantly, CA models allowed the inclusion of urban theory (e.g., spatial interaction) considering that cities exhibit several characteristics of complexity such as fractal dimensionality and self-similarity across scales, self-organization, and emergence (Torrens 2002). While the past decades have witnessed the development and application of many urban CA land change models (Pijanowski et al. 2005; Mundia and Aniya 2007; Yeh and Li 2009), most of the models have not been adopted by urban planners and policy makers (Sante et al. 2010). Nevertheless, urban CA models have provided deep insights into urban growth dynamics as well as "laboratories" to explore "what if" urban growth scenarios.

1.4 Summary of Book Chapters

As we will see from the various examples provided in this book, urbanization in Africa and Asia is creating metropolitan areas whose boundaries are constantly changing beyond the defined administrative boundaries. Therefore, it is critical to map land use/cover in an accurate, consistent and timely manner in order to understand the constantly evolving urban spatial developments beyond the defined formal city administrative boundaries. The combined methodological approach—based on remote sensing, spatial metrics and LCMs—adopted in this book has great potential to improve urban land use/cover mapping and modeling, particularly in complex urban areas in Africa and Asia.

This book is organized into four main parts. Part I presents the introduction, which is covered in Chaps. 1–3. This chapter outlines the importance of remote sensing and LCMs for urbanization studies, while Chap. 2 describes the overall methodological framework used to map, analyze and model land use/cover changes. Chapter 3 provides an overview of the rapid urbanization in Africa and Asia.

Part II covers Chaps. 4–12 and focuses on the major cities in Asia, including Beijing, Manila, Jakarta, Hanoi, Bangkok, Yangon, Dhaka, Kathmandu, and Tehran metropolitan areas. Part III covers Chaps. 13–18 and focuses on the major cities in Africa, including Dakar, Bamako, Nairobi, Lilongwe, Harare, and Johannesburg metropolitan areas. These chapters trace the origin and brief history of each of the metropolitan areas. The urban primacy of these cities and the spatiotemporal patterns and changes of their urban land use/cover are examined. The factors driving their urban development, as well as the implications of the observed and projected urban land use/cover changes for future sustainable urban development, are discussed.

Finally, Part IV provides the overall book summary and conclusions. More specifically, Chap. 19 focuses on the comparative analysis of the trends and spatial patterns of urbanization in Africa and Asia. Chapter 20 discusses the future of metropolitan areas in developing Africa and Asia.

References

Aplin P (2003) Comparison of simulated IKONOS and SPOT HRV imagery for classifying urban areas. In: Mesev V (ed) Remotely sensed cities. Taylor and Francis, London and New York, pp 23–45

Bagan H, Yamagata Y (2012) Landsat analysis of urban growth: how Tokyo became the world's largest megacity during the last 40 years. Remote Sens Environ 127:210–222

Batty M (1998) Urban evolution on the desktop: simulation with the use of extended cellular automata. Environ. Plann. B. 30:1943–1967

Briassoulis H (2000) Analysis of land use change: theoretical and modeling approaches. Accessed on 14 May 2005 from www.rri.wvu.edu/WebBook/Briassoulis/contents.html

Brown A (2001) Cities for the urban poor in Zimbabwe: urban space as a resource for sustainable development. Develop Pract 11:263–281

Cao X, Chen J, Imura H, Higashi O (2009) A SVM-based method to extract urban areas from DMSP-OLS and SPOT VGT data. Remote Sens Environ 113:2205–2209

Collier P (2016) African urbanization: an analytic policy guide. Accessed on 26 Feb 2016 from http://www.theigc.org/wp-content/uploads/2016/01/African-UrbanizationJan2016_Collier_Formatted-1.pdf

Conitz MW (2000) GIS applications in Africa: introduction. Photogram Eng Remote Sens 66:672–673

Couclelis H (1995) Cellular worlds: a framework for modeling micro-macro dynamics. Environ. Plann. A 17:585–596

de la Barra T (1989) Integrated land use and transport modeling. Cambridge University Press, Cambridge

de la Barra T, Perez B, Vera N (1984) TRANUS-J: putting large models into small computers. Environ Plann B 11:87–101

Engelen G (1988) The theory of self-organization and modeling complex urban systems. Eur. J. Oper. Res. 37: 42–57

Estoque RC, Murayama Y (2015) Classification and change detection of built-up lands from Landsat-7 ETM+ and Landsat-8 OLI/TIRS imageries: a comparative assessment of various spectral indices. Ecol Ind 56:205–217

Foody GM (2000) Estimation of sub-pixel land cover composition in the presence of untrained classes. Comput Geosci 26:469–478

Ghosh A, Richa Sharma R, Joshi PK (2014) Random forest classification of urban landscape using Landsat archive and ancillary data: combining seasonal maps with decision level fusion. Appl Geogr 48:31–41

Gong P, Howarth PJ (1990) The use of structural information for improving land-cover classification accuracies at the rural-urban fringe. Photogram Eng Remote Sens 56:67–73

Guindon B, Zhang Y, Dillabaugh C (2004) Landsat urban mapping based on a combined spectral-spatial methodology. Remote Sens Environ 92:218–232

Harris PM, Ventura SJ (1995) The integration of geographic data with remotely sensed imagery to improve classification in urban area. Photogram Eng Remote Sens 61:993–998

Hamburg JR, Creighton RL (1959) Predicting Chicago's land use pattern. J Am Inst Plan 25:67–72

Herold M, Scepan J, Clarke KC (2002) The use of remote sensing and landscape metrics to describe structures and changes in urban land uses. Environ Plan A 34:1443–1458

Herold M, Goldstein NC, Clarke KC (2003) Spatiotemporal form of urban growth: measurement, analysis and modeling. Remote Sens Environ 86:286–302

Herold M, Couclelis H, Clarke KC (2005) The role of spatial metrics in the analysis and modelling of urban land use change. Comput Environ Urban Syst 29:369–399

International Federation of Surveyors (2010) Rapid urbanization and mega cities: the need for spatial information management. Research study FIG Commission 3

Landis J (1994) The California urban futures model: a new generation of metropolitan simulation models. Environ Plann B 21:399–420

Landis J (1995) Imagining land use futures: applying the California urban futures model. J Am Plann Assoc 61:438–457

Liu Y (2009) Modelling urban development with geographical information systems and cellular automata. CRC Press, Taylor & Francis Group, New York

Lo CP, Choi J (2004) A hybrid approach to urban land use/cover mapping using landsat 7 enhanced thematic mapper plus (ETM+) images. Int J Remote Sens 25:2687–2700

Lu D, Weng Q (2007) A survey of image classification methods and techniques for improving classification performance. Int J Remote Sens 28:823–870

Masser I (2001) Managing our urban future: the role of remote sensing and geographic information systems. Habitat Int 25:503–512

Moller-Jensen L (1990) Knowledge-based classification of classification of an urban area using texture and context information in Landsat-TM imagery. Photogram Eng Remote Sens 56:899–904

Mundia CN, Aniya M (2007) Modeling urban growth of Nairobi city using cellular automata and geographical information systems. Geogr Rev Jpn 80:777–788

Pacione M (2007) Sustainable urban development in the UK: rhetoric or reality? Geography 92:246–263

Phinn S, Stanford M, Scarth P, Murray AT, Shyy PT (2002) Monitoring the composition of urban environments based on the vegetation-impervious surface-soil (VIS) model by subpixel analysis techniques. Int J Remote Sens 23:4131–4153

Pijanowski BC, Pithadia S, Shellito BA, Alexandridis K (2005) Calibrating a neural network-based change model for two metropolitan areas of the Upper Midwest of the United States. Int J Geogr Inf Sci 19:197–215

Putman SH (1983) Integrated urban models. Pion, London

Rakodi C (1995) Harare—inheriting a settler-colonial city: change or continuity?. Wiley, Chichester, UK

Ridd K (1995) Exploring a V-I-S (vegetation-impervious surface-soil) model for urban ecosystem analysis through remote sensing: comparative anatomy for cities. Int J Remote Sens 16:2165–2185

Sante I, Garcia AM, Miranda D, Crecente R (2010) Cellular automata models for the simulation of real-world urban processes: a review and analysis. Landscape Urban Plann 96:108–122

Seto KC, Liu W (2003) Comparing ARTMAP neural network with the maximum-likelihood classifier for detecting urban change. Photogram Eng Remote Sens 69:981–990

Stefanov WL, Ramsey MS, Christensen PR (2001) Monitoring urban land cover change: an expert system approach to land cover classification of semiarid to arid centers. Remote Sens Environ 77:173–185

The State of Land Change Modeling (2014) Advancing land change modeling: opportunities and research requirements. The National Academies Press, Washington, DC

Tobler W (1979) Cellular Geography. In: Gale S, Olsson G (eds). Philosophy in Geography, Reidel, Dordrecht pp 379–386

Torrens PM (2002) Cellular automata and multiagent systems as planning support tools. In: Geertman S, Stillwell J (eds) Planning support systems in practice, Springer, London, pp 208–222

Torrens P (2003) Automata-based models of urban systems. In: Longley PA, Batty M (eds) Advanced spatial analysis: the CASA book of GIS. ESRI Press, Redlands, CA, pp 61–81
UN (2010) World urbanization prospects: the 2009 revision. Highlights. United Nations Population Division, New York
United Nations (2006) State of the world's cities 2006/7. Accessed on 20 September 2008 from http://www.unhabitat.org/content.asp?cid=3397&catid=7&typeid=46&subMenuld=0
United Nations (2012) World urbanization prospects: the 2011 revision. Accessed on 25 July 2015 from http://esa.un.org/unpd/wup/index.htm
United Nations (2015) World urbanization prospects: the 2014 revision. Highlights (ST/ESA/SER. A/352). Accessed on 28 Sept 2015 from http://esa.un.org/unpd/wup/Highlights/WUP2014-Highlights.pdf
Ward D, Phinn SR, Murray AT (2000) Monitoring growth in rapidly urbanizing areas using remotely sensed data. Prof Geogr 52:371–386
White R, Engelen G (1993) Cellular automata and fractal urban form: a cellular modeling approach to the evolution of urban land-use patterns. Environ Plann A 25:1175–1199
White R, Engelen G (1997) Cellular automata as the basis of integrated dynamic regional modeling. Environ. Plann. B. 24:235–246
Wolfram S (1984) Cellular automata as models of complexity. Nature 311:419–424
Wu C, Murray AT (2003) Estimating impervious surface distribution by spectral mixture analysis. Remote Sens Environ 84:93–505
Wu F, Webster CJ (1998) Simulation of land development through the integration of cellular automata and multicriteria evaluation. Environ. Plann. B. 25:103–126
Xian G, Crane M (2005) Assessments of urban growth in the Tampa Bay watershed using remote sensing data. Remote Sens Environ 97:203–215
Xu H (2007) Extraction of urban built-up land features from Landsat imagery using a thematic-oriented index combination technique. Photogram Eng Remote Sens 73:1381–1391
Yeh AGO, Li X (2009) Cellular automata and GIS for urban planning. In: Madden M (ed) Manual of geographic information systems. American Society for Photogrammetry and Remote Sensing, Bethesda, MD, USA, pp 591–619
Yuan F, Saway KE, Loeffelholz BC, Bauer ME (2005) Land cover classification and change analysis of the Twin Cities (Minnesota) metropolitan area by multitemporal Landsat remote sensing. Remote Sens Environ 98:317–328
Zha Y, Gao J, Ni S (2003) Use of normalized difference built-index in automatically mapping urban areas from TM imagery. Int J Remote Sens 24:583–594

Chapter 2
Methodology

Courage Kamusoko

Abstract Remote sensing, GIS, and land change models (LCMs) are critical for mapping urban land use/cover and simulating "what if" urban growth scenarios, particularly in developing countries experiencing rapid urbanization. The purpose of this chapter is to describe briefly the methodology used to produce land use/cover maps, and simulate land use/cover changes for selected metropolitan areas in Asia and Africa. Land use/cover maps were classified from Landsat imagery for 1990, 2000, 2010, and 2014 using the random forest (RF) classifier. Quantitative accuracy assessment was not conducted for the 1990 land use/cover maps due to lack of reference data. However, qualitative and quantitative accuracy assessment was performed for the 2000, 2010, and 2014 land use/cover maps based on Google Earth imagery. Overall land use/cover classification accuracy for all land use/cover maps ranged from 70 to 90%. Land use/cover changes were simulated based on the boosted regression trees-cellular automata (BRT-CA) and RF-CA LCMs. We evaluated the goodness-of-fit of transition potential maps, and validated the simulated land use/cover changes based on robust statistical measures. Generally, the BRT-CA and RF-CA LCMs for all metropolitan areas in Asia and Africa performed relatively well. In particular, the BRT-CA and RF-CA LCMs for metropolitan areas in Africa had the best performance. The modeling and simulation results presented in this chapter provide an initial exploration of BRT-CA and RF-CA LCMs in Asia and Africa. This chapter demonstrates the significance of robust calibration, validation, and simulation of spatial LCMs for all metropolitan areas in Asia and Africa.

C. Kamusoko (✉)
Asia Air Survey Co., Ltd, Kawasaki, Japan
e-mail: kamas72@gmail.com

Y. Murayama et al. (eds.), *Urban Development in Asia and Africa*,
The Urban Book Series, DOI 10.1007/978-981-10-3241-7_2

2.1 Introduction

The past decades have witnessed tremendous development of land change models (LCMs) due to availability of remote sensing data, advances in geographical information and social sciences as well as theoretical developments of complexity and self-organizing systems (Tobler 1979; Wolfram 1984; Couclelis 1985; Engelen 1988; Batty 1998, 2005; Wu and Webster 1998; Torrens 2008; The Sate of Land Change Modeling 2014). To date, numerous LCMs have been developed to model and simulate land use/cover changes (Wu and Webster 1998; Verburg et al. 1999; Messina and Walsh 2001; Soares-Filho et al. 2002), deforestation (Lambin 1997; Geoghegan et al. 2001; Mas et al. 2004), urban growth (Couclelis 1989; Clarke et al. 1997; Cheng and Masser 2004; Yeh and Li 2009), climate change (Dale 1997), and hydrology (Matheussen et al. 2000).

While LCMs have highlighted significant insights into landscape change processes, most of these models have been criticized for lacking robust calibration and validation procedures (Pontius and Malanson 2005; Vliet et al. 2011). For example, previous studies show that transition potential maps—which are key inputs of LCM—have been validated using the relative operating characteristic (ROC) area under the curve (AUC) statistic (Eastman et al. 2005). However, the AUC statistic has limitations, especially for validating transition potential maps (Mas et al. 2013; Pontius and Parmentier 2014; Pontius and Si 2014) since it includes persistence areas (Eastman et al. 2005). For example, Kamusoko and Gamba (2015) demonstrated that the AUC can be large due to correctly predicted persistence not correctly predicted change (The Sate of Land Change Modeling 2014). Furthermore, percent correct and the standard Kappa statistics have been widely used to validate LCM (Verbug et al. 2004, Pontius and Malanson 2005; Vliet et al. 2011). However, the use of standard Kappa statistic for validating LCM has been criticized given its tendency to overestimate the agreement between the simulated and observed (reference) maps (Hagen 2002; Pontius et al. 2002). It has also been noted that the standard Kappa statistic neither reveals the components of agreement and disagreement between the simulated and observed (reference) maps nor accounts for persistence (that is, land use/cover classes that do not change during the simulation) (Pontius et al. 2007, 2008).

More recently, numerous statistical measures for calibrating and validating LCMs have been developed to overcome limitations of the ROC statistic and standard Kappa. For example, Pontius and Si (2014) developed the total operating characteristic (TOC) statistic to validate transition potential maps. The TOC statistic provides information such as misses and correct rejections in addition to ROC statistic such as hits (hits plus misses) and false alarms (false alarms plus correct rejections) (Pontius and Si 2014).

More importantly, the TOC statistic shows the actual units in the contingency table (e.g., square kilometers) instead of a unitless statistic such as AUC (Pontius and Si 2014). Furthermore, Visser and de Njis (2006) and Vliet et al. (2011) developed additional accuracy assessment statistics, which take into account information contained in the initial land use/cover map and the proportion of persistent land use/cover classes during the simulation period. The KSimulation expresses the agreement between the simulated land use/cover transitions and reference land use/cover transitions, while KTranslocation measures the degree to which the transitions agree in terms of allocations (Vliet et al. 2011). The KTransition captures the agreement in terms of quantity of built-up and non-built-up transitions (Vliet et al. 2011). The KSimulation, KTransition, and KTranslocation statistics are available in the Map Comparison Kit software by Visser and de Njis (2006). Pontius et al. (2007, 2008) also introduced the Figure of Merit (FoM), which expresses agreement between the observed and simulated changes for validating simulated land use/cover changes.

While these novel statistics have provided a new paradigm for validation, to date, few studies (Kamusoko and Gamba 2015) have applied these robust statistical measures for validating LCMs. Therefore, more research is needed to better understand uncertainty of LCMs based on the above-mentioned validation statistics. This is critical since LCMs are being considered as useful procedures or tools to establish business-as-usual baselines for urban growth and other land use/cover change studies (The Sate of Land Change Modeling 2014; Kamusoko and Gamba 2015). The purpose of this chapter is to describe briefly the methodology used to produce the land use/cover maps, calibrate and validate LCMs (in this case, both transition potential and simulated land use/cover maps). The specific objectives of this chapter are to evaluate the goodness-of-fit of transition potential maps, validate the simulated land use/cover maps, and elucidate components of agreement and disagreement. Validation statistics developed by Pontius and Si (2014), Pontius and Malanson (2005), Visser and de Njis (2006), and Vliet et al. (2011) as well as simple GIS overlay analysis are used in this chapter.

This chapter is organized as follows: Sect. 2.2 provides an overview of the image processing and change analysis; Sect. 2.3 describes land change modeling implementation procedures for all the metropolitan areas in Asia and Africa; Sect. 2.4 presents the results and discussions; while Sect. 2.5 provides the summary and conclusion of the chapter.

2.2 Image Processing and Change Analysis

2.2.1 Satellite Imagery and Reference Data

Landsat 4 and 5 Thematic Mapper (TM), Landsat 7 Enhanced Thematic Mapper Plus (ETM+), and Landsat 8 datasets were used for land use/cover classification

(Tables 2.1 and 2.2). All the Landsat datasets were acquired between 1988 and 2014 (Tables 2.1 and 2.2). The selection of the image data was based on the availability of high-quality satellite imagery with minimal cloud cover. Landsat 8 (originally called Landsat Data Continuity Mission) was launched on February 11, 2013, as the eighth

Table 2.1 Summary of Landsat imagery used for Metropolitan Areas in Asia

Metropolitan area	Landsat sensor	Path/row	Acquisition date
Bangkok	L4 TM	129/50	30/03/1988
		129/51	30/03/1988
	L7 ETM+	129/50	20/12/1999
		129/51	20/12/1999
	L5 TM	129/50	19/01/2009
		129/51	19/01/2009
	L8	129/50	17/01/2014
		129/51	17/01/2014
Beijing	L4 TM	123/32	25/12/1988
		123/33	25/12/1988
	L7 ETM+	123/32	30/04/2000
		123/33	30/04/2000
	L5 TM	123/32	14/03/2009
		123/33	14/03/2009
	L8	123/32	29/04/2014
		123/33	29/04/2014
Dhaka	L4 TM	137/44	13/02/1989
	L7 ETM+	137/44	28/02/2000
	L5 TM	137/44	15/03/2010
	L8	137/44	30/03/2014
Hanoi	L5 TM	127/45	11/09/1988
	L7 ETM+	127/45	20/12/1999
	L5 TM	127/45	05/11/2009
	L8	127/45	19/01/2014
Jakarta	L5 TM	122/64	03/05/1989
	L7 ETM+	122/64	16/08/2001
	L5 TM	122/64	21/05/2010
	L8	122/64	25/08/2013
Kathmandu	L5 TM	141/41	24/01/1989
	L7 ETM+	141/41	04/11/1999
	L5 TM	141/41	11/02/2010
	L8	141/41	26/03/2014
Manila	L5 TM	116/50	02/04/1993
	L7 ETM+	116/50	26/11/2001
	L5 TM	116/50	05/03/2009
	L8	116/50	07/02/2014

(continued)

Table 2.1 (continued)

Metropolitan area	Landsat sensor	Path/row	Acquisition date
Tehran	L5 TM L4 TM	164/35 165/35	19/09/1988 16/09/1987
	L7 ETM+	164/35	18/07/2000
		165/35	25/07/2000
	L5 TM	164/35	22/07/2010
	L8	164/35	08/12/2014
		165/35	13/11/2014
Yangon	L5 TM	132/48	26/02/1989
	L7 ETM+	132/48	21/11/1999
	L5 TM	132/48	24/01/2009
	L8	132/48	23/02/2014

Table 2.2 Summary of Landsat imagery used for Metropolitan Areas in Africa

Metropolitan area	Sensor	Path/row	Acquisition date
Bamako	L4 TM	199/51	22/03/1990
	L7 ETM+	199/51	30/12/2000
	L5 TM	199/51	16/01/2010
	L8	199/51	16/03/2014
Dakar	L4 TM	205/50	15/10/1989
	L7 ETM+	205/50	04/11/1999
	L5 TM	205/50	25/10/2010
	L8	205/50	17/03/2013
Harare	L5 TM	170/72	23/06/1990
	L7 ETM+	170/72	30/09/2000
	L5 TM	170/72	26/05/2009
	L8	170/72	24/05/2014
Johannesburg	L5 TM	170/78	25/07/1990
	L7 ETM+	170/78	28/07/2000
	L5 TM	170/78	26/05/2009
	L8	170/78	25/06/2014
Lilongwe	L5 TM	168/70	11/07/1990
	L5 TM	168/70	02/06/1999
	L5 TM	168/70	22/08/2011
	L8	168/70	26/07/2013
Nairobi	L5 TM	168/61	17/10/1988
	L7 ETM+	168/61	21/02/2000
	L5 TM	168/61	19/08/2010
	L8	168/61	03/02/2014

satellite in the Landsat program (NASA 2013; USGS 2013). Landsat 8 consists of the Operational Land Imager (OLI) and the Thermal Infrared Sensor (TIRS), which provide images at a spatial resolution of 15 m (panchromatic), 30 m (visible, NIR, SWIR), and 100 m (thermal) (NASA 2013; USGS 2013).

2.2.2 Random Forest Classification

A modified land cover classification scheme was used for image classification. Three land use/cover classes were considered in this study: (1) built-up; (2) non-built-up; and (3) water. Detailed descriptions of the land use/cover classes are provided in Table 2.3. Land use/cover maps were produced from the classification of Landsat imagery for 1990, 2000, 2010, and 2014 using the (RF) classifier, an ensemble decision tree machine learning method (Breiman 2001). The RF classifier combines bootstrap sampling to construct many individual decision trees, from which a final class assignment is produced (Breiman 2001). This machine learning classifier can be used to learn nonlinear relationships, particularly in heterogeneous urban landscapes. The RF classifier has been demonstrated to be effective for accurate land cover mapping across complex and heterogeneous landscapes (Rodriguez et al. 2012). All the Landsat imagery for all metropolitan areas were classified using the "randomForest" package (Liaw and Wiener 2002), which is available in R (R Development Core Team 2005).

Quantitative accuracy assessment for the 1990 land use/cover maps was not conducted because of the unavailability of reference data such as aerial photographs and high-resolution satellite imagery. However, the Atlas of Urban Expansion developed by the Lincoln Institute of Land Policy (Angel et al. 2010) was used to visually check the quality of land use/cover maps for the 1990 epoch (that is, Landsat imagery acquired between 1988 and 1993). Qualitative and quantitative accuracy assessment was conducted for land use/cover maps from 2000, 2010, and 2014 epochs. The primary reference data for accuracy assessment was obtained from very high-resolution images (e.g., QuickBird image) in Google Earth™

Table 2.3 Land use/cover classes

Class	Description
Built-up	Residential, commercial and services, industrial, transportation, communication and utilities, construction sites, and landfills
Non-built-up	All wooded areas, riverine vegetation, shrubs and bushes, grass cover, golf courses, parks, cultivated land, fallow land, land under irrigation, bare exposed areas and transitional areas
Water	Rivers, reservoirs, and other water bodies

(Google Earth 2015). Overall land use/cover classification accuracy for all land use/cover maps (from 2000 to 2014) ranged from 70 to 90% for all the metropolitan areas.

2.3 Land Change Modeling

2.3.1 Data

We used land use/cover maps and driving factors to develop spatial LCMs for all metropolitan areas (Table 2.4). Major roads were obtained from OpenStreetMap data, while city center was digitized from Google Earth. Elevation was derived from ASTERGDEM, while population density data were acquired from the LandScan data (Bhaduri et al. 2007). We used built-up areas (extracted from the 1990 and 2010 land cover maps), major roads, and city center data to compute "distance to built-up areas", "distance to major roads", and "distance to city center" using the Euclidean distance procedures available in ArcGIS 10.2. We computed "distance to built-up areas" for 1990 and 2010, and "distance to major roads" because built-up areas and roads are dynamic driving factors that change over time. Furthermore, we used "distance to built-up areas" as the driving factor because previous urban form influences future urban patterns (Liu 2009). Finally, all driving factors were resampled to 30 m × 30 m spatial resolution in order to match the spatial resolution of the Landsat-derived land use/cover maps.

2.3.2 Model Calibration and Simulation

We used the following procedures to implement the LCMs for all metropolitan areas: (I) computing transition rates, (II) transition potential modeling, and (III) CA simulation. Machine learning and statistical algorithms available in R were used to model transition potential, while functions available in Dinamica Environment for Geoprocessing Objects (EGO) were used to compute transition rates and simulate land use/cover changes. R is a free and open-source statistical and computer graphic

Table 2.4 Input data for calibrating and simulating land use/cover change

Variable	Source
Land use/cover maps (1990, 2000, and 2010)	Classified from landsat data
Distance to built-up areas (1990, 2000, 2010)	Derived from land use/cover
Distance to major roads (1990–2000, 2000–2010)	Open street map
Distance to city center	Digitized from Google Earth
Elevation	ASTER GDEM
Population density (2000, 2010)	LandScan data

software (R Core Development Team 2005), while **Dinamica EGO** is a freeware that was developed by Soares-Filho et al. (2009). **Dinamica EGO** consists of a sophisticated platform for developing dynamic spatial models, which involve nested iterations, multiple-step transitions, dynamic feedbacks, and multiscale approaches (Soares-Filho et al. 2009).

(I) Computation of transition rates

We used land use/cover maps for 1990, 2000, and 2010 to compute multiple-step transition rates in Dinamica EGO. Multiple-step transition rates refer to transition rates that are computed at annual time step. Therefore, the "1990–2000", "2000–2010", and "1990–2010" multiple-step transition rates for all the metropolitan areas were used as input for the final CA simulation run following the methodology described in Kamusoko and Gamba (2015).

(II) Computation of transition potential maps

In order to compute the "non-built-up to built-up" transition potential maps, "non-built to built-up" change map from 1990 to 2010, biophysical and socioeconomic driving factors were combined based on two machine learning procedures. First, the RF model (Breiman 2001) was used to compute transition potential maps for all metropolitan areas. RF is a machine learning approach, which builds regression trees to describe the relationship between the response and predictor variables (Breiman 2001). In general, multiple trees are built, each based on a bootstrap sample of the data and a random subset of the predictors. The final model predictions are an average prediction across component trees. Previous studies have shown that the RF model is effective for modeling transition potential maps (Kamusoko and Gamba 2015). However, preliminary transition potential calibration results indicated overfitting problems for some metropolitan areas such as Beijing, Bamako, Dhaka, Hanoi, Johannesburg, Kathmandu, and Nairobi. Therefore, an alternative method based on boosted regression trees (BRT) (Friedman 2002; Elith et al. 2008) was employed. BRT is also a machine learning approach, which forms a relationship between a response variable and its predictors without a priori specification of a data model (Friedman 2002; Elith et al. 2008). Generally, a large number of simple models are combined to form a final model (Elith et al. 2008). The main advantage of the BRT model is that it uses a sequential model-fitting algorithm, which reduces both bias and variance and therefore improves model accuracy.

In this study, approximately 2000 training points randomly sampled from "non-built-up to built-up" and "no change" (that is, built-up and non-built-up persistence) areas between 1990 and 2010 were used to fit the BRT and RF models. Generally, 70% of the training areas were used for model development, while 30% were used for cross-validation. The gbm and dismo packages (Ridgeway 2006; Elith et al. 2008) available in R were used to fit the BRT model. The BRT model was optimized by changing the learning rate, tree complexity, and number of trees parameters. The learning rate controls the weight that is given to each component tree, while the complexity controls the number of nodes within each tree (Ridgeway 2006; Elith et al. 2008). We set the initial number of trees to five, learning rate to a maximum of 0.001, and bagging fraction to 0.5 (that is, at each iteration 50% of the data is drawn at random, without replacement from the full training set) for each metropolitan area. After many iterations, the best model was selected to compute a "non-built-up to built-up" transition potential map for each metropolitan area.

The RF model was used to compute "non-built-up to built-up" transition potential maps for Bangkok, Jakarta, Manila, Tehran, Yangon, Dakar, Harare, and Lilongwe. The "randomForest" (Liaw and Wiener 2002) package available in R was used to fit the RF model. The RF model parameters were adjusted by changing the number of input variables selected at each node split and the total number of trees included in the model (25, 50, 100, and 500) in order to achieve optimum model performance. After calibration, between 100 and 500 trees were used to construct the final RF model and then compute the "non-built-up to built-up" transition potential maps.

Figures 2.1 and 2.2 show "non-built-up to built-up" transition potential maps for metropolitan areas in Asia and Africa, respectively. Visual analysis revealed that the BRT and RF models produced relatively accurate transition potential maps. In particular, the BRT and RF models were relatively good at modeling built-up areas near previous built-up areas (from 1990 to 2010). In general, the transition potential maps have identified the areas where a change is likely to occur. As a result, the transition potential maps can be used as a useful input to the CA models.

(III) Cellular automata (CA) simulations

The initial land use/cover map (1990), the transition potential maps (1990–2010), and the three multiple-state transition rates were used to simulate land use/cover up to 2014 based on cellular automata (CA) functions available in Dinamica EGO. The expander transition function expands or contracts previous land use/cover class patches, while the patcher transition function forms new patches (Soares-Filho et al. 2009). The expander and patcher transition functions are composed of an allocation mechanism responsible for identifying cells with the highest transition potential for each transition (Soares-Filho et al. 2009). In order to simulate land use/cover changes, both transition functions use a stochastic selecting mechanism (Soares-Filho et al. 2009). The sizes of new land use/cover patches are set according to a lognormal probability function, whose parameters are defined by the mean patch size (MPS), patch size variance (VAR), and isometry (ISO). The CA model for each metropolitan area was calibrated by changing the parameters of the expander and patcher transition functions using trial and error. The initial simulation year was set to 1990, while the final year was set to 2014.

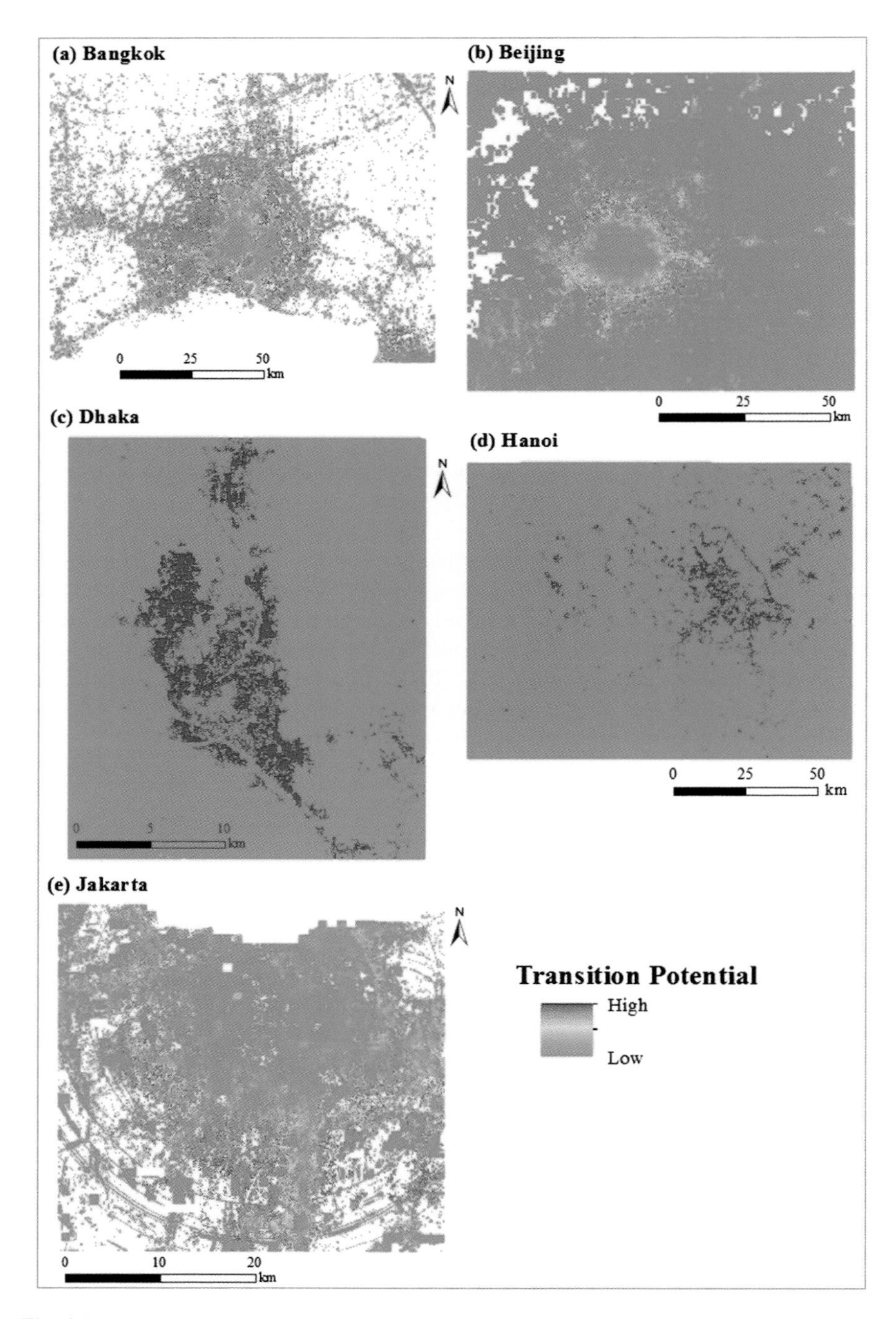

Fig. 2.1 Transition potential maps for Metropolitan Areas in Asia

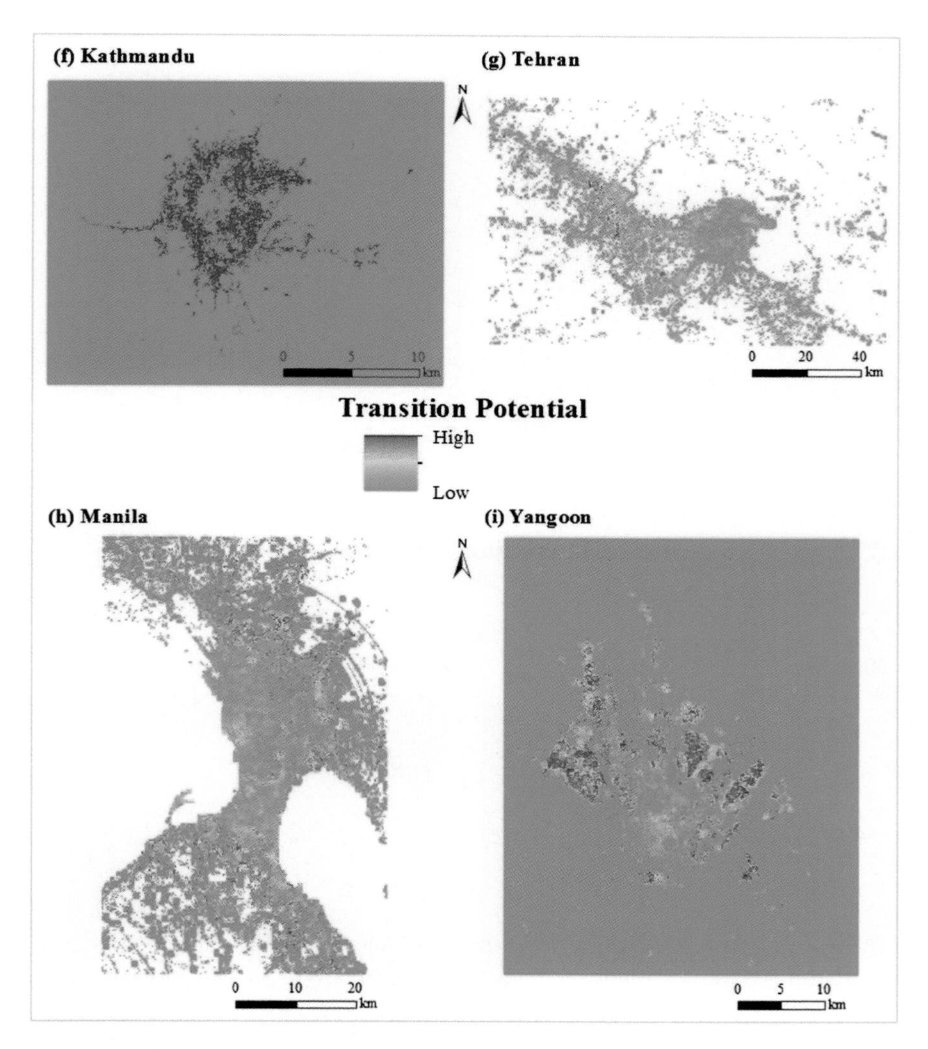

Fig. 2.1 (continued)

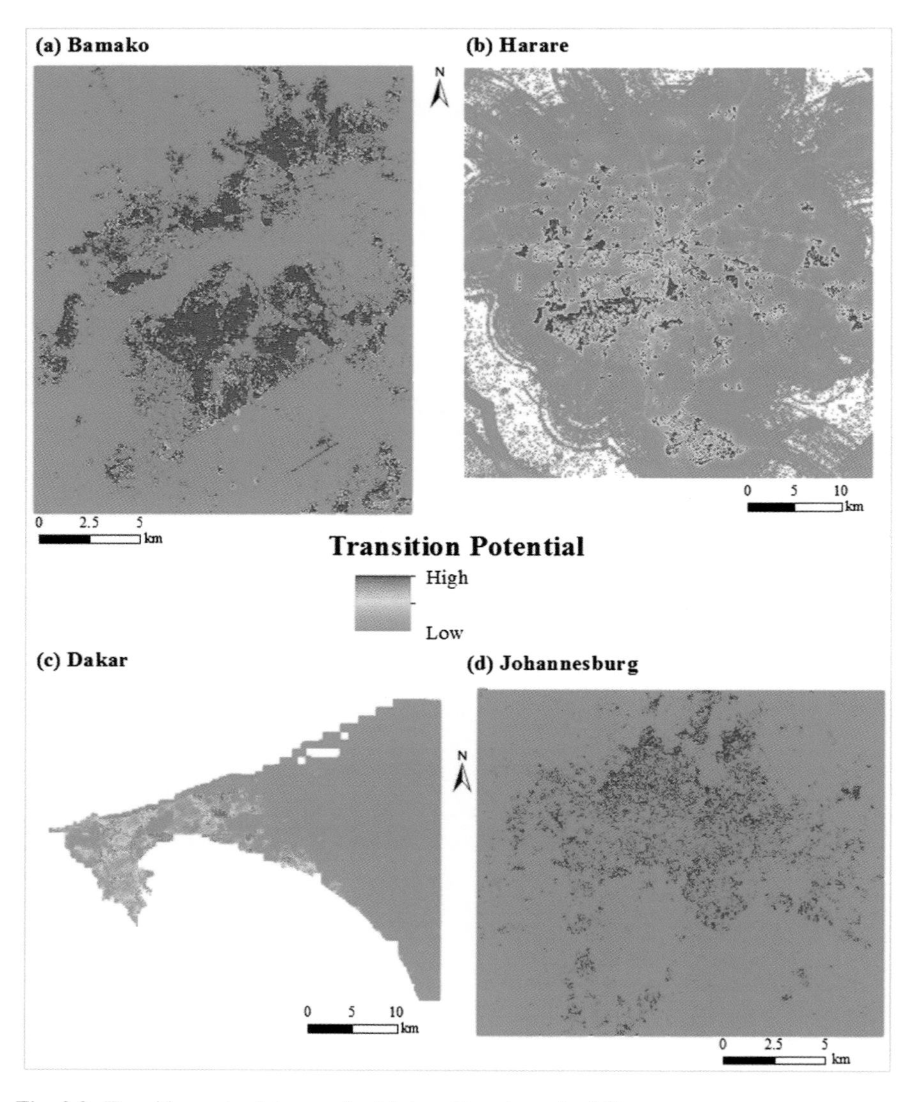

Fig. 2.2 Transition potential maps for Metropolitan Areas in Africa

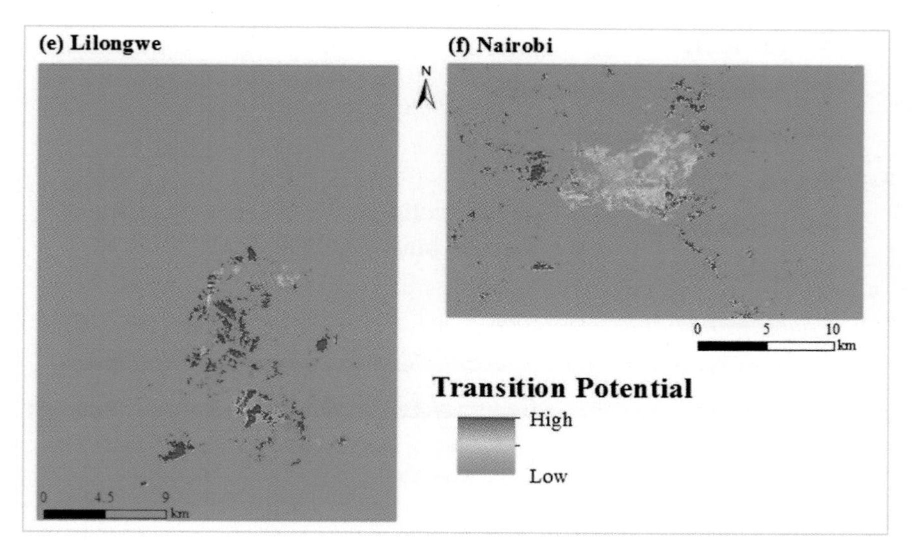

Fig. 2.2 (continued)

2.4 Results and Discussion

2.4.1 Evaluating the Goodness-of-Fit of Transition Potential Maps

2.4.1.1 Metropolitan Areas in Asia

Figure 2.3 shows the TOC graphs for all transition potential models in Asia. The TOC statistic (Pontius and Si 2014) is an excellent method to assess the validity of a model, which predicts the location of the occurrence of a class by comparing a transition potential map depicting the likelihood of that class occurring (that is, the input map) and a reference image showing where that class actually exists (that is, the non-built-up to built-up change between 1990 and 2010). In particular, TOC offers a statistical analysis that shows how the class of interest is concentrated at the locations of relatively high transition potential for that class (Pontius and Si 2014). Therefore, TOC was used to evaluate the goodness-of-fit of calibration for transition potential maps derived from the BRT and RF models.

We focused our analysis on hits (that is, the correct "non-built-up to built-up" change), which were derived from the TOC statistic (Pontius and Si 2014). Generally, the Bangkok metropolitan area had 684.5 km^2 hits (representing 74% of the correctly predicted "non-built-up to built-up" changes) compared to 930.9 km^2 of the observed "non-built-up to built-up" changes between 1990 and 2010. For Beijing metropolitan area, the TOC statistics revealed that out of the 887.3 km^2 "non-built-up to built-up" changes that occurred between 1990 and 2010, only

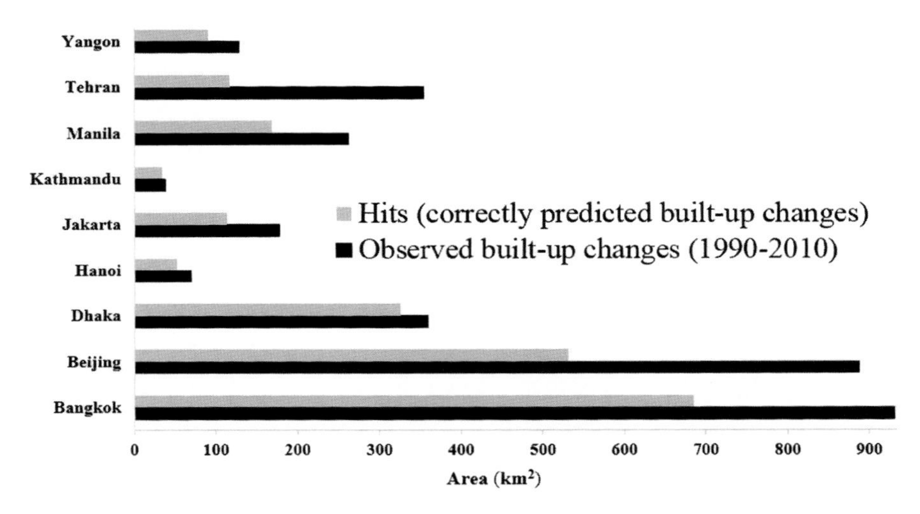

Fig. 2.3 Hits versus observed changes for Asia based on TOC statistics

531.7 km^2 representing 60%, were correctly predicted. However, Dhaka metropolitan area had approximately 325.7 km^2 hits representing 91% compared to 359.4 km^2 of the observed "non-built-up to built-up" changes between 1990 and 2010. For Hanoi metropolitan area, the TOC statistics revealed that out of the 70.4 km^2 "non-built-up to built-up" changes that occurred between 1990 and 2010, 52.8 km^2 representing 75% were correctly predicted. The Jakarta metropolitan area had approximately 114 km^2 hits representing 64% compared to 177.2 km^2 of the observed "non-built-up to built-up" changes between 1990 and 2010 (validation period). For Kathmandu metropolitan area, the TOC statistics revealed that out of the 38.3 km^2 "non-built-up to built-up" changes that occurred between 1990 and 2010, only 34.1 km^2 representing 89% were correctly predicted. The Manila metropolitan area had approximately 169.2 km^2 (representing 65% of the correctly predicted "non-built-up to built-up" changes) compared to 262.2 km^2 of the observed "non-built-up to built-up" changes between 1990 and 2010. However, Tehran had the lowest hits. The TOC statistics revealed that out of the 354.7 km^2 observed "non-built-up to built-up" changes that occurred between 1990 and 2010, only 117.6 km^2 representing 33% were correctly predicted. The Yangon metropolitan area had approximately 91 km^2 hits representing 71% compared to 129 km^2 of the observed "non-built-up to built-up" changes between 1990 and 2010.

Generally, all models expect Tehran produced relatively good transition potential maps. However, all models were excellent at predicting the allocation of built-up and non-built-up persistence since built-up and non-built-up persistence accounts for approximately 70% of the metropolitan areas. This is reflected by the relatively high TOC values, which are all above 86% for all metropolitan areas.

2.4.1.2 Metropolitan Areas in Africa

Figure 2.4 shows the TOC graphs for all transition potential models in Africa. The Bamako metropolitan area had 50.7 km^2 hits (representing 78% of the correctly predicted "non-built-up to built-up" changes) compared to 65 km^2 of the observed "non-built-up to built-up" changes between 1990 and 2010. For Dakar metropolitan area, the TOC statistics revealed that out of the 66.8 km^2 "non-built-up to built-up" changes that occurred between 1990 and 2010, only 48.3 km^2 representing 72% were correctly predicted. The Harare metropolitan area had 170.8 km^2 hits representing 72% compared to 236.9 km^2 of the observed "non-built-up to built-up" changes between 1990 and 2010. Johannesburg metropolitan area had the highest number of hits. The TOC statistics revealed that out of the 648.2 km^2 "non-built-up to built-up" changes that occurred between 1990 and 2010, 578.8 km^2 representing 89% were correctly predicted. However, Lilongwe metropolitan area had the lowest number of hits. For example, 8.1 km^2 hits representing 36% compared to 22.7 km^2 of the observed "non-built-up to built-up" changes between 1990 and 2010 were observed. For Nairobi metropolitan area, the TOC statistics revealed that out of the 74.9 km^2 "non-built-up to built-up" changes that occurred between 1990 and 2010, only 43.3 km^2 representing 58% were correctly predicted.

Generally, five models (except Lilongwe) produced relatively good transition potential maps. However, all models were excellent at predicting the allocation of built-up and non-built-up persistence. This is reflected by the relatively high TOC values since built-up and non-built-up persistence is dominant in the metropolitan areas of Africa.

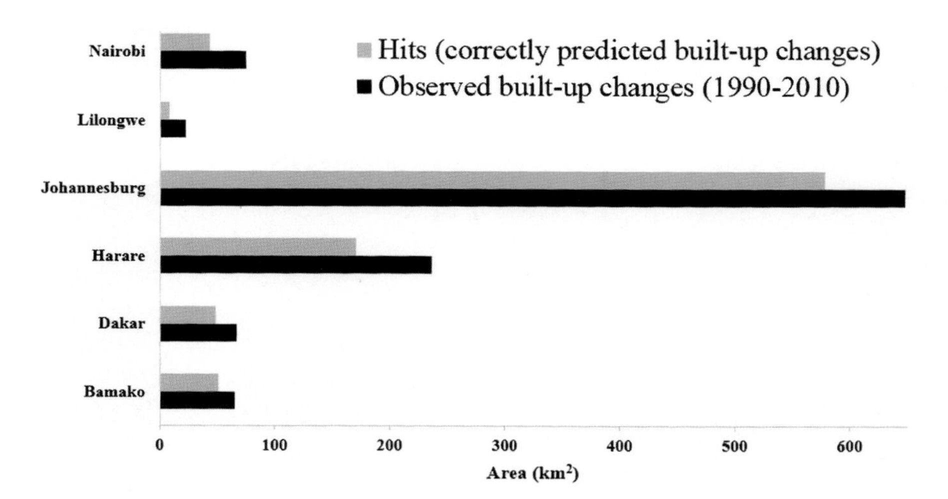

Fig. 2.4 Hits versus observed changes for Africa based on TOC statistics

2.4.2 Validation of the Simulated Land Use/Cover Changes

2.4.2.1 Metropolitan Areas in Asia

The observed and simulated land use/cover maps for all metropolitan areas in Asia are shown in Figs. 2.5, 2.6, 2.7, 2.8, 2.9, 2.10, 2.11, 2.12, and 2.13. Visual analysis shows that the BRT-CA and RF-CA models for Bangkok, Beijing, Dhaka, Hanoi, Jakarta, Kathmandu, Manila, and Yangon had good correspondence between the observed and simulated land use/cover maps for 2014 (Figs. 2.5, 2.6, 2.7, 2.8, 2.9, 2.10, 2.11, 2.12, and 2.13). The spatial patterns of built-up areas simulated by the BRT-CA and RF-CA models resemble the observed built-up patterns to a large

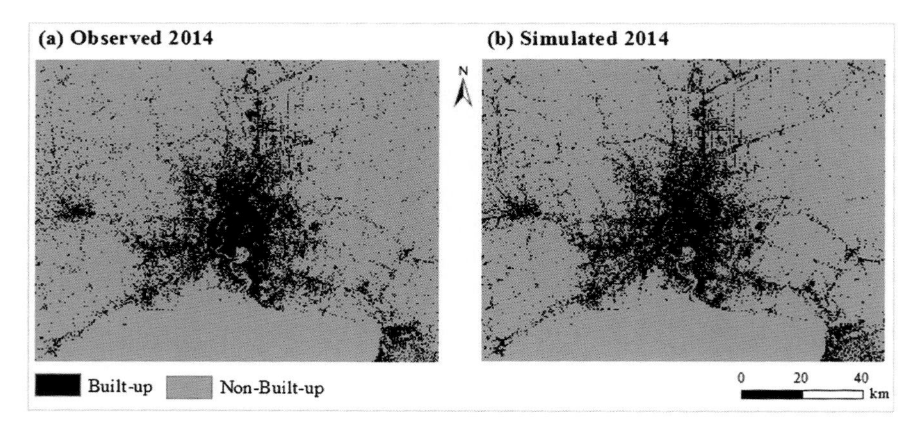

Fig. 2.5 Comparison of observed versus simulated land use/cover for Bangkok

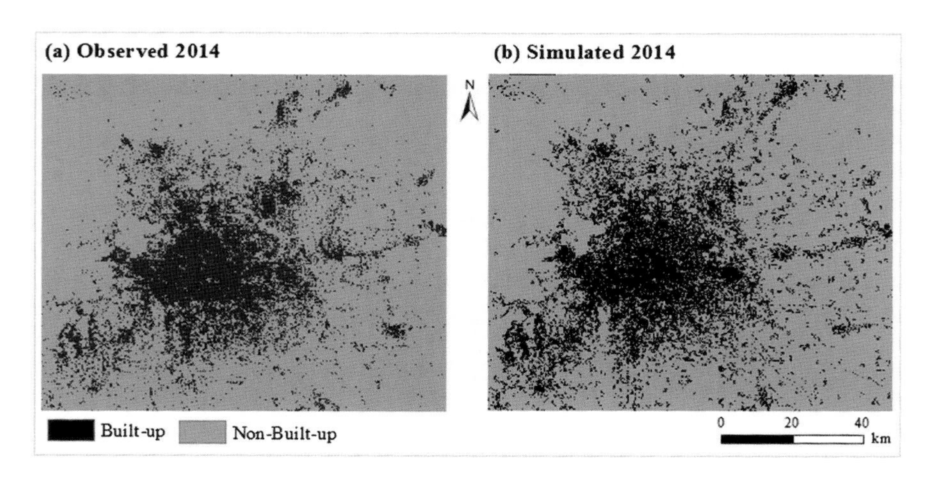

Fig. 2.6 Comparison of observed versus simulated land use/cover for Beijing

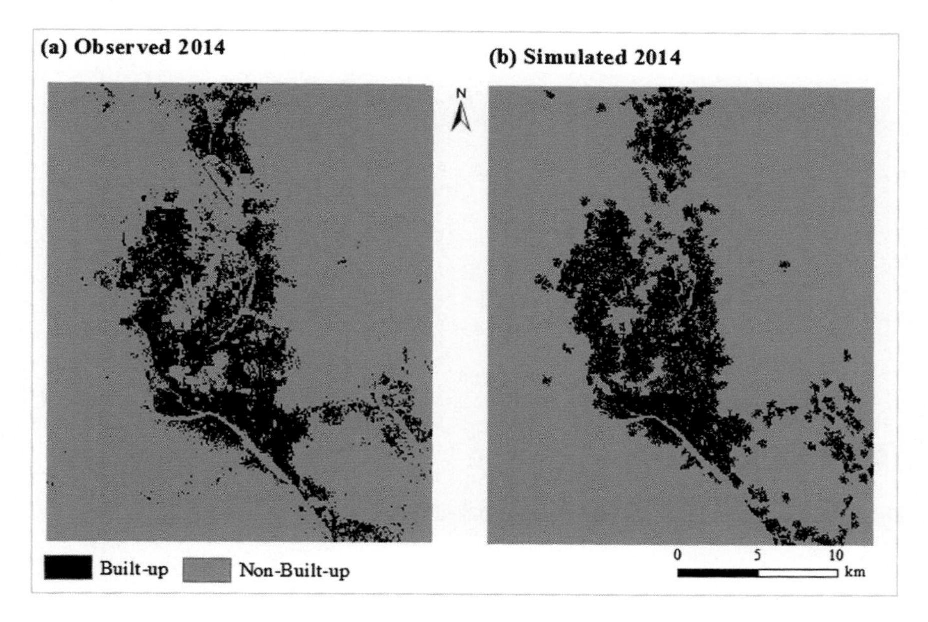

Fig. 2.7 Comparison of observed versus simulated land use/cover for Dhaka

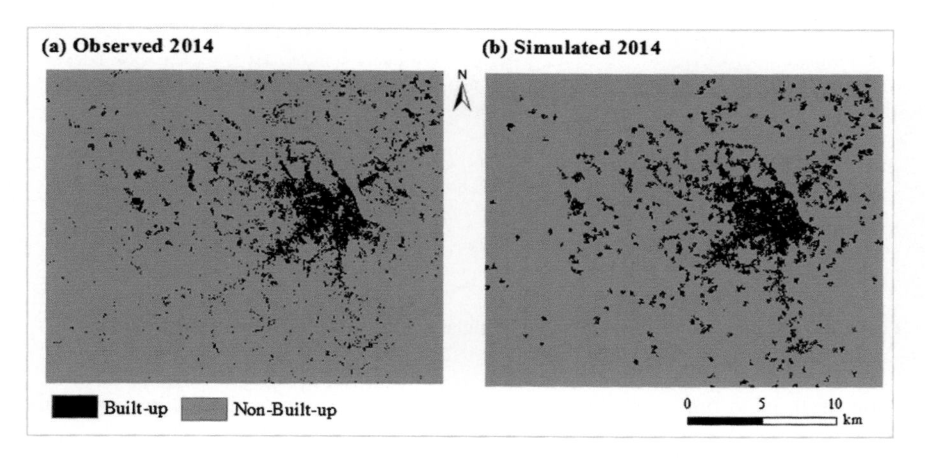

Fig. 2.8 Comparison of observed versus simulated land use/cover for Hanoi

extent. This suggests that the BRT-CA and RF-CA models were relatively accurate at allocating "non-built-up to built-up" changes.

However, Fig. 2.12 shows relatively medium correspondence between the simulated built-up patterns and the observed built-up patterns for Tehran. Generally, there was an underprediction of the built-up class for Tehran metropolitan area. This is partly attributed to lower spatial allocation of

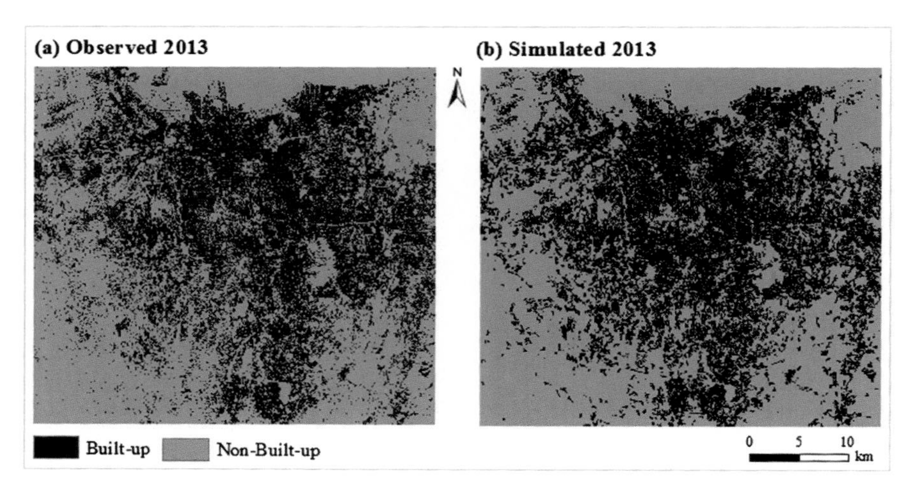

Fig. 2.9 Comparison of observed versus simulated land use/cover for Jakarta

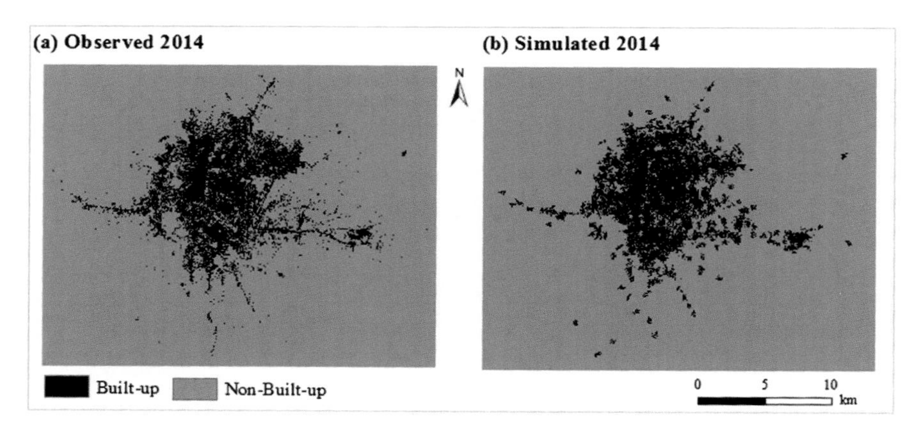

Fig. 2.10 Comparison of observed versus simulated land use/cover for Kathmandu

"non-built-up to built-up" changes that was observed during the calibration of the RF model for Tehran metropolitan area (Fig. 2.1g). In addition, it should be noted that Tehran metropolitan area has quite a unique setting with very high landscape fragmentation and scattered urban development, which makes it challenging to develop a robust simulation model. Nonetheless, it is also important to note that some of the allocated built-up areas had strong agreement between the simulated and observed land use/cover maps, which means that the CA model can be used to simulate future land use/cover changes.

For quantitative model validation, we used the observed (initial) land use/cover map for 1990, the observed (reference) land use/cover map for 2014, and the simulated land use/cover map for 2014. Table 2.5 shows the validation statistics

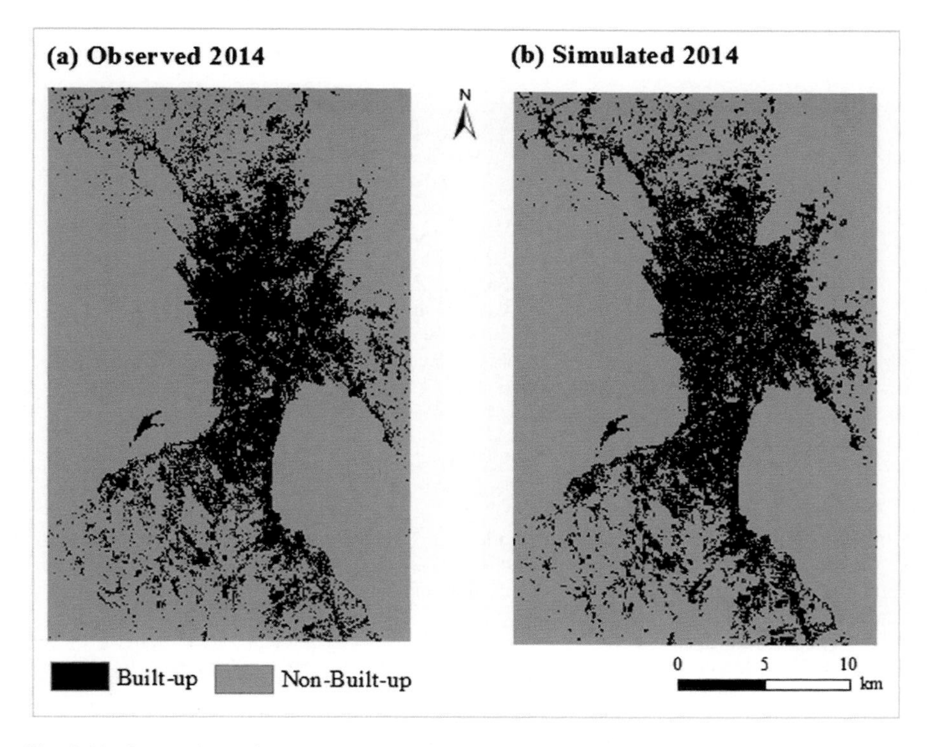

Fig. 2.11 Comparison of observed versus simulated land use/cover for Manila

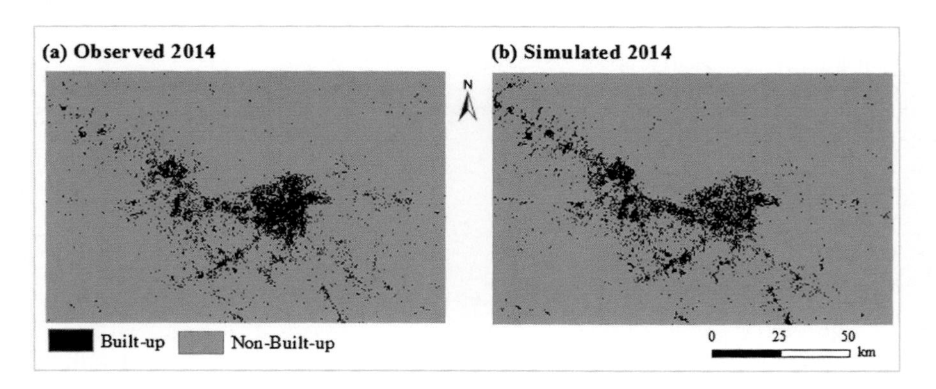

Fig. 2.12 Comparison of observed versus simulated land use/cover for Tehran

based on KSimulation, KTranslocation, KTransition, and the FoM for all metropolitan areas in Asia. KSimulation values range from 42 to 69%, while the KTranslocation values range from 44 to 71% (Table 2.5). Tehran had the lowest KSimulation and KTranslocation scores, while Dhaka had the highest KSimulation and KTranslocation scores. The FoM statistics also follow the similar pattern as the

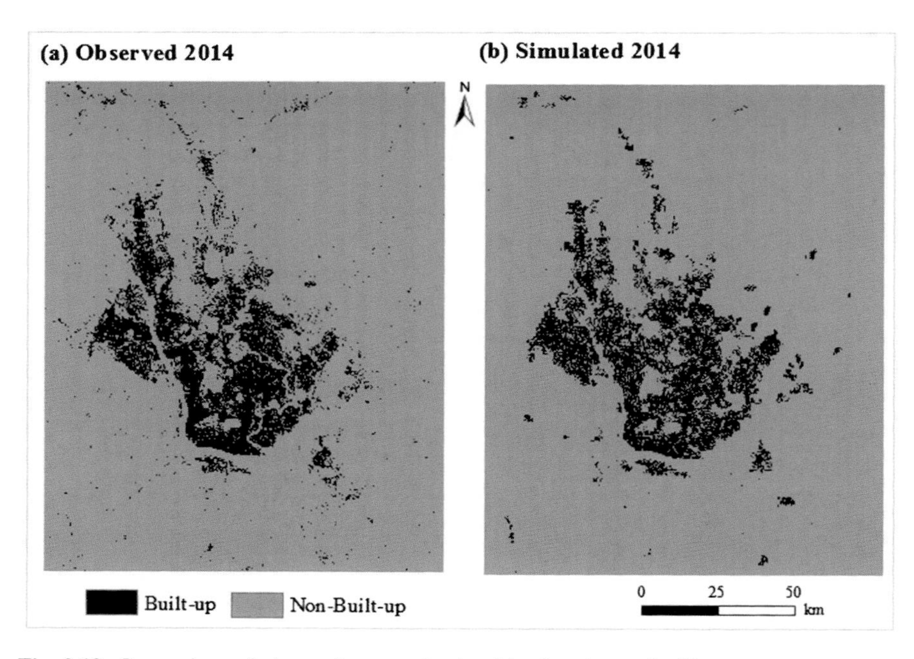

Fig. 2.13 Comparison of observed versus simulated land use/cover for Yangon

Table 2.5 Validation statistics for all simulation models in Asia

Metropolitan area	KSimulation	KTranslocation	KTransition	Figure of merit (%)
Bangkok	0.62	0.61	0.95	51
Beijing	0.51	0.55	0.93	45
Dhaka	0.69	0.71	0.97	58
Hanoi	0.56	0.67	0.83	42
Jakarta	0.47	0.49	0.96	43
Kathmandu	0.68	0.68	0.98	55
Manila	0.60	0.64	0.94	48
Tehran	0.42	0.44	0.96	29
Yangon	0.62	0.63	0.98	48

KSimulation and KTranslocation statistics. These results are in agreement with the goodness-of-fit transition potential results (Fig. 2.2), which suggest that transition potential maps have more influence in the overall accuracy of the CA simulation models. A study by Pontius et al. (2007, 2008) revealed that the FoM observed in other LCMs ranged from 1 to 59%. Therefore, the accuracy of the BRT-CA and RF-CA models are relatively high since the FoM is within the upper range of previously observed LCMs (Kamusoko and Gamba 2015).

Generally, all the simulation models had high KTransition score, which are above 83%. This is supported by the quantitative analysis between the simulated and observed land use/cover changes for all metropolitan areas in Asia. For example, a quantitative comparison for Bangkok revealed that the observed and projected quantities of built-up were 2292.4 and 2428.5 km^2, respectively. For Beijing, the observed built-up class was 2173.7 km^2, whereas the corresponding simulated class was 2364.1 km^2. However, the observed built-up class was 118.5 km^2, while the corresponding simulated class was 171.8 km^2 for Dhaka. For Hanoi, the observed built-up class was 133.3 km^2, whereas the corresponding simulated class was 171.8 km^2. In the case of Jakarta, the observed built-up class was 623.5 km^2, while the corresponding simulated class was 606.8 km^2. For Kathmandu, the observed built-up class was 75.9 km^2, whereas the corresponding simulated class was 76 km^2. Nevertheless, the observed built-up class was 848.6 km^2, while the corresponding simulated class was 984.6 km^2 for Manila. For Tehran, the observed built-up class was 923.3 km^2, whereas the corresponding simulated class was 969.1 km^2. Last but not least, the observed built-up class was 228.9 km^2, while the corresponding simulated class was 223.1 km^2 for Yangon. These results show that all simulation models were relatively accurate for simulating land use/cover quantity.

2.4.2.2 Metropolitan Areas in Africa

The observed and simulated land use/cover maps for all metropolitan areas in Africa are shown in Figs. 2.14, 2.15, 2.16, 2.17, 2.18, and 2.19. Visual analysis

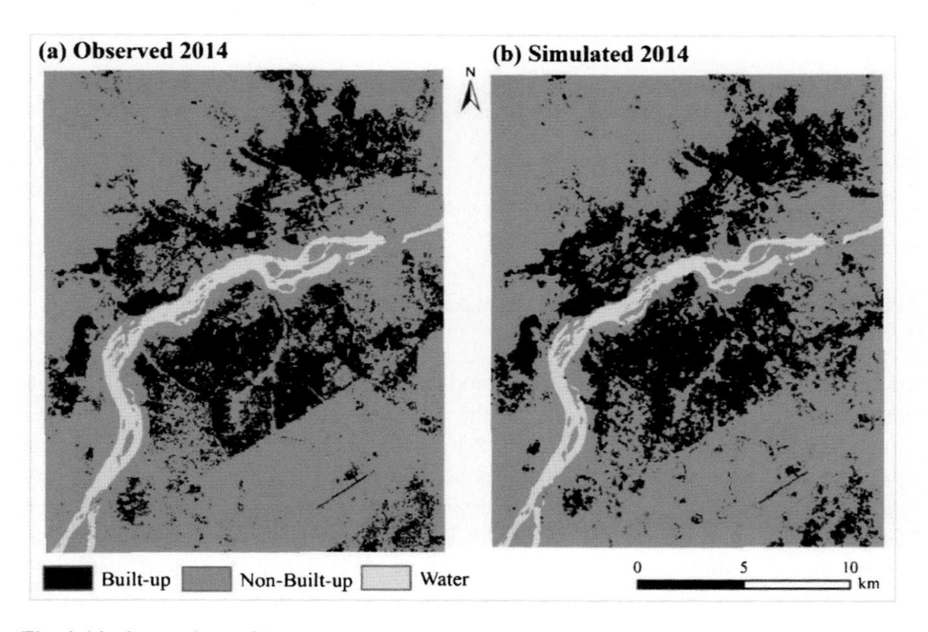

Fig. 2.14 Comparison of observed versus simulated land use/cover for Bamako

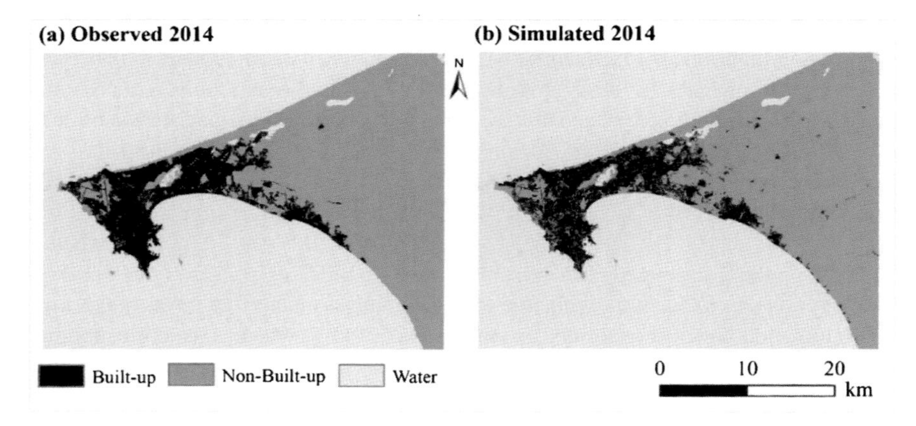

Fig. 2.15 Comparison of observed versus simulated land use/cover for Dakar

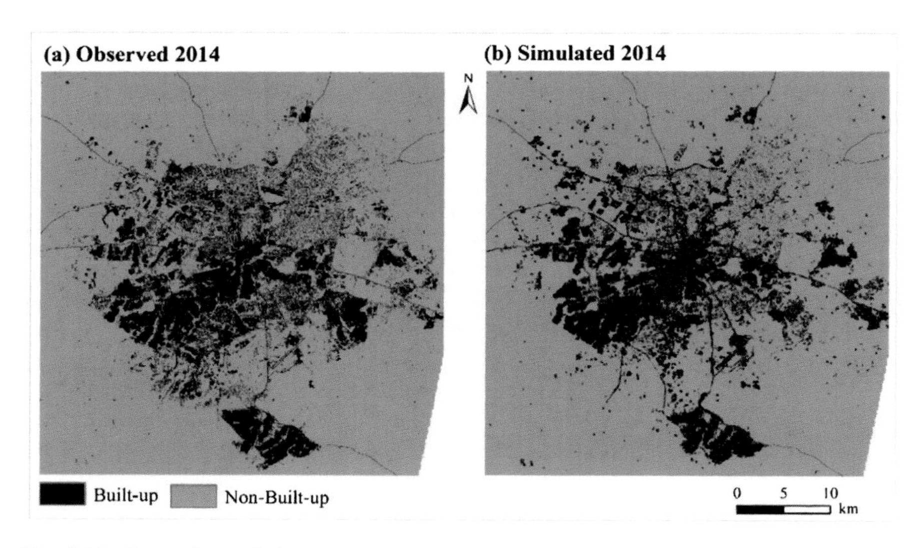

Fig. 2.16 Comparison of observed versus simulated land use/cover for Harare

shows that the BRT-CA and RF-CA models for all metropolitan areas had a relatively high correspondence between the observed and simulated land use/cover maps for 2014 (Figs. 2.14, 2.15, 2.16, 2.17, 2.18, and 2.19). Generally, the spatial patterns of the simulated built-up areas closely match the observed built-up patterns. While a slight degree of clumpiness is noted in some metropolitan areas such as Harare, in general the BRT-CA and RF-CA models were relatively good at allocating "non-built-up to built-up" changes and simulating land use/cover (Figs. 2.14, 2.15, 2.16, 2.17, 2.18, and 2.19). This attributed to the rigorous calibration of the transition potential maps, which were computed using BRT and RF

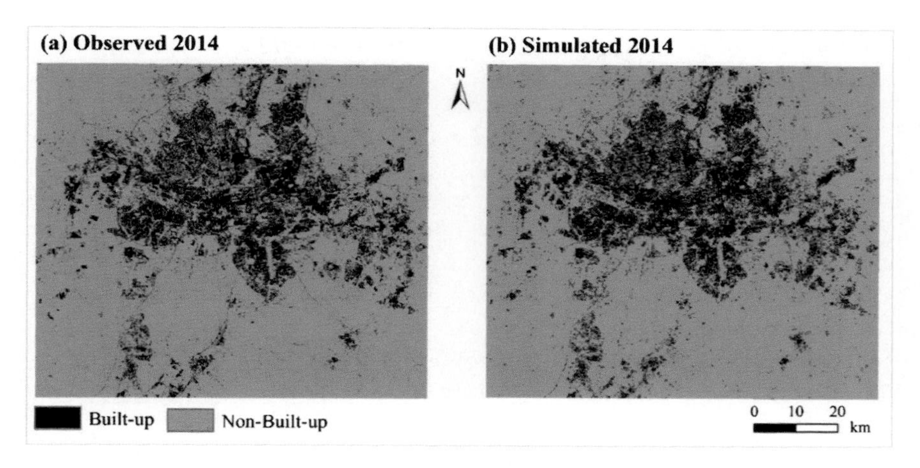

Fig. 2.17 Comparison of observed versus simulated land use/cover for Johannesburg

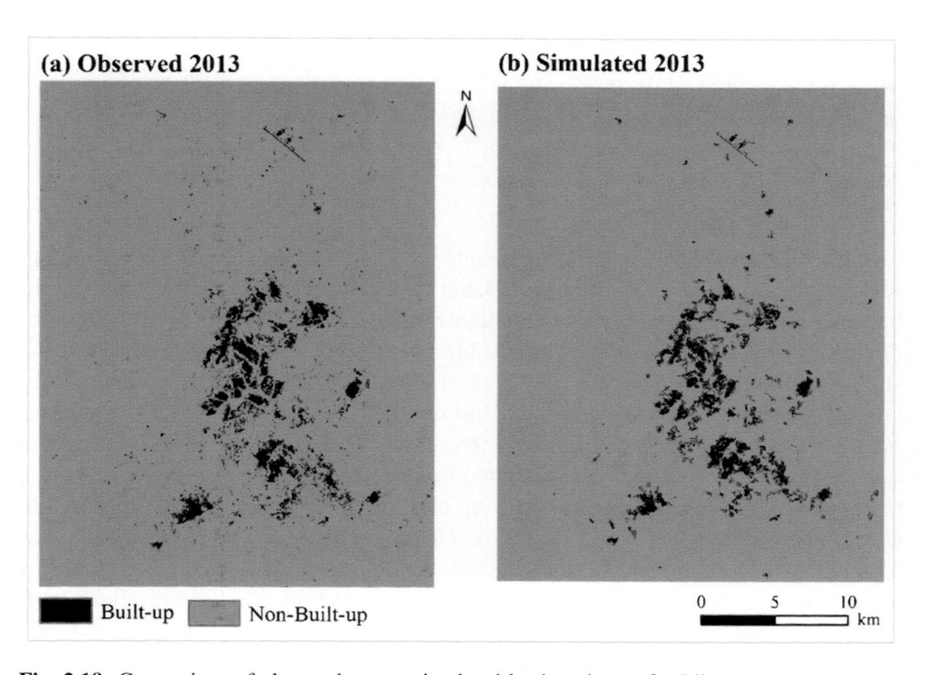

Fig. 2.18 Comparison of observed versus simulated land use/cover for Lilongwe

models (Fig. 2.3). These simulation results are significant given that all the BRT and RF models only incorporated a limited number of driving factors. Therefore, BRT-CA and RF-CA models can be used to simulate future land use/cover changes.

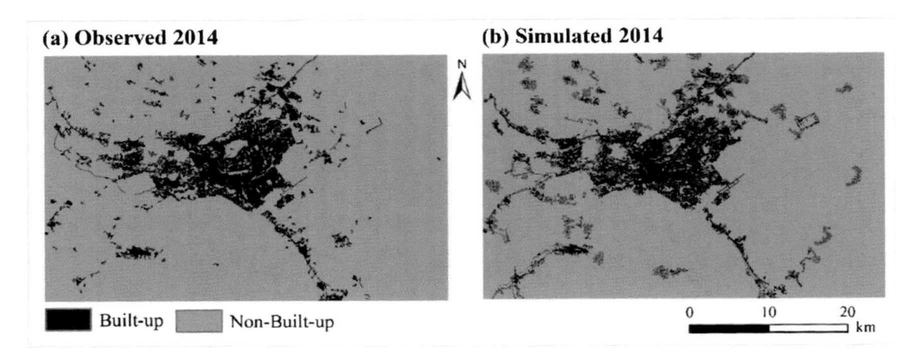

Fig. 2.19 Comparison of observed versus simulated land use/cover for Nairobi

Table 2.6 Validation statistics for all simulation models in Africa

Metropolitan area	KSimulation	KTranslocation	KTransition	Figure of merit (%)
Bamako	0.72	0.75	0.96	63
Dakar	0.75	0.79	0.94	62
Harare	0.49	0.50	0.99	47
Johannesburg	0.66	0.70	0.96	55
Lilongwe	0.68	0.70	0.97	53
Nairobi	0.56	0.60	0.93	43

Table 2.6 shows the validation statistics based on KSimulation, KTranslocation, KTransition, and the FoM. Bamako, Dakar, Johannesburg, Lilongwe, and Nairobi had KSimulation and KTranslocation scores above 50% (Table 2.6). All the simulation models had high KTransition score (above 93%), which is higher than the simulation models in metropolitan areas of Asia. This is supported by the quantitative analysis between the simulated and observed land use/cover changes for all metropolitan areas in Africa. For example, a quantitative comparison for Bamako revealed that the observed and simulated built-up class was 104.7 and 111.9 km^2, respectively. While the observed built-up class was 118.6 km^2, the corresponding simulated class was 109.7 km^2 for Dakar. However, the observed built-up class was 358.6 km^2, while the corresponding simulated class was 360.4 km^2 for Harare. For Johannesburg, the observed built-up class was 1418.6 km^2, whereas the corresponding simulated class was 1494.6 km^2. Although Lilongwe had a relatively lower accuracy for the transition potential (Table 2.6), the observed built-up class was 34.7 km^2, while the corresponding simulated class was 32.9 km^2. For Nairobi, the observed built-up class was 182.8 km^2, whereas the corresponding simulated class was 198.9 km^2. These results indicate that all simulation models were relatively accurate for simulating land use/cover quantity. This is supported by the high FoM, which was above 43% for all simulation models in Africa.

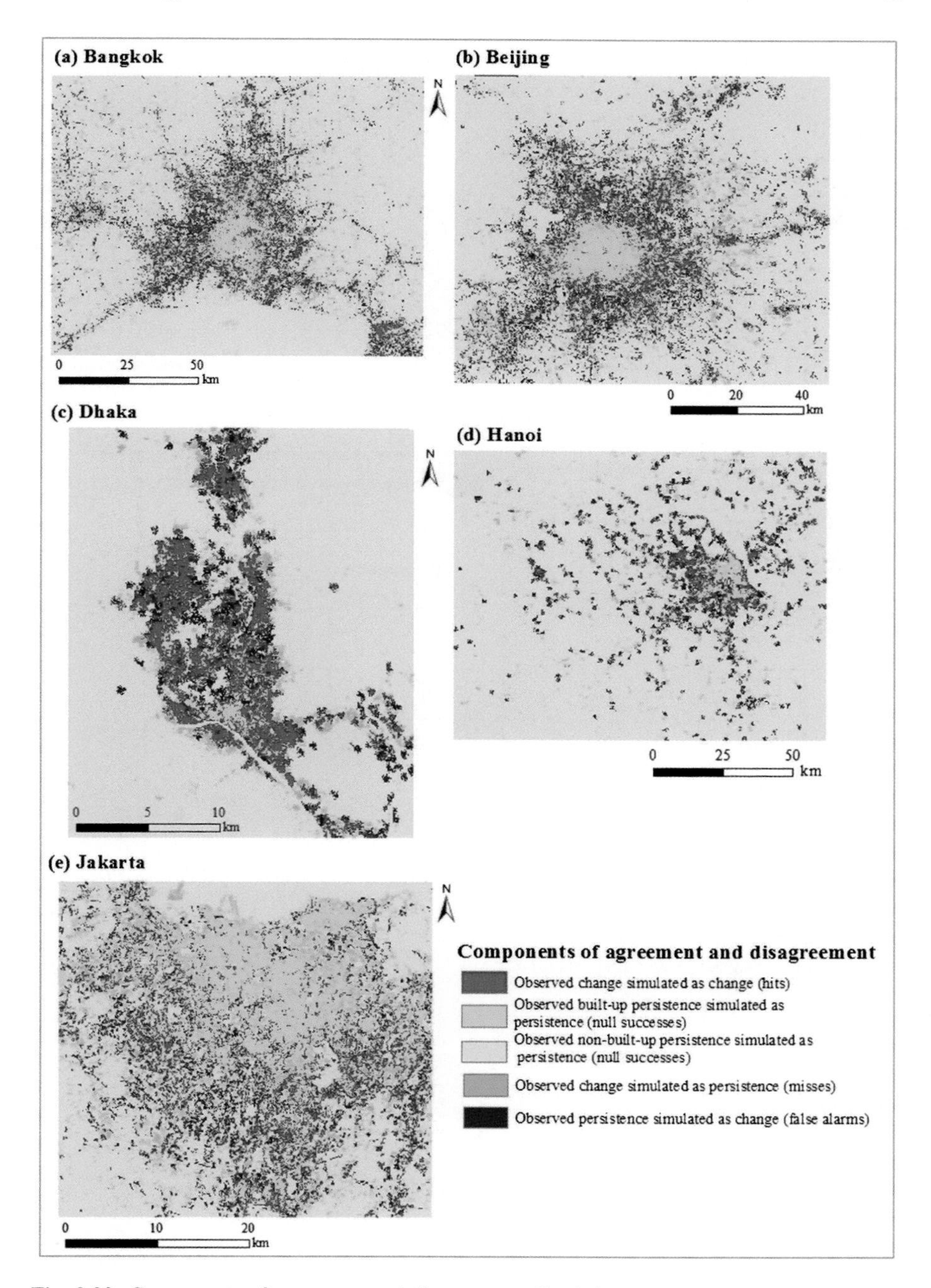

Fig. 2.20 Components of agreement and disagreement for Asia

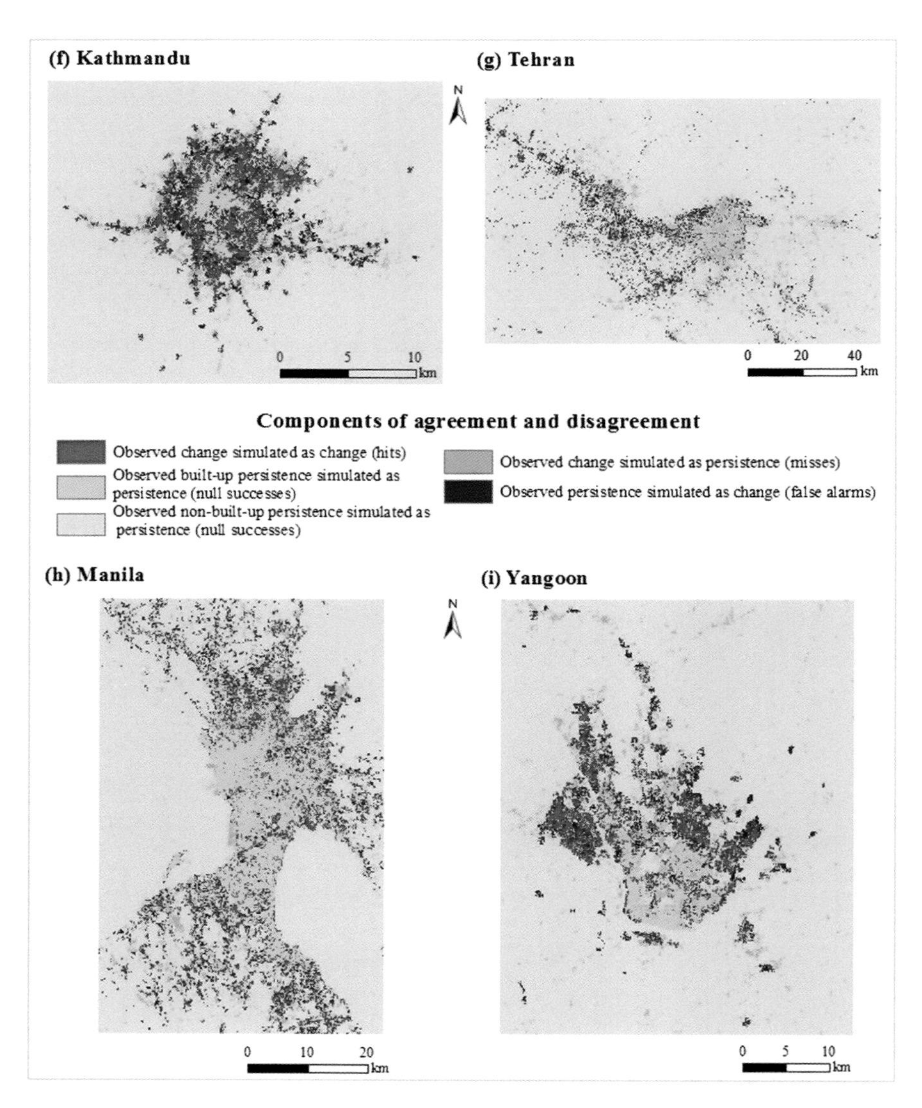

Fig. 2.20 (continued)

2.4.3 Analysis of Components of Agreement and Disagreement

2.4.3.1 Metropolitan Areas in Asia

Figures 2.20 and 2.21 show the components of agreement and disagreement based on the overlay of the initial (1990), the observed (2014), and simulated land

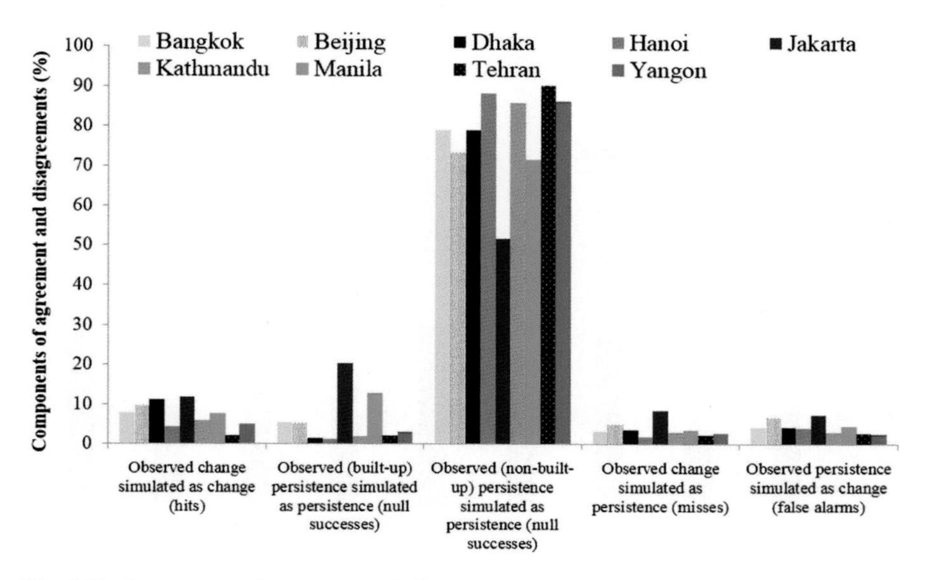

Fig. 2.21 Components of agreement and disagreement expressed as a percentage (Asia)

use/cover maps (2014) for all models. The components of agreement and disagreement reveal information such as: (1) observed change simulated correctly as change (hits); (2) observed persistence (built-up and non-built-up) simulated correctly as persistence (null successes); (3) observed change simulated wrongly as persistence (misses); and (4) observed persistence simulated wrongly as change (false alarms).

Results show that non-built-up persistence had the largest components of agreement for all the models in Asia (Figs. 2.20 and 2.21). This is because non-built-up persistence occupied about 60% of the metropolitan areas between 1990 and 2010. However, there are variation in terms of the hits, false alarms, and misses. For example, Bangkok, Dhaka, Kathmandu, Manila, and Yangon had the same or slightly more hits than the combined misses and false alarms (Figs. 2.20 and 2.21). This is encouraging since it shows that BRT-CA and RF-CA models performed relatively well. While Beijing, Hanoi, Jakarta, and Tehran had slightly more combined misses and false alarms than hits, the BRT-CA and RF-CA models also performed relatively well.

While the results show improvement in the LCMs, it must be noted that the simulated land use/cover maps include uncertainty of the original land use/cover maps, especially the 1990 land use/cover map (which was not quantitatively validated). In addition, it was observed that the BRT-CA and RF-CA models failed to simulate unconnected newly built-up areas, which is clearly apparent in the components of agreement and disagreement (Fig. 2.21). This is attributed to spatial and temporal nonstationarity in the built-up change process. Figure 2.22 shows the normalized observed built-up change rate between different epochs in Asia, which

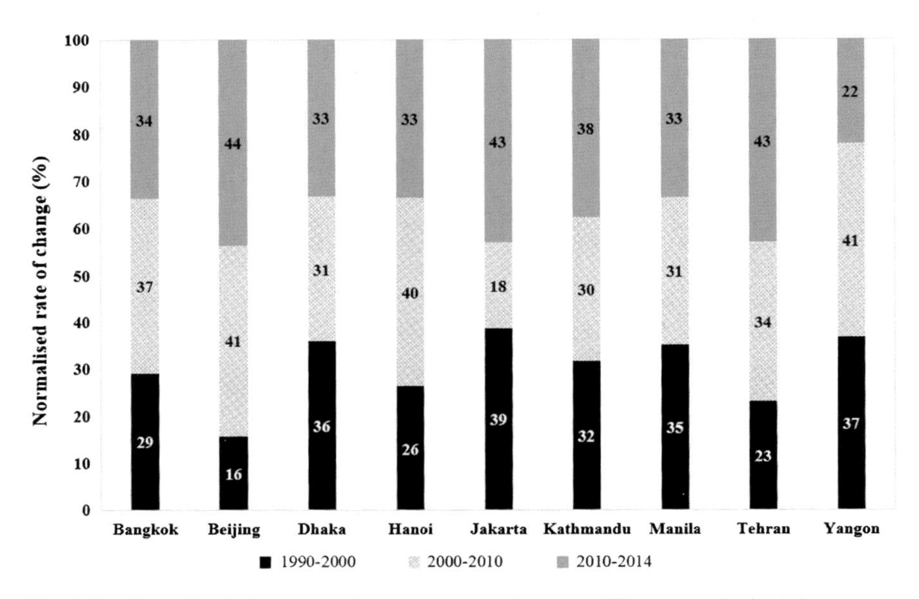

Fig. 2.22 Normalized observed built-up change rate between different epochs in Asia

indicates clearly that built-up changes for all metropolitan areas were nonstationary. The combination of rapid and slow urban growth developments between different time periods (e.g., "1990–2000" and "2000–2010" periods) is challenging for simulating unconnected newly built-up areas based on BRT-CA and RF-CA models. This is because statistical or machine learning algorithms have difficulty in handling nonstationarity (The Sate of Land Change Modeling 2014).

2.4.3.2 Metropolitan Areas in Africa

For Africa, non-built-up persistence had the largest components of agreement for all the models as was observed in Asia (Figs. 2.23 and 2.24). This is because non-built-up persistence occupied more than 50% of the metropolitan areas between 1990 and 2010. Furthermore, Bamako, Dakar, Johannesburg, Lilongwe, and Nairobi metropolitan areas had slightly more hits than the combined misses and false alarms (Figs. 2.23 and 2.24), indicating that the RF-CA model performed relatively well. However, Harare metropolitan area had more combined misses and false alarms than hits. This is because the RF model failed to predict unconnected newly built-up areas, particularly in unplanned and illegal settlement areas. In addition, uncertainty is increased due to the high-temporal nonstationarity (Fig. 2.25). Consequently, the RF model had difficulty modeling the unbalanced land outcomes, namely the combination of rapid and slow urban growth

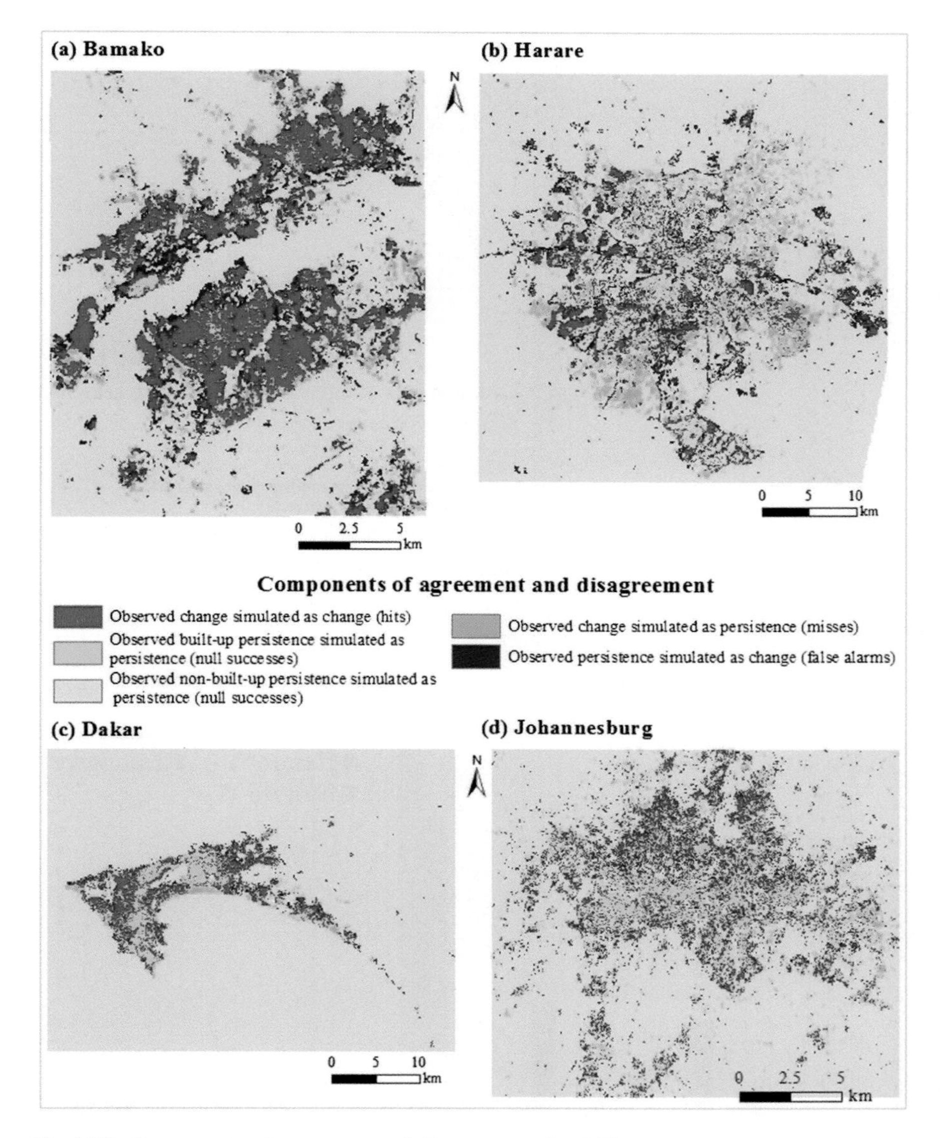

Fig. 2.23 Components of agreement and disagreement for Africa

developments, which occurred during the "1990–2000" and "2000–2010" periods (Fig. 2.25). For example, the rate of "non-built-up to built-up" change between 1990 and 2000 was approximately 114.4 km^2, while the "non-built-up to built-up" change slowed to 69.8 km^2 between 2000 and 2010 (Kamusoko and Gamba 2015).

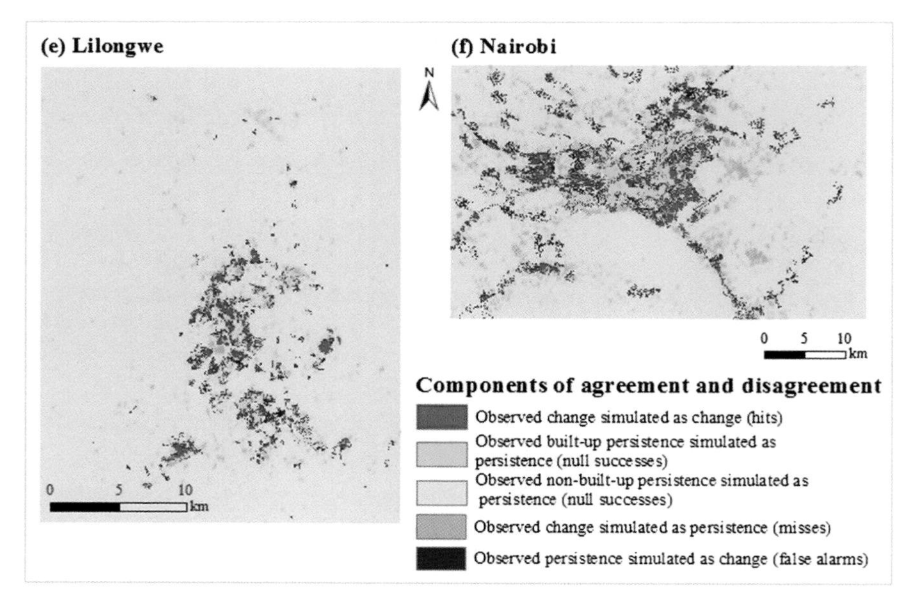

Fig. 2.23 (continued)

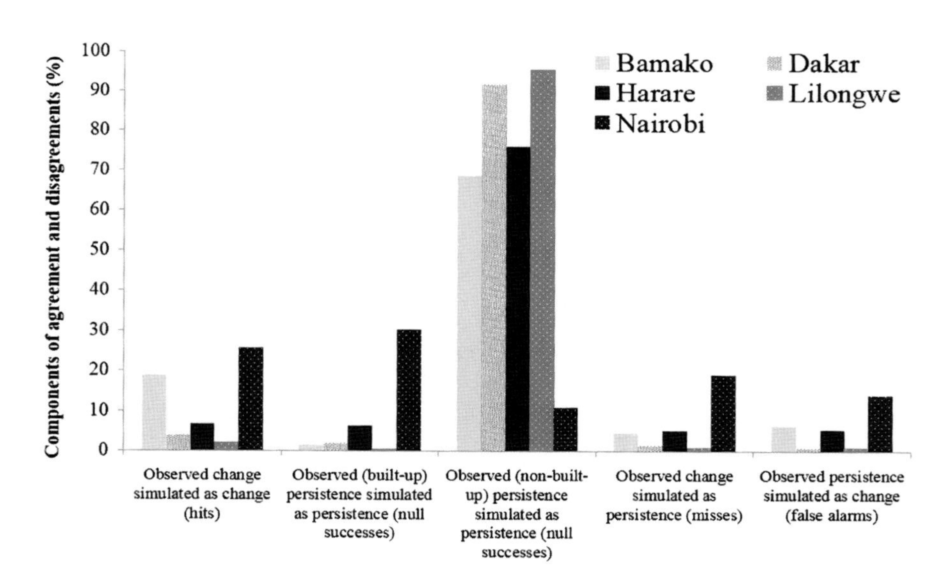

Fig. 2.24 Components of agreement and disagreement expressed as a percentage (Asia)

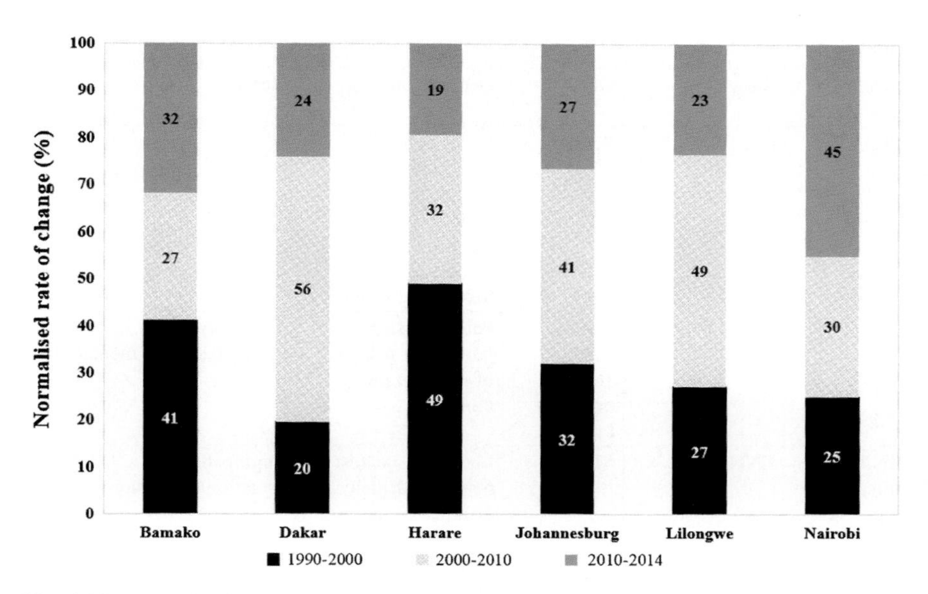

Fig. 2.25 Normalized observed built-up change rate between different epochs in Africa

2.5 Capturing Spatial Pattern of Urbanization

In order to capture, examine, and compare the spatial pattern of urbanization in all the cities, we used five spatial metrics. These include PLAND, PD, ENN, CIRCLE, and SHAPE (Table 2.7). This set of spatial metrics was also used recently by the Asian Development Bank (ADB) in a study entitled "Urban Metabolism of Six Asian Cities" (ADB 2014).

In this analysis, all these five spatial metrics were used at the class level (built-up). The 8-cell neighbor rule was used to determine the membership of each pixel to a patch. In this rule, all the four orthogonal and four diagonal neighbors of the focal cell are used. In the 8-cell neighbor rule, two cells of the same LULC class that are diagonally touching are considered as part of the same patch, but in the case of the 4-cell neighbor rule, these are considered separate patches (McGarigal et al. 2012). We selected the 8-cell neighbor as it has been used in various studies (e.g., Townsend et al. 2009; Estoque and Murayama 2013, 2016; Estoque et al. 2014). The five spatial metrics were computed using FRAGSTATS 4.2 (McGarigal et al. 2012).

Table 2.7 List and details of the class-level (built-up) spatial metrics used

Spatial metrics	Range	Unit	Description	Measure
PLAND	$0 < \text{PLAND} \leq 100$	Percent	Percentage of landscape; percentage of built-up relative to the whole landscape (excluding water)	Area
PD	$\text{PD} > 0$	Number per 100 ha or per km^2	Patch density; number of patches of built-up per unit area	Aggregation (subdivision/fragmentation)
ENN (mean)	$\text{ENN} > 0$	Meter	Euclidean nearest neighbor distance; distance to the nearest neighboring patch of the same type, based on shortest edge-to-edge distance	Aggregation (isolation/dispersion)
CIRCLE (mean)	$0 \leq \text{CIRCLE} < 1$	None	Related circumscribing circle; provides a measure of overall patch elongation; measures the circularity of built-up patches. Low values represent circular patches	Shape (geometry)
SHAPE (mean)	$1 \leq \text{SHAPE} \leq \infty$	None	Shape index; measures the complexity of patch shape compared to a standard shape (square) of the same size; measures the irregularity of built-up patches. Low values represent low complexity	Shape (complexity)

Source McGarigal et al. (2012). *Note* The water class was not included in the derivation of the metrics

2.6 Summary and Conclusions

Taking the selected metropolitan areas in Asia and Africa, the purpose of this chapter was to describe the methodology used to produce the land use/cover maps, calibrate and validate LCMs. The specific objectives of this chapter were to evaluate the goodness-of-fit of transition potential maps, validate the simulated land use/cover maps as well as elucidate components of agreement and disagreement. Land use/cover maps were classified from Landsat imagery for 1990, 2000, 2010, and 2014 using the (RF) classifier. Finally, land use/cover changes were simulated based on the BRT-CA and RF-CA models.

Quantitative accuracy assessment for the 1990 land use/cover maps was not conducted because of the unavailability of reference data such as aerial photographs and high-resolution satellite imagery. However, the Atlas of Urban Expansion developed by the Lincoln Institute of Land Policy (Angel et al. 2010) was used to visually check the quality of land use/cover maps for the 1990 epoch (that is, Landsat imagery acquired between 1988 and 1993). Qualitative and quantitative accuracy assessments were conducted for land use/cover maps from 2000, 2010, and 2014 epochs based on very high-resolution images (e.g., QuickBird image) from Google Earth™. Overall land use/cover classification accuracy for all land use/cover maps ranged from 70 to 90% for all the metropolitan areas.

Generally, the BRT-CA and RF-CA models for all metropolitan areas in Asia and Africa performed relatively well. However, the BRT-CA and RF-CA models for metropolitan areas in Africa performed better than the BRT-CA and RF-CA models for metropolitan areas in Asia. The modeling and simulation results presented in this chapter—however limited to selected case studies in Asia and Africa—provide an initial exploration of the machine learning-cellular (ML-CA) models for land change modeling. While urban expansion in Asia and Africa has been acknowledged, to-date spatial LCMs have not been rigorously explored and validated. Of particular importance here is the possibility of improving transition potential modeling using machine learning models. Consequently, this chapter highlights the value and significance of robust calibration, validation and simulation of spatial LCMs—in particular ML-CA models—for the 15 metropolitan areas in Asia and Africa. Therefore, this chapter provides a foundation for calibrating and validating spatial LCMs.

While the simulation results are encouraging, it is also important to acknowledge that all the BRT-CA and RF-CA models fail to simulate newly developed or built-up areas, which are not connected to existing urban built-up areas. Previous studies revealed that statistical or machine learning models underpredict the location of new patches, which are not connected to existing built-up areas (Pontius and Malanson 2005) due to spatial or temporal nonstationarity (Estoque and Murayama 2014; The State of Land Change Modeling 2014). Therefore, issues related to nonstationarity need to be addressed using more temporal land use/cover data (e.g., at 5 year intervals) or combining BRT-CA and RF-CA models with other LCMs. Although some model uncertainties remain, the BRT-CA and RF-CA models developed in this study have potential to improve land change modeling in general, and urban growth modeling and simulation in particular. Given the broader implications of the results from this chapter, further studies should be carried out to test the BRT-CA and RF-CA models using multiple temporal land use/cover maps at a shorter time interval in order to minimize the effects on spatial and temporal nonstationarity.

References

ADB (Asian Development Bank) (2014) Republic of the Philippines national urban assessment. Asian Development Bank, Metro Manila, Philippines

Angel SJ, Parent DL, Civco Blei AM (2010) Atlas of urban expansion. Lincoln Institute of Land Policy, Cambridge, MA

Batty M (1998) Urban evolution on the desktop: simulation with the use of extended cellular automata. Environ Plan B 30:1943–1967

Batty M, Xie Y (2005) Urban growth using cellular automata models. In: Maguire DJ, Batty M, Goodchild MF (eds) GIS, spatial analysis, and modelling. ESRI Press, CA, USA, pp 151–172

Bhaduri B, Bright E, Coleman P, Urban M (2007) LandScan USA: a high resolution geospatial and temporal modeling approach for population distribution and dynamics. Geo J 69:103–117

Breiman L (2001) Random forests. Mach Learn 45:5–32

Cheng J, Masser I (2004) Understanding spatial and temporal processes of urban growth: cellular automata modelling. Environ Plan B 31:167–194

Clarke KC, Hoppen S, Gaydos L (1997) A Self-modifying cellular automaton model of historical urbanization in the San Francisco Bay Area. Environ Plan B 24:247–261

Couclelis H (1985) Cellular worlds: a framework for modeling micro-macro dynamics

Couclelis H (1989) Macrostructure and microbehavior in a metropolitan area. Environ Plann 16:141–154

Dale VH (1997) The relationship between land-use change and climate change. Ecol Appl 17:753–769

Eastman JR, Solorzano LA, Van Fossen ME (2005) Transition potential modeling for land-cover change. In: Maguire DJ, Batty M, Goodchild MF (eds) GIS, spatial analysis, and modelling. ESRI Press, CA, USA, pp 357–385

Elith J, Leathwick JR, Hastie T (2008) A working guide to boosted regression trees. J Anim Ecol 77:802–813

Engelen G (1988) The theory of self-organization and modeling complex urban systems. Eur J Oper Res 37:42–57

Estoque RC, Murayama Y (2013) Landscape pattern and ecosystem service value changes: implications for environmental sustainability planning for the rapidly urbanizing summer capital of the Philippines. Landscape Urban Plan 116:60–72

Estoque RC, Murayama Y (2014) A geospatial approach for detecting and characterizing non-stationarity of land change patterns and its potential effect on modeling accuracy. GISci Remote Sens 51:239–252

Estoque RC, Murayama Y (2016) Quantifying landscape pattern and ecosystem service value changes in four rapidly urbanizing hill stations of Southeast Asia. Landscape Ecol 31:1481–1507

Estoque RC, Murayama Y, Kamusoko C, Yamashita A (2014) Geospatial analysis of urban landscape patterns in three major cities of Southeast Asia. Tsukuba Geoenviron Sci 10:3–10

Friedman JH (2002) Stochastic gradient boosting. Comput Stat Data Anal 38:367–378

Geoghegan J, Villar SC, Klepeis P, Mendoza PM, Ogneva-Himmelberger Y, Chowdhury RR, Turner BL II, Vance C (2001) Modeling tropical deforestation in the Southern Yucatan peninsular region: comparing survey and satellite data. Agri Ecosyst Environ 85:25–46

Google Earth (2015) Google Earth. Accessed on 2 Nov 2013 from http://www.earth.google.com

Hagen A (2002) Multi-method assessment of map similarity. In: 5th conference on geographic information science, 25–27 Apr 2002, Palma de Mallorca, Spain

Kamusoko C, Gamba J (2015) Simulating urban growth using a random-forest cellular automata (RF-CA) model. ISPRS Int J Geo-Inf 4(2):447–470

Lambin EF (1997) Modelling and monitoring land-cover change processes in tropical regions. Prog Phys Geogr 21:375–393

Liaw A, Wiener M (2002) Classification and regression by randomForest. R News 2:18–22

Liu Y (2009) Modelling urban development with geographical information systems and cellular automata. CRC Press, Taylor & Francis Group, Boca Raton

Mas JF, Puig H, Palacio J, Sosa-Lopez C (2004) Modelling deforestation using GIS and artificial neural networks. Environ Model Softw 19:461–471

Mas J, Soares-Filho B, Pontius R, Farfan Gutierrez M, Rodrigues H (2013) A suite of tools for ROC analysis of spatial models. ISPRS Int J Geo-Inf 2:869–887

Matheussen B, Kirschbaum RL, Goodman IA, O'Dennel GM, Lettenmaier DP (2000) Effects of land cover change on streamflow in the interior Columbia river basin (USA and Canada). Hydrol Process 14(5):867–885

McGarigal K, Cushman SA, Ene E (2012) FRAGSTATS v4: spatial pattern analysis program for categorical and continuous maps. Computer software program produced by the authors at the University of Massachusetts Amherst. Accessed 1 July 2015 from http://www.umass.edu/landeco/research/fragstats/fragstats.html

Messina J, Walsh S (2001) 2.5D morphogenesis: modeling landuse and landcover dynamics in the Ecuadorian Amazon. Plant Ecol 156:75–88

NASA (2013) Landsat 8 overview. Accessed on 25 May 2013 from http://www.nasa.gov/mission_pages/landsat/overview/index.html

Pontius RG Jr (2002) Statistical methods to partition effects of quantity and location during comparison of categorical maps at multiple resolutions. Photogram Eng Remote Sens 68:1041–1049

Pontius RG Jr, Malanson J (2005) Comparison of the structure and accuracy of two land change models. Int J Geogr Inf Sci 19:243–265

Pontius RG Jr, Parmentier B (2014) Recommendations for using the relative operating characteristic (ROC). Landscape Ecol 29:367–382

Pontius RG Jr, Si K (2014) The total operating characteristic to measure diagnostic ability for multiple thresholds. Int J Geogr Inf Sci 28:570–583

Pontius RG Jr, Walker R, Yao-Kumah R, Arima E, Aldrich S, Caldas M, Vergara D (2007) Accuracy assessment for a simulation model of Amazonian deforestation. Ann Assoc Am Geogr 97:677–695

Pontius RG Jr, Boersma W, Castella JC, Clarke K, de Nijs T, Dietzel C, Duan Z, Fotsing E, Goldstein N, Kok K, Koomen E, Lippit CD, McConnel W, Sood AM, Pijanowski B, Pithadia S, Sweeney S, Trung TN, Veldkamp AT, Verbug PH (2008) Comparing the input, output, and validation maps for several models of land change. Ann Reg Sci 42:11–37

R Development Core Team (2005) R: a language and environment for statistical computing. R Found Stat Comput. Accessed on 3 Apr 2014 from http://r-prject.kr/sites/default/files/2%EA%B0%95%EA%B0%95%EC%A2%8C%EC%86%8C%EA%B0%9C_%EC%8B%A0%EC%A2%85%ED%99%94.pdf

Ridgeway G (2006) Generalized boosted regression models. Documentation on the R package "gmb", version 1.5.7. Accessed on 3 Apr 2014 from http://www.i-pensieri.com/gregr/gmb.shtml

Rodriguez-Galiano VF, Chica-Olmo M, Abarca-Hernandez F, Atkinson PM, Jeganathan C (2012) Random forest classification of Mediterranean land cover using multi-seasonal imagery and multi-seasonal texture. Remote Sens Environ 121:93–107

Soares-Filho BS, Cerqueira GC, Pennachin CL (2002) Modeling the spatial transition probabilities of landscape dynamics in an Amazonian colonization frontier. BioScience 51:1059–1067

Soares-Filho BS, Rodrigues HO, Costa WLS (2009) Modeling environmental dynamics with Dinamica EGO. Accessed on 3 Aug 2009 from http://www.csr.ufmg.br/dinamica/

The State of Land Change Modeling (2014) Advancing land change modeling: opportunities and research requirements. The National Academies Press, Washington, DC

Tobler W (1979) Cellular geography. In: Gale S, Olsson G (eds) Philosophy in geography. Reidel, Dordrecht, pp 379–386

Torrens PM (2008) Simulating sprawl. Ann Assoc Am Geogr 96:248–275

Townsend PA, Lookingbill TR, Kingdon CC, Gardner RH (2009) Spatial pattern analysis for monitoring protected areas. Remote Sens Environ 113:1410–1420

USGS (2013) Landsat 8 data product information. Accessed on 25 May 2013 from http://landsat.usgs.gov/LDCM_DataProduct.php

Verburg PH, Veldkamp A, Koning GHJ, Kok K, Bouma J (1999) A spatial explicit allocation procedure for modelling the pattern of land use change based upon actual land use. Ecol Model 116:45–61

Verburg PH, Schot PP, Dijst MJ, Veldkamp A (2004) Land use change modelling: current practice and research priorities. Geo J 61:309–324

Visser H, de Nijs T (2006) The map comparison kit. Environ Model Softw 21:346–358

Vliet J, Bregt AK, Hagen-Zanker A (2011) Revisiting Kappa to account for change in the accuracy assessment of land-use change models. Ecol Model 222:1367–1375

Wolfram S (1984) Cellular automata as models of complexity. Nature 311:419–424

Wu F, Webster CJ (1998) Simulation of land development through the integration of cellular automata and multicriteria evaluation. Environ Plan B 25:103–126

Yeh AGO, Li X (2009) Cellular automata and GIS for urban planning. In: Madden M (ed) Manual of geographic information systems. American Society for Photogrammetry and Remote Sensing, Bethesda, MD, USA, pp 591–619

Chapter 3
Rapid Urbanization in Developing Asia and Africa

Akio Yamashita

Abstract This chapter gives an overview of urbanization in Asia and Africa in the late twentieth century. Furthermore, rapid urbanization in the 15 countries that are covered in this book is analyzed using several indices related to population and economics. Many Asian and African countries had long been under Western colonial rule. Most of these countries gained independence after World War II. Following independence, most of them have rapidly developed, and urbanization has progressed. Given the colonial historical background, population and urban functions were dominantly concentrated in primate cities. While the increase in population and industry in the primate cities are considered as symbols of economic growth, rapid urban development resulted in social and environmental problems. To date, these problems have not yet been solved, and are still serious in Asian and African cities. Nevertheless, rapid economic growth has been taking place since the 1980s in Asia, and since 2000 in Africa. While Asia is at a more advanced phase of urban development and its ensuing problems, Africa is following Asia's tracks. In this chapter, a time series of population and GDP growth data is used to gain important insights of rapid urbanization in Asia and Africa.

3.1 Overview of Urbanization

Urbanization can be defined with various indices depending on the field of study. In geography, land use/cover, population, and economic data are often used as indices for explaining urbanization. In this book, land use/cover changes are placed at the core of the analysis of urbanization in Asia and Africa. Details of the selected cities in Asia and Africa are described in Parts 2 and 3, respectively. In this chapter, an overview of urbanization in Asia and Africa in the late twentieth century is provided. In addition, various population and economic indices are used to analyze and discuss rapid urbanization in the 15 countries that are covered in this book.

A. Yamashita (✉)
Faculty of Life and Environmental Sciences, University of Tsukuba, Tsukuba, Japan
e-mail: akio@geoenv.tsukuba.ac.jp

Y. Murayama et al. (eds.), *Urban Development in Asia and Africa*,
The Urban Book Series, DOI 10.1007/978-981-10-3241-7_3

Many Asian and African countries had long been under colonial rule of Western countries, but after World War II, they gained independence one after another. Table 3.1 summarizes the year these countries gained independence (excluding Central Asia). In Asia, many countries became independent from 1945 to 1949, while others attained independence by the 1970s. In Africa, many countries became independent in the 1960s, a little later than the Asian countries. Following independence, most of these Asian and African countries have rapidly developed, and urbanization has progressed.

Due to such a historical background, population and urban functions in these countries were dominantly concentrated in the primate cities where the Western countries established their bases in the colonial period. The superiority of the primate cities was rapidly encouraged even after independence. This resulted in the successional emergence of megacities, which are metropolitan areas with a population of over 10 million. According to UN (2015b), only Tokyo, Osaka, New York, Mexico City, and São Paulo metropolitan areas had a population of over 10 million in 1980. This grew to 23 cities by 2010, including cities in Asia (Tokyo and Osaka in Japan; Beijing, Chongqing, Shanghai, and Shenzhen in China; Manila in the Philippines; Dhaka in Bangladesh; Delhi, Kolkata, and Mumbai in India; Karachi in Pakistan; and Istanbul in Turkey) and Africa (Cairo in Egypt and Lagos in Nigeria). Six more cities joined megacities in 2015, five from Asia (Guangzhou and Tianjin in China; Jakarta in Indonesia; Bangalore in India) and Kinshasa in the Democratic Republic of the Congo in Africa. Although the capitals of the other Asian and African countries do not meet the requirement for a megacity, excess concentrations of population and industry are progressing.

Although population increase and industrial development in primate cities are considered as symbols of economic growth, rapid urbanization or over-urbanization has resulted in social and environmental problems such as increased poverty, water and air pollution. Over-urbanization is a phenomenon referred to as urbanization without industrialization, or population growth without employment creation. More specifically, a large amount of labor flows from the surrounding rural areas into the primate cities due to their excessively concentrated urban function and industry. However, the industry is not yet developed enough to absorb all labor. As a result, there is high unemployment and an increase of the urban poor coupled with insufficient infrastructure development (e.g., transportation and sanitation).

These problems have not been solved yet, and are still serious in Asian and African cities. However, rapid economic growth has been taking place since the 1980s in Asia, and since 2000 in Africa, through Official Development Assistance (ODA) and Foreign Direct Investment (FDI) from developed countries.

Table 3.1 Number of countries by independence year in Asia and Africa

	Before 1945	1945–49	1950s	1960s	1970s	After 1980
Asia	14	11	3	3	4	3
Africa	4	0	6	32	8	4

Source Ministry of Foreign Affairs of Japan

Accordingly, the availability of more employment opportunities has resulted in increased incomes, thus an increase in middle class population. The expanding middle class will play an important role for urban socioeconomies in Asia and Africa in the future.

3.2 Increase of Urban Population

This section contains a comparative analysis based on the data from UN (2015b), of the 15 countries in Asia and Africa that are covered in this book.

First, the total population increases of Asia and Africa are compared with the world average. The total population of Asia was about 1.4 billion in 1950, and increased to about 4.2 billion by 2010 (accounting for about 60% of the global population). The population growth rate in Asia had a similar trend to world population growth rate. Nonetheless, the total population of Africa in 2010 was about 1.0 billion, accounting to less than 15% of the global population. Yet the population growth rate already exceeded 60% between 1950 and 1970. In the 60 years from 1950 to 2010, the population growth rate in Africa was about twice as fast as the global average (Table 3.2). Urban population analysis reveals that the growth rate in Asia and Africa was much higher than that of the total population, which indicates rapid urbanization after the late twentieth century. The growth rates for Asia (654%) and Africa (1114%) from 1950 to 2010 were remarkable, which are about double and triple the global urban population growth rate, respectively (Table 3.3). The ratio of the urban population to the total population has subsequently increased through the years. In 1950, the global urban population ratio was about 30%, while that of Asia and Africa were less than 20%. However in 2010, the urban population ratio in Asia and Africa increased faster than the global ratio, though the increasing global ratio. This also indicates rapid urbanization in Asia and Africa (Table 3.4). Nonetheless, the population concentration in Asian cities is more significant than in the African cities.

Second, demographic data for the 15 countries are analyzed (Table 3.5). Among the nine Asian countries, China had the highest urban population in 1950. Nonetheless, the urban population accounted for only 11.8% of the total population,

Table 3.2 Total population and growth rate of Asia and Africa

	Total population (million)				Growth rate (%)			
	1950	1970	1990	2010	1950–1970	1970–1990	1990–2010	1950–2010
Asia	1403.4	2135.0	3199.5	4164.3	52.1	49.9	30.2	196.7
Africa	229.9	368.1	635.3	1022.2	60.1	72.6	60.9	344.7
World	2532.2	3696.2	5306.4	6895.9	46.0	43.6	30.0	172.3

Source UN (2015b)

Table 3.3 Urban population and growth rate of Asia and Africa

	Urban population (million)				Growth rate (%)			
	1950	1970	1990	2010	1950–1970	1970–1990	1990–2010	1950–2010
Asia	245.1	505.7	1032.3	1847.7	106.4	104.1	79.0	654.0
Africa	33.0	86.6	203.4	400.7	162.3	134.9	97.0	1114.0
World	745.5	1352.4	2281.4	3558.6	81.4	68.7	56.0	377.3

Source UN (2015b)

Table 3.4 Ratio of urban population (%)

	1950	1970	1990	2010
Asia	17.5	23.7	32.3	44.4
Africa	14.4	23.5	32.0	39.2
World	29.4	36.6	43.0	51.6

Source UN (2015b)

Table 3.5 Total and urban populations in 15 countries (million)

	Country	1950	1970	1990	2010
Asia					
	China	550.8	814.6	1145.2	1341.3
		65.0	141.7	302.8	660.3
		11.8	17.4	26.4	49.2
	Philippines	18.4	35.5	61.6	93.3
		5.0	11.7	29.9	45.4
		27.1	33.0	48.6	48.6
	Indonesia	74.8	118.4	184.3	239.9
		9.3	20.2	56.4	119.8
		12.4	17.1	30.6	49.9
	Vietnam	28.3	44.9	67.1	87.8
		3.3	8.2	13.6	26.7
		11.6	18.3	20.3	30.4
	Thailand	20.6	36.9	57.1	69.1
		3.4	7.7	16.8	23.3
		16.5	20.9	29.4	33.7
	Myanmar	17.2	26.2	39.3	48.0
		2.8	6.0	9.7	15.4
		16.2	22.8	24.6	32.1
	Bangladesh	37.9	66.9	105.3	148.7
		1.6	5.1	20.9	41.5
		4.3	7.6	19.8	27.9
	Nepal	8.2	11.9	19.1	30.0
		0.2	0.5	1.7	5.0
		2.7	4.0	8.9	16.7
	Iran	17.4	28.7	54.9	74.0
		4.8	11.8	30.9	51.0
		27.5	41.2	56.3	68.9

(continued)

Table 3.5 (continued)

	Country	1950	1970	1990	2010
Africa					
	Senegal	2.4	4.1	7.2	12.4
		0.4	1.2	2.8	5.3
		17.2	30.0	38.9	42.3
	Mali	4.6	6.0	8.7	15.4
		0.4	0.9	2.0	5.3
		8.5	14.3	23.3	34.3
	Kenya	6.1	11.3	23.4	40.5
		0.3	1.2	3.9	9.5
		5.6	10.3	16.7	23.6
	Malawi	2.9	4.5	9.4	14.9
		0.1	0.3	1.1	2.3
		3.5	6.1	11.6	15.5
	Zimbabwe	2.7	5.2	10.5	12.6
		0.3	0.9	3.0	4.8
		10.6	17.4	29.0	38.1
	South Africa	13.7	22.5	36.8	50.1
		5.8	10.8	19.1	30.9
		42.2	47.8	52.0	61.5

Note upper row—total population; middle row—urban population; lower row—proportion of urban population (%). *Source* UN (2015b)

indicating that more than 80% was rural population at that time. Meanwhile, the Philippines and Iran had higher urban population of 27.1 and 27.5%, respectively. This implies high population concentration in cities during that period. Note that even in 2010, Iran had the highest urban population rate. Nonetheless, the urban population rates in China, the Philippines, and Indonesia were about 50% in 2010, indicating urbanization in these countries was relatively advanced. Among the six countries in Africa, the urban population ratio of South Africa is outstanding. For example, South Africa had an urban population of about 30 million out of a total population of approximately 50 million in 2010 (that is 60% urban population ratio). In contrast, Kenya—the second most populated African country—had an urban population ratio of approximately 23.6%. While the total populations for Malawi, Mali, Senegal, and Zimbabwe range between 12 and 16 million, their degrees of urbanization are different. Senegal had the highest urban population ratio, followed by Zimbabwe, Mali, and Malawi.

Figure 3.1 shows the growth index (derived by dividing the population in n year by the population in 1950) of total population and urban population in nine Asian countries (populations after 2015 are estimated). Generally, urban population

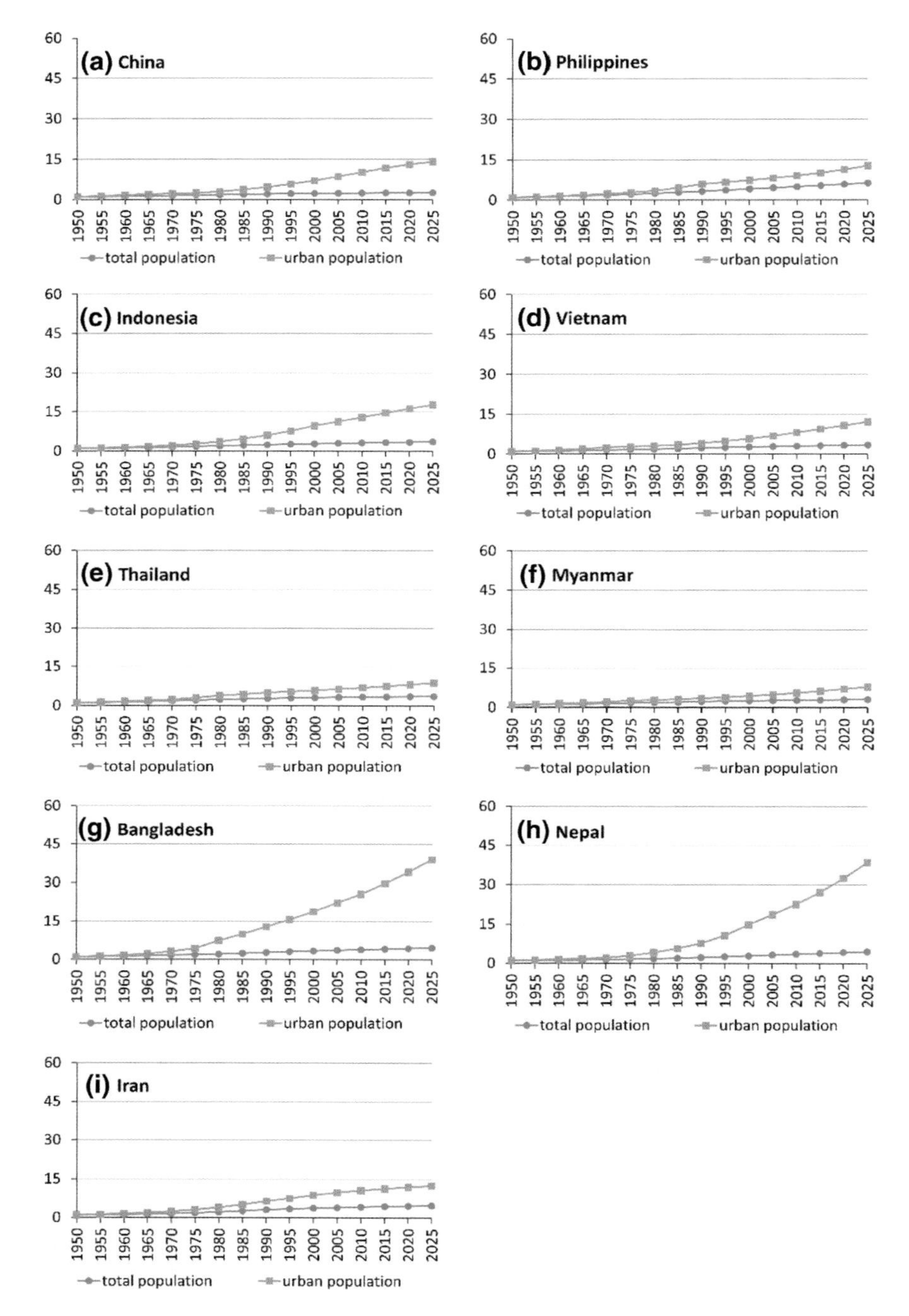

Fig. 3.1 Population growth index (derived by dividing the population in n year by the population in 1950) in nine Asian countries. *Source* UN (2015b)

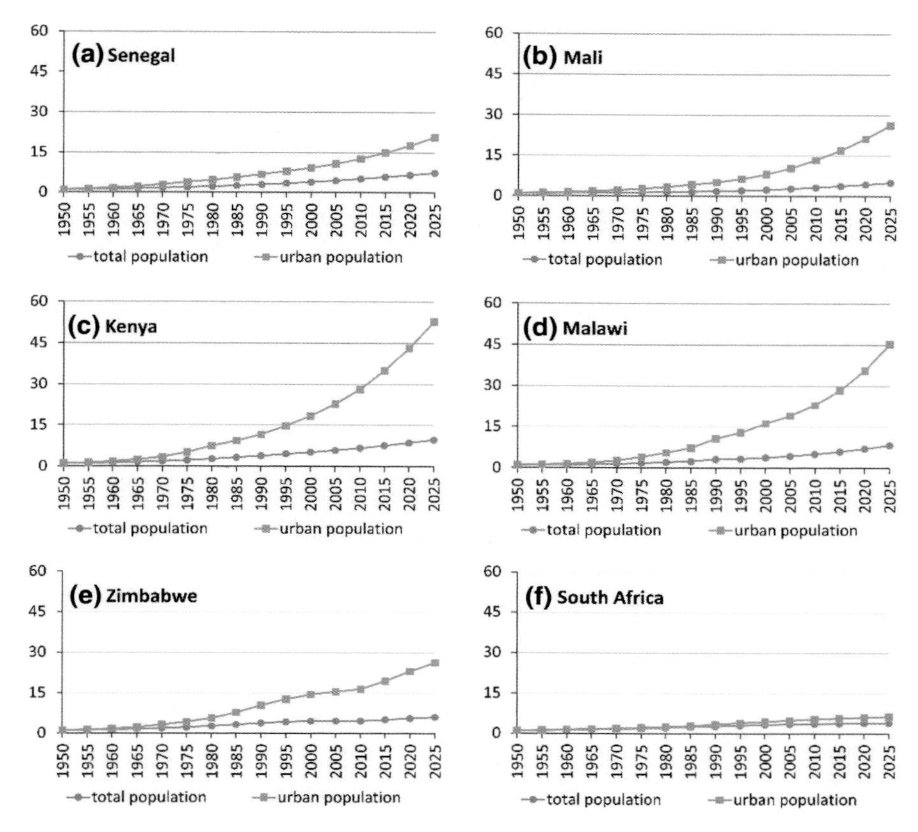

Fig. 3.2 Population growth index (derived by dividing the population in *n* year by the population in 1950) of six African countries. *Source* UN (2015b)

growth rate is higher than the total population growth rate. This suggests that rural–urban migration contributed to urban population growth. Such a tendency is especially strong in Bangladesh and Nepal, where the urban population growth index was 25.6 and 22.7, respectively in 2010. A comparison of demographic data from China, the Philippines, Indonesia, Vietnam, and Iran shows that the total population growth rate is relatively high in the Philippines and Iran. However, the urban population growth rate is higher in China and Indonesia. The urban population growth rate is similar in Thailand and Myanmar, albeit rather slow. The urban population trend shows earlier urbanization in Thailand. However, Myanmar's growth rate is expected to be faster in the future.

Figure 3.2 shows the total population and urban population growth indices for six African countries. In 2010, the total population growth index ranged from 3.3 in

Mali to 6.7 in Kenya. This indicates that there was no significant difference among the six countries in the growth index of total population. However, the urban population growth index varied among the six African countries. Table 3.5 shows that South Africa had a high urban population in 1950, compared to the other five countries. Nonetheless, the urban population in the other five countries increased later due to rapid urbanization. In particular, the urban population growth rates for Kenya and Malawi were higher than others. For example, the urban population growth indices for Kenya and Malawi were 28.1 and 22.9, respectively in 2010. These countries are projected to remain the highest in the future. The urban population growth index for Zimbabwe was 16.4, followed by Mali with 13.4, and Senegal with 12.6.

Table 3.6 Demographics of the 15 countries

Country	Natural increase rate (%)		Birth rate (%)		Mortality rate (%)	
	1985–1990	2005–2010	1990	2010	1990	2010
Asia						
China	18.5	5.7	25.2	12.2	6.7	6.5
Philippines	27.3	19.5	34.1	25.6	6.9	6.1
Indonesia	19.2	14.1	27.5	21.3	8.3	7.2
Vietnam	23.2	11.7	29.8	17.3	6.6	5.6
Thailand	14.8	5.1	20.5	12.2	5.6	7.2
Myanmar	18.4	12.6	29.1	21.2	10.7	8.6
Bangladesh	26.5	16.6	37.8	22.5	11.3	5.9
Nepal	25.3	18.3	39.6	25.3	14.3	7.0
Iran	29.1	13.0	38.2	18.1	9.1	5.2
Africa						
Senegal	32.6	30.6	44.8	38.9	12.2	8.3
Mali	27.7	34.0	48.7	47.1	20.9	13.1
Kenya	35.1	27.3	45.1	37.9	10.0	10.6
Malawi	31.0	29.9	51.5	42.5	20.5	12.5
Zimbabwe	31.0	19.2	39.5	35.6	8.5	16.3
South Africa	22.6	7.5	31.1	22.1	8.5	14.7
Country	Total fertility rate		Rate of youth (%)		Rate of elderly (%)	
	1990	2010	1990	2010	1990	2010
Asia						
China	2.8	1.5	28.8	17.4	5.3	8.2
Philippines	4.5	3.3	40.9	33.6	3.1	4.2
Indonesia	3.4	2.5	36.4	28.9	3.8	4.9
Vietnam	3.9	1.9	37.4	23.7	5.7	6.5
Thailand	2.3	1.6	30.2	19.2	4.5	8.9
Myanmar	3.8	2.6	37.6	29.8	4.2	5.0

(continued)

Table 3.6 (continued)

Country	Total fertility rate		Rate of youth (%)		Rate of elderly (%)	
	1990	2010	1990	2010	1990	2010
Bangladesh	5.0	2.5	42.3	32.1	3.1	4.7
Nepal	5.3	3.0	42.5	37.2	3.5	5.0
Iran	5.6	1.8	45.4	23.5	3.3	4.9
Africa						
Senegal	6.9	5.2	47.1	43.7	3.0	3.1
Mali	7.2	6.7	46.7	47.1	3.8	2.7
Kenya	6.5	4.8	49.0	42.6	2.7	2.7
Malawi	7.3	5.8	45.8	46.2	2.7	3.2
Zimbabwe	5.7	4.0	46.1	41.5	3.0	3.2
South Africa	4.0	2.6	38.9	30.9	3.2	5.1

Source UN (2015a)

Another demographic data (between 1990 and 2010) from UN (2015a) was analyzed for the 15 countries. The rate of natural increase (RNI), birth rate, mortality rate as well as youth and aging populations were examined (Table 3.6).

The RNI for Malawi, Kenya, Senegal, and Zimbabwe exceeded 30% between 1985 and 1990. While the RNI in Asia was mainly 20%, the RNI for China, Indonesia, Thailand, and Vietnam were relatively low. Between 2005 and 2010, the RNI in Asian countries significantly decreased. In contrast, the RNI remained high in Senegal, Mali, Kenya, and Malawi.

Most Asian countries had an approximate birth rate of 30% in 1990 (with highest rates in Bangladesh, Nepal, and Iran), while the birth rate in African countries exceeded 40% with the exception of South Africa. In 2010, the birth rate in Asian countries decreased to 20% (highest in the Philippines and Nepal), while African countries such as Mali and Malawi had a birth rate above 40%. However, South Africa experienced a significant decrease. Although the mortality rate in 1990 was high in Myanmar, Bangladesh, and Nepal, the mortality rate in the remaining Asian countries was below 10%. In 1990, the mortality rate for Malawi, Mali, Kenya, and Senegal was above 10%, while it was below 10% for Zimbabwe and South Africa. In 2010, the mortality rate in all Asian countries (with the exception of Thailand) decreased, while the rate in African countries remained higher than 10% (except for Senegal). The total fertility rate decreased in all Asian and African countries during the 20-year period. However, the rate was lower in Asia than in Africa. The total fertility rate for China, Vietnam, Thailand, and Iran was below 2.0 in 2010. Generally, Asian countries shifted from a high birth and infant mortality rate phase to a low birth and premature death rate phase. To date, African countries still have high birth and infant mortality rates.

In general, Africa has a higher youth population (under 15 years old) and a lower population of elderly people (over 65 years old) than Asia. From 1990 to 2010, most of Asian and African countries experienced a decrease in the youth population ratio (with a slight increase in Mali and Malawi), and an increase in the elderly population ratio (except for Mali and Kenya). Nevertheless, in Asian countries, the youth population rate had remained above 20% in 2010, except in China and Thailand, and the elderly population rate was around 5%, except for China and Thailand where the rate was slightly higher. At this point, the problem of declining birth rates and aging populations is not observed.

3.3 Developing Industry and Economy

Indices related to the industry and economy of each country, employment structure and GDP are investigated.

From an examination of the employment structure in Asian countries around 1990 (Table 3.7), the primary industries were the key industries in except for Iran. The highest employment ratio in the primary industries was found in Nepal, followed by Vietnam and Myanmar. Meanwhile, the secondary industries were relatively advanced in China, and the tertiary industries in the Philippines and Indonesia. In Iran, a lower employment ratio in the primary industries and a higher employment ratio in the secondary and tertiary industries were found. These are perhaps due to the fact that agriculture was not well developed since the country is

Table 3.7 Employment structure of the 15 countries

Country	Primary industry (%)		Secondary industry (%)		Tertiary industry (%)	
	Around 1990	Around 2010	Around 1990	Around 2010	Around 1990	Around 2010
Asia						
China	54.8	39.6	17.6	27.2	9.9	33.2
Philippines	45.8	33.2	14.4	15.0	39.7	51.8
Indonesia	56.6	40.6	13.0	17.5	30.2	41.9
Vietnam	70.3	47.3	10.3	20.7	19.4	32.0
Thailand	64.1	41.1	13.8	19.3	22.0	39.5
Myanmar	70.2	–	8.7	–	21.0	–
Bangladesh	66.4	47.8	12.9	17.5	16.2	35.3
Nepal	81.2	74.1	2.6	10.6	14.9	15.2
Iran	24.5	19.8	26.6	31.7	44.6	48.6
Africa						
Senegal	–	47.2	–	17.0	–	22.2
Mali	–	42.0	–	16.0	–	41.9
Kenya	–	–	–	–	–	–
Malawi	85.8	–	4.4	–	8.1	–
Zimbabwe	–	67.8	–	7.1	–	24.9
South Africa	12.9	7.0	29.3	22.5	53.7	70.1

Note "–" means data is not available. *Source* ILOSTAT

in arid region, while the ratio of oil-related industries was high as an oil-producing nation. By around 2010, the employment ratio in the primary industries had decreased in all countries, and the employment ratio in the tertiary industries was the highest in the Philippines and Indonesia. In the nine Asian countries, the employment ratio of the secondary industries still tends to be smaller than the tertiary industries. This is perhaps the reason for the immaturity of the secondary industries with a higher employment capacity rather than the expansion of the tertiary industries in the urban area. In any case, the employment ratio of the primary industries in 2010 was also much higher than in developed countries, with the exception of Iran. Although much of the data on African countries in the ILOSTAT is unorganized, the employment ratio in the primary industries stood out in Malawi around 1990, and the rates in both secondary and tertiary industries were below 10%. In South Africa, however, the tertiary industries were most developed. The country had an employment structure similar to that of developed countries, and the employment ratio of the primary industries was only 12.9% even around 1990. Around 2010, the employment ratio of the primary industries was very high in Zimbabwe, and its ratio of secondary industries was less than 10%. The secondary and tertiary industries in Senegal and Mali are developed to a degree similar to that in Asian countries. South Africa is increasingly on the road toward a service economy: the employment ratio in the tertiary industries exceeded 70%, but the ratio of the primary industries is only 7.0%.

Next, GDP was examined. GDP in Asia was around 500 billion USD in 1970, and increased to around 20 trillion USD by 2010, accounting for about 30% of global GDP. The growth rate per 10 years shows a trend of rapid increase in the 1970s and slight stagnation in the 1990s, similar to the global trend, though the growth rate was much higher in Asia. In Africa, GDP in 2010 was 1.7 trillion USD,

Table 3.8 GDP and its growth rate of Asia and Africa

	GDP (billion USD)					Growth rate (%)			
	1970	1980	1990	2000	2010	1970–1980	1980–1990	1990–2000	2000–2010
Asia	506.1	2489.6	5511.5	9218.0	20,279.3	391.9	121.4	67.2	120.0
Africa	90.3	434.4	495.0	599.6	1747.9	381.1	13.9	21.1	191.5
World	3355.6	12,043.2	22,595.4	32,859.7	64,406.4	258.9	87.6	45.4	96.0

Source UN National Accounts of Main Aggregate Database

Table 3.9 GDP per capita and its growth rate of Asia and Africa

	GDP per capita (USD)					Growth rate (%)			
	1970	1980	1990	2000	2010	1970–1980	1980–1990	1990–2000	2000–2010
Asia	243	966	1715	2480	4869	297.4	77.5	44.5	96.4
Africa	247	909	787	743	1698	268.4	−13.5	−5.6	128.5
World	910	2709	4249	5363	9315	197.8	56.9	26.2	73.7

Source UN National Accounts of Main Aggregate Database

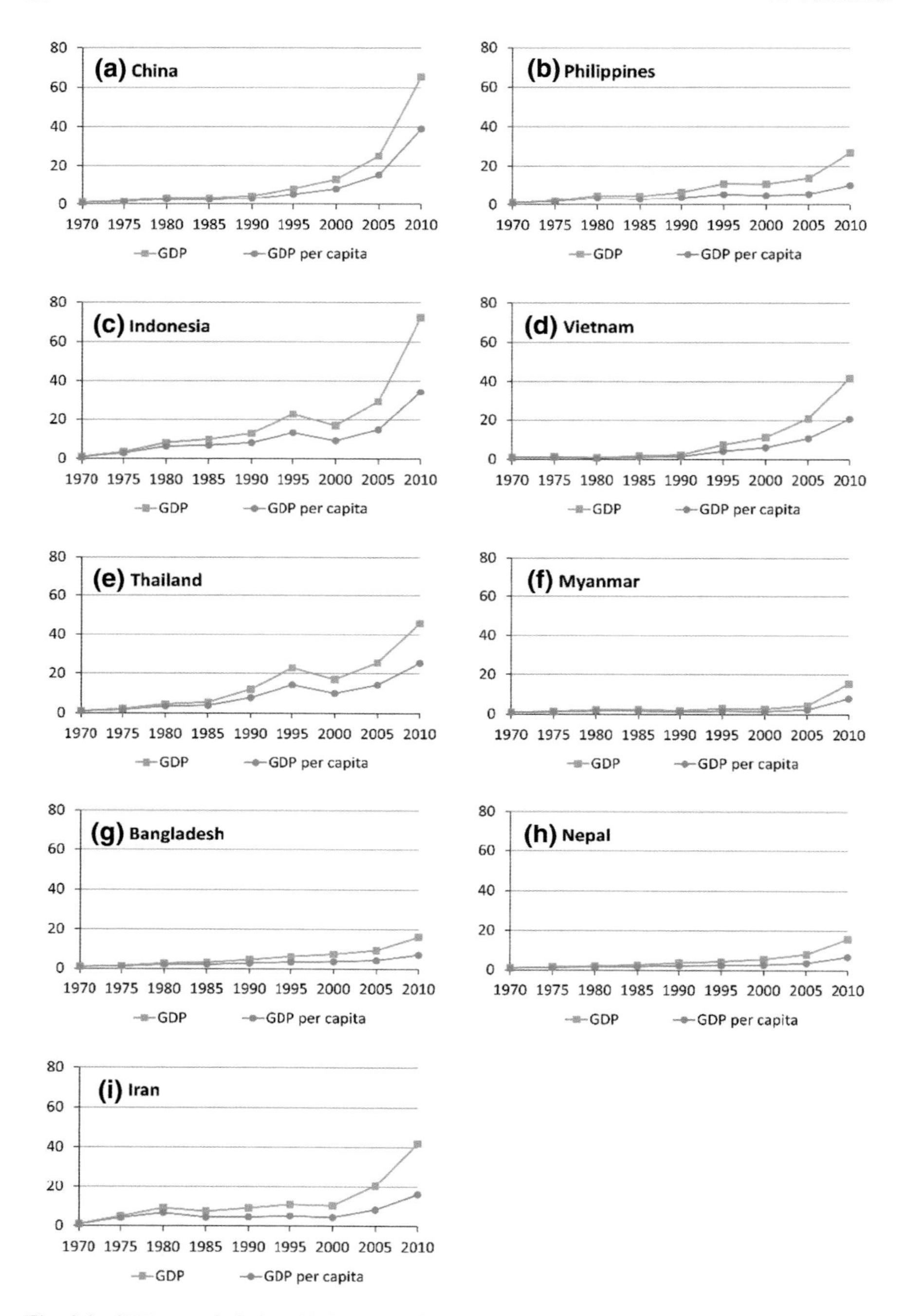

Fig. 3.3 GDP growth index (derived by dividing the GDP/GDP per capita in *n* year by the GDP/GDP per capita in 1970) in nine Asian countries. *Source* UN National Accounts of Main Aggregate Database

Table 3.10 GDP and GDP per capita of the 15 countries

Country	1970	1980	1990	2000	2010
Asia					
China	91,039	306,520	404,494	1192,836	5949,786
	112	311	347	932	4375
Philippines	7413	35,954	49,095	81,026	199,591
	207	759	793	1043	2136
Indonesia	9805	79,636	125,720	165,021	709,191
	86	545	701	790	2947
Vietnam	2775	2396	6472	31,173	115,932
	63	44	94	385	1302
Thailand	7374	33,467	88,299	126,148	338,001
	200	707	1561	2023	5090
Myanmar	2692	5905	5172	7275	41,518
	99	171	123	150	799
Bangladesh	6987	18,866	31,730	51,277	112,412
	105	229	295	387	744
Nepal	1041	2089	3780	5730	16,305
	90	145	209	247	607
Iran	10,032	91,950	91,036	104,016	421,716
	351	2364	1615	1578	5663
Africa					
Senegal	952	3254	6205	4680	12,886
	226	584	826	475	995
Mali	294	1517	2510	2655	9400
	51	225	315	259	672
Kenya	2195	9165	11,037	12,604	32,181
	195	563	471	403	787
Malawi	546	2108	2985	2970	6752
	121	338	316	262	450
Zimbabwe	2023	7148	11,738	7549	7433
	388	981	1122	604	568
South Africa	17,907	80,544	112,014	132,878	363,241
	796	2770	3044	2963	7060

Note upper row—GDP (million USD); lower row—GDP per capita (USD). *Source* UN National Accounts of Main Aggregate Database

accounting for less than 3% of global GDP. However, the growth rate after 2000 was higher than in Asia, and its economies are rapidly growing at a rate twice as fast as the world average (Table 3.8). In terms of GDP per capita (Table 3.9), in Asia, it increased from 243 USD in 1970, which was around 25% of the world average, to 4869 USD in 2010, which was about half of the world average at that time. In Africa, it was about the same as Asia in 1970, but decreased from 1980 to

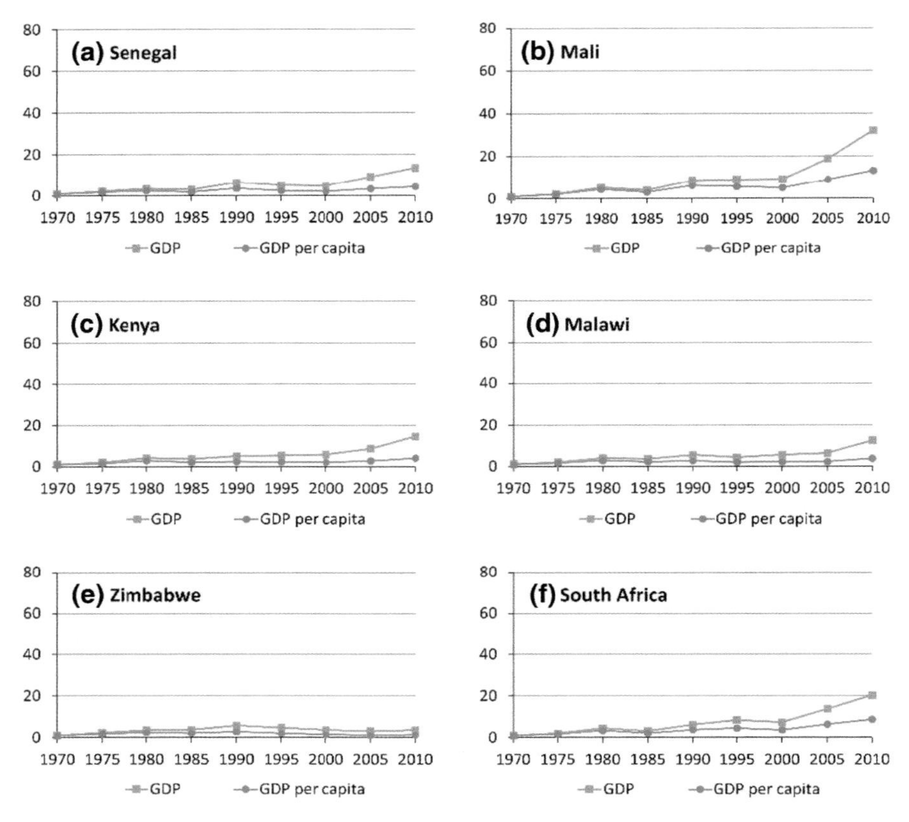

Fig. 3.4 GDP growth index (derived by dividing the GDP/GDP per capita in *n* year by the GDP/GDP per capita in 1970) in six African countries. *Source* UN National Accounts of Main Aggregate Database

2000. Although it quickly increased after 2000, it was about one-third of that in Asia in 2010.

Through individual investigation of GDP for the 15 countries (Table 3.10), China was found to have an extremely high GDP among Asian countries. It remarkably grew by about 6400% in 40 years, from 91 billion USD in 1970 to 5.9 trillion USD in 2010. However, in terms of GDP per capita, oil producer Iran is the highest, followed by Thailand. Vietnam's GDP per capita increased from below 100 USD in 1990 to above 1000 USD in 2010. When focusing only on the growth rates of GDP and GDP per capita, Vietnam's growth rates exceeded China's during these 20 years. Figure 3.3 shows the growth indices of GDP and GDP per capita (derived by dividing the GDP/GDP per capita in *n* year by the GDP/GDP per capita in 1970) for the nine Asian countries. According to this figure, the growth rates of China and Indonesia were the highest, followed by Vietnam, Thailand, and Iran. Economic growth in the Philippines, Indonesia, and Thailand began in the late

1980s, but subsided due to the impact of the Asian Financial Crisis in 1997. The rapid growth was resumed after 2000. Rapid economic growth occurred after 2005 in countries such as Vietnam and Iran.

South Africa is outstanding among the six African countries, however, the average GDP and GDP per capita are lower compared with Asia, and the growth rate is also not as high. In Zimbabwe, GDP decreased after 1990 due to economic turmoil, and GDP per capita decreased by half in the 20 years from 1990 to 2010 (Table 3.10). Figure 3.4 shows economic growth began after 2000 in Senegal, Mali, Kenya, and South Africa. The growth rate in Mali and South Africa was particularly high. Economic growth of Malawi began after 2005.

3.4 Concluding Remarks

Many Asian and African countries went independent after World War II, and experienced a rapid population increase and economic growth in the late twentieth century, particularly after the 1980s. This resulted in excess concentrations of population and industry in the primate cities. The speed of such transition was much faster than any Western developed countries had ever experienced. However, this also caused various social and environmental problems. Asia is at a more advanced phase in terms of such urban development and occurrence of problems, and Africa is following Asia's track with a given time lag. These trends can be grasped from the analyses discussed in this chapter, such as analysis of the time series changes of population increase and GDP growth. On one hand, economic growth in Asia stagnated due to the Asian Financial Crisis in the late 1990s, but resumed after 2000. On the other hand, Africa is in the midst of a growth period at this moment in the twenty-first century, and its rate of population increase and economic growth is even faster than Asia. While the connection between Asian and African countries, and Western developed countries will be enhanced due to economic globalization and advancement of information technologies, the role of Asian and African cities in global politics, economics, and society will become increasingly greater in the future.

References

International Labor Organization (ILO) ILOSTAT database. Accessed on 17 Dec 2015 from https://www.ilo.org/ilostat/

UN (United Nations) (2015a) World population prospects, the 2015 revision. Accessed on 17 Dec 2015 from http://esa.un.org/unpd/wpp/

UN (United Nations) (2015b) World urbanization prospects: the 2014 revision. Accessed on 17 Dec 2015 from http://esa.un.org/unpd/wup/

UN (United Nations) National accounts of main aggregate database. Accessed on 16 Dec 2015 from http://unstats.un.org/unsd/snaama/introduction.asp

Part II
Urbanization in Asia

Chapter 4
Beijing Metropolitan Area

Chiaki M. Akiyama

Abstract This chapter traces the origin and brief history of Beijing Metropolitan Area, one of the megacities of Asia and the national capital region of the People's Republic of China. The urban primacy, urban land use/cover changes, and the driving forces that influence the rapid urbanization of Beijing Metropolitan Area are examined. Their implications for future sustainable urban development of this megacity are also discussed. Beijing Metropolitan Area is one of the global cities. It has a long history as well as unprecedented urbanization process. Over the past 25 years from 1989 to 2014, Beijing Metropolitan Area has urbanized rapidly due to various interrelated factors. The changes in land management system and city functions (that is, change from industrial center to political, cultural, and science and technological center) as well as population and economic growth contributed to urban development. However, key urban issues such as the gap between urban and rural areas, environmental pollution, etc., should be considered for future development. Both the local and national government as well as other stakeholders need to push for the sustainable urban development of Beijing Metropolitan Area.

4.1 Origin and Brief History

Beijing City is the capital city of the People's Republic of China. The city center has played a leading role in the history, politics, culture, and other aspects of civilization in China. In this chapter, the study area is Beijing Metropolitan Area (BMA), approximately measuring 100 km by 100 km. Beijing City covers approximately 80% of BMA.

C.M. Akiyama (✉)
National Institute for Environmental Studies, Tsukuba, Japan
e-mail: mizutani.chiaki@nies.go.jp

Y. Murayama et al. (eds.), *Urban Development in Asia and Africa*,
The Urban Book Series, DOI 10.1007/978-981-10-3241-7_4

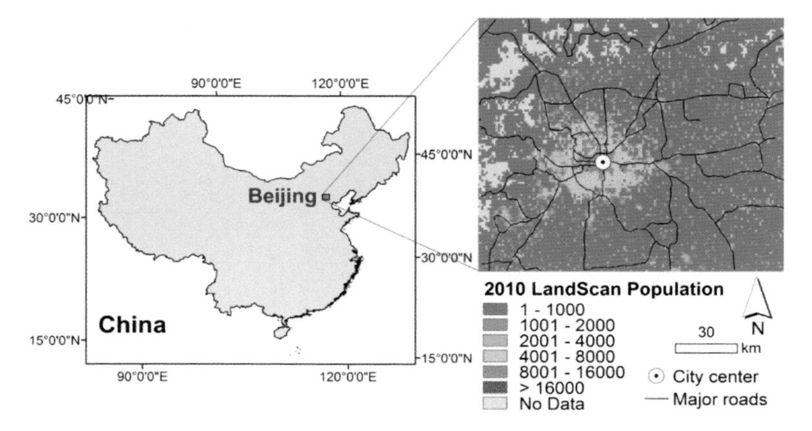

Fig. 4.1 Location and LandScan population of Beijing Metropolitan Area, the People's Republic of China

Beijing is the last of the Four Great Ancient Capitals of China [Beijing, Nanjing, Luoyang, and Chang'an (Xi'an)], which has three millennia of history. Before 1403, Beijing, which was named by Emperor Yongle, was called Dadu, Yangching, and Beiping. Emperor Yongle constructed the Forbidden City for his family. Today, the Forbidden City is one of the seven UNESCO World Heritage Sites in Beijing City.

Beijing City is located in northern China (Fig. 4.1). The north and west parts of the city are hilly, while the south and east parts are flat. Approximately 62% of the city area is mountainous. The highest mountain is East Lingshan (elevation 2303 m) in the north ridge, where the Great Wall is located. The city center lies on low and flat land, with elevation generally between 20 and 60 m above sea level. While the city area is a part of the Hai river basin, the Yongding and Chaobai rivers, which are also located in the city, are drying out. Consequently, water shortage is a serious problem in BMA.

Beijing City is a direct-controlled municipality under the central government. The city has been known as "The People's Government of Beijing Municipality" since 1949, and is characterized by a complex hierarchy of politics. It has to immediately and strictly adopts new policies or laws introduced by the central government.

Figure 4.1 shows the population density in 2010 based on the 1-km resolution LandScan population data. The population is concentrated approximately within the 20 km radius area of the city center, where the population density is more than 4000 people per km^2 (Fig. 4.1). The population density is more than 8000 people per km^2 along major roads, 30 km or more from the city center. Beijing is considered a major transport hub, with the presence of national highways, expressways, railways, high-speed rail networks, and the Beijing Capital International Airport.

4.2 Primacy in the National Urban System

Beijing City is a multifunctional city, and BMA is one of the most urbanized areas in China (Chen et al. 2014). Figure 4.2 shows the (a) Urban area ratio (%), or the percentage of urban area in provincial-level administrative area in 2012, (b) Gross Regional Product (GRP) per person (1000 Chinese Yuan/person), or the economic index for each provincial-level administrative division in 2013, and (c) Tertiary sector ratio of GRP (%), or the percentage of tertiary sector in whole industrial sectors of GRP of each provincial-level administrative division in 2013. Beijing City had an urban area ratio of 72.5%, the second highest in all the 31 provincial-level municipalities in 2013 (Fig. 4.2a). The most urbanized province was Shanghai City with an urban area ratio of 100%. Shanghai City, located on the flat Yangzi River delta, has been historically a business center in China. In contrast to Beijing, topography supported Shanghai's urban development. Tianjin City is the third most urbanized, with an urban area ratio of 20.7%. Although Tianjin City is also a direct-controlled municipality under the national government (as Beijing City and Shanghai City), its urban area ratio is much lower than Shanghai and Beijing.

In China, urban area ratio is high in coastal areas and low in inland areas. While inland areas are bigger than coastal areas, their urban area ratio is less than 1%. The GRP per person of Tianjin, Beijing, and Shanghai are 97.6, 92.2, and 89.4 thousand Chinese Yuan/person, respectively (Fig. 4.2b). Note that one Chinese Yuan was equivalent to 0.16 US dollar in 2015. The GRP per person shows a similar trend, that is, high in coastal area and low in inland areas. The tertiary sector ratio of GRP for Beijing, Shanghai, and Xizang are 76.9, 62.2, and 53%, respectively (Fig. 4.2c). Generally, the main sector has transitioned from primary to tertiary sector in tandem with economic development. Therefore, tertiary sector ratio of GRP represents the degree of urbanization.

Beijing City, the core of BMA, consists of two county-level prefectures, 14 districts, and 289 towns and villages. The city area and the city center area (districts No. 1 and 2) are 16,801.3 and 87.1 km^2, respectively. In 2014, the city area and city center had a population of 21.5 million and 2.2 million, respectively. Although the city center occupies 0.52% of the total city area, it accounts for 10.3% of the total population (Fig. 4.3). Figure 4.3b shows the county-level population density of Beijing City in 2010. The population density decreases with distance from the city center. Districts No. 1 and 2, located at the center of the city, are the most populated. Based on population distribution pattern, 16 divisions are classified into three categories: urban area, suburban area, and rural area. Boundaries of these three areas are characterized by a road network that spreads concentrically from the 2nd Ring Road, the castle wall surrounding the city center (districts No. 1 and 2).

"Urban area" is the most populated area among three categories and includes districts No. 1–6. Both districts No. 1 and 2, where the Forbidden Palace and the Great Hall of the People are located, have since been the nation political center. Additionally, these districts have many tourist attractions, including the Imperial Garden, which is a world heritage site, and the traditional courtyard residence,

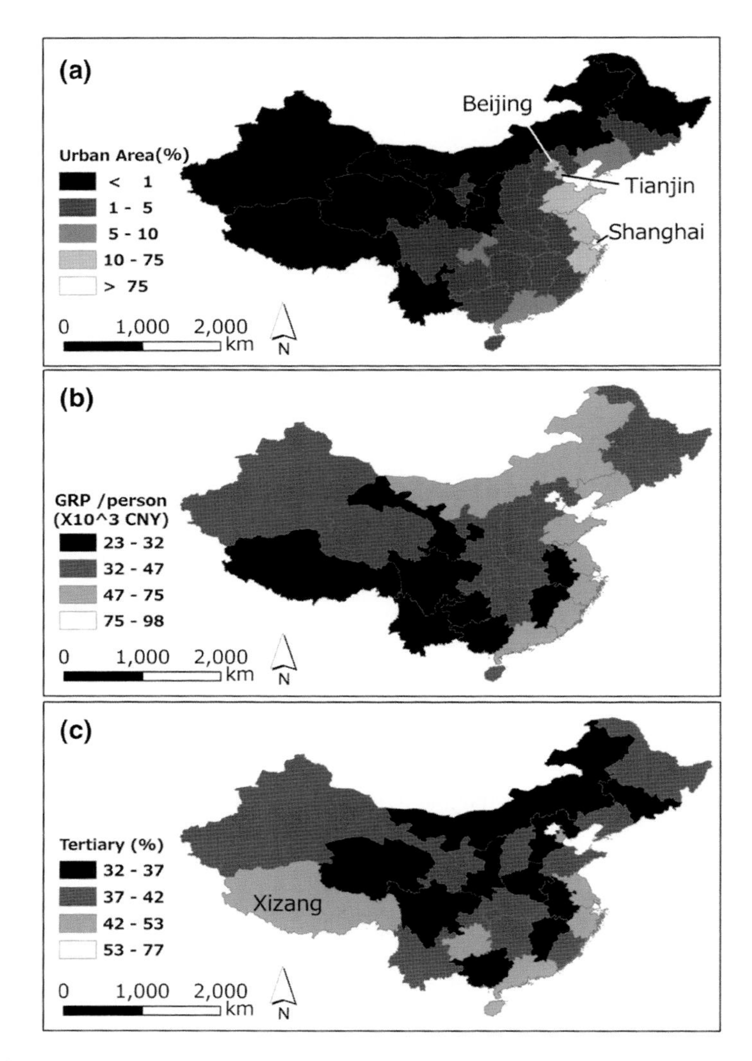

Fig. 4.2 **a** Urban area ratio in 2012 (%), **b** Gross regional product (GRP) per person in 2013 (1000 Chinese Yuan/person), and **c** Tertiary sector ratio of GRP in 2013 (%). *Source* China Statistics Yearbook 2012–2013

named "hutong". For the purpose of preserving the city's historical landscape, it is strictly prohibited to construct high-rise buildings (Fig. 4.4a–c) (Qian and Okazaki 2008). Districts No. 3–6 are located between the 2nd Ring Road and the 5th Ring Road. In these districts, there are many top-ranked educational, cultural, and art institutions: Peking University, Tsinghua University, Chinese Academy of Sciences, the National Stadium designed for the 2008 Summer Olympics and Paralympics Games, and others. In 1988, based on the accumulation of knowledge,

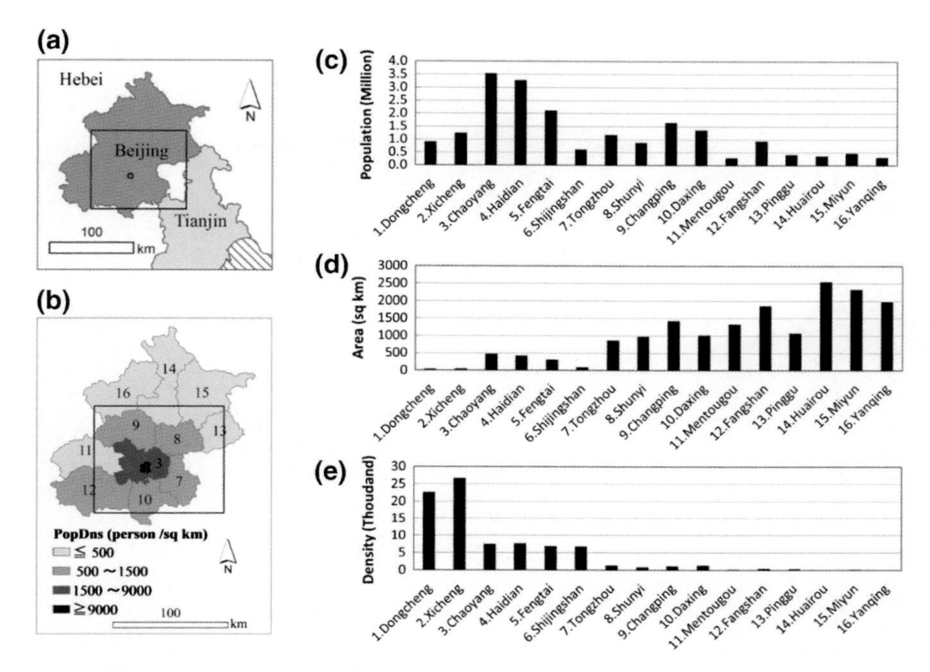

Fig. 4.3 **a** Location of Beijing Metropolitan Area; **b** Population density of Beijing City in 2010 (people per km^2); **c** Population in 2010 (million); **d** Area (km^2); and **e** Population density in 2010 (1000 people per km^2). *Data Source* Population and area [Beijing City (http://www.bjstats.gov.cn)]

the Zhongguancun Science Park (ZSP) at the northeast of the 4th Ring was established as the center of Information and Communication Technology. Today, the ZSP is referred to as "China's Silicon Valley" because of its leadership in information communication technology. Many international technology companies have built their headquarters and research centers in this area.

The primacy of BMA in many fields is supported by its rich transportation system, composed of the international airport, ring roads, and subway network. In 2015, the Beijing Capital International Airport, located in district No. 3, was ranked as the world's second busiest airport by the Airports Council International. The subway network in Beijing City has 18 routes and is planned to increase to 30 routes by 2020. Once the subway network plan is completed, 95% of the residents living inside the 4th Ring Road (Fig. 4.4d) will be 15 min walking distance from the nearest subway station.

Fig. 4.4 Landscape of BMA: **a** Inner city (District No. 1), **b** 2nd Ring Road, **c** Hutong district, **d** Line 5, Lishuiqiao South Station and its surroundings, **e** Changping line, Nanshao station, **f** Construction site in Changping district, **g** Construction site along the highway, and **h** New residential area for rural residents. *Source* (**a**)–(**h**) Author's fieldwork (2015)

"Suburban area" is located around the 6th Road Ring. Districts No. 7–12 are categorized under "suburban area". Many new subway stations have opened in the "suburban area" along with the development of residential area (Fig. 4.4g and h) due to relatively affordable residential spaces.

Districts No. 13–16 represent the "rural area" located outside of the 6th Ring Road. The Great Wall lies on the steep mountain ridge and attracts many tourists from the whole country and all over the world.

4.3 Urban Land Use/Cover Patterns and Changes (1990–2030)

4.3.1 Observed Changes (1990–2014)

The results of the urban land use/cover mapping and change detection (Figs. 4.5 and 4.6) show that the built-up land of BMA has developed and expanded more than three times from 1988 to 2014. With an area of 579.57 km^2 in 1988, BMA's built-up land increased to 2153.92 km^2 in 2014 (Table 4.1) at an annual rate of change (increase) of 60.55 km^2/year. The annual rate of change increased over the observed periods: 20.68 km^2/year (1988–2000 period), 71.06 km^2/year (2000–2009 period), and 137.35 km^2/year (2009–2014 period) (Table 4.2).

In the context of landscape pattern, the percentage of landscape (PLAND) metric measures the proportion of a particular class at a certain point in time relative to the whole landscape. In 1988, the PLAND of BMA's built-up lands was 5.35%. It increased to 7.71, 13.52, and 19.95% in 2000, 2009, and 2014, respectively

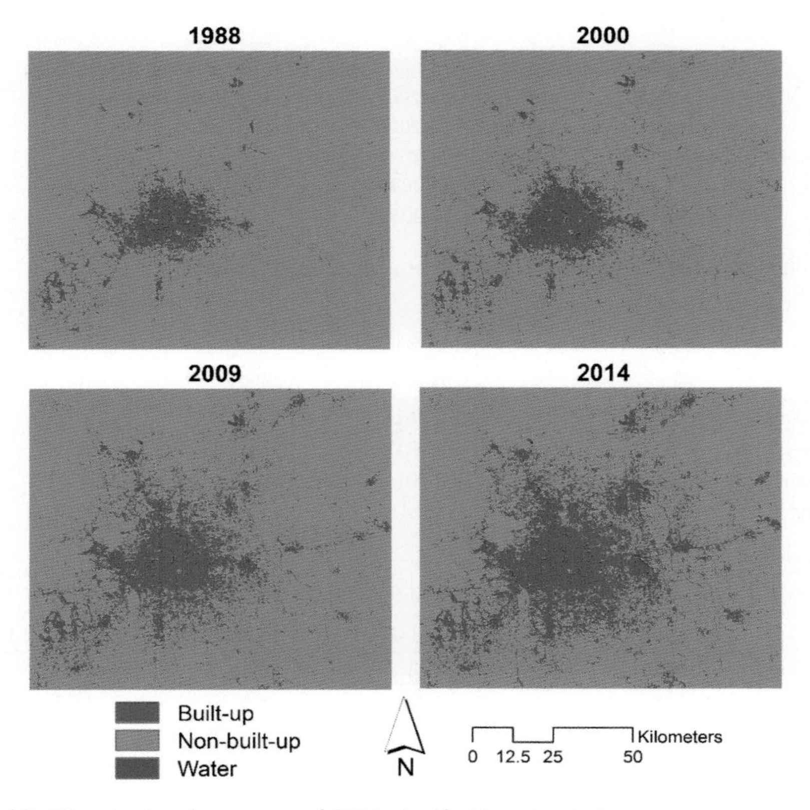

Fig. 4.5 Urban land use/cover maps of BMA classified from Landsat imagery

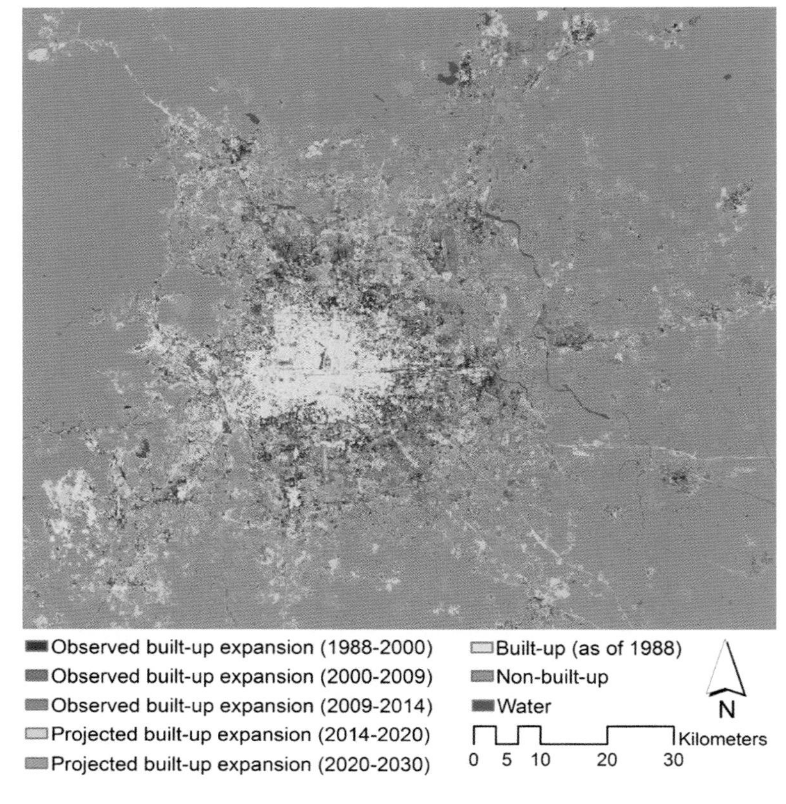

Fig. 4.6 Observed and projected urban land use/cover changes in BMA

Table 4.1 Observed urban land use/cover of BMA (km^2)

	1988	2000	2009	2014
Built-up	579.57	827.67	1467.19	2153.92
Non-built-up	10248.70	9904.11	9386.31	8640.29
Water	86.52	183.01	61.28	120.57
Total	10914.78	10914.78	10914.78	10914.78

Table 4.2 Observed urban land use/cover changes in BMA (km^2)

	1988–2000	2000–2009	2009–2014
Built-up	248.10	639.52	686.73
Annual rate of change (km^2/year)	*20.68*	*71.06*	*137.35*
Non-built-up	−344.59	−517.80	−746.02
Annual rate of change (km^2/year)	*−28.72*	*−57.53*	*−149.20*
Water	96.49	−121.73	59.29
Annual rate of change (km^2/year)	*8.04*	*−13.53*	*11.86*

Table 4.3 Observed landscape pattern of BMA

Class-level (built-up) spatial metrics	1988	2000	2009	2014
PLAND (%)	5.35	7.71	13.52	19.95
PD (number per km^2)	0.73	0.99	2.02	2.18
ENN (mean) (m)	200.31	183.45	148.58	141.97
CIRCLE (mean) ($0 \leq$ CIRCLE < 1)	0.44	0.44	0.45	0.46
SHAPE (mean) ($1 \leq$ SHAPE $\leq \infty$)	1.23	1.23	1.24	1.24

(Table 4.3). The patch density (PD) metric is a measure of fragmentation based on the number of patches per unit area (in this case per 100 ha or 1 km^2), in which a patch is based on an 8-cell neighbor rule. For the case of BMA, the PD of its built-up lands was 0.73 in 1988 and increased to 0.99, 2.02, and 2.18 in 2000, 2009, and 2014, respectively. The increase in PD during 1988–2014 (Table 4.3) can be due to the diffusion and fragmentation of patches of built-up lands.

The Euclidean nearest neighbor distance (ENN) metric is a measure of dispersion based on the distance of a patch to the nearest neighboring patch of the same class. For BMA, the mean ENN of its built-up patches decreased from 200.31 m in 1988 to 183.45 m, 148.58 m and 141.97 m in 2000, 2009 and 2014, respectively. The decrease in mean ENN from 2009 to 2014 can be due to the expansion of the old built-up patches and the development of new patches of built-up lands in between the old patches, as indicated by the slight increase in PD during the same period.

The related circumscribing circle (CIRCLE) metric measures the circularity of patches. The CIRCLE value ranges from 0 for circular or one cell patches to 1 for elongated, linear patches one cell wide. For BMA, the mean CIRCLE value of its built-up lands showed a slightly increasing trend during the 1988–2014 period (Table 4.3), indicating that built-up patches became more elongated, exhibiting a linear pattern. It can also be noted that the urban expansion of BMA, especially during the later periods, followed the road and rail networks (Figs. 4.5 and 4.6). The increasing trend can be due to the aggregation of much smaller, circular isolated patches.

The shape index (SHAPE) metric is a measure of complexity. This metric has a value of 1 when the patch is square and increases without limit as patch shape becomes more irregular. BMA's built-up lands had at least 1.23 mean SHAPE value across the whole 1988–2014 period (Table 4.3), indicating complexity in the shape of the built-up patches. Although the mean SHAPE value increased slightly to 1.24 between 2000 and 2009, it was stable between 2009 and 2014.

Figure 4.7 shows spatial metrics for the built-up lands of BMA (1988–2014) along the gradient of the distance from the city center. The figure shows that PLAND decreases as the distance from the city center increases, indicating that the proportion of built-up lands near the city center was relatively higher. PD increases first as it approaches the 15 km distance from the city center.

The annual maximum of PD moved far from the city center—15 km in 1988 to 35 km in 2014 (Fig. 4.7). This indicates expansion of built-up lands outside of city

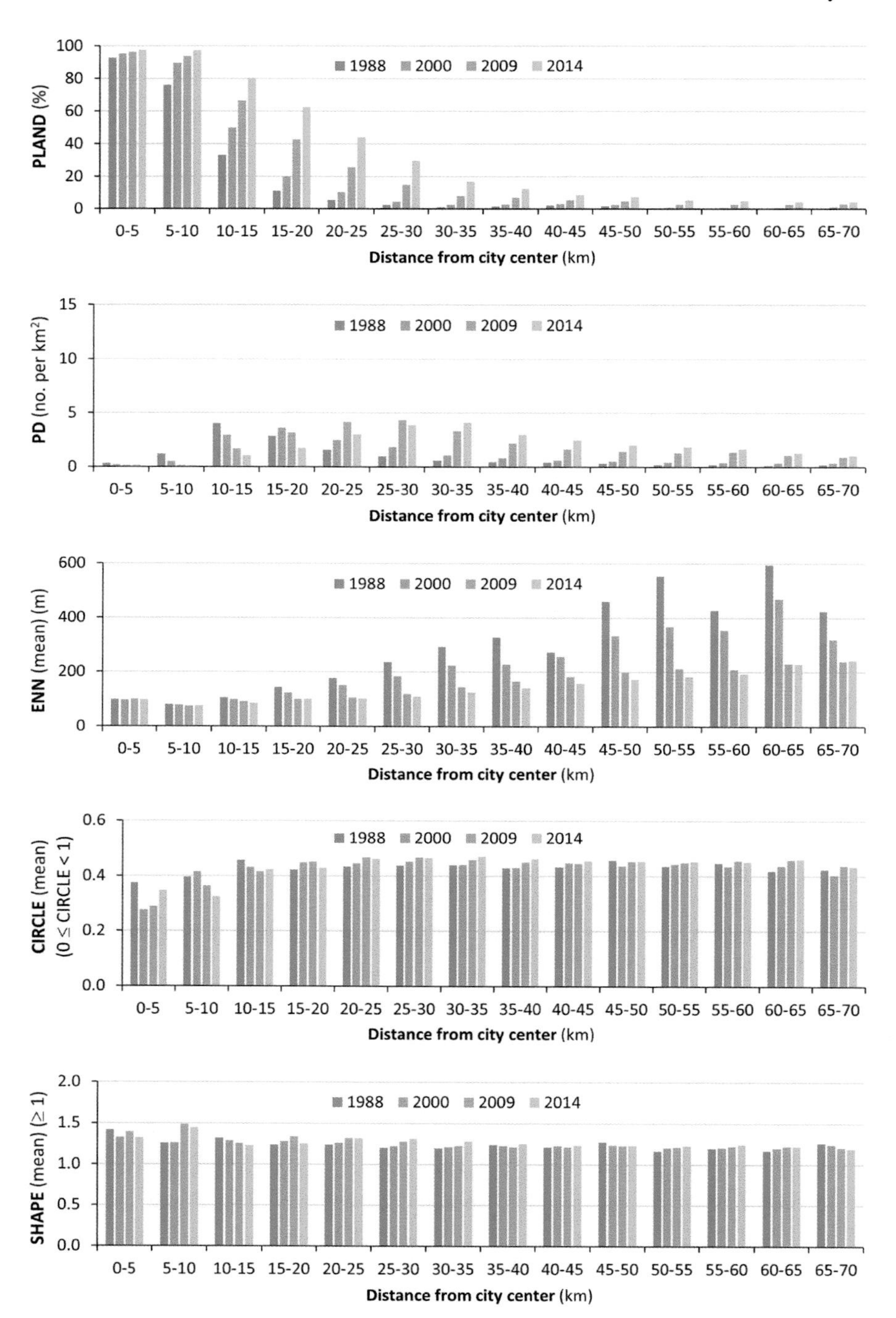

Fig. 4.7 Observed class-level spatial metrics for built-up along the gradient of the distance from city center of BMA. *Note* The y-axis values are plotted in the same range as those in Fig. 4.9

center. The mean ENN also increased with distance from the city center, indicating that patches were dispersed. The mean CIRCLE value increased slightly along the gradient of the distance from the city center. However, despite the variability of these metrics along the gradient of the distance from the city center, the complexity of the built-up patches of BMA was more or less uniform or stable as indicated by the SHAPE metric.

4.3.2 Projected Changes (2014–2030)

The simulation results revealed that the area of built-up lands would increase from 2153.9 km^2 in 2014 to 2530.9 and 3029.4 km^2 by 2020 and 2030, respectively (Table 4.4; Fig. 4.8). The annual rate of change was estimated to be at 62.83 and 49.85 km^2/year, respectively (Table 4.5).

PLAND would increase from 19.95% in 2014 to 23.45 and 28.06% in 2020 and 2030, respectively (Table 4.6). The rate of built-up increase would be, in general, higher in middle distances, between 10 km and 40 km as shown by the simulated increase in PLAND along the gradient of the distance from city center (Fig. 4.9).

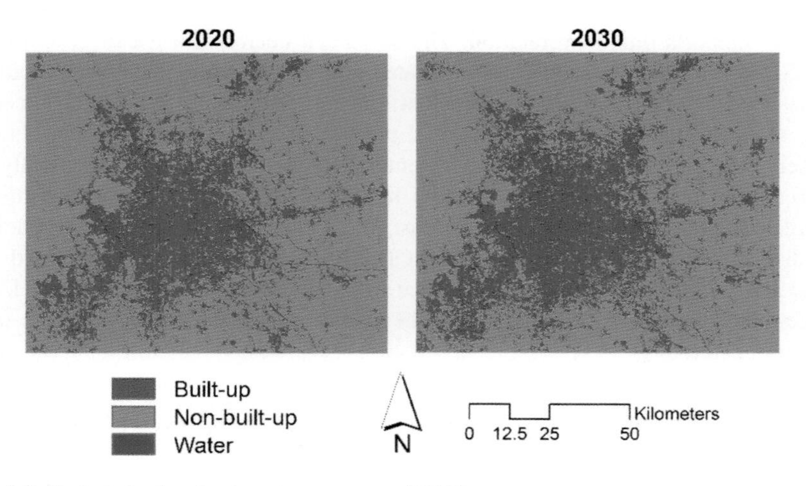

Fig. 4.8 Projected urban land use/cover maps of BMA

Table 4.4 Projected urban land use/cover of BMA (km^2)

	2020	2030
Built-up	2530.91	3029.37
Non-built-up	8263.30	7764.84
Water	120.57	120.57
Total	10914.78	10914.78

Table 4.5 Projected urban land use/cover changes in BMA (km^2)

	2014–2020	2020–2030
Built-up	376.99	498.46
Annual rate of change (km^2/year)	*62.83*	*49.85*
Non-built-up	−376.99	−498.46
Annual rate of change (km^2/year)	*−62.83*	*−49.85*
Water	0.00	0.00
Annual rate of change (km^2/year)	*0.00*	*0.00*

Table 4.6 Projected landscape pattern of BMA

Class-level (built-up) spatial metrics	2020	2030
PLAND (%)	23.45	28.06
PD (number per km^2)	4.86	7.56
ENN (mean) (m)	76.39	74.67
CIRCLE (mean) ($0 \leq$ CIRCLE < 1)	0.25	0.23
SHAPE (mean) ($1 \leq$ SHAPE $\leq \infty$)	1.10	1.08

The simulated increase in PD from 2014 to 2020 and 2030 suggests that built-up patches would be more fragmented in the future. The simulated decrease in mean ENN from 2014 to 2020 and 2030 indicates development of new built-up patches in between, but not connected to, the old patches. The simulated increase in PLAND and PD supports this interpretation. The simulated decrease in the mean CIRCLE and SHAPE values indicates less elongated and low-complex patches of built-up lands, respectively. Along the gradient of the distance from the city center (Fig. 4.9), the PLAND of the simulated patches of built-up lands would also be higher at distances closer to the city center. PD would increase dramatically by 2020 and 2030 from the city center to 35 km, but would still be relatively higher in middle distances (30–40 km). By contrast, the mean ENN would be stable across the distance from city center. The mean CIRCLE value would decrease slightly by 2020 and 2030 beyond the 5 km distance from the city center. The mean SHAPE value would also decrease slightly and be stable beyond the 10 km distance from the city center.

4.4 Driving Forces of Urban Development

During the study period, BMA experienced drastic changes in urban development policy and planning. Urban development policy is deeply related to land use management and socioeconomic development. As a result, "the agriculture to built-up" conversions in Beijing City was due to the demand for land for urban residential space, new public facilities (including transportation network) and business/industrial districts (Long et al. 2013). The following section focuses on the socioeconomic factors such as the functional change, population, and economic growth in BMA.

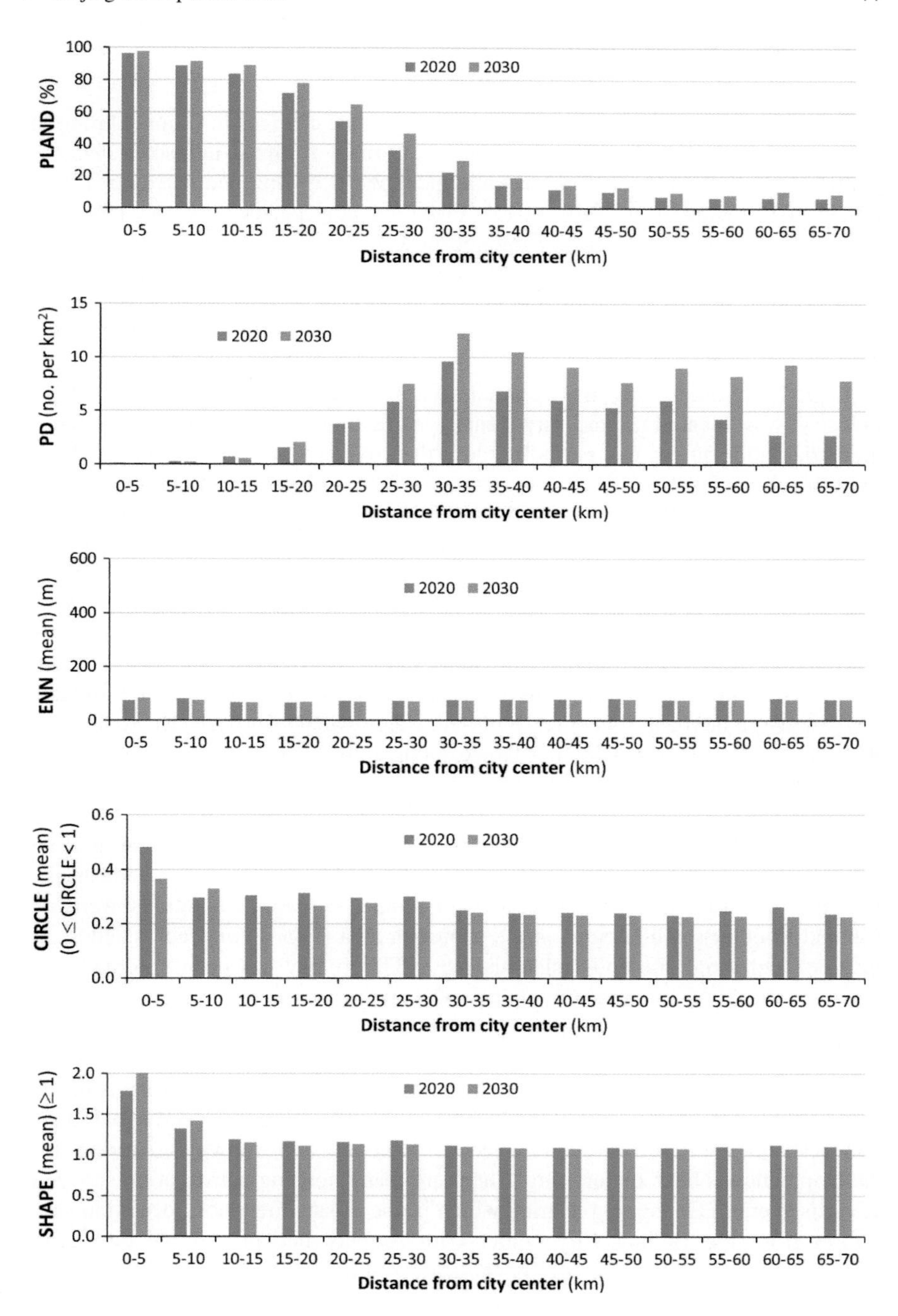

Fig. 4.9 Projected class-level spatial metrics for built-up along the gradient of the distance from city center of BMA. *Note* The y-axis values are plotted in the same range as those in Fig. 4.7

4.4.1 Functional Change

Beijing City has been strongly influenced by dynamic changes in China. During the 26-year study period, Beijing City has transformed from an industrial center to political, cultural, scientific, and technological center (Wang and Nakayama 2008; Zhang 2012; Yang et al. 2013, 2015). From a spatial perspective, this means the transformation of Beijing from a monocentric to a polycentric city. The concentration of all economic activities within the business center of the core city is a sign of a functionally monocentric urban region. However, functionally polycentric regions are determined by the distribution of economic activities in a lot of hubs in the city or region.

Until 1988, urban growth was based on a monocentric city model, whereby the industry was located in the city center. Agricultural lands were converted to industrial lands during this period because the government supplied industrial sites to factory owners free of charge. As a result, low land use efficiency became widespread. Agricultural areas were transformed to wastelands during the "Great Leap", while many industrialized development projects suffered setbacks around 1960. Consequently, there were many vacant lands in areas that were planned as industrial sites.

In order to promote land market development, the Land Administration Law was amended in 1988 to separate land use rights and land ownership (Zhang 1997; Ding 2003). The amendment allowed land users to rent and transfer land use rights as well as charge land users whether they were state-owned or private (Du et al. 2014). As a result, Beijing municipality earned income and incentive to convert agricultural land to industrial area.

After 1988, urbanization policy changed from a monocentric city model to polycentric city model (Beijing Municipal Commission of Urban Planning 1993–2010). As a result, the urban functions in the BMA changed from industrial to political, cultural, art, and scientific, while other economic functions were distributed in other satellite cities in the suburban area (Fig. 4.10). Each satellite city included industrial site and residential area. This policy has been enforced since 2001 for the 2008 Summer Olympics and Paralympics Games (Zhang and Zhao 2009). With economic development, rent in Beijing City center increased rapidly (Ding and Zhao 2014). Many residents have moved to suburban area to secure residential spaces at an affordable price, even though their working place is located at the city center. Moving out of the city brought traffic congestion and expansion of new transportation infrastructure such as highway and subway. Housing land development has been undergoing. There are many housing construction sites with advertisement in Changping district, which is one of satellite cities, located in about 34 km from the Beijing City center with subway network to the city center (Fig. 4.4e–f). Changping district is characterized as having high-tech R & D and tourism services.

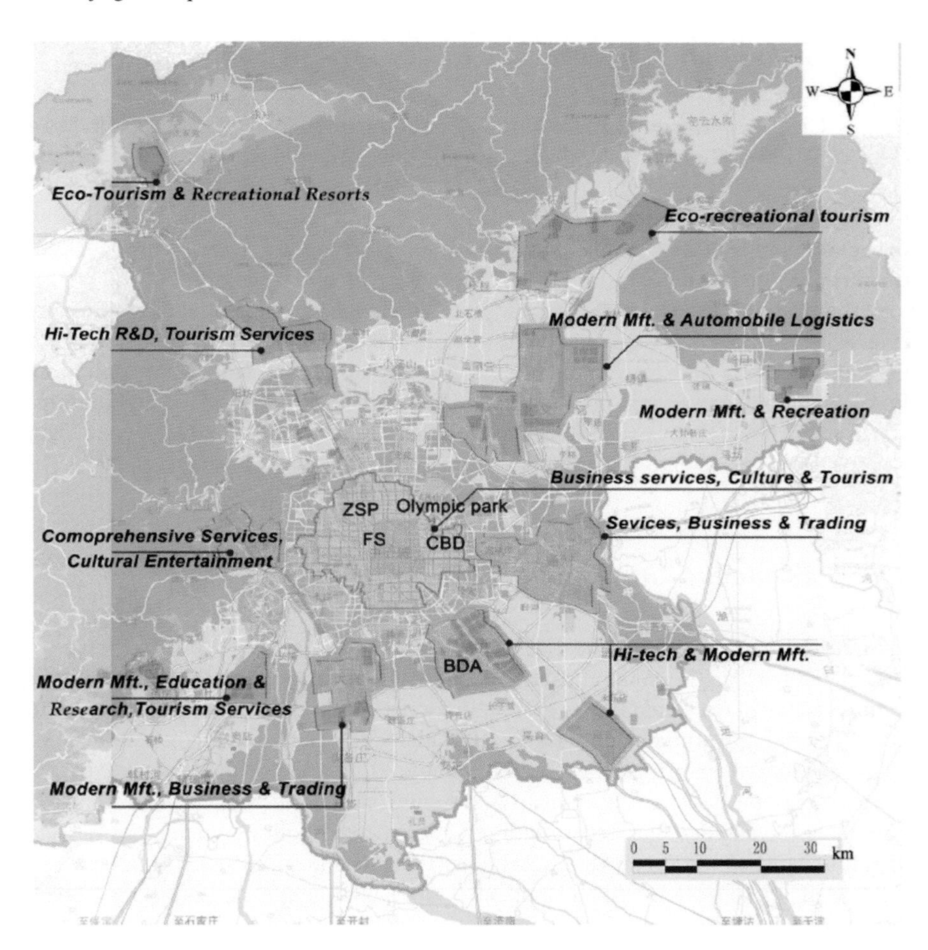

Fig. 4.10 Urban master plan for Beijing City. *Source* Yang et al. (2015)

These changes are reflected in landscape metrics. Between 2000 and 2014, the PLAND and PD of built-up lands increased substantially (Table 4.3), while the mean ENN decreased substantially after the 20 km distance from the city center (Fig. 4.7). These metrics indicate development of new satellite cities with expansion and scattering of patches of built-up lands. According to Fig. 4.9, this trend may continue in the future. These results support the following suggestion by Burger and Meijers (2012): in morphologically monocentric cities, the decrease of density can be observed with increasing distance from the city center; In polycentric form, cities of similar size can be located close to each other (Burger and Meijers 2012).

4.4.2 Population Growth

Beijing City's population increased from 8.7 million in 1978 to 21.5 million in 2014 (Fig. 4.11 left). The rapid population increase was mainly attributed to migrants since Beijing City strictly enforced China's "one child policy". Consequently, migrants increased the demand for residential areas and infrastructure. Originally, the developing area was mainly concentrated in the city center. However, population pressure resulted in the construction of high-rise residential buildings.

Urbanization has expanded to the rural area to release population pressure (Fig. 4.11 right). High demand for residential lands resulted in the conversion of agricultural lands and green spaces as well as increased air and water pollution. Farmers who were living in the rural area were moved to high-rise buildings constructed by the government or developers. These landless farmers were then assigned low-level service works in the urban areas.

4.4.3 Economic Growth

Figure 4.12 (right) presents the Gross Regional Production (GRP) of Beijing City from 1978 to 2014 in nominal price. GRP continues to grow rapidly. The ratio of tertiary sector in the total GRP is gradually increasing year by year. This resulted to both national government and Beijing Municipality encouraging industrial shift from secondary sector to tertiary sector (Zhang 2012). One of the reasons why the governments pushed for the industrial structure shift is to reduce environmental pollutants. To achieve this target, Beijing City relocated polluting industries outside its administrative territory. Although industrial shift and industrial relocation have succeeded, these policies forced not only rapid population increase (Geng et al. 2014) but also expansion of the gap between urban area and rural area (Fig. 4.11 right). Rapid industrialization and urbanization caused many environmental issues such as air pollutants, water pollutants, soil erosion, food security, etc.

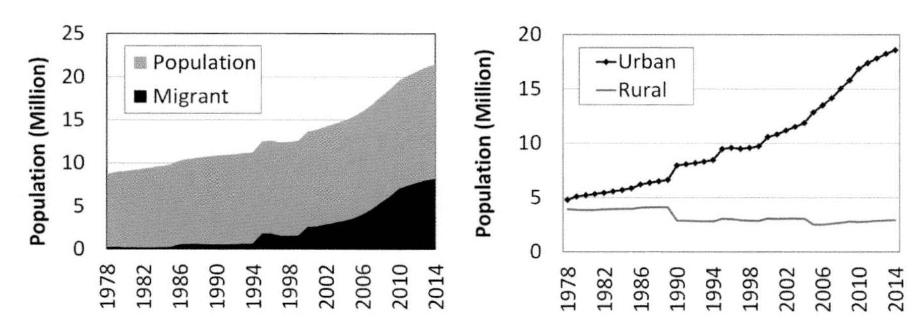

Fig. 4.11 Beijing City: (*left*) Population and migrant; (*right*) Urban and rural population. *Source* Beijing statistical information net (http://www.bjstats.gov.cn/rkjd/shcx/)

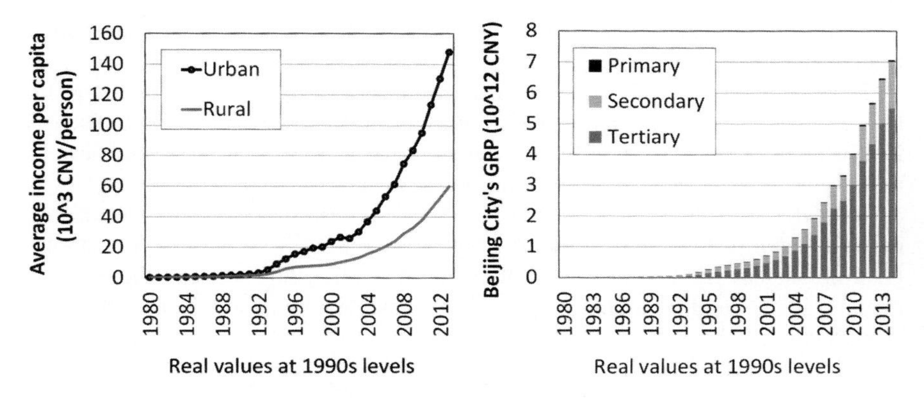

Fig. 4.12 Economic growth in Beijing City and a comparison between urban and rural areas (real values at 1990s levels): (*left*) Urban and rural income per person, (*right*) Beijing's GRP in three sectors. *Source* Beijing statistical information net (http://www.bjstats.gov.cn/), IMF cross country macroeconomic statistics (https://www.quandl.com/data/ODA/CHN_NGDP_D-China-GDP-Deflator)

4.5 Implications for Future Sustainable Urban Development

The results of the urban land use/cover change simulation indicated that the area of built-up lands in BMA would continue to increase in 2020 and 2030. In addition, the PD of built-up lands would also increase. As a result, built-up patches would be more isolated in the future. This implies that urban expansion under the current scenario would continue in the future. Therefore, current environmental problems confronting Beijing City such as air and water pollution as well as water scarcity will also continue to persist based on the urban land use/cover change patterns. Beijing City and the national government are faced with these environmental issues associated with urban development.

Air pollution impacts on both health of residents and business activities. The origins of air pollutants are listed such as yellow sand, automobile emissions, and coal fuel for business use (power generation) and consumer use (heating and cocking). Currently, the national government intends to shift to a low-carbon society and reduce coal fuel use by adopting air pollution-control measures and global warming mitigation measures (Khanna et al. 2014).

BMA suffers from a water shortage because the water resource in the northern part of China is very limited. The increasing demand for water aggravates the problem. Therefore, it is important for BMA to have a stable water resource. The "South-North Water Diversion" project has been in operation, supplying water from the Yangtze River to Beijing, Tianjin and other cities with relatively scarce water resources. It started to supply water to Beijing City in 2014. However, the amount of water supply is less than what was planned due to water quality deterioration; the cost is also high (Zhao et al. 2015). Large-scale project for water resource development faced difficulties to be sustainable in terms of environment and economy.

Considering sustainability, the policy is changing from supply-based management to demand-based management.

Sustainable urban development is characterized by a well-balanced relationship between environmental quality, social justice, and economic progress (Estoque 2017). Rapid economic growth has brought huge urban development and sustainable urban development issues related to social and physical environment management (Cao et al. 2014). At present, the relationship between environmental quality, social justice, and economic progress is getting imbalanced. The national government has tried to keep a well-balanced relationship between the environmental, social, and economic aspects of urban development, even with lower growth target as "the new normal". It is a big challenge and an indispensable target to maintain BMA as ecumene.

4.6 Concluding Remarks

Beijing Metropolitan Area has been growing remarkably. The result of the urban land use/cover change simulation indicated that built-up lands would expand even further from the central area in the near future. Its rapid urbanization has been driven by various interrelated factors. Its land management system has changed. Its main function has changed from an industrial center to political, cultural, and science and technological center. Its population and economic growth together played a key role in its urban development. However, there are a number of key urban issues to be considered in its future development, such as the increasing gap between urban and rural areas, environmental pollution, etc. (Wang et al. 2012). Both the local and national governments, including all sectors of the society, need to push for the sustainable urban development of Beijing Metropolitan Area. If this is resolved, this capital city will be a useful compass for sustainable development.

References

Burger M, Meijers E (2012) Form follows function? Linking morphological and functional polycentricity. Urban Stud 49:1127–1149

Cao S, Lv Y, Zheng H, Wang X (2014) Challenges facing China's unbalanced urbanization strategy. Land Use Policy 39:412–415

Chen M, Huang Y, Tang Z, Lu D, Liu H, Ma L (2014) The provincial pattern of the relationship between urbanization and economic development in China. J Geog Sci 24:33–45

Ding C (2003) Land policy reform in China: assessment and prospects. Land Use Policy 20: 10–120

Ding C, Zhao X (2014) Land market, land development and urban spatial structure in Beijing. Land Use Policy 40:83–90

Du J, Thill JC, Peiser RB, Feng C (2014) Urban land market and land-use changes in post-reform China: a case study of Beijing. Landscape Urban Plan 124:118–128

Estoque RC (2017) Manila metropolitan area. In: Murayama Y, Kamusoko C, Yamashita A, Estoque RC (eds) Urban development in Asia and Africa—geospatial analysis of metropolises. Springer Nature, Singapore, pp 85–110
Geng Y, Wang M, Sarkis J, Xue B, Zhang L, Fujita T, Yu X, Ren W, Zhang L, Dong H (2014) Spatial-temporal patterns and driving factors for industrial wastewater emission in China. J Clean Prod 76:116–124
Khanna N, Fridley D, Hong L (2014) China's pilot low-carbon city initiative: a comparative assessment of national goals and local plans. Sustain Cities Soc 12:110–121
Long Y, Han H, Lai SK, Mao Q (2013) Urban growth boundaries of the Beijing Metropolitan Area: comparison of simulation and artwork. Cities 31:337–348
National Bureau of Statistics of the People's Republic of China. National statistical yearbook. Accessed on 14 Oct 2017 from http://www.stats.gov.cn/tjsj/ndsj/
Qian W, Okazaki A (2008) The formation process of the historical environment conservation system and its present structure in Beijing. J Archit Planning (Transactions of AIJ) 73:1007–1013 (in Japanese with English abstract)
Wang F, Nakayama T (2008) Over the urban development of Beijing and its plan—the study on the development from satellite city to new city. J Archit Planning (Transactions of AIJ) 73:1521–1528 (in Japanese with English Abstract)
Wang J, Chen Y, Shao X, Zhang Y, Cao Y (2012) Land-use changes and policy dimension driving forces in China: present, trend and future. Land Use Policy 29:737–749
Yang Z, Cai J, Otens HFL, Sliuzas R (2013) Beijing. Cities 31:491–506. http://www.sciencedirect.com/science/article/pii/S0264275111000989
Yang Z, Hao P, Cai J (2015) Economic clusters: a bridge between economic and spatial policies in the case of Beijing. Cities 42:171–185
Zhang XQ (1997) Urban land reform in China. Land Use Policy 14:187–199
Zhang J (2012) Historical changes in the land use regulation policy system in Beijing since 1949. J Appl Sci 12:2202–2214
Zhang L, Zhao SX (2009) City branding and the Olympic effect: a case study of Beijing. Cities 26:245–254
Zhao ZY, Zuo J, Zillante G (2015) Transformation of water resource management: a case study of the South-to-North water diversion project. J Cleaner Prod. doi:10.1016/j.jclepro.2015.08.066

Chapter 5
Manila Metropolitan Area

Ronald C. Estoque

Abstract Metro Manila, the national capital region of the Philippines, is one of the megacities in Asia. This chapter traces its origin and examines its urban primacy. It also examines the recent (1993–2014) and potential future (2014–2030) urban land changes, i.e., changes from non-built-up to built-up lands, in Metro Manila and its surrounding areas using geospatial tools and techniques. Some of the possible key factors influencing the urban development of Metro Manila and the potential implications of its rapid population growth and urban land changes to its future sustainable urban development are discussed. The analysis showed compelling evidence for Metro Manila's urban primacy over the other metropolitan areas and regions in the country based on population and gross domestic product. Over the past 21 years (1993–2014), the area of built-up lands has increased almost twofold, transforming the landscape of Metro Manila and its surrounding areas. The relatively small land area of Metro Manila, its geographic characteristics and population and economic growth, the concentration of key urban functions/services and opportunities in the area, and its accessibility are hypothesized to be among the key factors influencing the spatiotemporal patterns of urban land changes and the overall urban development of the region. The simulated urban land changes indicated that built-up lands would continue to expand in the future (2014–2030) under the influence of infill and sprawl development patterns. The intensifying pressure of urbanization due to rapid population growth and urban land changes poses many challenges that need to be considered in sustainable urban development and landscape planning.

5.1 Origin and Brief History

Metropolitan Manila (*Filipino: Kalakhang Maynila, Kamaynilaan*), commonly known as Metro Manila and the National Capital Region (NCR) of the Philippines (Fig. 5.1), is the seat of government and the economic and political center of the

R.C. Estoque (✉)
Faculty of Life and Environmental Sciences, University of Tsukuba, Tsukuba City, Japan
e-mail: estoque.ronald.ga@u.tsukuba.ac.jp; rons2k@yahoo.co.uk

Y. Murayama et al. (eds.), *Urban Development in Asia and Africa*,
The Urban Book Series, DOI 10.1007/978-981-10-3241-7_5

country. It is one of the megacities in Asia. Metro Manila is composed of Manila, the capital city of the country, Quezon City, the country's most populous city, the Municipality of Pateros, and the cities of Caloocan, Las Piñas, Makati, Malabon, Mandaluyong, Marikina, Muntinlupa, Navotas, Parañaque, Pasay, Pasig, San Juan, Taguig, and Valenzuela (Fig. 5.2).

There are various accounts on the origin of the word "Manila". One of which is that Manila was derived from two Tagalog words: "may", meaning "there is"; and "nilad", the name of a shrub that originally grew abundantly along the shores of the Pasig River and Manila Bay (www.aenet.org/philip/manila.htm) (see Fig. 5.2a).

The history of Manila may date back to the year 900 AD based on the Laguna Copperplate Inscription, from which the first reference to "Tondo" is found and thought to be referring to the present-day district of Manila City (Fig. 5.2a) (Postma 1992). Before Miguel Lopez de Legaspi established the capital of Spanish colonization in Asia for the next 300 years (1565–1898), Muslim people were living at the mouth of the Pasig River by the Manila Bay, located in the present-day Metro Manila (Porio 2009) (Fig. 5.2a). The Spanish built the walled city known as *Intramuros* (within the walls) to serve as the seat of their colonial government (Fig. 5.2a).

Due to its central location in the vital Pacific sea trade routes, the richness of the country, its vast undeveloped resources and the opportunities for profitable investment, Manila was called "*The Pearl of the Orient*" (Manila Merchants' Association 1908). The Manila-Acapulco galleon was among the first known commercially traveled trade routes in the world, which lasted from 1565 to 1815

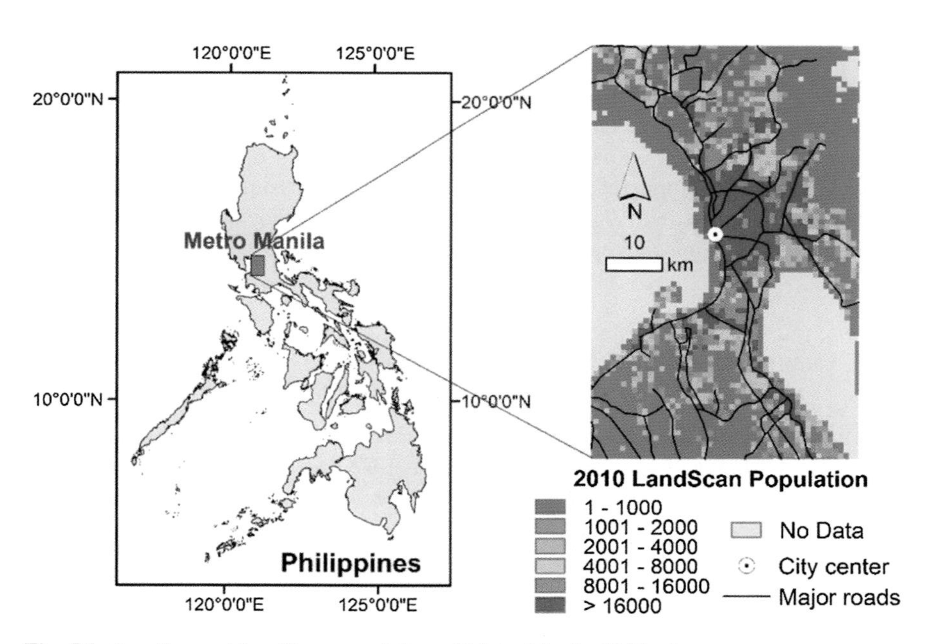

Fig. 5.1 Location and LandScan population of Metro Manila, Philippines

(Fish 2011). This makes Manila as one of the original "global cities". Today, Tondo and Intramuros are among the six districts of the city of Manila (Fig. 5.2a).

The British occupied Manila from 1762 to 1764. At this time, Bacolor, Pampanga became the headquarters of the exiled Spanish colonial government. The British occupation ended by virtue of the Treaty of Paris of 1763. The signing of the Treaty of Paris of 1898 ended the Spanish–American War in the same year (1898). It also ended the Spanish colonial rule in the Philippines. During the Philippine Revolution against Spain (1896–1897) and the Philippine–American War (1898–1902), several places became the headquarters of the revolutionary government, including San Miguel and Malolos in the Province of Bulacan. In 1898, Marikina also became the capital of the then Province of Manila. The American occupation in the Philippines started in 1898 and ended in 1946 (Caoili 1988; Porio 2009).

On August 7, 1901, Manila became a chartered city. In 1905, due to the small area of Intramuros, the American colonial government commissioned the famous American architect and urban planner Daniel H. Burnham to design a plan for the development of Manila. Both the Spanish and American colonial governments initiated the development of Manila as a primate city (Caoili 1988; Porio 2009). However, according to Felino Palafox, Jr., a prominent Filipino architect, an urban planner and environmentalist, Burnham's grand plan for Manila, i.e., grand scale, wide radial boulevards, landscaped parks, and pleasant vistas, was not followed, and this is one of the major loopholes in the urban development of Manila (Macas 2014). In Burnham's grand plan, Manila was envisioned to cope with population growth to an anticipated level of 800,000 people (Alcazaren 2004; Morley 2011). However, in 1948, this population level had already been surpassed when Manila recorded a population of 983,906 people (Stinner and Bacol-Montilla 1981).

During the Japanese occupation (1942–1945), Manila remained the capital city of the country. However, on July 17, 1948, the Congress approved Republic Act No. 333, declaring Quezon City as the capital of the Philippines in place of Manila. On May 29, 1976, the stature of the nation's capital was transferred back to the city of Manila by virtue of Presidential Decree No. 940 under President Ferdinand E. Marcos, Sr. This was made a few months after the creation of the present-day Metropolitan Manila under Presidential Decree No. 824 (November 7, 1975). In 2012, a Quezon City councilor passed a resolution urging the Philippine Congress to enact a law, declaring Quezon City as the "new" capital of the Republic of the Philippines. As of writing, Manila City remains as the capital of the country.

Since the creation of Metro Manila 40 years ago, its member local government units (LGUs) have grown and all of them, except one (Pateros), have been converted into cities. The cityhood of each of these LGUs provides an overview of the spatiotemporal pattern of the socioeconomic growth and urban development of Metro Manila (Fig. 5.2a). Figures 5.3, 5.4, and 5.5 provide a glimpse of the present-day Metro Manila.

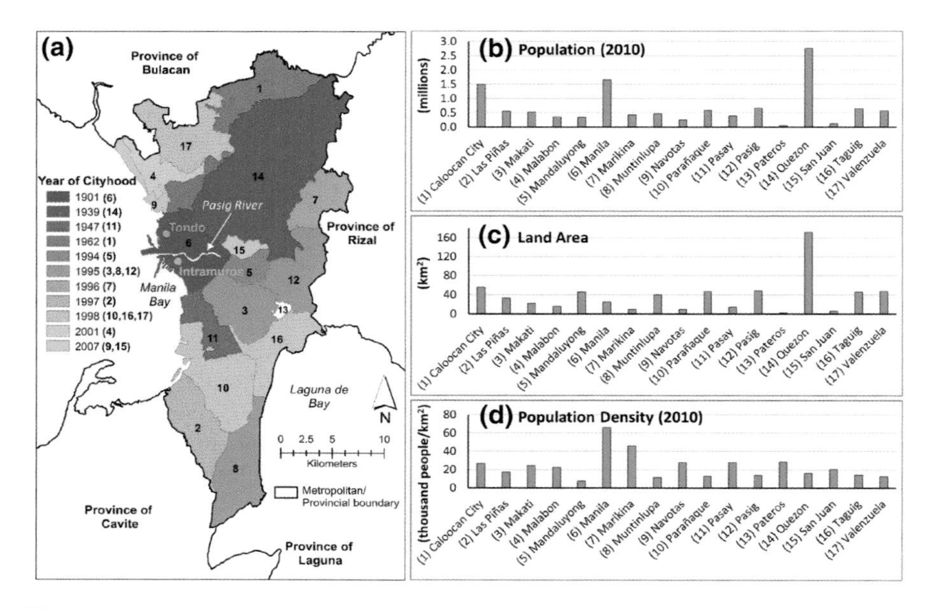

Fig. 5.2 **a** Cityhood of the member local government units (LGUs) of Metro Manila; **b** 2010 population; **c** land area; and **d** 2010 population density. *Source* The data on land area (2007) and population (2010) were consolidated from the Philippine Statistics Authority-National Statistical Coordination Board (http://nap.psa.gov.ph/; http://www.nscb.gov.ph); the boundary map was downloaded from the Philippine GIS Data Clearinghouse (http://philgis.org/), with some updates from the author on the boundary of Manila City on the bay area, i.e., western part

5.2 Primacy in the National Urban System

5.2.1 The Philippines and Its National Urban System

The Philippines is one of the 11 sovereign states or countries in the southeastern region of Asia. In terms of land area, it is ranked 73rd worldwide and 6th in Southeast Asia (UN 2013). With a population of 100.10 million as of 2014, the Philippines is ranked 12th worldwide and second in Southeast Asia (World Bank 2015a). In terms of population density, the Philippines, with 336 people per km^2, is ranked 32nd worldwide and second in Southeast Asia (World Bank 2015b). In general, the Philippines is considered a highly urbanized nation (ADB 2014), with 45.3% of the country's 2010 total population of 92.34 million living in urban areas (UN 2015).

The Philippines is an archipelago composed of 7107 islands, with a total land area of approximately 300,000 km^2. The country is divided into three island groups—Luzon (north), Visayas (middle), and Mindanao (south) (Fig. 5.6a), and into 18

Fig. 5.3 Inside Binondo district (the world's oldest Chinatown) in Manila City, Metro Manila, Philippines. *Source* Author's fieldwork (2015)

regions[1] based on cultural, ethnological, and geographical characteristics (Fig. 5.7). Under a democratic-presidential form of government, the Philippines is politically and administratively divided into 81 provinces, 144 cities, 1490 municipalities, and 42,029 barangays (PSA 2015a). The provinces are the primary administrative and political divisions in the country, while the barangays are the smallest administrative units. As provided by the 1987 Philippine Constitution and the subsequent laws enacted by the Philippine Congress, the country is also divided into legislative districts allocated among the provinces, cities, the Metropolitan Manila area, and the registered national, regional, and sectoral parties and organizations.

In the National Framework for Physical Planning: 2001–2030, 12 metropolitan areas are identified as the country's leading industrial, financial, and technological centers that serve as the main hubs for international trade (NEDA 2002, 2007; ADB

[1] Region I—Ilocos Region; Region II—Cagayan Valley; Region III—Central Luzon; Region IV-A—CALABARZON; Region IV-B—MIMAROPA; Region V—Bicol Region; Region VI—Western Visayas; Region VII—Central Visayas; Region VIII—Eastern Visayas; Region IX—Zamboanga Peninsula; Region X—Northern Mindanao; Region XI—Davao Region; Region XII—SOCCSKSARGEN; XIII—Caraga; ARMM—Autonomous Region in Muslim Mindanao; CAR—Cordillera Administrative Region; NCR—National Capital Region (Metro Manila); NIR—Negros Island Region. See also Fig. 5.7.

Fig. 5.4 Inside Makati City, Metro Manila, Philippines. *Source* Author's fieldwork (2015)

2014). These include Metro Manila, Metro Cebu, Metro Davao, Metro Cagayan de Oro, Metro Angeles, Metro Iloilo-Guimaras, Metro Bacolod, Metro Naga, Metro Baguio (BLISTT[2]), Metro Batangas, Metro Dagupan (CAMADA[3]), and Metro Olongapo (Fig. 5.6).

According to the United Nations (UN), the territorial spread of cities of different sizes across the whole territory of one country constitutes a national urban system (UN 2015). Such a system can be linked to the organization of the government at various levels: national, regional, and local levels (Kim and Law 2012; UN 2015). For the case of the Philippines, the spatial distribution of its 12 metropolitan areas (Fig. 5.6a) provides an overview of its national urban system and the regional, social, and economic agglomerations in the country.

[2]BLISTT—Acronym for the city of Baguio and municipalities of La Trinidad, Itogon, Sablan, Tuba, and Tublay in the province of Benguet (see Estoque and Murayama 2013a). See also the 'note' section in Fig. 5.6.

[3]CAMADA—Acronym for the municipalities of Calasiao, Mangaldan, and Dagupan in the province of Pangasinan.

Fig. 5.5 A glimpse of the urban development (high-rise buildings and urban green spaces) of Bonifacio Global City (also known as The Fort), Metro Manila, Philippines. *Source* Author's fieldwork (2015)

5.2.2 Primacy of Metro Manila

Urban primacy indicates the degree of dominance of one urban area (e.g., city or region) based on population, economy, and urban functions and services. This section examines the urban primacy of Metro Manila by comparing its population, which has been classified as 100% *urban population* (PSA 2013), and its gross regional domestic product (GRDP)[4] with the other metropolitans and regions in the country.

Figure 5.6b shows that Metro Manila is not the "largest" metropolitan in the Philippines in terms of land area. It has a land area of 638.55 km^2, which is much smaller than Metro Cebu, Metro Iloilo-Guimaras, Metro Naga, and Metro Baguio, but more especially Metro Davao and Metro Cagayan de Oro (Fig. 5.6b). However,

[4]Gross Regional Domestic Product (GDP) is the aggregate of gross value added of all resident producer units in a region. It includes regional estimates on the three major sectors and their subsectors, namely the (a) agriculture, fishery, and forestry sector, (b) industry sector (mining and quarrying, manufacturing, construction, and electricity, and water), and (c) service sector (transport, communication and storage, trade, finance, ownership of dwellings and real estate, private services, and government services) (http://nap.psa.gov.ph/).

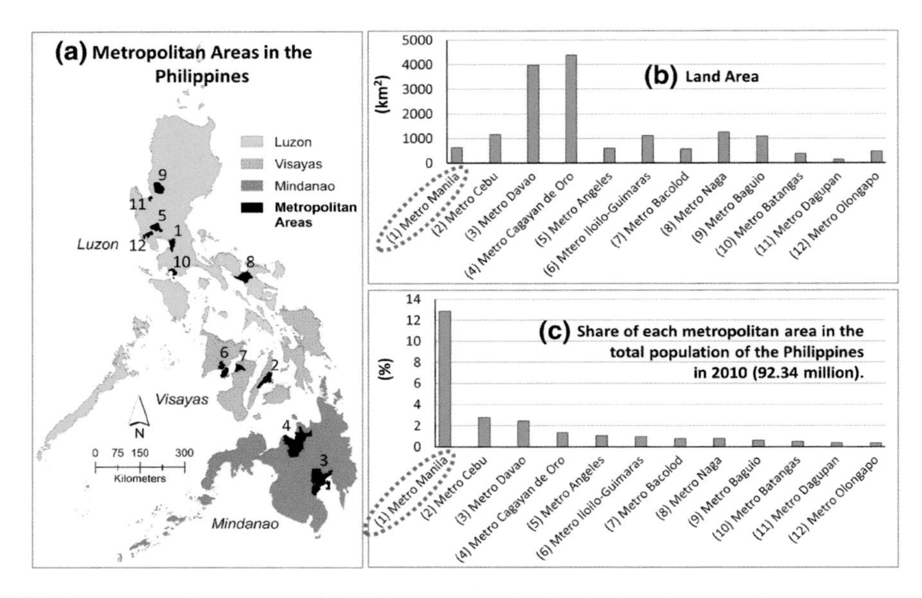

Fig. 5.6 Metropolitan areas in the Philippines. **a** spatial distribution of metropolitan areas across Luzon (with seven), Visayas (with three) and Mindanao (with two); **b** land area distribution among the metropolitan areas; and **c** individual share of the metropolitan areas in the total population of the country (2010). *Source* The data on land area (2007) and population (2010) were consolidated from the Philippine Statistics Authority-National Statistical Coordination Board (http://nap.psa.gov.ph/; http://www.nscb.gov.ph). *Note* In this article, the cities and municipalities included in each metropolitan are based on latest updates. In NEDA (2002, 2007), the municipality of Tublay in the province of Benguet was not yet included in Metro Baguio, and the municipality of Carmen in the province of Davao del Norte was not yet included in Metro Davao

in terms of population, as of 2010 it has the largest percentage share, accounting for 12.84% (11.86 million) of the total population of the country (92.34 million). Its share is more than four times the share of the second highest, that is Metro Cebu with 2.76% (Fig. 5.6c). In fact its share is even higher than the share of all other metropolitan areas combined with 11.95%. This shows how dominant Metro Manila is among the metropolitans in the country in terms of population, despite having a much smaller land area.

As mentioned earlier, Metro Manila is also called the National Capital Region (NCR), one of the 18 regions in the country. Figure 5.7a shows the population density of the Philippines based on regional boundary. In 2010, Metro Manila (NCR) had the highest population density with 18,567 people/km^2, followed by CALABARZON (IV-A) (770 people/km^2) and Central Luzon (III) (471 people/km^2). The Cordillera Administrative Region (CAR), where the summer capital of the country, Baguio City, is located, and MIMAROPA (IV-B) had the lowest population density with 84 and 93 people/km^2, respectively. This shows that Metro Manila's density in 2010 was 24 and 220 times the density of CALABARZON and CAR, respectively.

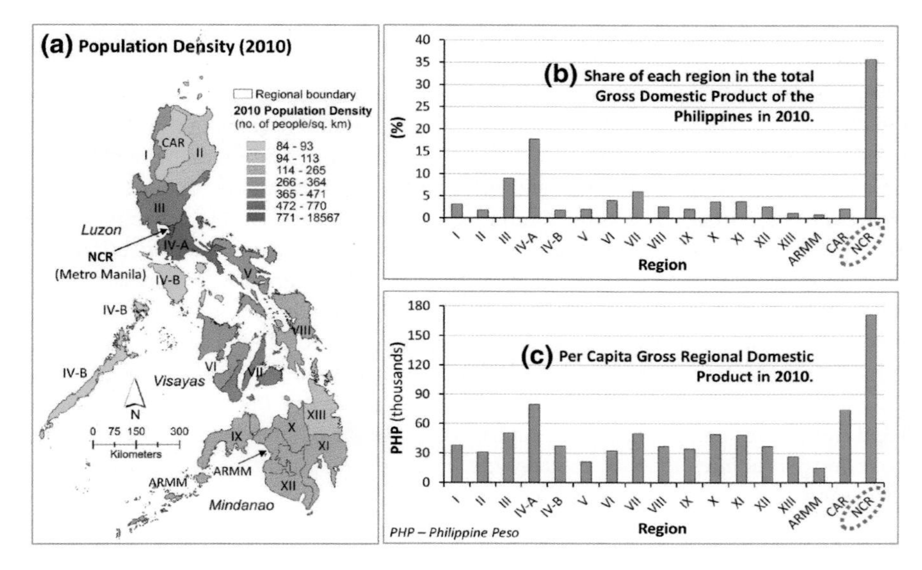

Fig. 5.7 Regions of the Philippines. **a** population density of the regions (2010); **b** individual share of the regions in the total gross domestic product of the country (2010); and **c** per capita domestic product of the regions (2010; at constant 2000 prices; In 2000, 1 USD = 29.47 PHP). *Source* The data on land area (2007), population (2010), and GRDP (2010) were consolidated from the Philippine Statistics Authority-National Statistical Coordination Board (http://nap.psa.gov.ph/; http://www.nscb.gov.ph); the boundary map was downloaded from the Philippine GIS Data Clearinghouse (http://philgis.org/). In this figure, the Negros Island Region (NIR), which was created on May 29, 2015, under Executive Order No. 183, is still merged with Regions VI and VII

In terms of GRDP, in 2010 Metro Manila (NCR) had the highest share, with 35.75%, followed by CALABARZON with 17.70% and Central Luzon with 8.96% (Fig. 5.7b). ARMM had the lowest share with 0.84%, followed by Caraga (XIII) with 1.13% and Cagayan Valley (II) with 1.75%. This shows that Metro Manila's (NCR's) GRDP share in 2010 was more than two times the share of CALABARZON and 42 times the share of ARMM. Metro Manila (NCR) also had the highest per capita GRDP in 2010 (171,442 PHP), followed by CALABARZON (79,699 PHP), CAR (74,104 PHP), and Central Luzon (50,207 PHP) (Fig. 5.7c). ARMM also had the lowest per capita GRDP (14,588 PHP), followed by Bicol Region (V) (21,004 PHP) and Caraga (26,504 PHP).

The land area of Regions IV-A (CALABARZON) and III (Central Luzon) is more than 16,000 and 21,000 km^2, respectively, that is about 25 and 33 times the area of Metro Manila (NCR). This is the reason why, as of 2010, CALABARZON, with 12.61 million people, and Central Luzon, with 10.14 million people, have the largest and third largest population among all regions in the country, respectively, with Metro Manila in between, with 11.86 million people. In fact, these two regions have been part of the various current development plans for Metro Manila because of their potential critical roles in the national urban system and in influencing the future development of the country's capital region (more discussion on this issue in Sect. 5.5).

Overall, based on population and GRDP, the urban primacy of Metro Manila (or the NCR) in comparison with the other metropolitans and regions in the Philippines is evident and overwhelming. The case of Metro Manila provides evidence for a positive relationship between concentration of people and economic growth. However, the primacy of Metro Manila creates an imbalance in the urban hierarchy and development processes in the country. Nevertheless, the identification and inclusion of other metropolitan areas in the national framework for physical planning for the whole country (NFPP: 2001–2030) promotes countryside developments and a more balanced national urban system. Metro Manila and its governing agency, the Metropolitan Manila Development Authority (MMDA), were both created by law. However, the other metropolitan areas in the country still lack a legal framework and governing structure. The absence of such legal framework and governing structure impedes metro-wide urban development planning and implementation (e.g., see Estoque and Murayama 2013a).

5.3 Urban Land Use/Cover Patterns and Changes (1993–2030)

This section discusses the observed and projected urban land changes, i.e., changes from non-built-up to built-up lands, in Metro Manila and its surrounding areas. Remote sensing-derived urban land use/cover maps and spatial metrics were used to detect the temporal and spatial patterns of urban land changes. The details of the urban land use/cover mapping, change detection and simulation modeling, and spatial pattern analysis are described in the methodology chapter (Kamusoko 2017). Estoque and Murayama (2017) provide a comparative analysis of the trends and spatial patterns of urbanization in Asia and Africa.

5.3.1 Observed Changes (1993–2014)

The urban land change analysis revealed that the area of built-up lands in Metro Manila and its surrounding areas has increased almost twofold over the past 21 years (1993–2014) (Fig. 5.8; Table 5.1). It increased from 455.5 km^2 in 1993 to 848.6 km^2 in 2014 (Table 5.1). This increase translates to an annual rate of change (increase) of 18.7 km^2/year. The annual rate of change during the 1993–2001 period was slightly higher than during the 2001–2009 period (Table 5.2). During the 2009–2014 period, it increased substantially. It is also worth noting that built-up expansions in the area have been occurring and "moving" beyond the boundary of Metro Manila (Figs. 5.8 and 5.9). This is indicative of a sprawl urban development pattern.

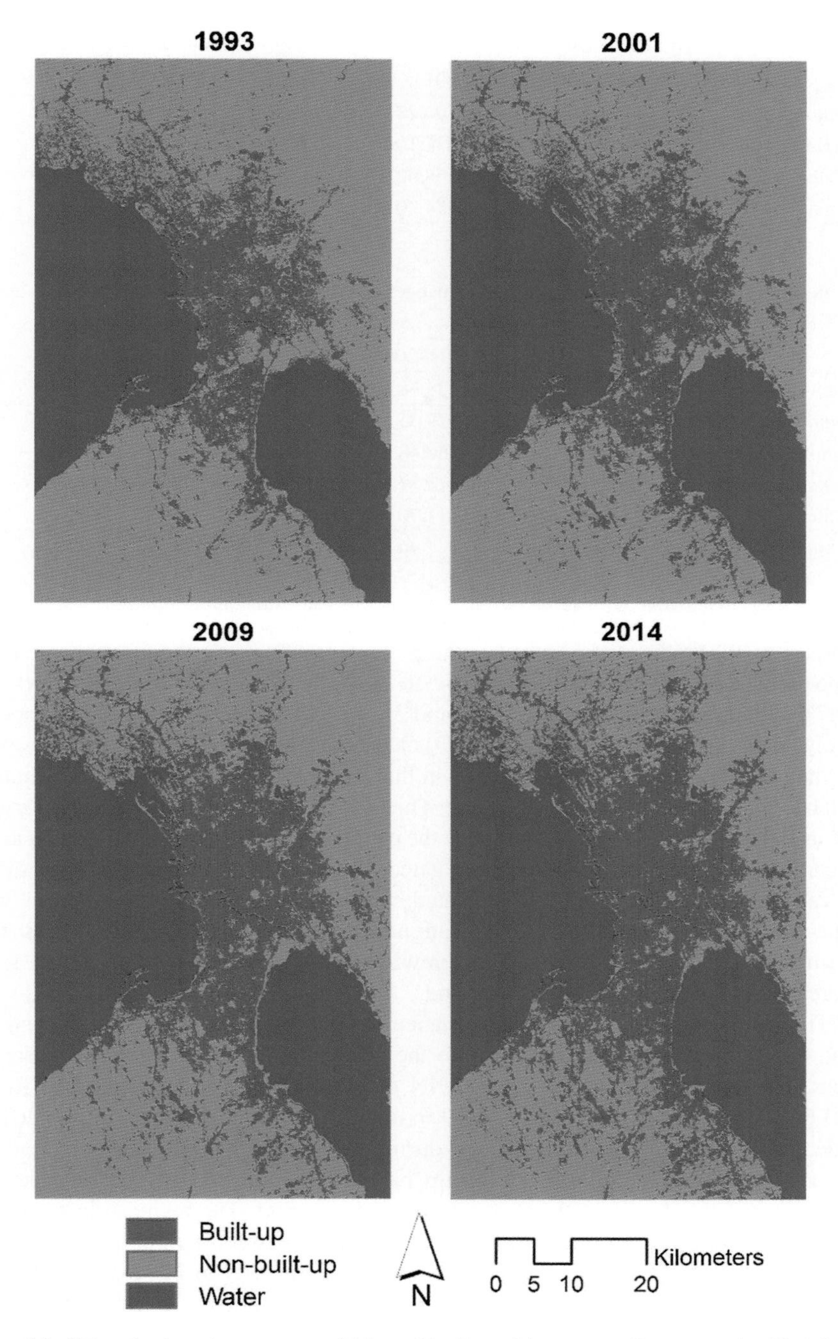

Fig. 5.8 Urban land use/cover maps of Metro Manila and its surrounding areas classified from Landsat imagery

Table 5.1 Observed urban land use/cover of Metro Manila and its surrounding areas (km^2)

	1993	2001	2009	2014
Built-up	455.50	594.28	716.94	848.63
Non-built-up	1981.58	1831.10	1704.43	1554.47
Water	1147.12	1158.82	1162.83	1181.10
Total	3584.20	3584.20	3584.20	3584.20

Table 5.2 Observed urban land use/cover changes in Metro Manila and its surrounding areas (km^2)

	1993–2001	2001–2009	2009–2014
Built-up	138.78	122.65	131.69
Annual rate of change (km^2/year)	*17.35*	*15.33*	*26.34*
Non-built-up	−150.48	−126.66	−149.96
Annual rate of change (km^2/year)	*−18.81*	*−15.83*	*−29.99*
Water	11.70	4.01	18.27
Annual rate of change (km^2/year)	*1.46*	*0.50*	*3.65*

Table 5.3 shows the spatial metrics for the built-up class. The percentage of landscape (PLAND) metric measures the proportion of a particular class at a certain time point relative to the whole landscape. In 1993, the built-up class had a PLAND of 18.7%, which increased to 24.5, 29.6, and 35.3% in 2001, 2009, and 2014, respectively (Table 5.3). The patch density (PD) metric is a measure of fragmentation based on the number of patches per unit area, in this case per 100 ha or 1 km^2, in which a patch is based on an 8-cell neighbor rule. The PD of the built-up class decreased from 3.8 in 1993 to 2.6 in 2009, indicating that the patches of built-up lands in the study area became less fragmented and more aggregated. However, built-up lands in 2014 were more fragmented than in 2009 as indicated by the increase in PD between these two time points. This suggests that densification/infill development was more dominant during the 1993–2009 period, while sprawl/diffusion of new built-up patches was more active during the 2009–2014 period.

The Euclidean nearest neighbor distance (ENN) metric is a measure of dispersion based on the distance of a patch to the nearest neighboring patch of the same class. For the study area, the mean ENN of the built-up patches increased from 103.62 m in 1993 to 105.65 and 111.19 m in 2001 and 2009, respectively. In 2014, it decreased to 103.07 m. The increase during the 1993–2009 period can be due to the aggregation of neighboring built-up patches as indicated by the increase in PLAND and the decrease in PD during the same period. The increase in PLAND and decrease in PD redefined the average distance between neighboring built-up patches (Table 5.3). The decrease in mean ENN from 2009 to 2014 can be due to the expansion of the old built-up patches and the development of new patches in between but not necessarily connected to the old patches, as indicated by the increase in PLAND and PD during the same period.

Fig. 5.9 Observed and projected urban land use/cover changes in Metro Manila and its surrounding areas

Table 5.3 Observed landscape pattern of Metro Manila and its surrounding areas

Class-level (built-up) spatial metrics	1993	2001	2009	2014
PLAND (%)	18.69	24.50	29.61	35.31
PD (number per km^2)	3.82	3.12	2.58	3.19
ENN (mean) (m)	103.62	105.65	111.19	103.07
CIRCLE (mean) ($0 \leq$ CIRCLE < 1)	0.33	0.34	0.37	0.35
SHAPE (mean) ($1 \leq$ SHAPE $\leq \infty$)	1.26	1.26	1.28	1.25

The related circumscribing circle (CIRCLE) metric measures the circularity of patches. The value of CIRCLE is 0 for circular or one cell patches and approaches 1 for elongated, linear patches one cell wide. For the study area, the mean CIRCLE value of the built-up patches showed an increasing trend during the whole 1993–2014 period (Table 5.3), indicative of the development of more elongated patches exhibiting a linear pattern. It can be noted that built-up expansion at the outer portions of Metro Manila, especially during the later periods, followed the road network (Figs. 5.8 and 5.9). The increasing trend can be due to the aggregation of much smaller, circular isolated built-up patches.

The shape index (SHAPE) metric is a measure of complexity. This metric has a value of 1 when the patch is square and increases without limit as patch shape becomes more irregular. For the study area, the mean SHAPE value of the built-up patches ranged from 1.25 to 1.28 across the whole 1993–2014 period (Table 5.3), indicating complexity in the shape of the built-up patches. Although the mean SHAPE value increased slightly between 2001 and 2009, it also decreased between 2009 and 2014, suggesting that the complexity of the built-up patches was more or less stable across the whole 1993–2014 period.

Figure 5.10 presents all the metrics for the built-up class of the study area along the gradient of the distance from the city center across all time periods from 1993 to 2014. PLAND decreases as the distance from the city center increases, indicating that the proportion of built-up lands near the city center was relatively higher. By contrast, PD increases first as it approaches the 25-km distance from the city center and then decreases in farther distances. This implies that there were more patches of built-up lands in middle distances. Built-up patches were relatively more dispersed in farther distances, as shown by the increasing trend of the mean ENN across the distance from the city center. The figure also shows a slightly increasing trend of the mean CIRCLE value along the gradient of the distance from the city center, indicating that the patches of built-up lands were slightly more elongated or linear in farther distances. However, despite the variability of these metrics along the gradient of the distance from the city center, the complexity of built-up patches was almost uniform or stable as indicated by the mean SHAPE value.

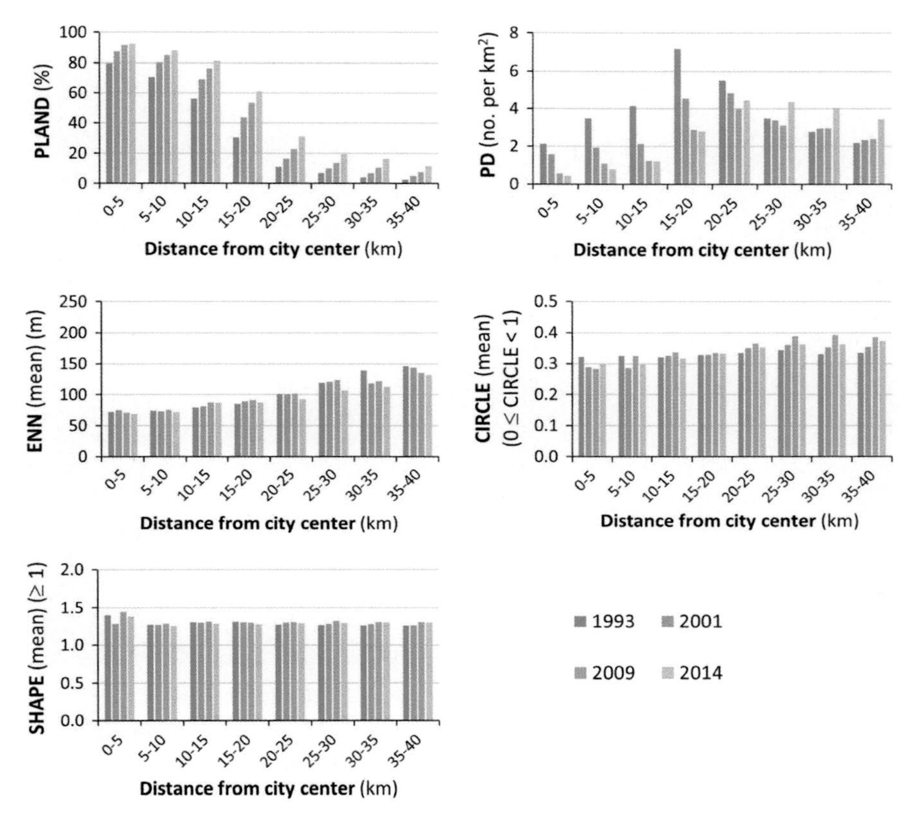

Fig. 5.10 Observed class-level spatial metrics for built-up along the gradient of the distance from city center of Metro Manila. *Note* The y-axis values are plotted in the same range as those in Fig. 5.12

5.3.2 Projected Changes (2014–2030)

The results of the urban land change simulation revealed that the area of built-up lands would increase from 848.63 km^2 in 2014 to 978.34 km^2 in 2020 and 1112.27 km^2 in 2030 (Figs. 5.9 and 5.11; Table 5.4). It would increase at the rate of 129.71 km^2/year from 2014 to 2020 and 133.92 km^2/year from 2020 to 2030 (Table 5.5). The spatial pattern analysis also revealed that the simulated built-up patches in 2020 and 2030 would be more aggregated as indicated by the simulated increase in PLAND and decrease in PD (Table 5.6). The simulated increase in mean ENN also indicates that more neighboring built-up patches would become connected. This simulated aggregation of built-up patches would redefine the average distance between neighboring built-up patches. The simulated increase in the average values of CIRCLE and SHAPE indicates more connected, elongated/linear, and complex patches of built-up lands, respectively.

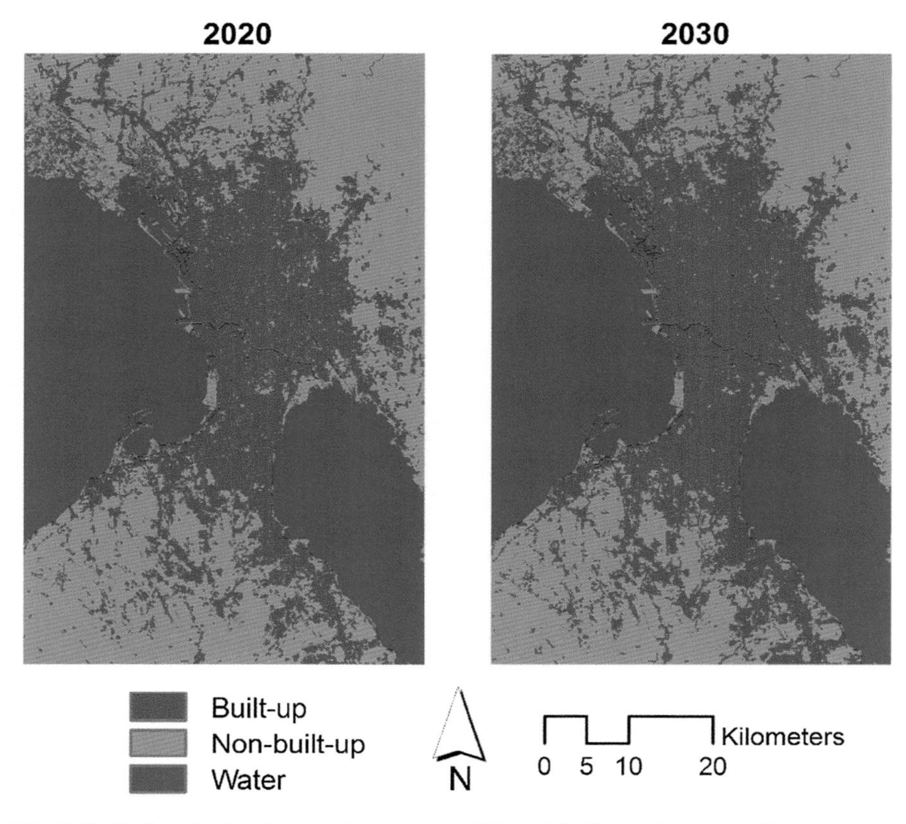

Fig. 5.11 Projected urban land use/cover maps of Metro Manila and its surrounding areas

Table 5.4 Projected urban land use/cover of Metro Manila and its surrounding areas (km^2)

	2020	2030
Built-up	978.34	1112.27
Non-built-up	1424.76	1290.84
Water	1181.10	1181.10
Total	3584.20	3584.20

Along the gradient of the distance from the city center (Fig. 5.12), the PLAND of the simulated built-up in 2020 and 2030 would also be higher at distances closer to the city center. PD would decrease dramatically in 2020 and 2030, but would still be relatively higher in middle distances. By contrast, the mean ENN would increase dramatically, though it would still follow the pattern, i.e., mean ENN increases along the gradient of the distance from the city center, especially for the 2020 simulated built-up patches. The mean CIRCLE value would also increase in 2020 and 2030, especially at 0–5 km and 20–40 km distances from the city center. The mean SHAPE value would also increase, but would also be relatively more uniform or stable along the gradient of the distance from the city center (Fig. 5.12).

Table 5.5 Projected urban land use/cover changes in Metro Manila and its surrounding areas (km^2)

	2014–2020	2020–2030
Built-up	129.71	133.92
Annual rate of change (km^2/year)	*21.62*	*13.39*
Non-built-up	−129.71	−133.92
Annual rate of change (km^2/year)	*−21.62*	*−13.39*
Water	0.00	0.00
Annual rate of change (km^2/year)	*0.00*	*0.00*

Table 5.6 Projected landscape pattern of Metro Manila and its surrounding areas

Class-level (built-up) spatial metrics	2020	2030
PLAND (%)	40.71	46.28
PD (number per km^2)	0.69	0.48
ENN (mean) (m)	168.37	181.92
CIRCLE (mean) ($0 \leq$ CIRCLE < 1)	0.37	0.42
SHAPE (mean) ($1 \leq$ SHAPE $\leq \infty$)	1.43	1.56

5.4 Driving Forces of Urban Development

As discussed above, the urban land changes in the study area over the past 21 years (1993–2014) have been remarkable, with most of the changes occurring at the outer parts of Metro Manila and in its surrounding areas (Sect. 5.3.1). It can be observed that during the early 1990s, a large portion of Metro Manila's landscape had already been covered with built-up (Fig. 5.8). Thus, the space for future development was very limited and located mostly in the northern and southern parts of the region. Metro Manila is located in between two main bodies of water: Manila Bay on the west and Laguna de Bay on the southeastern side (Fig. 5.2a). Thus, since the 1990s, although there were some infill developments in the central area, the spatial expansions of built-up lands in the region have been mostly toward the northern and southern directions because of this geophysical feature (Figs. 5.8 and 5.9).

Population growth is probably the most important common driver of urbanization elsewhere in the world. Since the formal creation of Metro Manila in 1975, the number of its member LGUs has remained the same; its population, however, had grown rapidly. In 1970, Metro Manila only had 3.97 million population, but in 2010 it already had 11.86 million (Stinner and Bacol-Montilla 1981; Ortega 2014; Estoque and Murayama 2015; https://psa.gov.ph), resulting in an almost threefold increase over a 40-year period. This population increase could have also raised the need for various urban services including housing and commercial and business centers, which means that more non-built-up lands had to be converted into built-up. The urban land change analysis from 1993 to 2014 (Sect. 5.3.1) provides some evidence for this proposition. The rapid population growth of Metro Manila is the result of the combined effect of natural birth and migration. Between 2005 and

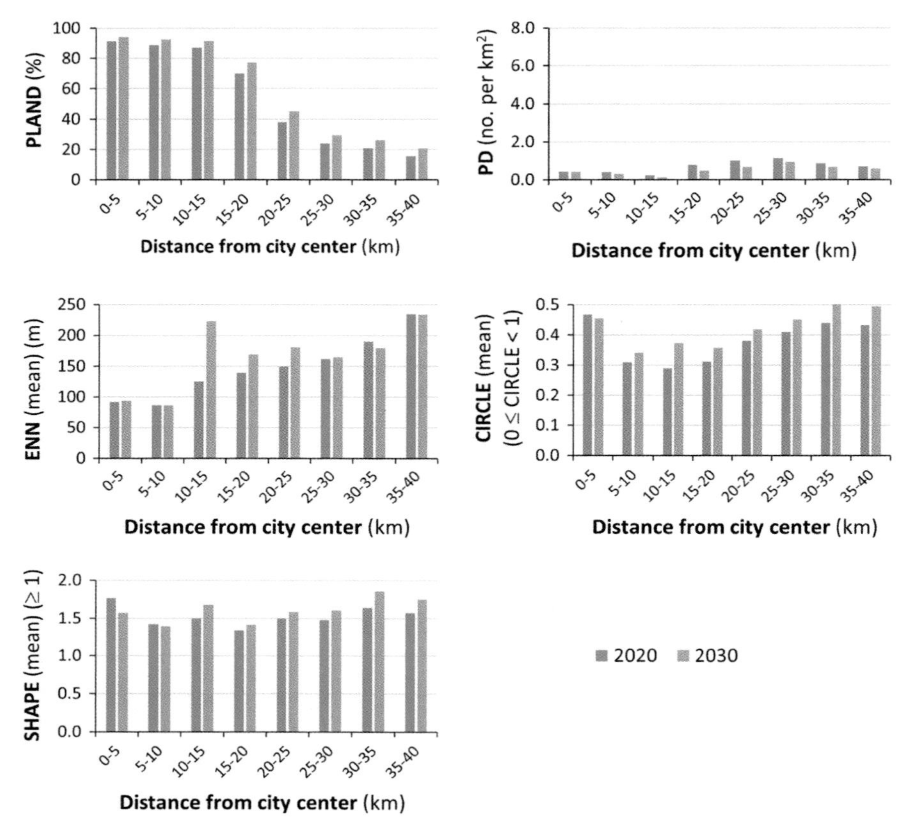

Fig. 5.12 Projected class-level spatial metrics for built-up along the gradient of the distance from city center of Metro Manila. *Note* The y-axis values are plotted in the same range as those in Fig. 5.10

2010 alone, Metro Manila had a total of 284,000 domestic and 19,000 foreign migrants (PSA 2012).

Metro Manila, being the country's capital region, houses most of the national government offices responsible for the three major sectors of the national economy: the agriculture, fishery, and forestry sector, the industry sector, and the service sector. The country's central business districts, international airport, and most of the Philippines' prime educational and research institutions, and cultural, sports and healthcare centers are also located in Metro Manila. The concentration of these urban functions and services and various socioeconomic opportunities such as employment in Metro Manila is a major factor why people flock to the area.

Economic growth is another important driver of rapid urbanization. In fact, population and economy influence each other. A large and healthy population can provide the manpower needed for the economy: production, distribution, consumption, and resource maintenance. In return, a vibrant and productive economy can create various socioeconomic opportunities for the people. It can also attract more investments and people. As of January 2015, Metro Manila has an employment rate of 90.7% (PSA 2015b); 3.5% higher than the July 2008 employment rate of 87.2% (NSO 2010).

Gross domestic product (GDP; called GRDP when it is measured per region) is among the most important indicators of the size and vigor of an economy. The time-series statistical data (see http://nap.psa.gov.ph) show that Metro Manila's GRDP and per capita GRDP have been continuously increasing over the past 14 years (2001–2014). Metro Manila's share of 30.59% to the country's total GDP in 2001 increased to 36.29% in 2014. Its per capita GRDP of 30,000 PHP in 2001 also increased to 42,000 PHP in 2008 (based on constant 1985 prices), and from 162,000 PHP in 2009 to 203,000 PHP in 2014 (based on constant 2000 prices).

Among some other physical factors, accessibility is an important physical factor for the urban development of Metro Manila. Metro Manila is accessible by land, air, and water. Its accessibility and relative position promotes and enables high interaction with neighboring provinces and regions in the country, as well as with other Asian cities. The pull factor of Metro Manila due to its primacy seems to have outweighed the risk factor inherent to its geographic location. With its location in a low-lying coastal area and the presence of the west valley fault on the eastern part, Metro Manila is vulnerable to various natural hazards such as floods, earthquake, and tsunami, among others. However, despite all these environmental hazards, Metro Manila has been continuously growing. People continue to flock to the area to seek for better opportunities and socioeconomic progress.

In general, the relatively small land area of Metro Manila, its geographical characteristics and population and economic growth, the concentration of key urban functions/services and opportunities in the area, and its accessibility are considered to be among the key factors influencing the spatiotemporal patterns of urban land changes (Figs. 5.8, 5.9, and 5.10; Tables 5.1, 5.2, and 5.3) and the overall urban development of the region. In addition, there are also a number of urban development-related legislations that guide urban development planning and implementation in the country, including Metro Manila, the country's prime urban center (see ADB 2014).

5.5 Implications for Future Sustainable Urban Development

5.5.1 Sustainability, Population Growth, and Urban Land Changes

Sustainability is one of the important concepts to emerge in the Earth's current geological epoch, the Anthropocene, also known as the "Age of Man". This concept encapsulates three important dimensions known as the triple bottom line, that is, people (social justice), planet (environmental quality), and profit (economic prosperity) (Elkington 1997; Estoque and Murayama 2014). Sustainable urban development, characterized by a well-balanced relationship between environmental quality, social justice, and economic progress, is an important component and an indispensable part of the sustainability goal of humankind. According to the Sustainable City Agenda of the International Council for Local Environmental Initiatives (ICLEI), "sustainable cities ensure an environmentally, socially, and economically healthy and resilient habitat for existing populations, without compromising the ability of future generations to experience the same" (http://www.iclei.org/activities/our-agendas.html).

The socioeconomic conditions in Metro Manila have improved over the years. However, its overall sustainability, including its environmental sustainability, remains an important issue. This section examines the potential implications of Metro Manila's population growth and built-up expansion pattern to its socioeconomic and urban environment. A brief overview of some of the current major urban development plans for Metro Manila is also provided.

To a great extent, the strength and primacy of a city lie on its population. At 18,567 people/km^2 density in 2010, Metro Manila is one of the densest urban agglomerations in the world. The large population of Metro Manila, which has also been projected to reach 16.8 million by 2030 (World Bank 2015a), has been a major factor to its primacy over all metro areas and regions in the Philippines (Figs. 5.6 and 5.7). However, its high and continuously increasing population density also causes various socioeconomic problems, such as congestion (including traffic congestion) and urban poverty (urban poor and slum areas, including those living in informal settlements or squatter areas) (Mathur 2013; ULI 2013; Porio 2015). To address these issues, the national government has to keep up the pace on the delivery of the needed urban services to the rapidly growing population of its prime urban region.

In Sect. 5.4, population increase and economic growth have been discussed as among the many possible factors that influence the rapid urbanization of Metro Manila. It has also been highlighted that the GRDP of Metro Manila has been increasing, and that despite the increase in population, the per capita GRDP has also been increasing. However, whether this economic growth has been translated into various basic socioeconomic services for the local population is another important issue.

In other words, Metro Manila's increasing population has a critical implication to its future sustainable urban development particularly to its per capita socioeconomic condition. Both the local and national governments need to make sure that the issue on congestion and urban poverty in the region are taken into consideration in landscape and urban planning.

The implications of the spatiotemporal patterns of built-up expansions in Metro Manila and its surrounding areas can be further examined using the diffusion-coalescence urban growth theory (Dietzel et al. 2005; Wu et al. 2011; Estoque and Murayama 2015, 2016). The theory suggests that urbanization exhibits a cyclic pattern in time and space driven by two alternating processes: *diffusion*, in which new urban patches are dispersed from the origin point or seed location, and *coalescence* or the union of individual urban patches, or the growing together of the individual patches into one form or group (Dietzel et al. 2005; Wu et al. 2011; Estoque and Murayama 2015, 2016).

As discussed in Sect. 5.3, the patches of built-up lands in the study area, in general, have become more aggregated over the years. This is one indication that built-up expansion, especially in Metro Manila, has been moving toward the coalescence phase, and this observation is consistent with previous findings (e.g., Estoque and Murayama 2015). From 2009 to 2014, the PD of the built-up class increased, and this was due to the diffusion of new built-up patches, especially on the outskirts of Metro Manila (Table 5.3; Figs. 5.8 and 5.9). The urban land change simulation indicated that built-up lands would continue to expand and again undergo the process of aggregation or coalescence in the near future (Tables 5.3 and 5.6; Figs. 5.9 and 5.11). The process of coalescence can result in an infilling growth pattern, whereas the process of continuous diffusion and expansion can result in a sprawl development pattern (Estoque and Murayama 2015). The urban land changes in Metro Manila and its surrounding areas are characterized by both infilling and sprawl urban development patterns.

While an infilling growth pattern has some potential advantages, e.g., the use of existing infrastructures, the promotion of walkable neighborhoods, and the prevention of the associated external costs of sprawl development, it also has some potential disadvantages, e.g., increased traffic congestion and pollution, limited open space, potential loss of urban green spaces, and crowded services (Estoque and Murayama 2015, 2016). For Metro Manila and its surrounding areas, infilling pattern poses a threat to their remaining urban green spaces, which are important sources of various urban ecosystem services[5] and valuable elements for sustainable urban development, especially if such pattern will continue undisrupted. That said, this should not be a problem if future urban development will follow the concept of "sustainable cities", where urban green spaces are kept, improved, restored, or introduced. On the other hand, the sprawling pattern detected in the area also has various important implications (e.g., higher urban development costs, greater disturbance or loss of natural habitat, etc.), and thus requires landscape and urban planning in a wider scale.

5.5.2 *Current Major Development Plans*

In a recently published project report, *Ten Principles for Sustainable Development of Metro Manila's New Urban Core*, by the Urban Land Institute (http://uli.org), a new urban core for Metro Manila has been identified. Metro Manila's urban core is composed of Manila City and Makati City, but in this newly identified urban core, Taguig City replaces Manila City (ULI 2013). Urban Land Institute (ULI) is a nonprofit research and education organization, whose mission is to provide

[5]Ecosystem services refer to the benefits that ecosystems (cropland, forest, etc.) generate for and provide to people. Such benefits can be tangible (goods, e.g., cropland for providing food) or intangible (services, e.g., forest for absorbing CO_2), large or small, and direct or indirect (MEA 2005; Estoque and Murayama 2013b, 2016).

leadership in the responsible use of land and in creating and sustaining thriving communities worldwide. In the said project report for Metro Manila, ULI proposes ten principles for the sustainable development of Metro Manila's new urban core (ULI 2013, p. 21):

1. *create one Metro Manila*, which should be the common goal and vision of its member LGUs and the national government;
2. *improve urban mobility*, where an integrated transport and infrastructure must be put in place;
3. *make beautiful places* by establishing business improvement districts and high-quality public spaces;
4. *work together* through collaboration and partnerships, such as with private sector;
5. *establish good governance* through a streamlined regulatory framework and effective development control;
6. *engage everyone* in an inclusive, participatory, and transparent process;
7. *empower people* by establishing community improvement districts to enhance education, awareness, and employment opportunities;
8. *be prepared*—disaster preparedness and resilience;
9. *restore human dignity* through affordable housing policy and delivery; and
10. *go beyond smart communities* by aiming for more livable and sustainable communities.

Landscape and urban development planning in a wider scale has also been the subject of various major development plans for Metro Manila. Such development plans include the *Metro Manila Greenprint 2030* by the Philippine national government through the MMDA in partnership with the World Bank, AusAID and Cities Alliance, and the *Mega Manila Dream Plan*, formally titled the *Roadmap for Transport Infrastructure Development for Metro Manila and Its Surrounding Areas*, by the Philippine national government through the National Economic Development Authority (NEDA) and Japan International Cooperation Agency (JICA).

The Greenprint is a 20-year development and spatial plan, envisioned to provide an overall framework and recommendations on the use of land and other resources in Metro Manila (MMDA 2012). The four main goals of the Greenprint are as follows: (i) to provide an urban environment that is more conducive for investors, entrepreneurs, and innovators, as well as creative minds that will enhance competitiveness vis-à-vis other cities in Asia; (ii) to improve coordination among key players, especially the 17 LGUs of Metro Manila; (iii) to guide the future urban form of Metro Manila, taking into consideration its neighboring areas in Region III (Central Luzon) and Region IV-A (CALABARZON) (see Figs. 5.7a and 5.13); and (iv) to be a primary infrastructure, providing green systems and the clustering of economic activities to improve livability (JICA and NEDA 2014; Estoque and Murayama 2015).

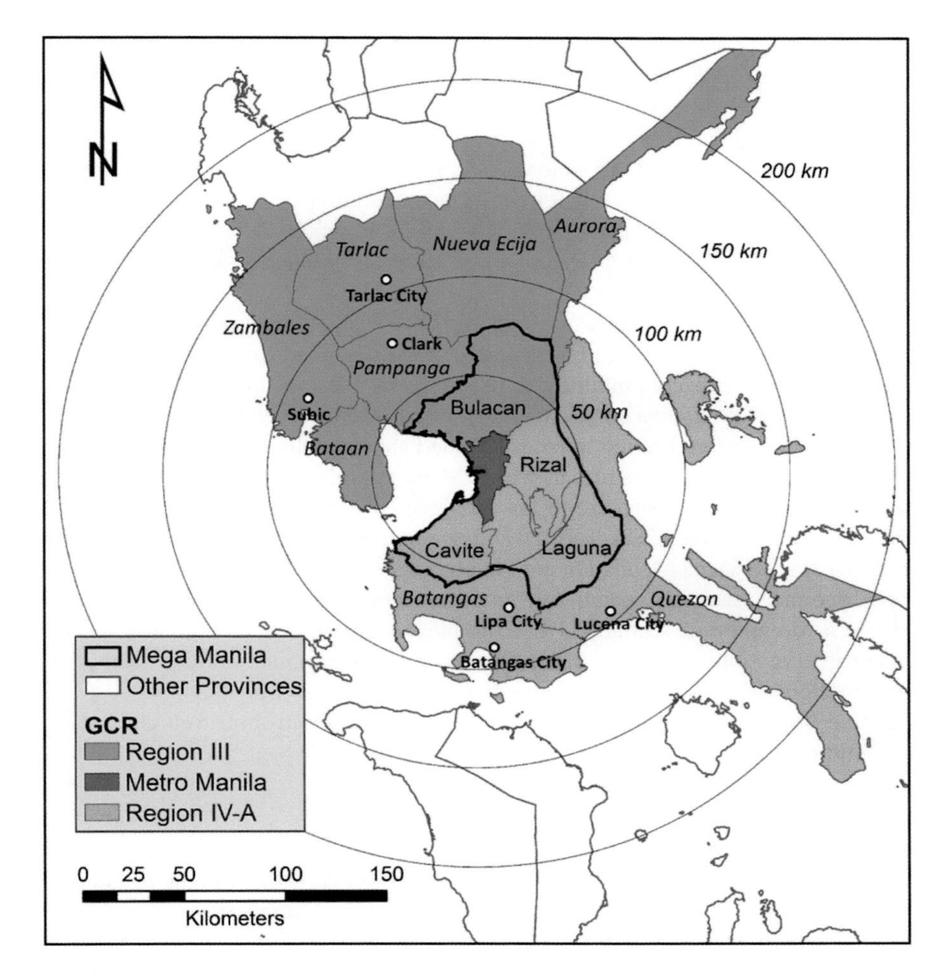

Fig. 5.13 Map of the proposed Mega Manila and Greater Capital Region (GCR). *Source* Author's own elaboration based from JICA and NEDA (2014)

In connection with the new urban core concept (Makati City and Taguig City), ULI (2013) suggests that, in order to achieve a more livable and sustainable community, Greenprint 2030 should include provisions for reducing pollution, improving the pedestrian environment, making Metro Manila more environment-friendly, and for promoting the use of public transportation, as well as walking and biking. However, ULI (2013) also points out that although these provisions are promising, a well-crafted interventions plan is needed.

The Mega Manila Dream Plan, on the other hand, is an integrated plan for improving the transport system in Metro Manila and its surrounding areas, and for addressing the pressing, interlinked problems on transportation, land use, and the environment (JICA and NEDA 2014). In this plan, Mega Manila will be composed of Metro Manila and the provinces of Bulacan (part of Region III), Rizal, Cavite and Laguna (parts of Region IV-A) (Fig. 5.13). In addition to the list of short term and medium-long term priority projects on transport system (including airports and seaports) and industrial development, the plan also presents a spatial development strategy, aiming to expand the national capital region (NCR) of the Philippines into a "Greater Capital Region" (GCR), with the integration of Metro Manila (or the NCR), Region III, and Region IV-A (Fig. 5.13). It can be noted that these three regions have the largest population and GRDP share among all regions in the country today (Fig. 5.7).

The Mega Manila Dream Plan highlights the need for regional integration in order to achieve a tri-engine growth for the GCR: (i) gate to wellspring of hope; (ii) place for livable communities; and (iii) space for dynamic business centers. Moving away from a monocentric to a polycentric urban development, five primary growth centers are envisioned to be established and spatially distributed from north to south: three in Mega Manila, in which Metro Manila will remain as the central function area (the other two are in the provinces of Bulacan, and Cavite-Laguna); one in the north (Subic-Clark-Tarlac); and one in the south (Batangas-Lipa-Lucena) (Fig. 5.13). The Mega Manila Dream Plan has been evaluated of its feasibility from the economic, financial, social, and environmental perspectives by comparing it against a "do nothing" scenario. JICA and NEDA (2014) concluded that if a set of proper interventions are made, traffic congestions can be removed from most of the road sections. Compared to the present situation, the overall transport cost will decrease by 13% and the air quality in Metro Manila will improve. The socioeconomic conditions in adjoining provinces will also improve.

5.6 Concluding Remarks

History tells us that Manila is one of the original global cities, i.e., as part of the Manila-Acapulco galleon, one of the first known commercially traveled trade routes in the world. The analysis showed compelling evidence for Metro Manila's urban primacy over the other metropolitan areas and regions in the country based on population and gross regional domestic product. Over the past 21 years (1993–2014), the area of built-up lands has increased almost twofold, transforming the landscape of Metro Manila and its surrounding areas. The relatively small land area of Metro Manila, its geographic characteristics and population and economic growth, the concentration of key urban functions/services and opportunities in the area, and its accessibility are hypothesized to be among the key factors influencing the spatiotemporal patterns of urban land changes and the overall urban development of the region.

The simulated future urban land changes indicated that built-up lands would continue to expand in the future (2014–2030) under the influence of infill and sprawl development patterns. The intensifying pressure of urbanization due to rapid population growth and urban land changes poses many challenges that need to be considered in sustainable urban development and landscape planning. The region's natural environment, an important asset toward sustainable urbanization and comfortable urban life, needs to be protected and conserved. There is also a need to address the other key urban issues, including urban poverty, congestion, limited urban green spaces, and disaster preparedness, among others. To overcome these issues, both the local and national governments, including all sectors of the society, need to implement and observe the ten principles outlined above (ULI 2013) and support development plans that can promote sustainable urban development, such as the Metro Manila Greenprint 2030 and Mega Manila Dream Plan. If these principles are observed and if these plans are realized, this primate city in the Orient will have a better prospect of becoming a high-ranking world-class city.

References

ADB (Asian Development Bank) (2014) Republic of the Philippines national urban assessment. Asian Development Bank, Metro Manila, Philippines

Alcazaren P (2004) The city: blueprint for a city's soul. Philippine Center for Investigative Journalism. Accessed on July 7, 2015 from http://pcij.org/stories/blueprint-for-a-citys-soul/

Caoili MA (1988) The origins of Metropolitan Manila: a political and social analysis. New Day Publishers, Quezon City, Philippines

Dietzel C, Oguz H, Hemphill JJ, Clarke KC, Gazulis N (2005) Diffusion and coalescence of the Houston metropolitan area: evidence supporting a new urban theory. Environ Planning B 32:231–236

Elkington J (1997) Cannibals with forks: the triple bottom line of the 21st century business. Capstone, Oxford

Estoque RC, Murayama Y (2013a) City profile: Baguio. Cities 30:240–251

Estoque RC, Murayama Y (2013b) Landscape pattern and ecosystem service value changes: implications for environmental sustainability planning for the rapidly urbanizing summer capital of the Philippines. Landscape Urban Planning 116:60–72

Estoque RC, Murayama Y (2014) Measuring sustainability based upon various perspectives: a case study of a hill station in Southeast Asia. AMBIO J Hum Environ 43:943–956

Estoque RC, Murayama Y (2015) Intensity and spatial pattern of urban land changes in the megacities of Southeast Asia. Land Use Policy 48:213–222

Estoque RC, Murayama Y (2016) Quantifying landscape pattern and ecosystem service value changes in four rapidly urbanizing hill stations of Southeast Asia. Landscape Ecol 31:1481–1507

Estoque RC, Murayama Y (2017) Trends and spatial patterns of urbanization in Asia and Africa: a comparative analysis. In: Murayama Y, Kamusoko C, Yamashita A, Estoque RC (eds) Urban development in Asia and Africa—geospatial analysis of metropolises. Springer Nature, Singapore, pp 393–414

Fish S (2011) The Manila-Acapulco Galleons: the treasure ships of the Pacific: with an annotated list of the transpacific Galleons 1565–1815. Author House UK Ltd., Central Milton Keynes, UK

JICA (Japan International Cooperation Agency) and NEDA (National Economic and Development Authority, Philippines) (2014) Roadmap for transport infrastructure development for Metro Manila and its surrounding areas (Regions III and Region IV-A). Accessed on July 15, 2015 from http://www.neda.gov.ph/?pageid=5061

Kamusoko C (2017) Methodology. In: Murayama Y, Kamusoko C, Yamashita A, Estoque RC (eds) Urban development in Asia and Africa—geospatial analysis of metropolises. Springer Nature, Singapore, pp 11–46

Kim S, Law MT (2012) History, institutions, and cities: a view from the Americas. J Reg Sci 52:10–39

Macas T (2014) Burnham's century-old ideas can still be used to improve Manila—architect. GMA Network, Metro Manila, Philippines

Manila Merchants' Association (1908) Manila, the pearl of the orient: guide book to the intending visitor. Bureau of Printing, Manila, Philippines

Mathur OP (2013) Urban poverty in Asia. Asian Development Bank, Metro Manila, Philippines

MEA (Millennium Ecosystem Assessment) (2005) Ecosystems and human well-being: current state and trends: findings of the condition and trends working group. Island Press, Washington, DC

MMDA (Metropolitan Manila Development Authority) (2012) 2011 Accomplishment report. Metropolitan Manila Development Authority, Metro Manila, Philippines

Morley IB (2011) America and the Philippines: modern civilization and city planning. Educ About Asia 16:34–38

NEDA (National Economic and Development Authority) (2002) National framework for physical planning: 2001–2030. National Economic and Development Authority Metro Manila, Philippines

NEDA (National Economic and Development Authority) (2007) Building globally competitive metro areas in the Philippines. National Economic and Development Authority, Metro Manila, Philippines

NSO (National Statistics Office) (2010) Special release: employment situation in National Capital Region (NCR). National Statistics Office-NCR, Metro Manila, Philippine

Ortega AAC (2014) Mapping Manila's mega-urban region: a spatio-demographic accounting using small-area census data. Asian Population Studies 10:208–235

Porio E (2009) Shifting spaces of power in Metro Manila, City: analysis of urban trends, culture, theory, policy, action. City 13:110–119

Porio E (2015) Sustainable development goals and quality of life targets: insights from Metro Manila. Curr Sociol 63:244–260

Postma A (1992) The Laguna copper-plate inscription: text and commentary. Philippines Studies 40:183–203

PSA (Philippine Statistics Authority) (2012) Domestic and international migrants in the Philippines (Results from the 2010 Census). Philippine Statistics Authority, Metro Manila, Philippines

PSA (Philippine Statistics Authority) (2013) Urban barangays in the Philippines (based on 2010 CPH). Accessed on July 15, 2015 from https://psa.gov.ph/content/urban-barangays-philippines-based-2010-cph

PSA (Philippine Statistics Authority) (2015a) Provincial summary: number of provinces, cities, municipalities and barangays, by region. Philippine Statistics Authority, Metro Manila, Philippines

PSA (Philippine Statistics Authority) (2015b) Employment situation in January 2015 (Final Results). Philippine Statistics Authority, Metro Manila, Philippines

Stinner WF, Bacol-Montilla M (1981) Population deconcentration in Metropolitan Manila in the twentieth century. J Developing Areas 16:3–16

ULI (Urban Land Institute) (2013) Ten principles for sustainable development for Metro Manila's new urban core. Urban Land Institute, Washington DC

UN (United Nations) (2013) 2012 Demographic yearbook. United Nations, New York

UN (United Nations) (2015) World urbanization prospects: the 2014 revision. United Nations, New York

World Bank (2015a) Population ranking. Accessed on July 15, 2015 from http://data.worldbank.org/data-catalog/Population-ranking-table

World Bank (2015b) Population density (people per sq. km of land area). Accessed on July 15, 2015 from http://data.worldbank.org/indicator/EN.POP.DNST

Wu J, Jenerette GD, Buyantuyev A, Redman CL (2011) Quantifying spatiotemporal patterns of urbanization: the case of the two fastest growing metropolitan regions in the United States. Ecol Complex 8:1–8

Chapter 6
Jakarta Metropolitan Area

Akio Yamashita

Abstract Jakarta Metropolitan Area expands outside of DKI Jakarta (*Daerah Khusus Ibukota Jakarta*: Special Capital Region of Jakarta) and includes Bogor and Bekasi in West Java Province and Tangerang in Banten Province. Land use/cover pattern and change analysis indicated that built up area in Jakarta expanded in all directions since 1989. The east–west development strategy, which was designed to prevent the deterioration of the aquifer environment, has not appeared to be thoroughly implemented. This may have caused, in part, escalation of land subsidence due to excessive groundwater pumping for urban water demand. Therefore, proper urban and land use planning at watershed or metropolitan scale is required for the sustainable development of Jakarta.

6.1 Origin and Brief History

Jakarta, the capital city of Indonesia, is located on the coast of Java Sea (Fig. 6.1). Jakarta originally prospered as a port city of the Hindu Sunda Kingdom and was called Sunda Kelapa. This name still remains as the name of a port in central Jakarta (Fig. 6.2). In 1527, the Muslim Banten Sultanate conquered the land and changed the name to Jayakarta.

In 1619, the Dutch East India Company occupied the land and changed the name to Batavia. In 1645, the construction of Batavia Fort was completed, and the fort was used as a base for the Asian colonial policy by Holland (Fig. 6.3). The town center of Batavia is the current Kota area, and there are still Dutch colonial squares and architecture (Fig. 6.4). In 1808, Daendels, the Governor General of the Dutch East Indies, moved the administrative center to Weltevreden (called Gambir at present), the city which is more on the South and inlying. The administrative center in Jakarta remains in Gambir at present, where the Indonesian Presidential Palace and Merdeka Square with the National Monument are located (Fig. 6.5).

A. Yamashita (✉)
Faculty of Life and Environmental Sciences, University of Tsukuba, Tsukuba, Japan
e-mail: akio@geoenv.tsukuba.ac.jp

Y. Murayama et al. (eds.), *Urban Development in Asia and Africa*,
The Urban Book Series, DOI 10.1007/978-981-10-3241-7_6

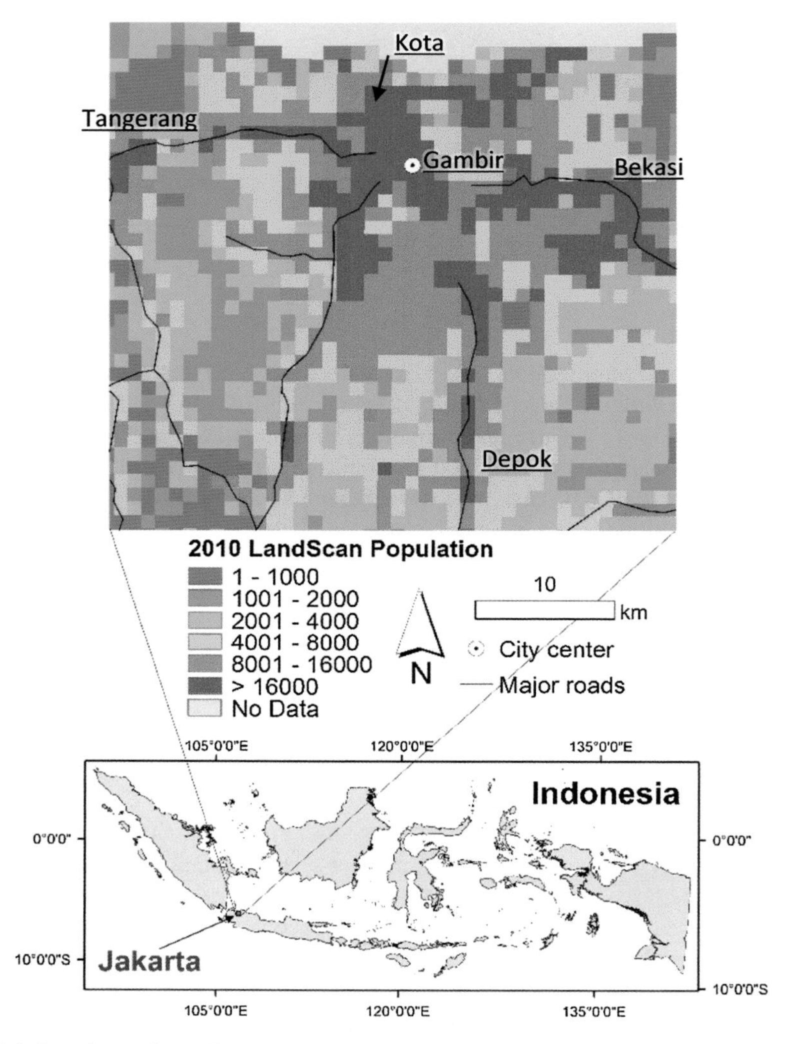

Fig. 6.1 Location and LandScan population of Jakarta Metropolitan Area, Indonesia

In 1942, Japan occupied the Dutch East Indies and changed the name of the city Batavia to Jakarta. After the Second World War, Indonesia declared its independence. However, countries like Holland refused to accept it, which led to the Indonesian War of Independence. In 1949, Indonesia finally gained independence. Jakarta became the capital city of Indonesia (Yamashita 2014).

During the early 1950s, the urban structure of Jakarta followed the Dutch colonial development pattern (Konagaya 1999). After the 1960s, new business

Fig. 6.2 Sunda Kelapa. *Source* Author's fieldwork (2008)

districts were formed along Thamrin Street and Sudirman Street on the south of Gambir (Figs. 6.6 and 6.7). The triangular-shaped area made of Sudirman Street, Kuningan Street, and Gatot Subroto Street is called the Golden Triangle, where the majority of major business offices in DKI Jakarta are located. By the 1980s, the urban area extended further south, beyond the southern Kebayoran Baru, an exclusive residential district.

During the latter part of the twentieth century, Jakarta has developed rapidly as Indonesia's capital city and seen its population significantly increased (see also Sect. 6.2.1). Population density now exceeds 16,000 people per km^2 in central Jakarta and is over 8000 people per km^2 in its environs (Fig. 6.1).

Land use analysis based on old topographic maps (Fig. 6.8) revealed that Jakarta urban area was formed around Kota area and Gambir in the 1930s (Yamashita 2011, 2014). The Kota area was an old inner-city district during Holland's occupation, while Gambir is the current center. Although the inner-city districts expanded slightly during the 1960s, the overall urban development pattern remained the same (Fig. 6.8). However, urbanization progressed rapidly in Jakarta, while the surrounding urban areas extended between 1960 and 2000 (Fig. 6.8 and see also Sect. 6.3.1). Today, Jakarta metropolitan area expands outside of DKI Jakarta (*Daerah Khusus Ibukota Jakarta*: Special Capital Region of Jakarta) and

Fig. 6.3 A former warehouse of the Dutch East India Company. *Source* Author's fieldwork (2010)

includes Bogor and Bekasi in West Java Province and Tangerang in Banten Province. Jakarta Metropolitan Area is now called "Jabotabek" (or "Jabodetabek" with the initial of Depok City in West Java State).

6.2 Primacy in the National Urban System

6.2.1 Population Growth of Indonesia and Jakarta

The total population in Indonesia (Fig. 6.9) increased more than three times from 75 million in 1950 to 240 million in 2010 (UN 2015). The population in Jakarta increased from 1,450,000 (that is 1.9% of the total population) in 1950 to 5,980,000 in 1980; while between 1980 and 2010, the population increased to 9,630,000 (which corresponds to 4.0% of the total population). The population in Jakarta increased approximately six times from 1950 to 2010 (Fig. 6.9). During the past 60 years, the population growth rate in Jakarta was double that of the population growth rate in Indonesia. Furthermore, following the expansion of urban areas in Jakarta, the population of Jakarta and its environs (called Jabotabek) has exceeded 20 million, and became one of the world's largest megacities (Yamashita 2014).

Fig. 6.4 Architectures in Kota area. *Source* Author's fieldwork (2008)

6.2.2 National Urban System in Indonesia

The Indonesian archipelago comprises more than 13,000 islands of various sizes. The main islands include: Java, with the capital Jakarta; Sumatra on the west side; Kalimantan, which borders Malaysia; New Guinea, which borders Papua New Guinea; and Timor that borders East Timor, world famous tourist attraction Bali, and Sulawesi. There are 28 cities with populations of 300,000 or more (UN 2015): 11 in Java, 7 in Sumatra, 4 in Kalimantan, 2 in Sulawesi, and one each in Batam, Bali, Lombok, and Ambon. These cities play central roles in each island politically and economically. Figure 6.10 shows that the country has a typical primate city type urban system, with the most prominent population in the capital Jakarta. In addition, this figure shows population accumulation in Java, where the second (Surabaya), third (Bandung), and fifth (Semarang) most populated cities exist as well as the capital Jakarta.

Such population accumulation in Java has been present since the period of Dutch rule. The Transmigrasi program promoted migration from densely populated islands to less densely populated islands in order to improve economic growth based on agriculture and natural resource development in local areas. Nevertheless, Java has more than half of the country's total population of 240 million. Led by the capital Jakarta, the economic dominance of Java remains unchanged till today.

Fig. 6.5 National monument in Merdeka Square. *Source* Author's fieldwork (2008)

6.3 Urban Land Use/Cover Patterns and Changes (1989–2030)

6.3.1 Observed Changes (1989–2013)

Figures 6.11 and 6.12 show that built-up areas extended to surrounding areas such as Bekasi on the east side, Tangerang on the west side, and further into the south side (that is, toward Bogor City and its environs, which is a suburban summer retreat). While some of these built-up areas were developed from natural green lands such as forests and grasslands, the majority of them were developed from farmlands such as rice paddy fields (Fig. 6.8).

Table 6.1 shows that built-up areas increased from 311.88 to 623.48 km^2 during the past 24 years. The built-up area increased at an annual rate of 10.08 km^2/year during the "1989–2001" period, while annual built-up rate decreased to 6.28 km^2/ year during the "2001–2010" period (Table 6.2). However, built-up areas expanded rapidly since 2010, at an annual rate of 44.71 km^2/year (Table 6.2).

The PLAND, PD, ENN, CIRCLE, and SHAPE landscape metrics (Table 6.3; Fig. 6.13) were analyzed to assess landscape pattern changes for the built-up lands (Estoque et al. 2014; Kamusoko 2017).

Fig. 6.6 Buildings along Thamrin street. *Source* Author's fieldwork (2008)

The percentage of landscape (PLAND) metric is a measure of proportion of built-up area to the whole area. PLAND increased substantially from 22.61 in 1989 to 43.76 in 2013 (Table 6.3). This is attributed to the rapid built-up area expansion of Jakarta, which occupied only about one fifth of the metropolitan area in 1989, but covered nearly 50% of Jakarta area in 2013. Figure 6.13 shows that PLAND increased within all distance buffer zones from city center. However, the percentage increase was higher near the city center, particularly within the 0–20 km distance buffer zones.

The patch density (PD) metric is a measure of fragmentation of built-up area patches based on the number of patches (regardless of the size) per unit area (in this analysis per km^2), in which a patch is based on an 8-cell neighbor rule. PD increased from 7.72 in 1989 to 9.24 in 2013 (Table 6.3). Generally speaking, if suburban small settlements were taken over expanding central urban area, PD should decrease because the expanding core urban center patch aggregated the suburban small patches. In Jakarta, however, PD actually shows temporal increase. The reason for this tendency may be due to new small settlements that continue to form outside the widespread central urban area in Jakarta. PD within distance zone 10–15 is lower in more recent years because built up area in central Jakarta has grown with aggregating existing fragmented patches. On the other hand, PD beyond distance zone 20–25 has become high with time, implying new built up area patches emerged mainly in this zone (Fig. 6.13).

Fig. 6.7 Buildings at the golden triangle. *Source* Author's fieldwork (2008)

The Euclidean nearest neighbor distance (ENN) metric expresses dispersion based on the distance from a patch to the nearest patch. The mean ENN decreased from 92.17 in 1989 to 82.13 in 2013 (Table 6.3). This trend is similar to the changes that were observed with PD. If the size of each patch increased, the mean ENN would also increase. In Jakarta, however, new small patches emerged while existing patches expanded. The mean ENN tends to show higher in farther distance zone (Fig. 6.13). It is because farther distance zone is larger in size.

The related circumscribing circle (CIRCLE) metric indicates the circularity of patches. This metric ranges from 0 for circular or one cell patches to 1 for elongated linear patches one cell wide. The mean CIRCLE value of Jakarta showed little change between 1989 and 2013 (Table 6.3). The shape index (SHAPE) metric indicates the complexity of patches. The value of 1 means a patch shape is square. The higher the value is, the more irregular a patch shape is. The mean SHAPE value of Jakarta gradually decreased between 1989 and 2013 (Table 6.3). The values of these two metrics mean that built up area in Jakarta sprawled to every direction, not always along some specific linear features such as railway and highway since 1989 (Fig. 6.12).

The land use/cover change and landscape metrics analysis confirmed that the built-up area in Jakarta expanded in all directions since 1989. However, the east–west development strategy was actually proposed in the urban development master plan for Jakarta, called Structure Plan, which began in 1985. Since the Dutch

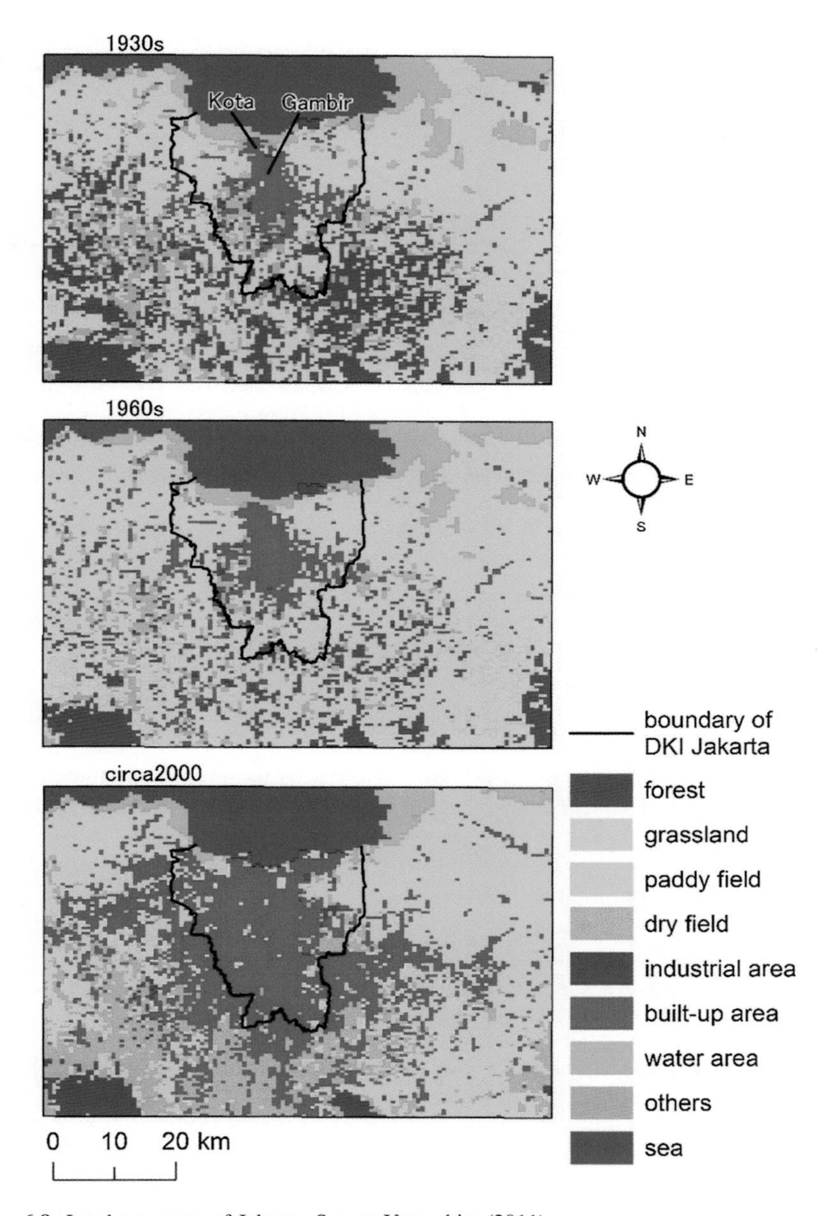

Fig. 6.8 Land-use maps of Jakarta. *Source* Yamashita (2011)

colonial rule, development in Jakarta progressed from the frequently flooding north coast toward the southern part, which has a better living environment. However, further development in the southern part could have caused deterioration of the aquifer environment because the hills and mountains around Bogor in the southern

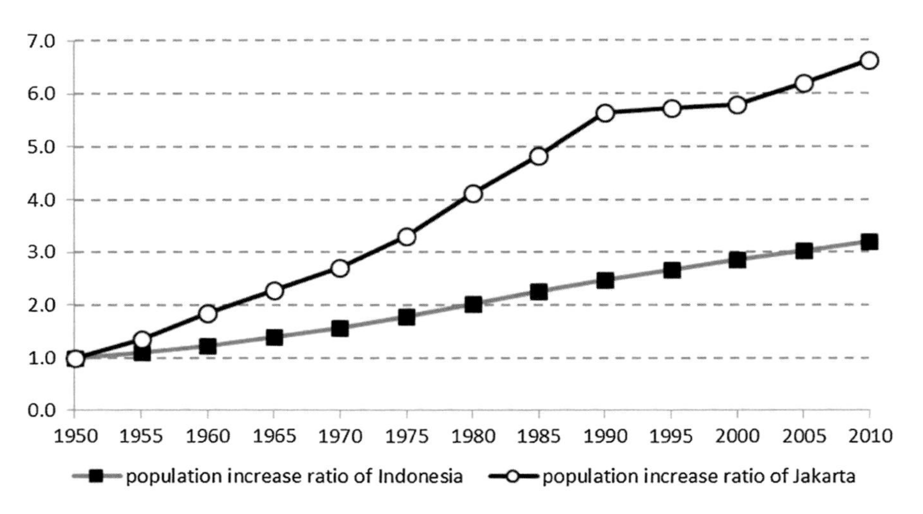

Fig. 6.9 Population increase ratio of Indonesia and Jakarta. *Source* UN (2015)

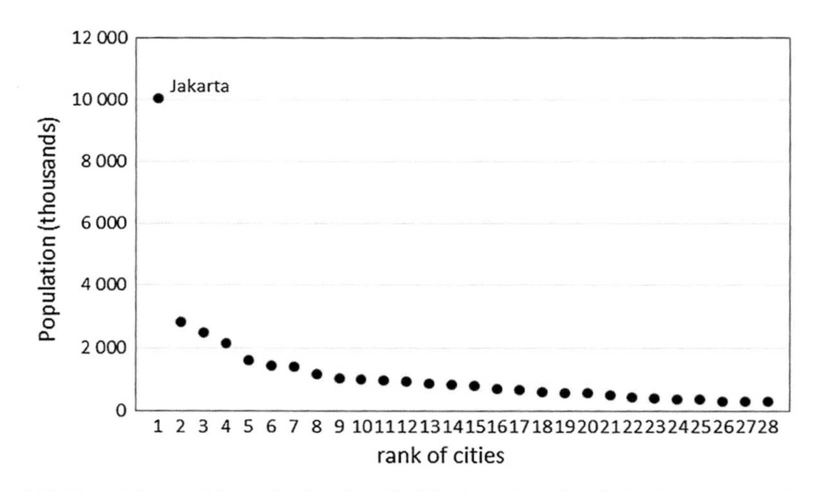

Fig. 6.10 Population and its rank of major 28 cities in Indonesia (2013). *Source* UN (2015)

part were the water reservoir of urban areas in Jakarta. Therefore, development progressed in the east–west direction along the highway stretching from the center of Jakarta, toward Bekasi in the east side and Tangerang in the west side, creating large housing estates and industrial districts (Konagaya 1997) (see also Sect. 6.4). Yet, based on the result of this analysis, the east–west development strategy has not appeared to be thoroughly implemented. This is because urban development—in particular the conversion of water catchment areas into built-up areas—partly contributed to the land subsidence due to excessive groundwater pumping for water consumption in Jakarta and its environs (see also Sect. 6.5).

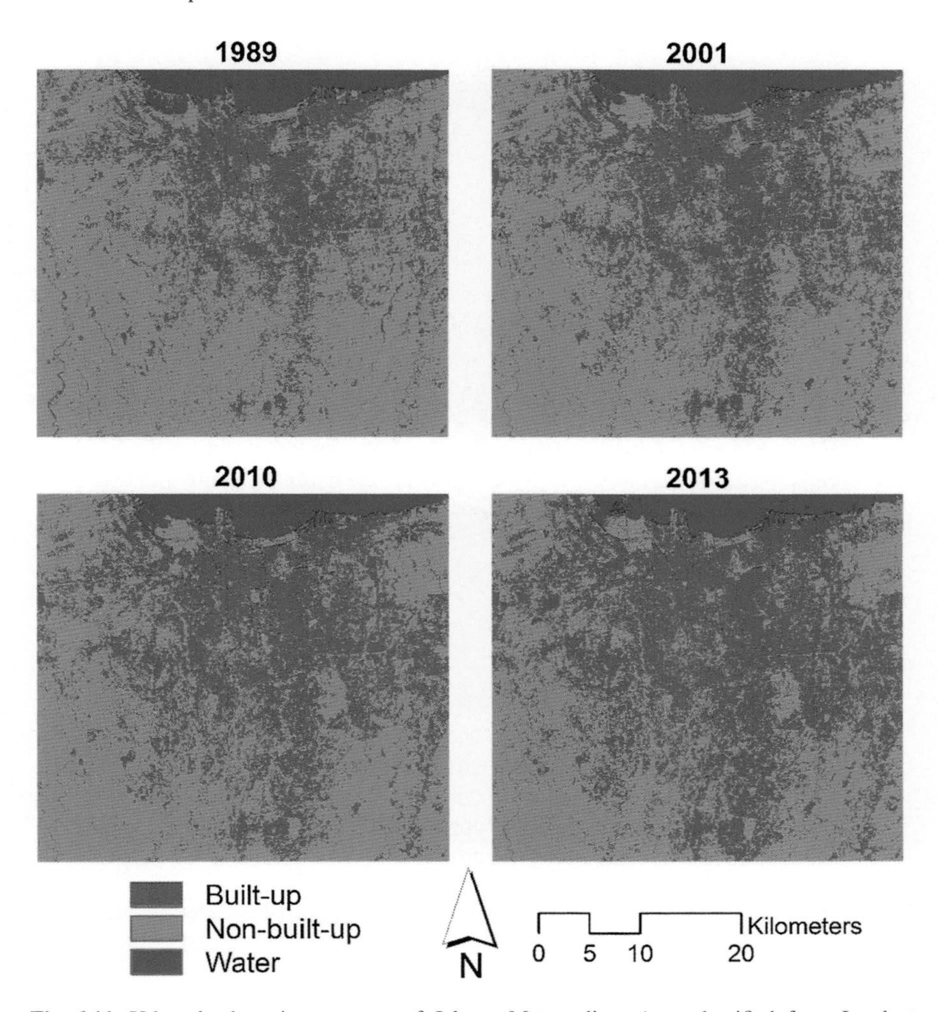

Fig. 6.11 Urban land use/cover maps of Jakarta Metropolitan Area classified from Landsat imagery

6.3.2 Projected Changes (2013–2030)

The projected land use/cover changes indicate that built up area will continue to expand in the future (Tables 6.4 and 6.5; Fig. 6.14, see also Fig. 6.12). Built-up area would increase from 623.48 in 2013 to 650.52 km^2 in 2020 at an annual rate of 3.86 km^2/year. However, built-up area is expected to increase further by 61.88 km^2, at an annual rate of 6.19 km^2/year between 2020 and 2030. This suggests that another wave of rapid urbanization would likely occur between 2020 and 2030 under the current simulation scenario.

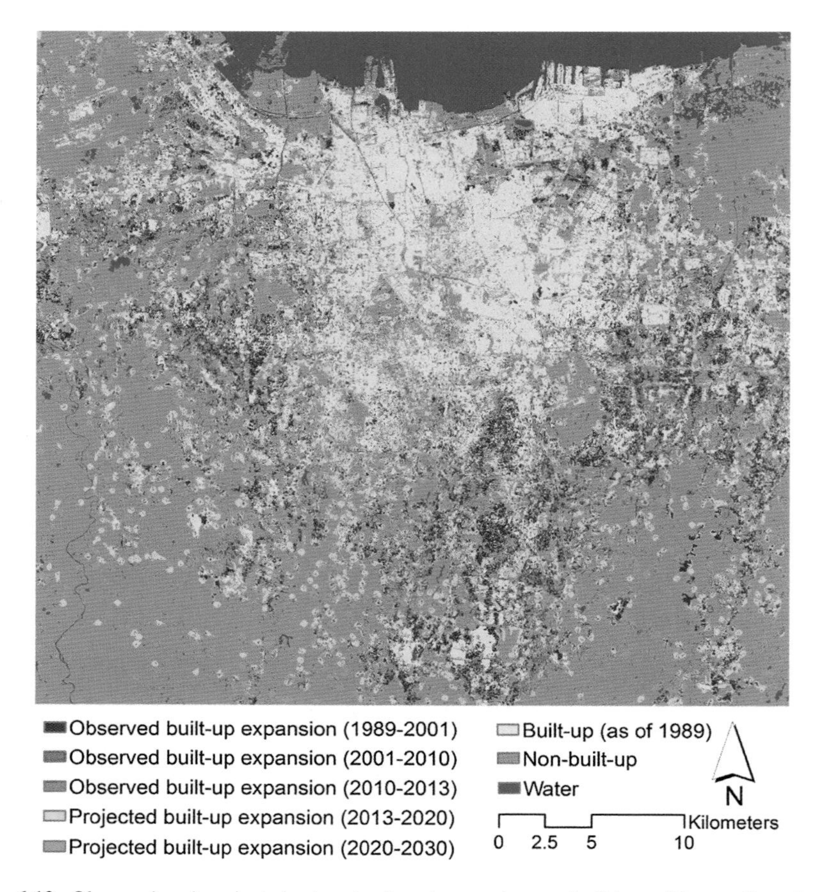

Fig. 6.12 Observed and projected urban land use/cover changes in Jakarta Metropolitan Area

Table 6.1 Observed urban land use/cover of Jakarta Metropolitan Area (km^2)

	1989	2001	2010	2013
Built-up	311.88	432.78	489.34	623.48
Non-built-up	1067.21	968.12	930.34	801.44
Water	148.82	127.00	108.23	102.99
Total	1527.91	1527.91	1527.91	1527.91

According to landscape pattern analysis (Table 6.6; Fig. 6.15), PLAND will increase within all distance buffer zones. The mean SHAPE value will also increase slightly, particularly in distance zones beyond 15 km from city center. The mean CIRCLE value has the similar tendency. These simulation results imply that built up area will fill most area within 15 km from the city center of Jakarta and that will also spread comparatively linearly and irregularly in the outer edge of Jakarta.

Table 6.2 Observed urban land use/cover changes in Jakarta Metropolitan Area (km^2)

	1989–2001	2001–2010	2010–2013
Built-up	120.91	56.56	134.14
Annual rate of change (km^2/year)	*10.08*	*6.28*	*44.71*
Non-built-up	−99.08	−37.79	−128.90
Annual rate of change (km^2/year)	*−8.26*	*−4.20*	*−42.97*
Water	−21.82	−18.77	−5.25
Annual rate of change (km^2/year)	*−1.82*	*−2.09*	*−1.75*

Table 6.3 Observed landscape pattern of Jakarta Metropolitan Area

Class-level (built-up) spatial metrics	1989	2001	2010	2013
PLAND (%)	22.61	30.89	34.47	43.76
PD (number per km^2)	7.72	8.10	8.78	9.24
ENN (mean) (m)	92.17	86.39	84.51	82.13
CIRCLE (mean) ($0 \leq \text{CIRCLE} < 1$)	0.35	0.34	0.32	0.32
SHAPE (mean) ($1 \leq \text{SHAPE} \leq \infty$)	1.27	1.26	1.25	1.22

6.4 Driving Forces of Urban Development

The major rivers flowing through Jakarta to the Java Sea are the Ciliwung River and the Cisadane River. The downstream basins of these rivers have rather a flat terrain, though gradually sloping from south to north, and are low-lying areas 100 m above sea level or lower. Therefore, when the urbanization progressed within and around Jakarta, there were few topographic restrictions, resulting in radial expansion of urban land use (Yamashita 2012, 2013).

According to Konagaya (1999), urban expansion progressed toward the south until the 1980s, since the area has a better living environment. During this time, districts with accumulated offices, shopping malls, condominiums, etc., were developed in many locations in DKI Jakarta. The Golden Triangle is a representative example of office and condominium accumulation, and the Block M near Kebayoran Baru in the south is a representative commercial center in a suburban area, while Mangga Dua in the north is an example of existing urban core redevelopment.

Up to 1997, when the economic crisis hit in Indonesia, many housing estates and industrial districts were developed along the highway stretching in the east–west direction (the Merak Highway toward the west and the Cikampek Highway toward the east), taking into account the east–west development strategy. In 1996, 13 giant housing estates with a 1000 ha or larger development area existed within Jabotabek (one in DKI Jakarta, two in Bogor Regency in the south, three in Bekasi Regency in the east, and seven in Tangerang Regency in the west) (Konagaya 1999). These massive developments were encouraged by a series of financial deregulation policies in Indonesia during the 1980s (Winarso and Firman 2002). This deregulation

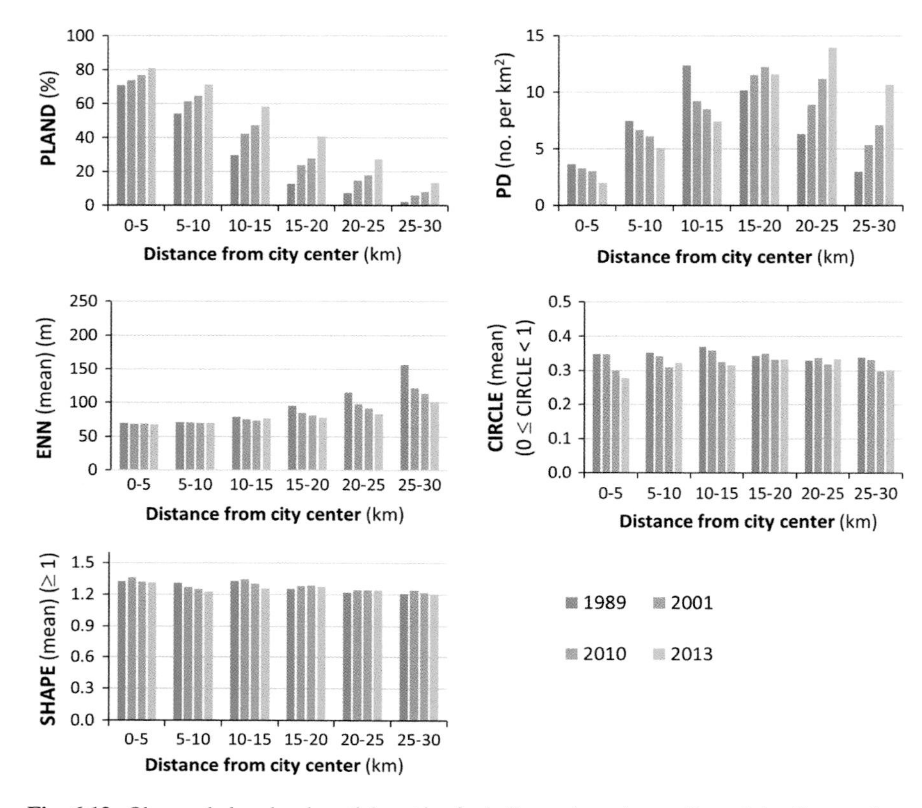

Fig. 6.13 Observed class-level spatial metrics for built-up along the gradient of the distance from city center of Jakarta Metropolitan Area. *Note* The y-axis values are plotted in the same range as those in Fig. 6.15

Table 6.4 Projected urban land use/cover of Jakarta Metropolitan Area (km^2)

	2020	2030
Built-up	650.52	712.39
Non-built-up	774.41	712.53
Water	102.99	102.99
Total	1527.91	1527.91

enhanced suburban housing markets for middle- and high-income classes and housing development by private sectors. One example is the development of Lippo Karawaci. It was begun in 1992 by Lippo Group, an Indonesian big business. It is a composite housing estate, built near an interchange of the Merak Highway, where high-rise office buildings, condominiums (Fig. 6.16), a large shopping mall, and a university are even located. A golf course is at the center of the housing estate, surrounded by an exclusive residential district (large sites with yards) for the wealthy class. Residential areas for the middle class are distributed at the outer rim

Table 6.5 Projected urban land use/cover changes in Jakarta Metropolitan Area (km^2)

	2013–2020	2020–2030
Built-up	27.03	61.88
Annual rate of change (km^2/year)	*3.86*	*6.19*
Non-built-up	−27.03	−61.88
Annual rate of change (km^2/year)	*−3.86*	*−6.19*
Water	0.00	0.00
Annual rate of change (km^2/year)	*0.00*	*0.00*

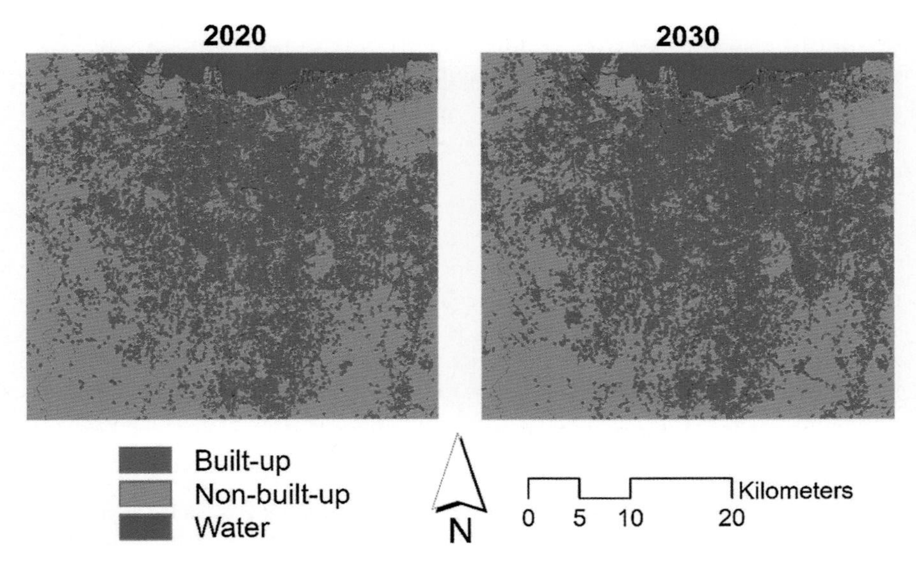

Fig. 6.14 Projected urban land use/cover maps of Jakarta Metropolitan Area

Table 6.6 Projected landscape pattern of Jakarta Metropolitan Area

Class-level (built-up) spatial metrics	2020	2030
PLAND (%)	45.65	50.00
PD (number per km^2)	1.85	1.37
ENN (mean) (m)	104.73	108.10
CIRCLE (mean) ($0 \leq$ CIRCLE < 1)	0.31	0.32
SHAPE (mean) ($1 \leq$ SHAPE $\leq \infty$)	1.29	1.30

of the housing estate (Fig. 6.17), and housing complexes and terraced houses for residential and commercial use can be seen.

The major industrial districts are mainly located in Bekasi Regency and Karawang Regency further east, along the Cikampek Highway on the east side (Konagaya 1999). The MM2100, one such district in Bekasi Regency, was

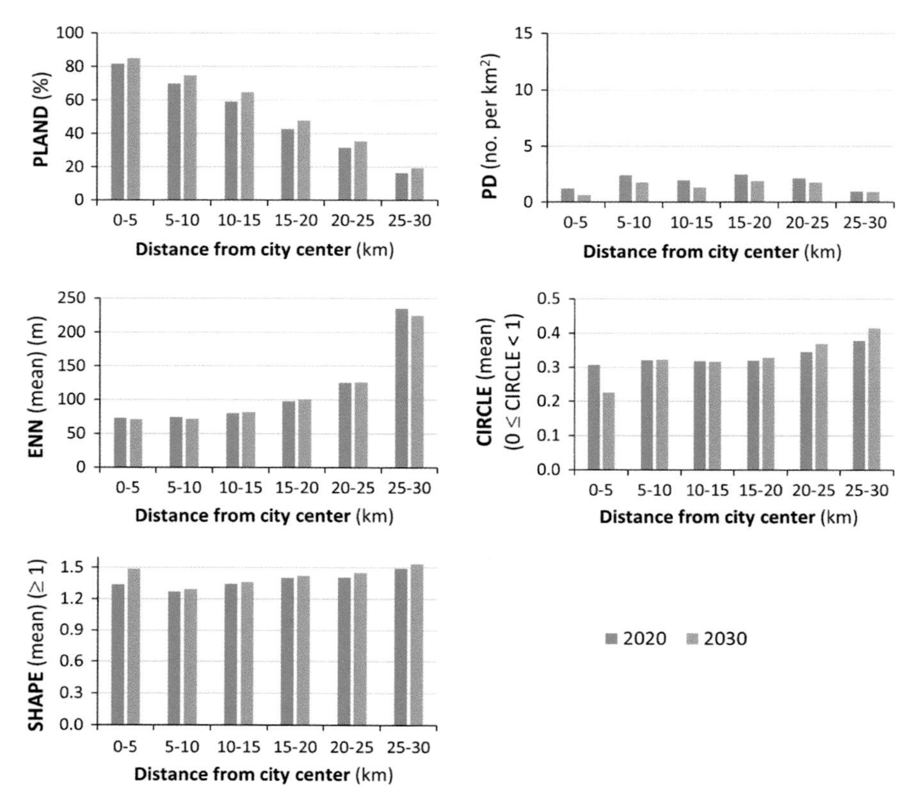

Fig. 6.15 Projected class-level spatial metrics for built-up along the gradient of the distance from city center of Jakarta Metropolitan Area. *Note* The y-axis values are plotted in the same range as those in Fig. 6.13

developed in the 1990s by a joint venture of Japanese and Indonesian companies (Fig. 6.18). About 170 factories and businesses are located in the 805 hectare site (as of 2008), including many Japanese-affiliated companies. There are several other industrial districts in the area, developed with capital from Japan.

Foreign direct investment (FDI) from countries such as Japan, the United States, Singapore, Korea, and China was a major driving factor for urban expansion of Jakarta. However, FDI for Jakarta decreased substantially due to the Asian Currency Crisis in 1997. On the other hand, investment from Korean companies has been increasing recently. The Japanese government has also provided financing through ODA since 2010, based on the Jobodetabek Metropolitan Priority Area for Investment and Industry, and development of infrastructure including the new international port, public transportation, and road networks, is in progress (Ikuta 2011).

Fig. 6.16 Condominiums in Lippo Karawaci. *Source* Author's fieldwork (2008)

6.5 Implications for Future Sustainable Urban Development

As a megacity in a developing country, Jakarta has the typical issues of excessive urbanization which are characterized by insufficient employment creation opportunities and infrastructure development (such as water supply and sewage systems) that is lagging behind the rapidly increasing population. This has resulted to various social and environmental problems. Here, according to Yamashita (2014), water environmental problem will be emphasized because it has become especially serious in Jakarta lately.

The Jakarta Metropolitan Area has experienced rapid urban growth over the past decades. Furthermore, the projected land use/cover changes indicate that future built-up expansion will likely continue in the southern part of Jakarta, where major aquifer zones for the metropolitan water demand are found. This implies that impervious surfaces (built-up areas) will continue to replace the natural landscape. As a result, water environment (rivers, canals, and groundwater) will be affected both in water quality and quantity.

Jakarta is built on low-lying marshlands and has an extensive waterway network. The waterway network in Jakarta seems to have been used to flow out rainwater and wastewater into the sea. Two things can be pointed out as a qualitative aspect of

Fig. 6.17 Houses for middle class in Lippo Karawaci. *Source* Author's fieldwork (2008)

water environmental problem in the canals and ditches in Jakarta: serious nonpoint source pollution caused by urbanization, and rubbish dumping by residents living alongside the river.

Impervious surfaces in Jakarta have accelerated the delivery of pollutants from the built-up environment to canals and ditches. For example, a large amount of dust and rubbish, which exists on the surface of the land flows into canals and ditches. In fact, some canals and ditches in central Jakarta turned grey, and in some cases it is smelly along many parts of the canals and ditches.

In addition, inadequate public services such as sewage systems and rubbish collection following rapid population growth have led to significant water environmental issues. People tend to drain household wastewater directly into the waterway and dump rubbish. With regard to measures to deal with pollution such as worsening nonpoint source pollution and rubbish dumping, it is important to improve public infrastructure, such as the sewage system and rubbish collection service.

On the other hand, change from permeable surface to impermeable surface is significant from a quantitative perspective of water environmental problem. Rainwater that runs off the land surface instead of penetrating into the soil suggests a decrease of the amount of groundwater recharge. Rapid urbanization and population growth drastically increase the requirement for the pumping of groundwater. This results in land subsidence. Jakarta is currently facing serious land subsidence

Fig. 6.18 Entrance of the MM2100 district. *Source* Author's fieldwork (2008)

and expanding areas below sea level (called "zero meter areas"). Such areas are also seen in large coastal cities in Japan. However, a problem unique to a large tropical city such as Jakarta is that there are frequent flood inundations during the wet season. Heavy storm water in a short period of time is not able to penetrate into the impermeable land surface and does not flow out into the sea through river or waterway due to land subsidence. As a result, whenever there is a rain downfall, roads become flooded.

In Jakarta, construction of discharge channels is under way. However, such a plan for channels alone is not adequate to deal with flooding. Comprehensive flood control including the land use plan in the wider areas at watershed scale is necessary. In Jakarta, included in the structure plan in the 1980s, the east–west development strategy was launched and measures were taken in order to slow down the development of the upper reaches of the river in the south, seen as water source cultivation areas (Konagaya 1997). Although such measures brought some success, they are not entirely satisfactory as the development from forests to farmlands continues in the south. There is development restriction in the highly elevated, steeply sloped areas for water source forest conservation, but these areas are relatively small for the overall watershed. In order to reduce flooding and slow down land subsidence, it is necessary not only to successfully complete construction of discharge channels, but also to build facilities to store rainwater and to cultivate groundwater at each catchment and to facilitate systems to make a good use of such water.

6.6 Concluding Remarks

Jakarta, a port town facing the Java Sea, was a hub of the Dutch East India Company and became a center of marine traffic for Asia. Rapid urbanization progressed after independence, for about 50 years in the late twentieth century. Especially after the 1980s, large scale development had been conducted also in suburban areas by FDI from Japan and other countries, resulting in making Jakarta one of the world's top megacities. But such rapid urban development caused serious environmental problems as well. For the sustainable development of Jakarta in the future, it is necessary to point toward urban policy and land use planning based on proper arrangements of population and industry on watershed scale or metropolitan area, while developing infrastructure in urban areas.

References

Estoque RC, Murayama Y, Kamusoko C, Yamashita A (2014) Geospatial analysis of urban landscape patterns in three major cities of Southeast Asia. Tsukuba Geoenvironmental Sci 10:3–10

Ikuta M (2011) Mega-cities in Southeast Asia: widening regional integration including Japan. Kokon-Shoin, Tokyo (Japanese)

Kamusoko C (2017) Methodology. In: Murayama Y, Kamusoko C, Yamashita A, Estoque RC (eds) Urban development in Asia and Africa—geospatial analysis of metropolises. Springer Nature, Singapore, pp 11–46

Konagaya K (1997) Urban planning and urban structure of DKI Jakarta. Q J Econ Stud 20(2):85–99 (Japanese)

Konagaya K (1999) Urban structure. In: Miyamoto K, Konagaya K (eds) Asian large city 2 Jakarta. Nippon Hyoron Sha, Tokyo, pp 87–116 (Japanese)

UN (United Nations) (2015) World urbanization prospects: the 2014 revision. United Nations, New York

Winarso H, Firman T (2002) Residential land development in Jabotabek, Indonesia: triggering economic crisis? Habitat Int 26:487–506

Yamashita A (2011) Comparative analysis on land use distributions and their changes in Asian mega cities. In: Taniguchi M (ed) Groundwater and subsurface environments: human impacts in Asian coastal cities. Springer, Tokyo, pp 61–81

Yamashita A (2012) Land use change and its correlation with topographic condition in Asian large cities in the 20th century: cases of Seoul, Taipei and Jakarta. Proceedings of geographical information systems association 21 (Japanese with English Abstract)

Yamashita A (2013) Land use change and its correlation with topographic condition in the Ciliwung-Cisadane River Basin, Indonesia. Proceedings of geographical information systems association 22 (Japanese with English Abstract)

Yamashita A (2014) Aspects of water environmental issues in Jakarta due to its rapid urbanization. Tsukuba Geoenvironmental Sci 10:43–50

Chapter 7
Hanoi Metropolitan Area

Duong Dang Khoi

Abstract Vietnam in general and Hanoi in particular, have experienced rapid urbanization due to economic reforms and open door policies in the last two decades. The purpose of this chapter is to examine the observed and projected urban land changes (i.e., changes from non-built-up to built-up lands) between 1989 and 2030, as well as to understand the driving forces of urban development in Hanoi Metropolitan Area (Metro Hanoi). Their implications for Metro Hanoi's future sustainable development are also discussed. History reveals that the urban development of Metro Hanoi is closely related to feudal dynasties and political status of Vietnam. The observed urban land changes show that Metro Hanoi has undergone rapid urbanization over the past two decades. This rapid urbanization was mainly driven by economic development, urban planning, population growth, and urban policies. While urbanization has produced positive economic outcomes, serious problems have since emerged. The detected spatiotemporal patterns of the observed and simulated urban land changes can be used to assist strategic urban planning in Metro Hanoi.

7.1 Origin and Brief History

Hanoi, the capital city of Vietnam, is located in the Red River delta in the northern region of Vietnam (Fig. 7.1). The capital city consists of 12 urban districts, one town, and 18 rural districts. The topography is characterized by deltas and mountainous areas. The altitude in the northern part of the city is lower than the southern part.

The origin and development of Metro Hanoi is closely related to the Ly-Tran (1010–1397), Le (1428–1789), and Nguyen feudal dynasties (Nguyen et al. 2010) as well as the French colonial rule (1802–1945) and post-second world war conflicts (1945 to present).

In the Ly-Tran dynasty, King Ly Cong Uan selected the Dai La region, now Hanoi, to construct the capital city of Vietnam. According to Nguyen et al. (2010),

D.D. Khoi (✉)
Hanoi University of Natural Resources and Environment, Hanoi, Vietnam
e-mail: khoi_tn@yahoo.com

Y. Murayama et al. (eds.), *Urban Development in Asia and Africa*,
The Urban Book Series, DOI 10.1007/978-981-10-3241-7_7

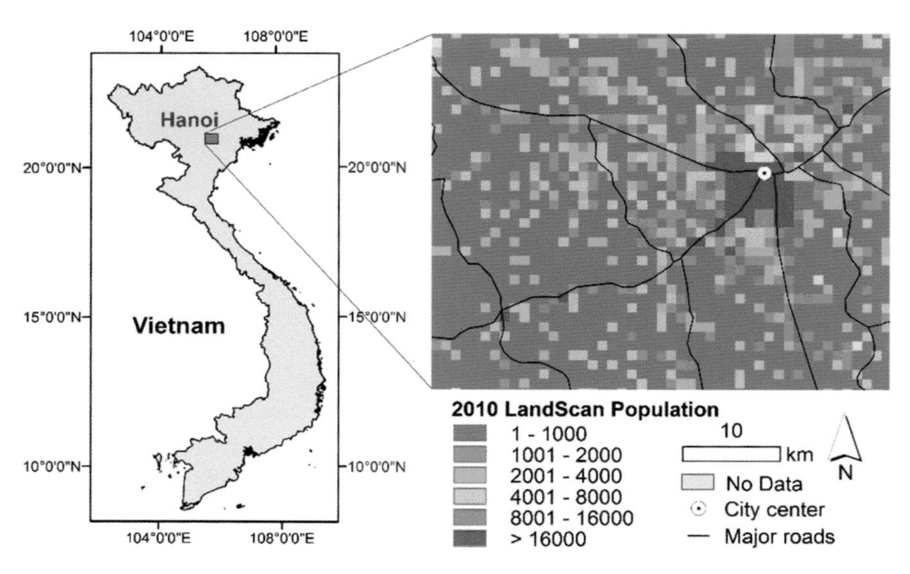

Fig. 7.1 Location and LandScan population of Metro Hanoi, Vietnam

when the King visited the Da La region, the King saw a cloud that looked like a flying dragon and named the capital, Thang Long (Dragon). Since then, Thang Long City was recognized as the center of politics, economy, and culture in Vietnam. In the early days, Thang Long was well known as the center of markets. Each market specialized in a particular product, e.g., rice noodle market and bamboo hat market. Craft workers across the Red River Delta of Vietnam left their native homeland to settle in the markets of Thang Long. Over time, urban infrastructures were gradually constructed. In 1029, the King Ly Thanh Tong constructed his royal palace. During the Ly dynasty, several architectural buildings were constructed in Thang Long. Thang Long continued to expand and grow until the Tran dynasty. During the Tran dynasty, Thang Long was divided into 61 wards. Each ward specialized in producing a handicraft product. During the 1258–1288 period, the Mongol Empire invaded Vietnam several times. The Tran dynasty failed to protect Thang Long from the Mongol invasion. The King then moved to An Ton district, Thanh Hoa province in 1397.

In 1406 under the Le dynasty, the Ming Empire of China attacked Vietnam. The King of Vietnam and his people lost against the Ming Empire. Thang Long was renamed Dong Quan. In 1418, the Ming Empire was defeated in a battle led by Le Loi. In 1430, Thang Long was renamed Dong Kinh, but its original name, Thang Long, was restored in 1527. Until 1749 (Le-Trinh dynasty), many buildings were constructed in Thang Long. In 1789, when Quang Trung became the King, Hue province was selected as the capital city of Vietnam.

Under the Nguyen dynasty and the French colonial rule (1802–1945), from 1803 to 1805, the King of the Nguyen, Gia Long, reconstructed new areas of Hanoi in the style of French architecture. When the French dominated northern Vietnam and settled in Hanoi, the population of Hanoi was less than 100,000 people. Hanoi City had three separate areas consisting of the castle of the King Nguyen, a small commercial area and rural areas around the wall castle. By 1902, Hanoi became the capital of the Indochina. French planners greatly transformed the appearance and functions of the city. In particular, roads, drainage, electricity networks, and grid street system were constructed.

Vietnam became independent in 1945. However, Hanoi was at war against the occupation of French and the United States until 1975. Between 1945 and 1975, the government moved people and industries from Hanoi to other regions due to the wars. In the late 1980s, Vietnam launched open door policies known as the "Doi Moi" (renovation policies). The population of Hanoi increased substantially to 3.2 million in 2007, attributing to rural–urban migration (HSO 2009). In 2008, due to rapid urbanization in the inner area of Hanoi (Figs. 7.2 and 7.3), the Vietnam National Assembly approved the extended boundary of Hanoi. To date, the city covers about 3329 km^2, with a population of about 6.4 million. A large area of the new Hanoi is classified as "rural areas" (HSO 2009). In terms of population size, Hanoi is second to Ho Chi Minh, which has a population of 6.8 million. The so-called "New Hanoi" is expected to become the center of politics, education, science, economy, and international exchanges. The other objective is to distribute inhabitants, industries, universities, hospitals, administration bodies, and overcrowded areas of the city to suburban areas (VET 2008). The city's GDP increased three times between 2000 and 2008. In 2008, Hanoi received about US$18.8 billion worth of foreign development investment, accounting for 7% of the total investment in Vietnam (HSO 2009).

Fig. 7.2 A view of a highly urbanized area in Tuliem district, Hanoi City. *Source* Author's fieldwork (2015)

Fig. 7.3 A view of a residential expansion in Tuliem district, Hanoi City. *Source* Author's fieldwork (2015)

7.2 Primary in the National Urban System

7.2.1 Vietnam Urban System

Urbanization is a global trend in the recent years. The United Nations (2008) estimated that about 50% of the world population lives in urban areas. This proportion may increase to over 72% by 2050 (United Nations 2012). Urbanization has taken place exponentially in the past two decades in Southeast Asia, especially in large urban areas. Many cities in the developing countries have radically shifted from agriculture- to industry-based economy (World Bank 2011). Similarly, since the launch of economic reforms and open door policies in the late 1980s, Vietnam has experienced a rapid urbanization rate. Vietnam's annual rate of urbanization has been estimated at 3.4% (World Bank 2011). There were 500 urban areas in the 1990s and this increased to 656 in 2003 (World Bank 2011). In particular, urbanization has occurred rapidly in large cities such as Hanoi, Ho Chi Minh, and Danang. Hanoi City has been one of the fastest growing cities in Vietnam (World Bank 2011).

Vietnam's economic development was centered on highly urbanized areas such as Hanoi, Ho Chi Minh, Danang, and Hue. In terms of economic geography, Hanoi is in the north, Ho Chi Minh is in the south and Danang and Hue are in the central region. The government has issued a series of policies to foster both economic growth and urban development. For example, Government Decision 10 (1998), which concerns the urban development strategy 2020, offered incentives for the development of medium and small cities while maintaining the growth of the largest cities. The 2011–2020 economic development strategy confirmed that urbanization was one of the most important strategies to achieve the country's goals of industrialization and modernization.

The Vietnam urban classification system, established in 2001 and revised in 2009 by the decree 42/2009/ND-CP, has served as the foundation of urban policies in Vietnam. According to the decree, Vietnam urban system is classified into six urban types based on economic activities, physical development, population density, the ratio of nonagricultural labors, and socioeconomic infrastructures. In terms of economic activities, the decree states that an urban center is a specialized center of national, interprovincial, provincial or district level, or a center of an intraprovincial region, which plays a role in promoting socioeconomic development of the whole country or a certain region. The town in the lowest level of the urban classification system must have at least 4000 inhabitants. Nonagricultural labor in the inner of urban must be more than 60% of total population. Urban infrastructures and facilities are social facilities (e.g., schools, hospitals, administrative building, and other public facilities) and technical infrastructures (e.g., roads, water supply system, and electricity system).

Hanoi and Ho Chi Minh are termed "special cities" because they considerably contributed to Vietnam development in terms of economic and political aspects. In 2009, Vietnam urban system was comprised of two special cities, five class-I cities, 12 class-II cities, 40 class-III towns, 47 class-IV provincial towns, and 625 class-V small towns. Local governments often expect to promote their urban status to a higher ranking level in the classification in order to receive a bigger amount of budget from central government. Therefore, the urban classification provided incentives for cities to try to move to a higher class (Coulhart et al. 2006). In 2009, special cities, class-I cities, class-II cities, class-III cities, and class-IV cities contributed to 30.5, 6.9, 5.2, 5.7, and 3%, respectively, of the total GDP of Vietnam. Annual city population growth of the urban classes from 1999 to 2009 varied from 1.1% in the central coast to 3.7% in the northern midlands (World Bank 2011).

7.2.2 *Primacy of Metro Hanoi*

The ranking of cities in a country or in the world can support investors in the choice of location and serves as an important guide for future city development. In this section, urban primacy is based on city area, economic activities, population, and the ratio of nonagricultural labor, and infrastructure.

The extent of Hanoi expanded exponentially from 461 km^2 in 1961 to 3329 km^2 in 2015. The designation of Hanoi as the development center for the northern region of Vietnam resulted in territorial expansion and rapid economic growth over the last decades. Recently, Hanoi is ranked as the biggest urban area of Vietnam, followed by Ho Chi Minh City with a total area of 2095.6 km^2.

In regard to urban economic activities, Hanoi and Ho Chi Minh City are major contributors of the total GDP of Vietnam (Fig. 7.4). For instance, in 2009, these two

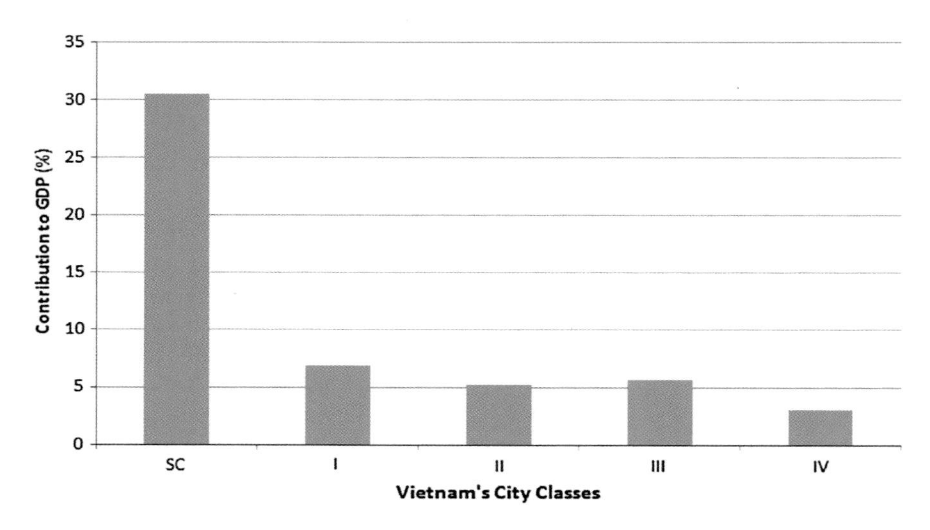

Fig. 7.4 Contribution of the city classes to the total GDP of Vietnam in 2009. *Note* SCs are special cities of Vietnam consisting of only Hanoi and Ho Chi Minh. *Source* General Statistic Office, Vietnam

cities contributed 30.5% of the total GDP. In terms of economic growth, Hanoi was ranked as the second after Ho Chi Minh City. The other urban areas in the city classes I, II, III, and IV accounted for 20.8% of the country's total GDP (GSO 2009). In general, nonagricultural activities were often concentrated in large cities. Hanoi and Ho Chi Minh City specialize in manufacturing and services industries. Many special economic zones, some of which are situated in urbanized areas, play important roles in the promotion of regional economic development in Vietnam. These are referred to as Industrial Clusters (IC), Industrial and Processing Zones (IZ), High-Tech Zones (HTZ), and Economic Zones (EZ). ICs consist of small-scale manufacturing enterprises located in one area in order to improve efficiency. In 2007, there were 333 ICs in Vietnam. IZs specialize in manufacturing industrial products and supporting services. In 2009, there were 264 IZs in the country, in which construction sector was a major fraction. HTZs are defined as high-tech enterprise and its supporting services. They are located only in Hanoi, Ho Chi Minh, and Danang City. The HTZs have been growing gradually. Lastly, EZ areas are border and coastal economic zones. By the end of 2010, there were 23 border gate economic zones and 15 coastal economic zones in Vietnam (World Bank 2011).

In terms of urban population, Metro Hanoi is ranked second to Ho Chi Minh City. Hanoi and Ho Chi Minh City accounted for almost 14% of the total urban population of the country, and all urban areas in the classes I, II, III and IV constituted 18% (Fig. 7.5) (GSO 2009). Income disparity between Metro Hanoi and its surrounding areas resulted in a large number of migrants into Hanoi. For example, in 2009, the average income varied from 11.0 million VND in the Northern Midlands to 23.1 million VND in the Red River Delta which includes

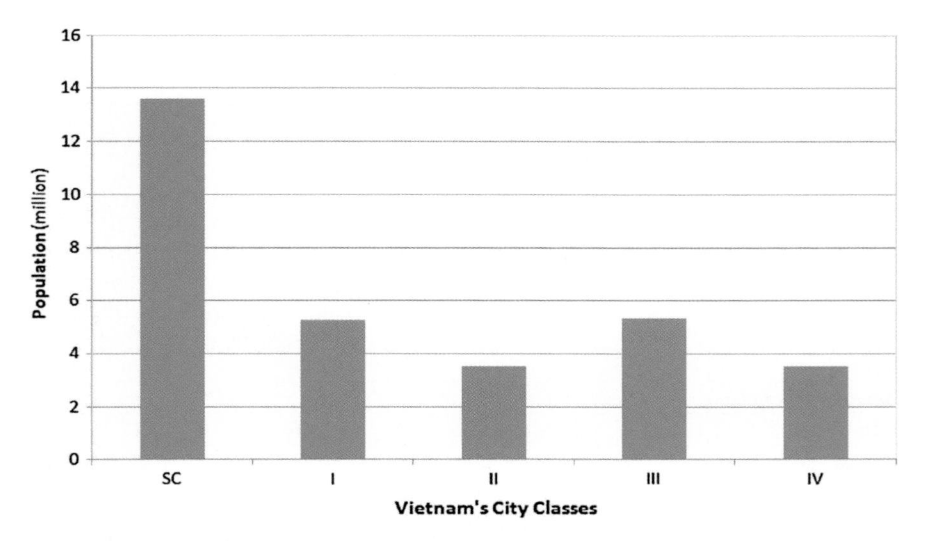

Fig. 7.5 Total population of the city classes of Vietnam in 2009. *Notes* SC are special cities consist of Hanoi and Ho Chi Minh City. *Source* General Statistic Office, Vietnam

Hanoi (GSO 2009). In addition, highly educated people and highly skilled workers often expect to settle in Hanoi because they can have better jobs with higher salary and access to better living facilities. Therefore, Hanoi has attracted a large number of highly qualified workers from rural areas across the country.

Regarding urban infrastructures and facilities, Hanoi offers better common facility services than other provincial cities and towns. In addition, the percentage of households that has access to safe water and electricity in Hanoi was higher than other provincial cities and towns (GSO 2009). However, the quality and the quantity of infrastructures in Hanoi, which are lower than international standards, are still a major constraint for rapid socioeconomic development of the city in the future. Currently, Metro Hanoi is urgently in need of modern infrastructures such as railway system, subway system, water supply and drainage system, electricity lines, and wastewater and solid waste processing facilities.

7.3 Urban Land Use/Cover Patterns and Changes (1989–2030)

7.3.1 Observed Changes (1989–2014)

Tables 7.1 and 7.2 and Figs. 7.6 and 7.7 show the urban land use/cover changes in Metro Hanoi. There was a substantial increase in built-up areas between 1989 and 2014, indicating rapid urbanization. Built-up area increased from 23.2 km^2 in 1989

Table 7.1 Observed urban land use/cover of Metro Hanoi (km^2)

	1989	1999	2009	2014
Built-up	23.22	51.36	94.10	129.63
Non-built-up	1568.54	1543.17	1533.51	1557.68
Water	168.86	166.08	133.01	73.31
Total	1760.62	1760.62	1760.62	1760.62

Table 7.2 Observed urban land use/cover changes in Metro Hanoi (km^2)

	1989–1999	1999–2009	2009–2014
Built-up	28.15	42.73	35.54
Annual rate of change (km^2/year)	*2.81*	*4.27*	*7.11*
Non-built-up	−25.36	−9.66	24.17
Annual rate of change (km^2/year)	*−2.54*	*−0.97*	*4.83*
Water	−2.78	−33.07	−59.70
Annual rate of change (km^2/year)	*−0.28*	*−3.31*	*−11.94*

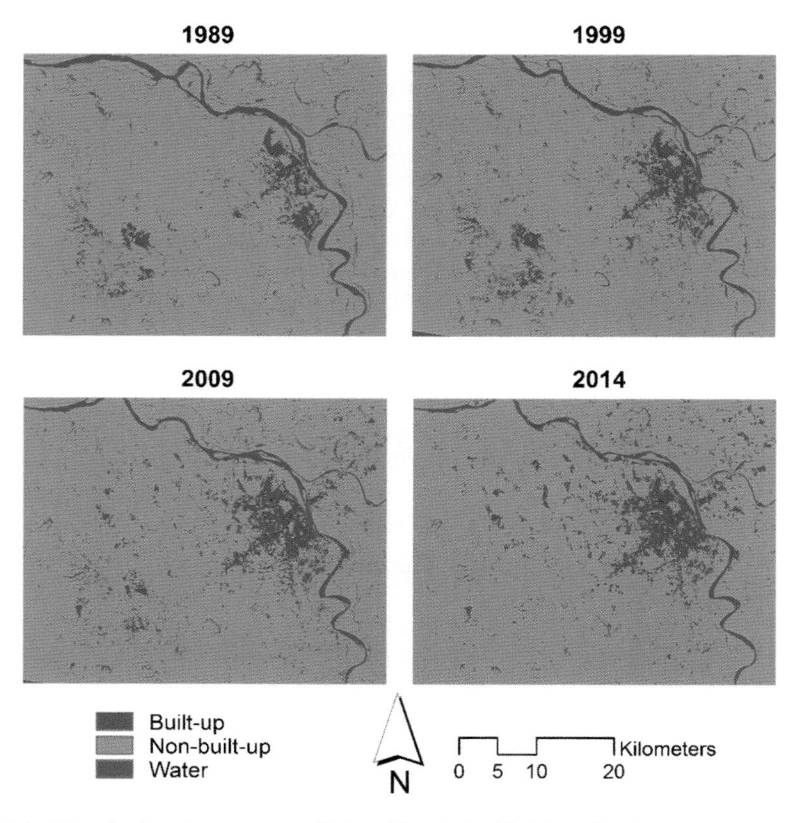

Fig. 7.6 Urban land use/cover maps of Metro Hanoi classified from Landsat imagery

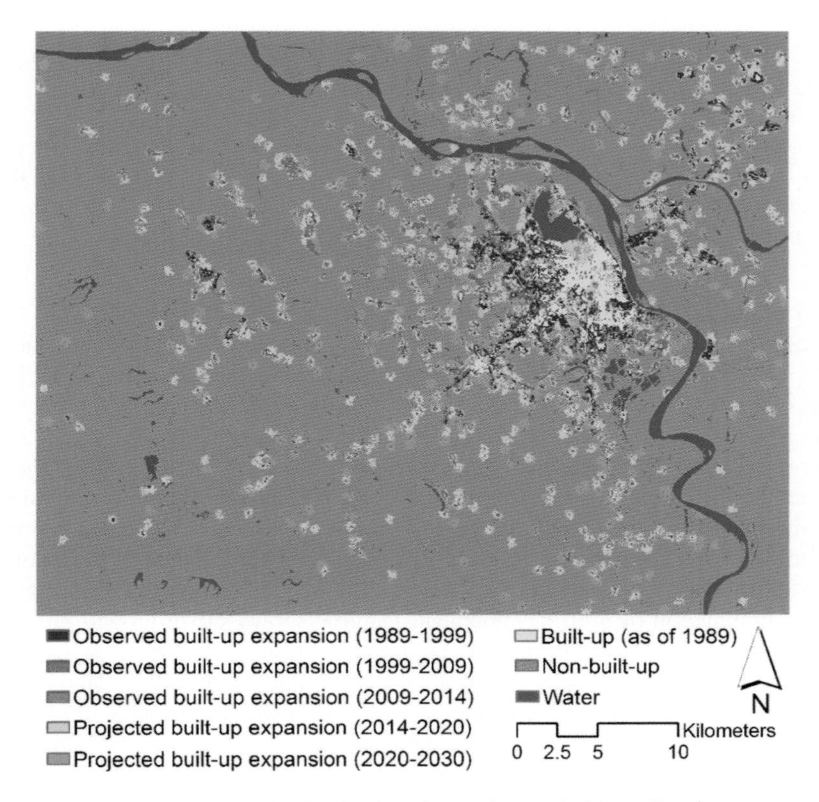

Fig. 7.7 Observed and projected urban land use/cover changes in Metro Hanoi

to 129.6 km^2 in 2014 (Table 7.1), while the non-built-up areas decreased from 1568.5 to 1557.7 km^2. The city expanded into its surrounding areas such as Thanh Xuan, Hoang Mai, Long Bien, Ha Dong, North Tu Liem, and South Tu Liem. In addition, Ha Tay province was merged into Hanoi City in 2008. Therefore, a large portion of the suburban districts still remains as agricultural land and plays an important role in supplying food and fresh vegetables to Hanoi. The built-up growth rate has accelerated over the study period. For example, during the 1989–1999 period, Metro Hanoi had an annual gain of built-up of 2.81 km^2/year. This rate increased to 4.27 km^2/year and 7.11 km^2/year during the 1999–2009 and 2009–2014 periods, respectively (Table 7.2).

Table 7.3 shows the landscape pattern metrics for the built-up class. The percentage of landscape (PLAND) metric measures the fraction of a class to the whole landscape at a specific time point, while the patch density (PD) metric is a measure of fragmentation based on the number of patches per unit area (per 100 ha or 1 km^2), in which a patch is based on an 8-cell neighbor rule. The Euclidean nearest neighbor distance (ENN) metric is a measure of dispersion based on the average distance of a patch to the nearest neighboring patch of the same class. The related circumscribing

Table 7.3 Observed landscape pattern of Metro Hanoi

Class-level (built-up) spatial metrics	1989	1999	2009	2014
PLAND (%)	1.46	3.22	5.78	7.68
PD (number per km^2)	0.50	0.84	1.78	3.36
ENN (mean) (m)	240.95	198.40	152.03	111.48
CIRCLE (mean) ($0 \leq$ CIRCLE < 1)	0.39	0.41	0.41	0.32
SHAPE (mean) ($1 \leq$ SHAPE $\leq \infty$)	1.22	1.24	1.27	1.25

circle (CIRCLE) metric measures the circularity of patches. The CIRCLE value ranges from 0 for circular or one cell patches to 1 for elongated, linear patches one cell wide. The shape index (SHAPE) metric is a measure of complexity. This metric has a value of 1 when the patch is square and increases without limit as patch shape becomes more irregular.

From 1989 to 2014, the PLAND metric increased from 1.46 to 7.68%. The PD metric significantly increased from 0.5 to 3.36, which indicates that the built-up became more fragmented over the years. However, the mean ENN decreased from 241 to 111.5 m. The decrease in mean ENN was attributed to the expansion of old built-up patches as well as the development of new built-up patches. This suggests that a combination of infill development, leapfrog/sprawl urban growth processes were dominant in Hanoi during this period. The mean CIRCLE value increased between 1989 and 2009, indicating that substantial patches of built-up land became more elongated or exhibited a linear pattern. However, the mean SHAPE value increased slightly between 1989 and 2009, and decreased between 2009 and 2014. This suggests that the complexity of the built -up patches was more or less stable.

Figure 7.8 shows all the metrics for the built-up class along the gradient of the distance from the city center between 1989 and 2014. PLAND decreases as the distance from the city center increases, indicating that the proportion of built-up land near the city center was relatively higher. In contrast, PD increases first as it approaches the 10-km distance from the city center and then decreases with distance. This indicates that there were more patches of built-up land in middle distances. The built-up land patches were relatively more dispersed in farther distances. This is shown by the increase in mean ENN from the city center. The mean CIRCLE value does not show a clear trend. However, the mean CIRCLE value was over 0.5 in the 25–30 km buffer zone in 1989, indicating that some built-up patches were more elongated or linear. The figure also shows that the mean SHAPE values were more or less stable over the years across the gradient of the distance from the city center. This suggests that the built-up patches had a more or less stable shape complexity, which was also not high.

7.3.2 Projected Changes (2014–2030)

The projected urban land use/cover changes show that built-up area would increase from 129.6 km^2 in 2014 to 230.82 km^2 in 2030 (Figs. 7.7, 7.9, and 7.10;

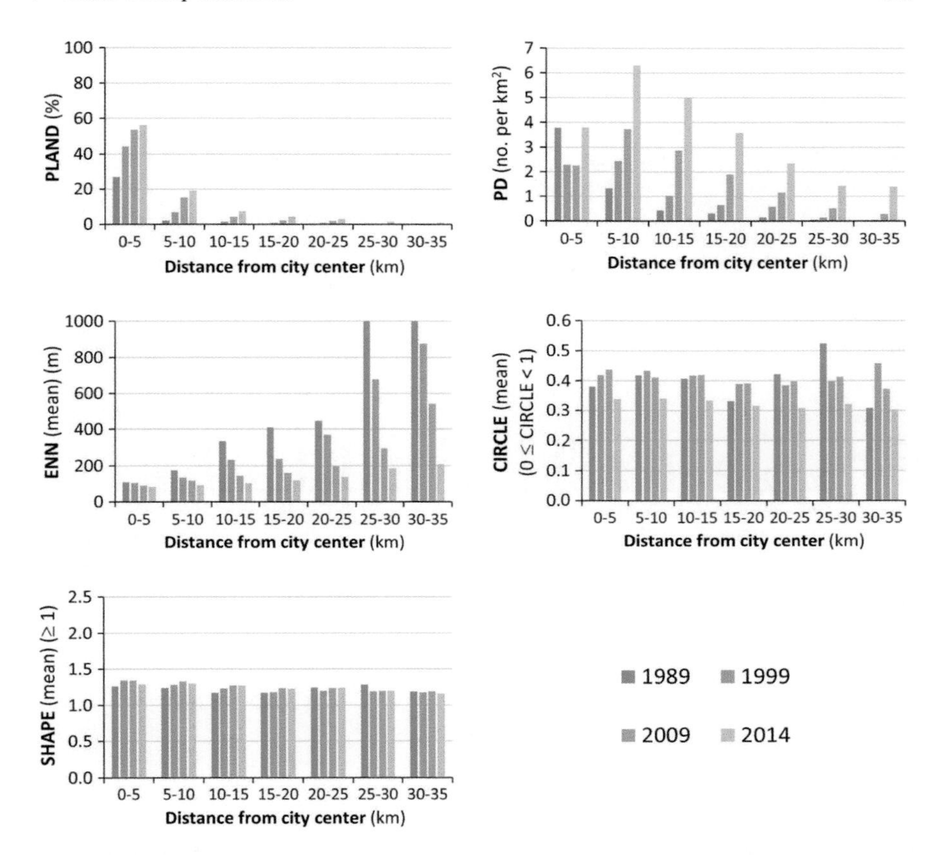

Fig. 7.8 Observed class-level spatial metrics for built-up along the gradient of the distance from city center of Metro Hanoi. *Note* The y-axis values are plotted in the same range as those in Fig. 7.10

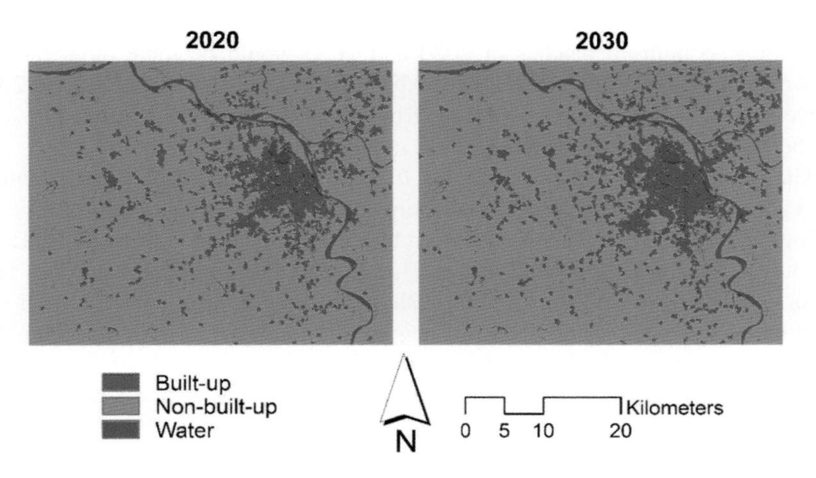

Fig. 7.9 Projected urban land use/cover maps of Metro Hanoi

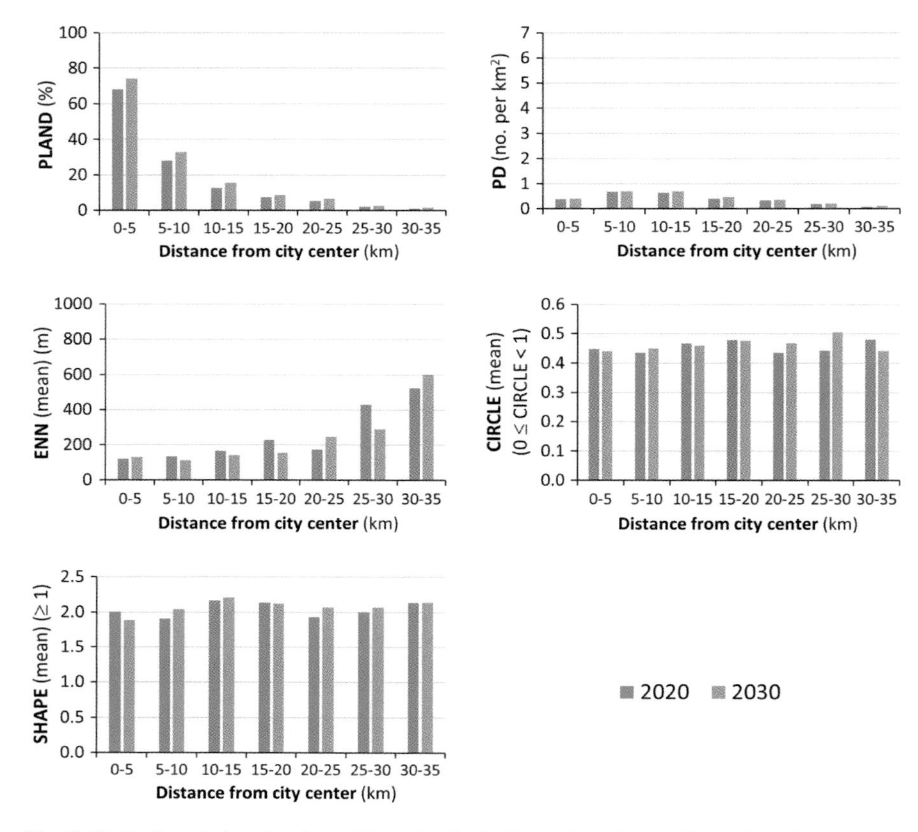

Fig. 7.10 Projected class-level spatial metrics for built-up along the gradient of the distance from city center of Metro Hanoi. *Note* The y-axis values are plotted in the same range as those in Fig. 7.8

Tables 7.4, 7.5, and 7.6). The landscape pattern analysis (Table 7.6) indicated that the simulated built-up patches in 2020 and 2030 would be more aggregated as shown by the simulated increase in PLAND and decrease in PD relative to 2014. However, PD would increase slightly from 2020 to 2030, indicating that some patches of built-up lands would become more fragmented.

The simulated increase in mean ENN from 2014 to 2020 is due to the simulated aggregation of built-up patches, redefining the average distance between neighboring built-up patches. This is supported by the simulated increase in PLAND and decrease in PD during the same period. The simulated decrease in mean ENN from 2020 to 2030 is due to the predicted slight increase in PD, which would also redefine the average distance between neighboring built-up patches. The simulated increase in the mean CIRCLE and SHAPE values from 2014 to 2030 indicates that built-up patches would become more connected, elongated, and complex in shape.

Table 7.4 Projected urban land use/cover of Metro Hanoi (km^2)

	2020	2030
Built-up	197.30	230.82
Non-built-up	1490.01	1456.49
Water	73.31	73.31
Total	1760.62	1760.62

Table 7.5 Projected urban land use/cover changes in Metro Hanoi (km^2)

	2014–2020	2020–2030
Built-up	67.67	33.53
Annual rate of change (km^2/year)	*11.28*	*3.35*
Non-built-up	−67.67	−33.53
Annual rate of change (km^2/year)	*−11.28*	*−3.35*
Water	0.00	0.00
Annual rate of change (km^2/year)	*0.00*	*0.00*

Table 7.6 Projected landscape pattern of Metro Hanoi

Class-level (built-up) spatial metrics	2020	2030
PLAND (%)	11.69	13.68
PD (number per km^2)	0.34	0.37
ENN (mean) (m)	181.29	161.92
CIRCLE (mean) ($0 \leq$ CIRCLE < 1)	0.44	0.45
SHAPE (mean) ($1 \leq$ SHAPE $\leq \infty$)	2.14	2.17

Along the gradient of the distance from the city center (Fig. 7.10), the PLAND of the simulated built-up patches in 2020 and 2030 would still be higher at distances closer to the city center. PD would first increase within the 15-km distance from the city center, before it would decrease in farther distances. The mean ENN would increase at distances farther from the city center. The mean CIRCLE and SHAPE values would have a wavy pattern along the gradient of the distance from the city center.

7.4 Driving Forces of Urban Development

The observed urban land changes (i.e., changes from non-built-up to built-up lands) in Metro Hanoi were due to a number of drivers and their interactions. Some of the most important driving forces are economic development, urban planning, population growth, and urban policies (World Bank 2011). Many studies have empirically confirmed the roles of socioeconomic factors in spatial urban growth (e.g., Verburg et al. 2004; Antrop 2005; Hersperger and Burgi 2007; Schneeberger et al. 2007). Generally, socioeconomic factors drive the spatial growth of a city, whereas biophysical factors are often constraints to city expansion. For instance, the spatial urban growth of Metro Hanoi has been limited by streams, water bodies, and wetlands (Ho and Shibayama 2009), and controlled by an urban green network (Uy and Nakagoshi 2008) and cultural heritages. In this study, only the socioeconomic drivers of urbanization are discussed.

7.4.1 Economic Development

Economic development is recognized to be the most important factor of urbanization in Vietnam. In 1986, the Vietnam government introduced liberal market mechanisms, and encouraged private sector initiatives. Since the inception of reform policies, the government has implemented a variety of policies in order to foster economic growth and urban development. For example, the 2010–2020 socioeconomic development strategy prioritizes industrialization and modernization goals and emphasizes that urban development is one of the most important strategies to promote regional economic development goals. This trend had been observed in China, India, and Indonesia in the late 1970s (World Bank 2011). The relationship between economic development and urban growth can be found in many studies (e.g., Verburg et al. 2004; Schneeberger et al. 2007; Cheshire et al. 2008). In particular, most foreign investments in Vietnam have been concentrated in urban areas. As a result, highly urbanized areas have attracted a large number of rural labor. Hanoi, Hai Phong, Ho Chi Minh City, Binh Duong, Dong Nai, Da Nang, and Can Tho are some examples of these highly urbanized areas.

7.4.2 Urban Planning

Urban planning has been an important factor affecting urban land change patterns in Metro Hanoi. The boundary of Hanoi City has changed many times from 1956 to 2008. Hanoi City occupied an area of about 70 km^2 between 1956 and 1960. Urban growth was mainly concentrated to the south area of the Red River. The core districts then consisted of Hoan Kiem and Ba Dinh. From 1960 to 1968, the city expanded to about 130 km^2. The city had four core districts and four suburban districts. During the same period, urban growth was concentrated mainly in the southern and northern region of the Red River. However, the area of Hanoi exponentially increased to 2123 km^2 between 1981 and 2000. In this period, urban growth was concentrated mainly in the northern region of the Red River consisting of Dong Anh and Gia Lam suburban districts. Ha Tay province was merged with Hanoi City. According to the Hanoi development plan toward 2030, Hanoi City is expected to increase to about 3344 km^2. Metro Hanoi is designed to have a central

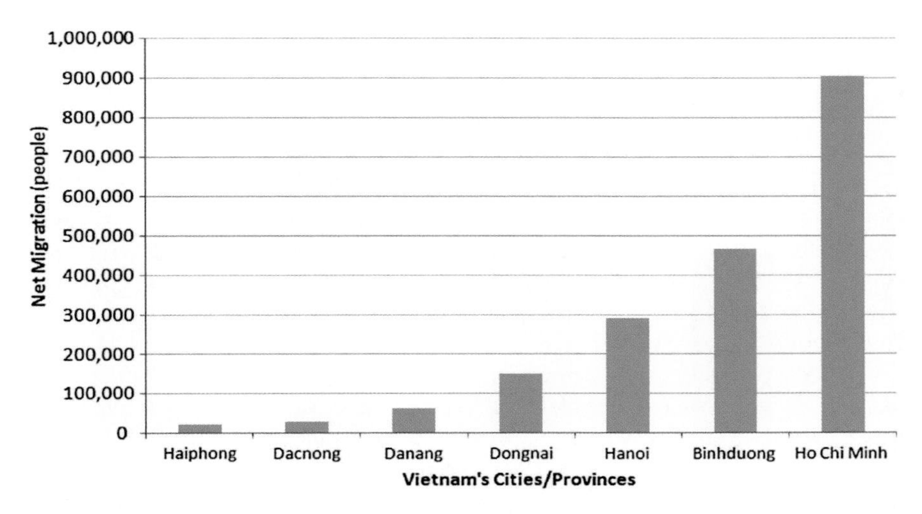

Fig. 7.11 Net migrants of major cities of Vietnam from 2004 to 2009. *Source* GSO (2009) and VMPI (2011)

city, five satellite cities, towns, and rural areas. Central city, satellite cities and towns will be separated by green corridors and ring roads.

7.4.3 Population Growth

The Urban proportion of the total population in the country was 10% in 1950 and 21.5% in 1975. In 2009, it further increased to 29.6% (GSO 2009). The increase was attributed to regional economic development policies and migration during the last two decades. Between 2004 and 2009, for instance, Metro Hanoi and the other major cities in the country had a high net migration (Fig. 7.11).

For Metro Hanoi, its rapid population growth is considered to be a very important factor to its urban expansion. In the 1989–1999 and 1999–2009 periods, the urban population of Hanoi increased by about 1.42 times and 1.7 times, respectively. But because of the inclusion of Ha Tay province, the urban population of Metro Hanoi decreased from 58.2 to 40.8% from 1999 to 2009 (Fig. 7.12). The new Hanoi covers the area of the old Hanoi and Ha Tay province terrestrial area.

7.4.4 Urban Policies

Urban policies have significantly affected the development of Metro Hanoi. Major urban policies in Vietnam include city boundary control, immigration control, urban finance, land market, and urban classification system (World Bank 2011).

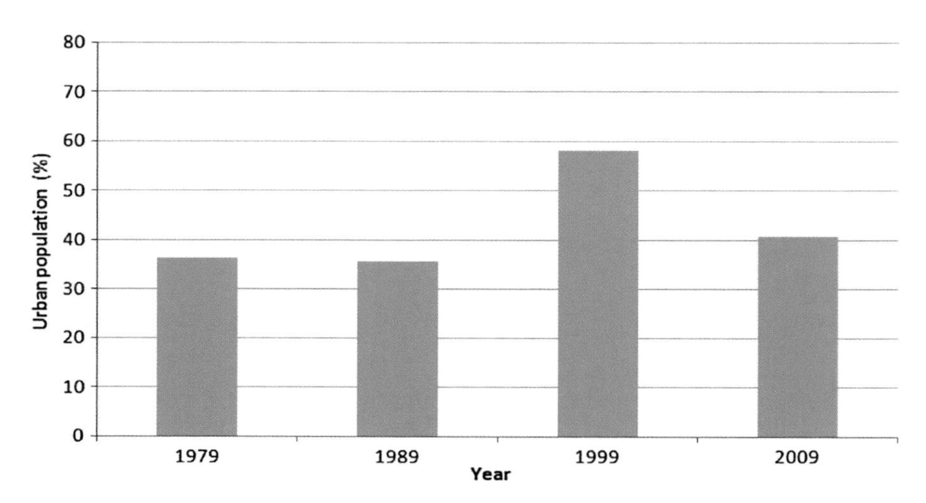

Fig. 7.12 Urban population proportion of Metro Hanoi from 1979 to 2009. *Source* GSO (2009)

Administrative boundary control policy means that the central government provides the approval for urban boundary expansion. Historically, this policy has been recognized as an effective tool for controlling city size and its sprawl. Immigration control policy refers to urban residency permission. This policy was effective from 1954 to 1990. After 1990, this policy was relaxed. As a result, the population of Hanoi increased rapidly. Urban finance means provision of finance for urban construction from central government budget. The urban finance policy strongly affected the growth rate of infrastructure and urbanization. Urban land markets were largely ignored from 1954 to 1990. The 1993 Land Law established land markets in Vietnam. This resulted to the rapid conversion of farmland to urban in many cities around the country. Vietnam's city and town classification system also enforced urbanization. This system relates to the allocation of the public budget from the central government for construction of urban infrastructure and other investments. The recent changes in urban policies such as industrial development in the rural fringe areas, the new dynamics of the private housing sector, and the commercial redevelopment of the inner city also contribute to urban expansion in the city.

7.5 Implications for Future Sustainable Development

Sustainable city (eco-city) is one of the sustainable development goals adopted by the United Nations. Cities cover only 3% of land surface of the earth, but they consume most energy and release most emissions. The sustainability of cities can be measured by a large number of indicators classified as social, economic, and environmental indicator groups. In terms of future city development, city development indicators can be used to measure different stages in the process of achieving sustainable city goals.

Rapid population growth and built-up expansion strongly influence the future sustainable development of Metro Hanoi. This section examines the relationships between population growth, built-up expansion, and the future sustainable development goals of Metro Hanoi. The current plan for Metro Hanoi is briefly reviewed.

7.5.1 Population Growth

In 2010, Vietnam had a population of 86.93 million (urban population: 26.22 million; rural population: 60.70). It has been projected that urban population will increase to 44 million in 2020 and 52 million in 2025 (MONRE 2011). Metro Hanoi is one of the biggest cities of Vietnam in terms of total population. The current population of Hanoi is about 6.4 million. It has been projected that Hanoi population will increase to about 9.2 million in 2030 (GSO 2009). Population growth will lead to rapidly increasing demands for lands for new residential areas and infrastructure development projects. This will lead to infilling of new built-up patches in the core districts of the city. The analysis of the PLAND and PD metrics indicates that between 1989 and 2014 more new built-up patches were established in the distance of about 10 km from the city center (Fig. 7.8). This implies that population growth clearly influenced infilling process in the core districts of the city.

7.5.2 Built-up Expansion

The results revealed that the observed built-up expansion in Metro Hanoi includes city boundary expansion and infilling process. The infilling process is characterized by expansion of old built-up patches and a rapid decrease in green space patches and public space patches in the core districts of the city such as Hoan Kiem, Tay Ho, Hai Ba Trung, and Ba Dinh. Figure 7.8 shows that the PD of built-up lands increased within the distance of the core district area between 1989 and 2014. The results of the urban land change simulation also showed that the area of built-up lands would increase substantially in 2020 and 2030. Also, the landscape pattern analysis suggested that the simulated built-up patches in 2020 and 2030 would be more aggregated as shown by the simulated increase in PLAND and decrease in PD. However, PD would increase slightly from 2020 to 2030, indicating that some patches of built-up lands would become more fragmented. These results suggest that urban expansion under the current scenario would continue in the future. Hence, current challenges will continue to persist. The expansion of the city boundary of Hanoi has resulted in a rapid increase of new built-up patches like new apartment buildings and new public facilities in the suburban districts of the cities of Tu Liem, Ha Dong, Thanh Xuan, Long Bien, Gia Lam, Dong Anh, and Soc Son.

The expansion of Hanoi has plays a great role toward the achievement of Vietnam's socioeconomic development goals. However, rapid urbanization has negatively affected the quality of life, and the quality of urban air and surface waters. Currently, Hanoi is dealing with air pollution, wastewater, and solid wastes. These environmental issues affect the quality of urban life and sustainability of the city. Air pollution in Hanoi is mainly caused by vehicles and construction activities, as well as by suspended particulates, SO_2, NO_x, CO, and volatile organic compounds from the manufacturing sector (Hung 2010). Wastewater is another serious source of pollution in Hanoi. It is estimated that Hanoi discharges about 670,000 m^3 of wastewater each day, while only a small volume (7%) is treated each day (MONRE 2012). Solid waste is also a major concern. In Hanoi, solid wastes consist mainly of domestic solid waste (6500 tons per day), industrial solid waste (1950 tons per day), and medical solid waste (15 tons per day) (MONRE 2011).

7.5.3 Current Development Plans

The infilling and expansion patterns of built-up lands in Metro Hanoi have resulted in various sustainable development challenges. In responding to the challenges of rapid urbanization, the Vietnam government approved the Hanoi region master plan 2020 and vision 2050 (Hanoi master plan). The master plan aims to control the infilling or densification process in the core districts of the city and support the future sustainable development goals of the Hanoi region (GOV 2008). The Hanoi region in the master plan includes Metro Hanoi and its surrounding provinces consisting of Ha Tay, Vinh Phuc, Hung Yen, Bac Ninh, Hai Duong, Ha Nam, and Hoa Binh. The region covers an area of 13,436 km^2. The total population of the region is predicted to reach 15 million in 2020 and 18.2 million in 2050. The master plan highlights the aim to reduce or minimize the infilling of new built-up patches in the core districts of Hanoi by building nearby provincial cities. This means that Metro Hanoi and its satellite cities in the neighboring provinces will be interconnected. In the master plan, Metro Hanoi will serve as the center for politics, finance, commerce, research institutes, and high-tech industries. Its satellite cities will serve as manufacturing centers. For example, Hai Duong will be designed as the center for light industries and food processing industries. Hoa Binh City will be designed as the center for tourism services, and Vinh Phuc as a center for industries and urban services. Bac Ninh, Hung Yen, and Ha Nam will be the centers for heavy industries, light industries, and agriculture production.

In addition, in 2013 the Hanoi plan 2030 and vision 2050 (Hanoi plan) were approved by the Government of Vietnam (GOV 2013). The Hanoi plan highlights the strict conservation of old built-up patches and restricts the establishment of new built-up patches as well as high-rise buildings in the core districts of the city. In terms of the spatial arrangement of the city, the Hanoi plan indicates that Metro Hanoi will include the core area (core districts of the city) and five satellite cities. Hoa Lac, Son Tay, Xuan Mai, Phu Xuyen, and Soc Son are designed as the satellite cities of Metro Hanoi. Each satellite city will play particular functions. Hoa Lac will be designed as the center for high-tech industries, research institutes, and national universities. Son Tay will serve as the center for tourism services, resort buildings, and cultural activities. Phu Xuyen will be designed as the center for industries and food processing industries, and Soc Son as the center for international gateway, hospital, and university clusters. Ring roads and highways will be upgraded and constructed to connect the central city with the satellite cities. The green buffers, which cover 70% of the total area of the new Hanoi, can reduce overloading in terms of population density, congestion, and urban services in the core area of Hanoi. This large green space and the satellite cities are expected to contribute significantly toward the achievement of the sustainable economic development goals of the new Hanoi. In terms of the sustainable environmental goals of the new Hanoi, the plan emphasizes on the strict protection of the city's green spaces which include natural forests, urban green parks, rivers, and lakes. The plan also emphasizes the building of solid waste recycling factories in satellite cities and urban wastewater treatment factories. It also aims to improve environmental monitoring activities concerning air pollution and water pollution.

7.6 Concluding Remarks

Based on the urban land use/cover change and landscape pattern analysis, it is found that Metro Hanoi has undergone rapid urbanization over the past two decades (1989–2014). The spatial expansion of built-up lands has been influenced by both infill and extension development patterns. The expansion of the built-up area was mainly driven by economic activities, population growth, urban planning, and urban policies. The analysis of landscape metrics along the gradient of the distance from the city center revealed that built-up patches were clearly more dense near the city center, with discontinuously scattered pattern at distances farther from the city center. The urban land change simulation indicated that built-up lands would continue to expand in the future (2014–2030). The landscape pattern analysis indicated that the simulated built-up patches would be more aggregated (2014–2020). However, PD would increase slightly from 2020 to 2030, indicating that some patches of built-up lands would become more fragmented.

In general, the rapid urbanization of Metro Hanoi, including its rapid built-up expansion, has brought positive and negative impacts. On the one hand, Metro Hanoi's contribution to the national GDP has increased and the living standards and urban services for most inhabitants in the city have improved. On the other hand, a number of urban-related issues have emerged, such as rising land and housing prices, traffic congestion, and urban pollution (air pollution and wastewater). Because of these issues, the government of Vietnam approved the Hanoi master plan and the Hanoi plan for future sustainable development. These plans aim to control the infilling or densification process in the districts near the city center. In particular, the Hanoi plan restricts the establishment of new built-up patches and high-rise buildings in the core districts near the city center. New built-up patches such as new residences are only allowed in suburban districts. If the Hanoi plan is realized, the infilling or densification process near the city center would be efficiently controlled.

References

Antrop M (2005) Why landscapes of the past are important for the future. Landscape Urban Plan 70:21–34

Cheshire P, Magrini S (2008) Urban growth drivers in a Europe of sticky people and implicit boundaries. J Econ Geogr 9:85–115

Coulhart A, Quang N, Sharp H (2006) Urban development strategy: meeting the challenges of rapid urbanization and the transition to market oriented economy. World Bank Office in Vietnam, Hanoi

GOV (Government of Vietnam) (2008) Hanoi region development master plan 2020 and vision 2050. The Government of Vietnam, Hanoi

GOV (Government of Vietnam) (2013) Hanoi capital construction master plan 2030 and vision 2050. The Government of Vietnam, Hanoi

GSO (General Statistic Office) (2009) Statistical book in 2009. General Statistic Office, Hanoi

Hersperger AM, Burgi M (2007) Driving forces of landscape change in the urbanizing Limmat Valley, Switzerland. In: Koomen E, Stillwell J, Bakema A, Scholten HJ (eds) Modelling land-use change. GeoJournal Library, vol 90, pp 45–60

Ho DD, Shibayama M (2009) Studies on Hanoi urban transition in the late 20th century based on GIS/RS. Southeast Asian Stud 46:532–546

HSO (Hanoi Statistical Organization) (2009) Hanoi statistical yearbook 2008. Hanoi Statistical Office, Hanoi

Hung NT (2010) Urban air quality modeling and management in Hanoi, Vietnam. Ph.D. dissertation. Aarhus University

MONRE (Ministry of Natural Resource and Environment) (2011) National environment assessment report in 2011. Ministry of Natural Resource and Environment, Hanoi

MONRE (Ministry of Natural Resource and Environment) (2012) National environment assessment report: current surface wastewaters in 2012. Ministry of Natural Resource and Environment, Hanoi

Nguyen VP, Le VL, Nguyen MT (2010) The history of Thang Long—Hanoi. The Times Publisher, Hanoi (in Vietnamese)

Schneeberger N, Burgi M, Hersperger AM, Ewald KC (2007) Driving forces and rates of landscape change as a promising combination for landscape change research—an application on the northern fringe of the Swiss Alps. Land Use Policy 24:349–361

United Nations (2008) Report of the meeting—urbanization: A global perspective. Proceedings of the expert group meeting on population distribution, urbanization, internal migration and development. New York, 21–23 Jan 2008

United Nations (2012) World urbanization prospects: the 2011 revision. New York

Uy PD, Nakagoshi N (2008) Application of land suitability analysis and landscape ecology to urban greenspace planning in Hanoi, Vietnam. Urban for Urban Greening 7:25–40

Verburg PH, Ritsema van Eck J, de Nijs T, Schot P, Dijst M (2004) Determinants of land-use change patterns in the Netherlands. Environ Plan B 31:125–150

VET (Vietnam Economic Times) (2008) Go west. Vietnam Economic Times (April), pp 20–21

VMPI (Vietnam Ministry of Planning and Investment) (2011) Migration and urbanization in Vietnam: patterns, trends and differentials. Vietnam population and housing census 2009, Hanoi, Vietnam

World Bank (2011) Vietnam urbanization review. Technical Assistance Report, Hanoi, Vietnam

Chapter 8
Bangkok Metropolitan Area

Akio Yamashita

Abstract Bangkok, the capital city of Thailand, has a high density population and viable economic prospects. According to the results of land use/cover change and landscape metrics analyses, built-up area in Bangkok increased substantially between 1988 and 2014. Built-up areas expanded almost evenly along the existing urban areas in all direction. Urban growth process in Bangkok was dominated by extension and densification. The land use/cover projection under the current scenario indicates that built-up area will fill most area within 30 km from the city center of Bangkok and that will also spread comparatively linearly and irregularly to the outer edge of Bangkok. After the 1980s, the influx of foreign companies including Japan promoted the development of Bangkok Metropolitan Region (BMR). Such prominent urban development of Bangkok brought various urban problems, such as traffic and water environmental problems. Countermeasures against these problems are highly important for sustainable development in the future.

8.1 Origin and Brief History

During the era of Ayutthaya Kingdom from 1351 to 1767, the capital Ayutthaya was located to the north of Bangkok, about 60 km on the upper stream of the Chao Phraya River. The capital city flourished as the trading center connecting India, the Arab world, and China. Trading goods from the west were transported by land from the west coast, Tenasserim, to Ayutthaya instead of taking a detour of Malay Peninsula by ship, and then traveling down the Chao Phraya River to Gulf of Thailand and heading for China. During those days, Bangkok was a front line base for Portuguese hired troops to protect Ayutthaya. “Bangkok,” which means “Makok-covered village” in local language, was an undeveloped wetland.

A. Yamashita (✉)
Faculty of Life and Environmental Sciences, University of Tsukuba, Tsukuba, Japan
e-mail: akio@geoenv.tsukuba.ac.jp

Y. Murayama et al. (eds.), *Urban Development in Asia and Africa*,
The Urban Book Series, DOI 10.1007/978-981-10-3241-7_8

In 1767, the Ayutthaya Kingdom was destroyed after the invasion of Burmese Troops, after which Ayutthaya fell into ruin. Phraya Tak who fought off the Burmese Troops ascended the throne as Taksin and set the capital in Thonburi located at the downstream of Ayutthaya and right coast of the Chao Phraya River (the Thonburi Kingdom). As the father of Taksin was a Chinese, he managed to fight off Burmese troops with assistance from China. Thonburi was then chosen as the capital, which was an advantageous place for trading with China. However, this Thonburi Kingdom did not last for long. In 1782, Taksin was executed during a coup inside the palace. Chao Phraya Chakri, one of military commanders, ascended the throne (Rama I). He later moved the capital to Bangkok (Fig. 8.1), located on the opposite side of Thonburi across the Chao Phraya River and named the place Krung Thep (capital of angel). This is the beginning of the current Thai Royal Family (Chakri Kingdom or Rattanakosin Kingdom) and local Thai people still call Bangkok "Krung Thep" to this day. The purpose of moving the capital to Bangkok was for protection from the threat of Burmese troops of the west. Bangkok was an ideal land for defense as the Chao Phraya River is a natural fort and the east side is a wetland.

By the early nineteenth century, Banglamphu Canal (Fig. 8.2) and Ong Ang Canal were built outside of the existing Khu Mueang Canal built during the Thonburi Kingdom, and along the canals, the castle walls were also built. Canals played a big role of draining water from swamps, securing daily water for residents, transportation, and defense of the capital. The area surrounded by these canals and the Chao Phraya River is called Rattanakosin Island and forms the center of Bangkok where the palace is also located. In 1853, Padung Krung Kasem Canal

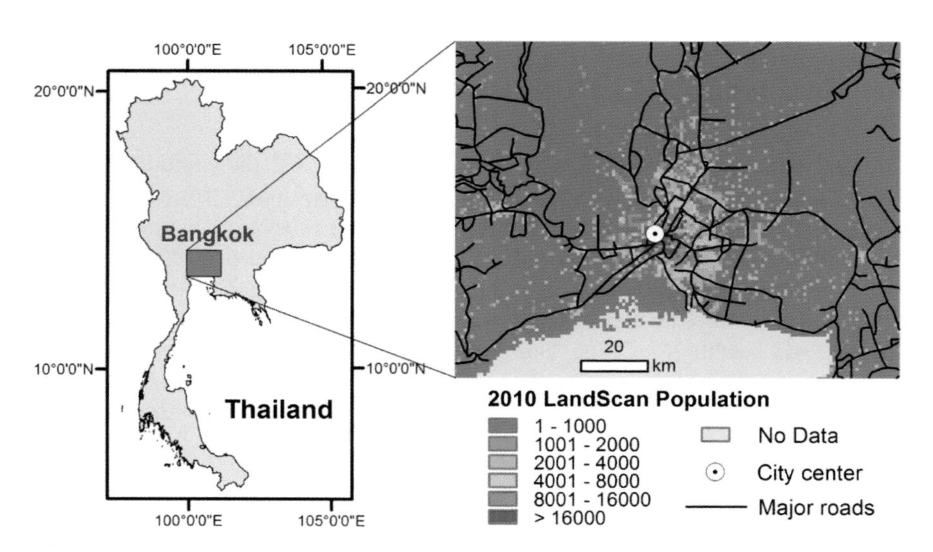

Fig. 8.1 Location and LandScan population of Bangkok Metropolitan Area, Thailand

Fig. 8.2 Banglumpoo Canal. *Source* Author's fieldwork (2006)

was built extending outside. As a result, the urban area of Bangkok extended twice. This new canal was for transportation. During this time, the construction of major roads including the current Rama IV Avenue became active. Given the swampy conditions of Bangkok, ships were the main mode of transport. However, the arrival of Europeans during the mid-nineteenth century promoted road construction in Bangkok. Nonetheless, shipping still remains as one of the main means of transportation in Bangkok today (Fig. 8.3).

Between the latter half of nineteenth century and the early twentieth century, the modernization policies such as emancipation of slaves and revolution of local administrative system were promoted. A centralized administrative framework centered on royal family was built in Bangkok. This modernization, which was led by the King, was to prevent Thailand from being colonized by European countries. In 1932, Siamese revolution occurred and Thailand became a constitutional monarchy (Tomosugi 1998, 2003). In the 1960s after World War II, rapid economic development increased the population in Bangkok dramatically (see also Sect. 8.2) and the urban area largely expanded (Fig. 8.4).

According to a temporal land use analysis based on old topographic maps (Yamashita 2011), the built-up area in Bangkok until the 1950s was within a radius of 5 km. However, the majority of the surrounding areas have already been converted to agricultural land. Therefore, the area had almost no natural green land. In the 2000s the built-up area expanded in a radial pattern to a radius of about 20 km, especially to the eastern side of the Chao Phraya River (Fig. 8.5).

Fig. 8.3 Commuting boat in Bangkok. *Source* Author's fieldwork (2006)

Fig. 8.4 Overview of urban area in Bangkok. *Source* Author's fieldwork (2006)

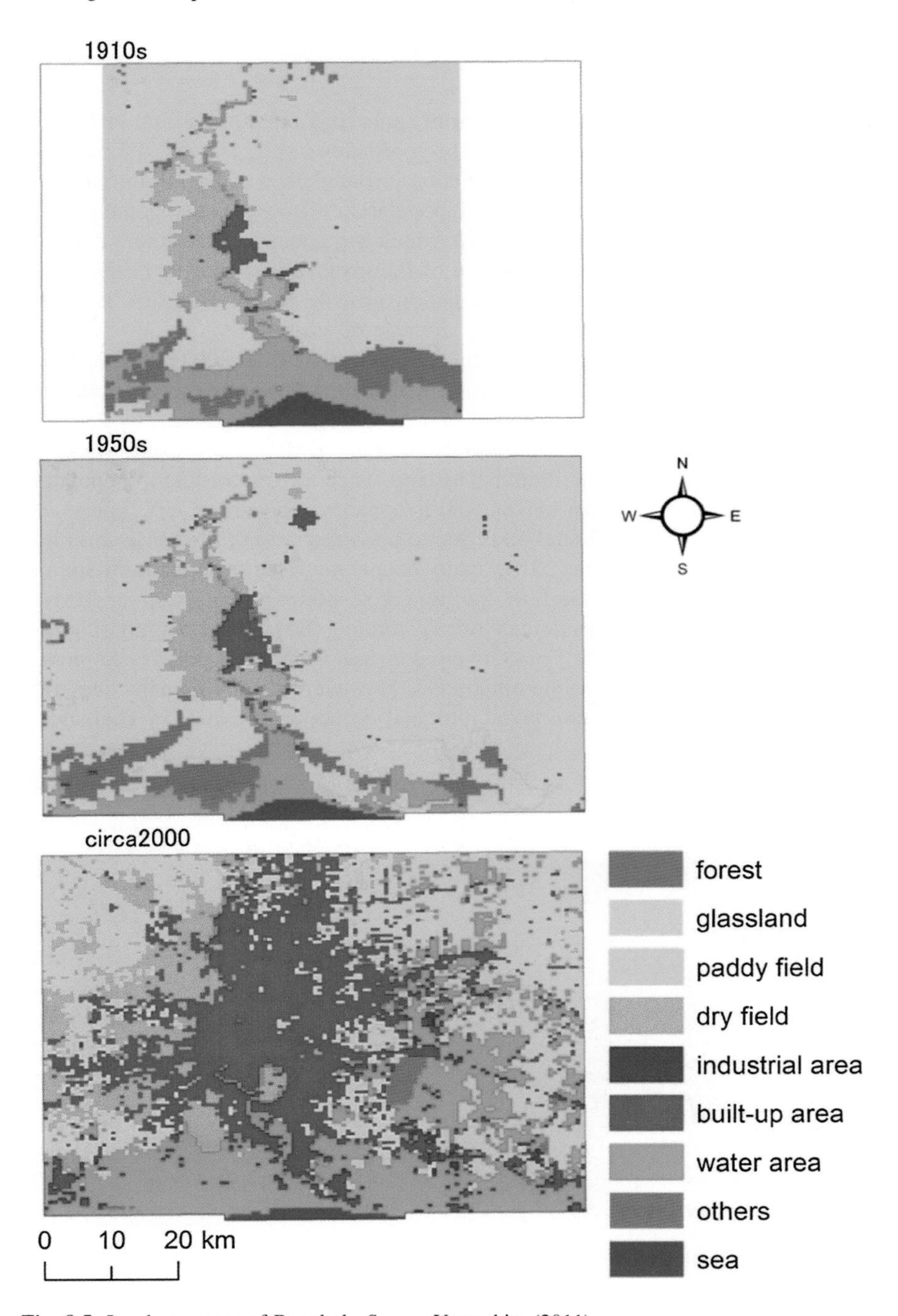

Fig. 8.5 Land-use maps of Bangkok. *Source* Yamashita (2011)

8.2 Primacy in the National Urban System

The total population in Thailand grew from approximately 20.6 million in 1950 to about 69.1 million in 2010 (UN 2015). The population has increased 3.4 times from 1950 to 2010 (Fig. 8.6). The total population in Bangkok was 1.4 million in 1950, corresponding to 6.6% of the national population. However, the population in Bangkok increased to 8.2 million in 2010, which was about six times more than that of 1950 (Fig. 8.6). The urban population of Bangkok corresponds to 11.9% of the total population in Thailand. The population growth rate of Bangkok is nearly double that of the entire country during the past 60 years.

Figure 8.7 ranks the cities with a population of more than 300,000. This figure shows that the population is mainly concentrated in the capital city Bangkok. The second city, Samut Prakan (population of 1,490,000), is located next to Bangkok and forms a part of Bangkok Metropolitan Region (BMR). Only these two cities have a population of over 1 million in Thailand. Therefore, according to population distribution, the urban system in Thailand is apparently a primate city type.

The large population in Bangkok is due to mass-migration from other areas into Bangkok. According to Niwa (2010), many people migrated from the northern and northeastern parts of Thailand to the Bangkok Metropolitan Region (BMR), including the Bangkok Metropolitan Administration (BMA) and neighboring districts. On the other hand, the Thai Government has introduced policies to redress disparities between BMR and the other areas. However, the city remains dominant given the high concentration of population and economic activities in Thailand.

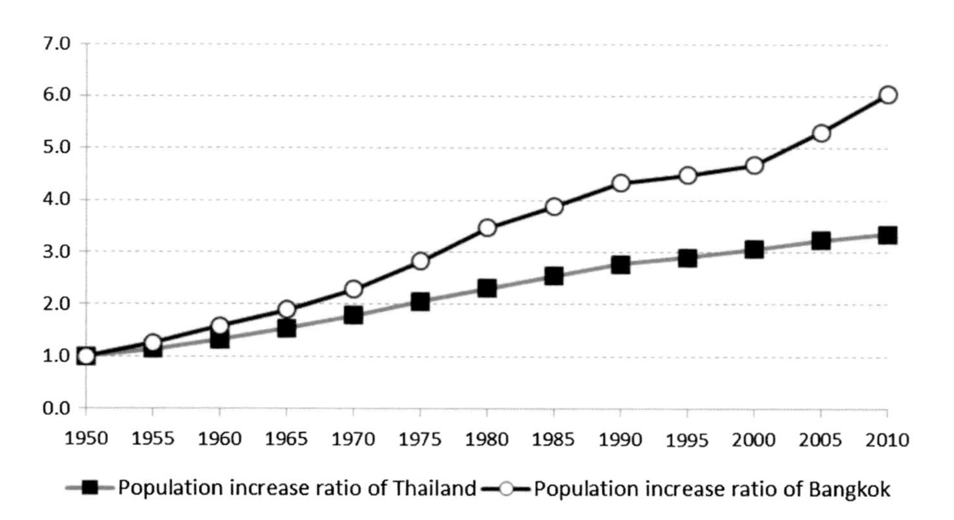

Fig. 8.6 Population increase ratio of Thailand and Bangkok. *Source* UN (2015)

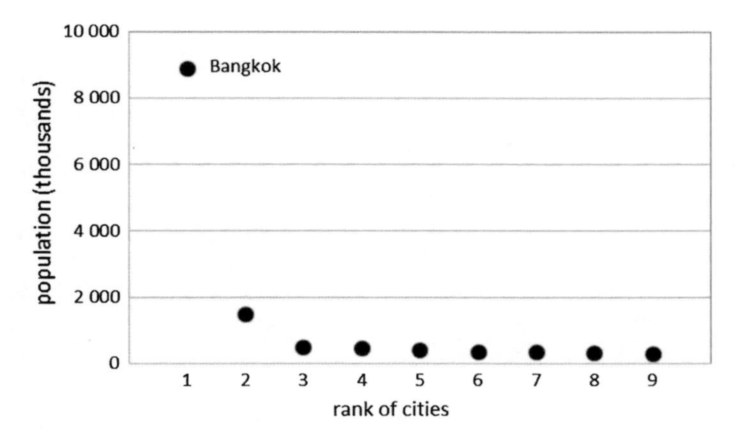

Fig. 8.7 Population and its rank of major nine cities in Thailand (2013). *Source* UN (2015)

8.3 Urban Land Use/Cover Patterns and Changes (1988–2030)

8.3.1 Observed Changes (1988–2014)

Figures 8.8 and 8.9 show that built-up area in Bangkok increased substantially between 1988 and 2014. The results of land use/cover change analysis indicate that built-up area grew from 706.73 km^2 in 1988 to 2107.62 km^2 in 2014 (Table 8.1). This means the built-up area of Bangkok almost tripled during the past 26 years. The annual rate of built-up area expansion was 37.27 km^2/year during the "1988–1999" period (Table 8.2). However, the annual rate of built-up area change significantly increased to 51.85 and 94.50 km^2/year during the "1999–2009" and "2009–2014" epochs, respectively (Table 8.2). The results of land use/cover change analysis show that rapid urbanization occurred in Bangkok during the study period.

According to Estoque et al. (2014) and Kamusoko (2017), various spatial metrics for the built-up class, including the percentage of landscape (PLAND), patch density (PD), Euclidean nearest-neighbor distance (ENN), related circumscribing circle (CIRCLE) and shape index (SHAPE), were analyzed (Table 8.3; Fig. 8.10).

The PLAND metric increased from 5.91 in 1988 to 17.98 in 2014. Note that built-up area occupied less than 20% of Bangkok even in 2014. Figure 8.10 shows that while PLAND increased in all distance buffer zones, the percentage was higher close to the city center.

The PD metric increased slightly from 1.11 in 1988 to 1.36 in 2009. However, PD decreased substantially from 1.36 in 2009 to 0.35 in 2014 (Table 8.3). This suggests that urban growth process was dominated by extension and densification. The built-up areas expanded along the existing roads, canals, and settlements, and

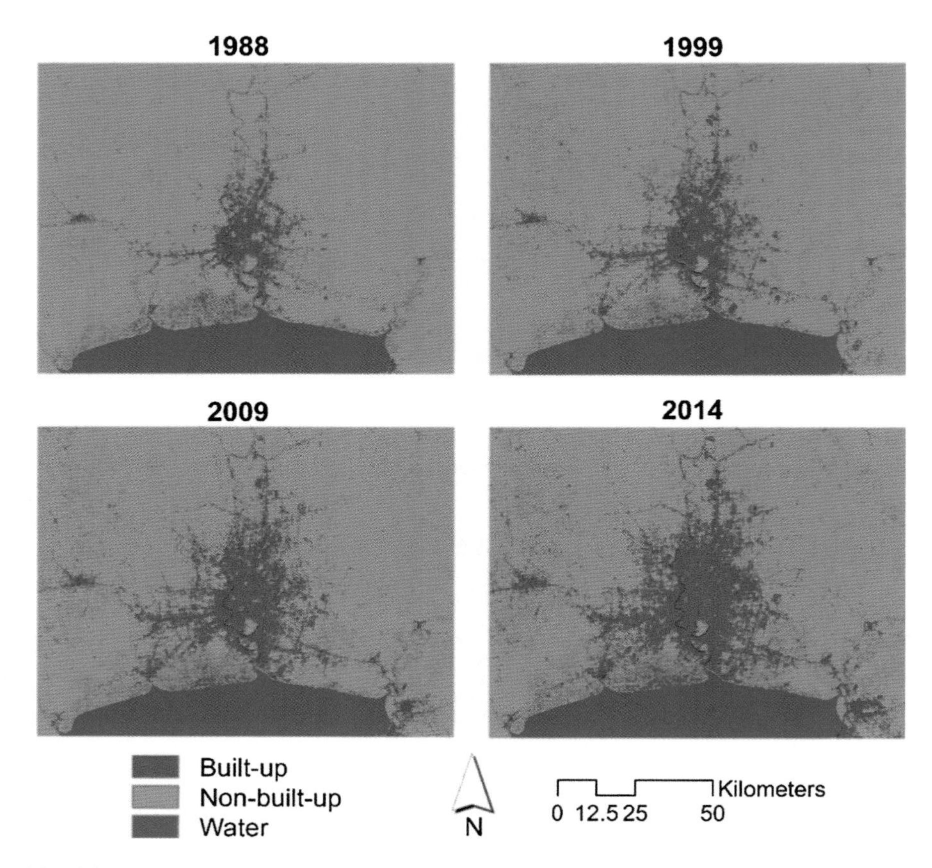

Fig. 8.8 Urban land use/cover maps of Bangkok Metropolitan Area classified from Landsat imagery

also merged with suburban small built-up patches. Nevertheless, the PD metric varied within the different buffer zones between 1988 and 2014. For instance, PD within 15–20 km zone is lower in more recent years because built-up area in central Bangkok has grown with aggregating existing fragmented built-up patches in this zone. On the other hand, PD beyond distance zone 30–35 km has become high until 2009 (Fig. 8.10). This implies new built-up patches emerged outside the central urban area in Bangkok.

The mean ENN decreased from 189.56 in 1988 to 173.64 in 2009. Nonetheless, the mean ENN increased substantially from 173.64 in 2009 to 373.38 in 2014 (Table 8.3). This trend is related to the change in PD, which suggests that urban growth process was dominated by extension and densification. For example, new small built-up patches emerged outside the 30–35 km distance buffer zone. Note that the mean ENN also tends to be higher with distance, particularly in 2014.

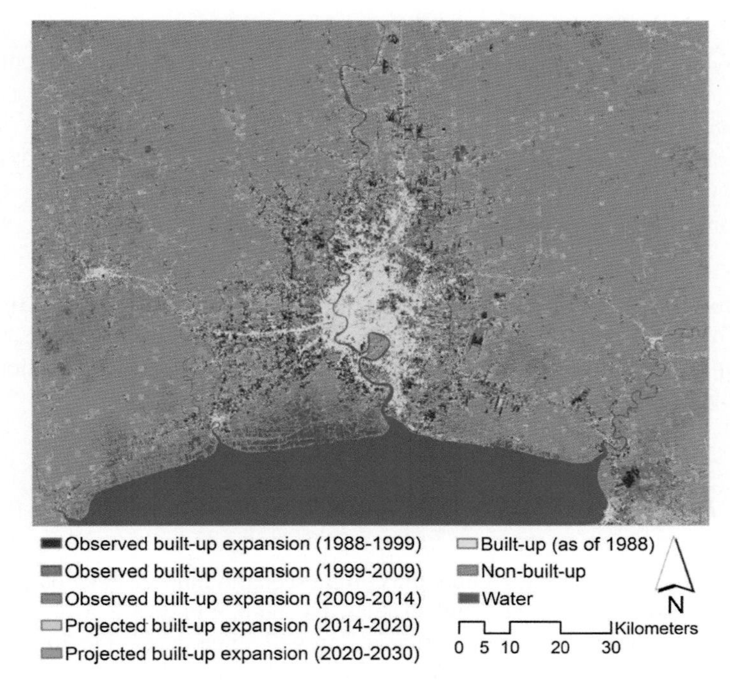

Fig. 8.9 Observed and projected urban land use/cover changes in Bangkok Metropolitan Area

Table 8.1 Observed urban land use/cover of Bangkok Metropolitan Area (km^2)

	1988	1999	2009	2014
Built-up	706.73	1116.68	1635.14	2107.62
Non-built-up	11,246.93	10,806.37	10,225.53	9612.69
Water	1875.11	1905.71	1968.10	2108.45
Total	13,828.77	13,828.77	13,828.77	13,828.77

Table 8.2 Observed urban land use/cover changes in Bangkok Metropolitan Area (km^2)

	1988–1999	1999–2009	2009–2014
Built-up	409.96	518.46	472.48
Annual rate of change (km^2/year)	*37.27*	*51.85*	*94.50*
Non-built-up	−440.56	−580.84	−612.84
Annual rate of change (km^2/year)	*−40.05*	*−58.08*	*−122.57*
Water	30.60	62.39	140.35
Annual rate of change (km^2/year)	*2.78*	*6.24*	*28.07*

Table 8.3 Observed landscape pattern of Bangkok Metropolitan Area

Class-level (built-up) spatial metrics	1988	1999	2009	2014
PLAND (%)	5.91	9.37	13.79	17.98
PD (number per km^2)	1.11	1.12	1.36	0.35
ENN (mean) (m)	189.56	185.55	173.64	373.38
CIRCLE (mean) ($0 \leq$ CIRCLE < 1)	0.41	0.41	0.42	0.46
SHAPE (mean) ($1 \leq$ SHAPE $\leq \infty$)	1.25	1.25	1.25	1.29

The mean CIRCLE and SHAPE values showed little change between 1988 and 2014. This suggests a stable and less complex growth process because built-up areas expanded almost evenly along the existing urban areas in all direction since 1988 (Fig. 8.9).

8.3.2 Projected Changes (2014–2030)

The land use/cover projection under the current scenario indicates that built-up area will continue to expand in the future (Tables 8.4 and 8.5; Figs. 8.9, 8.11, and 8.12). Especially, another wave of rapid urbanization will come during the 2020s. While built-up area will increase from 2107.62 km^2 in 2014 to 2211.56 km^2 in 2020 with annual rate of 17.32 km^2/year, more 433.96 km^2 built-up area will expand between 2020 and 2030 with annual rate of 43.40 km^2/year.

According to landscape pattern analysis (Table 8.6; Fig. 8.12), PLAND would increase in all distance zones. The PD metric also shows the same tendency. However, the mean SHAPE and CIRCLE values would likely decrease between 2020 and 2030 in the 5–30 km distance buffer zones. These simulated results imply that built-up area will fill most area within 30 km from the city center of Bangkok and that will also spread comparatively linearly and irregularly to the outer edge of Bangkok.

8.4 Driving Forces of Urban Development

Bangkok is located in the Chao Phraya River delta, which is fertile. In the early twentieth century, most of the land was used for cropping of rice (refer to Fig. 8.5). On the other hand, the natural environment in Bangkok often caused flood damage during the wet season. In the wake of the large flood in 1980, the Greenbelt was set up in the east and west outskirts of Bangkok to restrict development activities and absorb flood flow. However, the construction of Suvarnabhumi International Airport (which opened in 2006) resulted in housing construction even in the Greenbelt of east suburb.

Primacy position of Bangkok in Thailand is extremely high (see also Sect. 8.2). This is mainly resulted from strong centralized administrative structure centered on the capital city Bangkok. Although, in Thailand, local authorities in cities, towns, and districts were established by the 1950s, their authorities are still limited (Tamada 2003). Since the 1990s, decentralization was promoted. However, political primacy of Bangkok where the central government is located is still high. This encouraged the prominent development of Bangkok. On the other hand, having a look at Bangkok as a local authority, its area of the day when Bangkok city was established in 1936 was 41 km^2, but the administrative area has gradually expanded according to the extension of urban area. The Bangkok Metropolitan Administration (BMA) area integrated surrounding areas in 1972 and covered 1569 km^2 (Hashimoto 1998). Such an extension of the administrative area and centralization has elevated Bangkok's financial and economic status. As a result, the population expanded. To date, Bangkok Metropolitan Region (BMR) has extended over its administrative boundaries including the neighboring provinces.

On the other hand, the increase of foreign investments to the suburban areas of BMR promoted the expansion of built-up areas. In Thailand, after the 1980s, influx of foreign companies including Japan has been active, and they focused on BMR as a location. However, after the 1990s, many Japanese car industries moved to suburban areas particularly east waterfront area, Samut Prakan Province or Chon Buri Province (Ikuta 2011). According to Une (2009), one of the reasons for gathering Japanese car parts industries to a suburban industrial estate is the zone system based on Investment Incentive Regulation which gives more favorable treatment to companies located farther from BMA. After all, such influx of factories and accompanying growth of office functions and service industries in the suburb enhances population migration from rural areas. As a result, it promotes the development of extending BMR.

8.5 Implications for Future Sustainable Urban Development

According to land use/cover change and landscape metrics analyses, the built-up areas in Bangkok substantially increased since 1988. Urban growth process was dominated by extension and densification. Furthermore, the projected land use/cover change results indicate that built-up area will continue to expand in the future. Such prominent urban development of Bangkok brought various urban problems. Countermeasures against traffic and water environmental problems are highly important for sustainable development of future Bangkok because the problems are particularly severe in Bangkok.

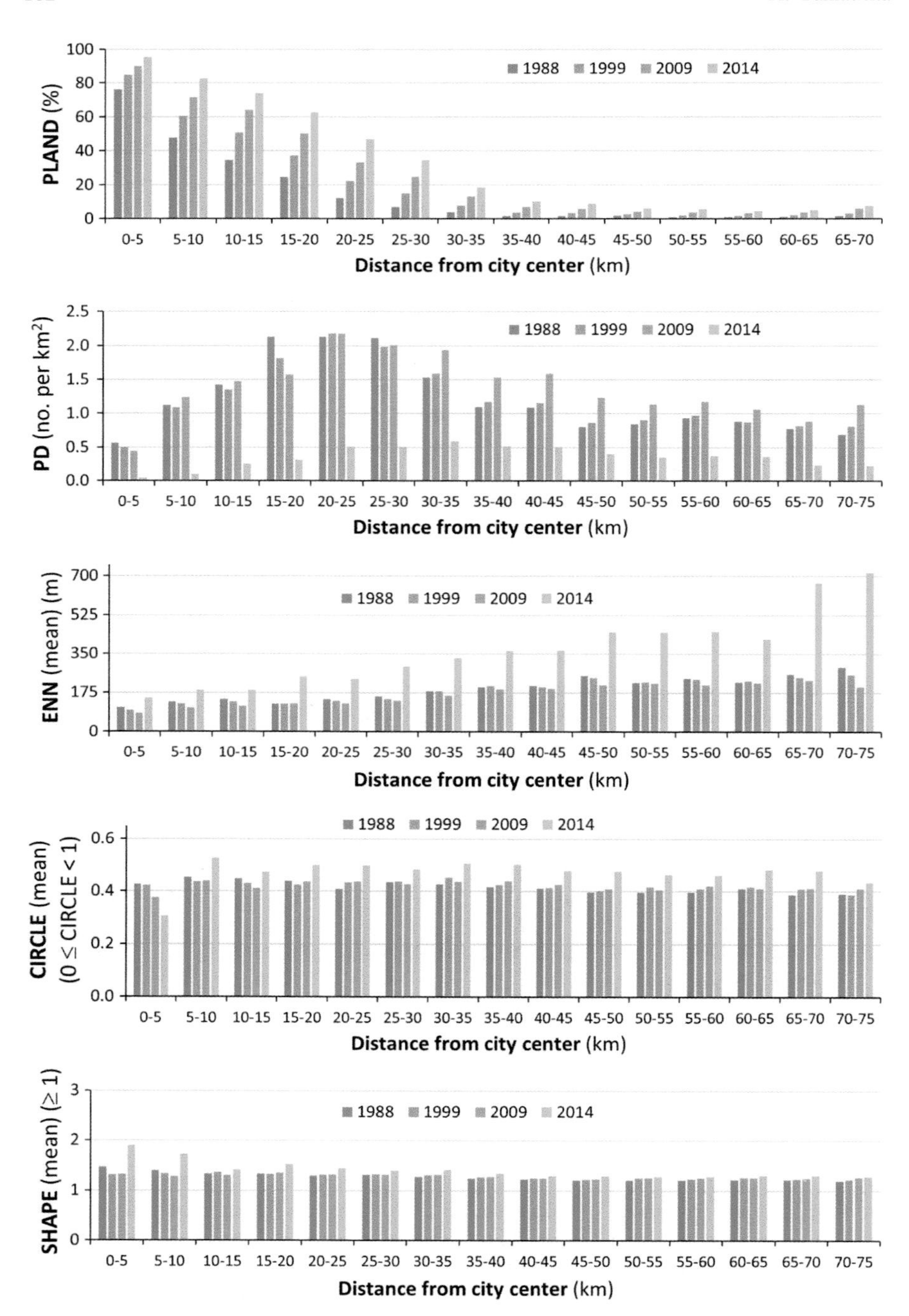

Fig. 8.10 Observed class-level spatial metrics for built-up along the gradient of the distance from city center of Bangkok Metropolitan Area. *Note* The y-axis values are plotted in the same range as those in Fig. 8.12

Table 8.4 Projected urban land use/cover of Bangkok Metropolitan Area (km^2)

	2020	2030
Built-up	2211.56	2645.52
Non-built-up	9508.76	9074.80
Water	2108.45	2108.45
Total	13,828.77	13,828.77

Table 8.5 Projected urban land use/cover changes in Bangkok Metropolitan Area (km^2)

	2014–2020	2020–2030
Built-up	103.93	433.96
Annual rate of change (km^2/year)	*17.32*	*43.40*
Non-built-up	−103.93	−433.96
Annual rate of change (km^2/year)	*−17.32*	*−43.40*
Water	0.00	0.00
Annual rate of change (km^2/year)	*0.00*	*0.00*

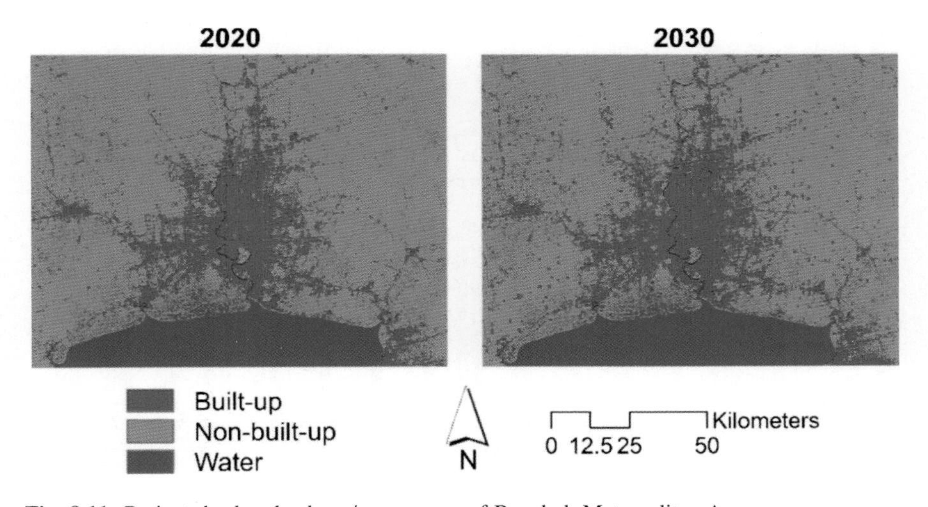

Fig. 8.11 Projected urban land use/cover maps of Bangkok Metropolitan Area

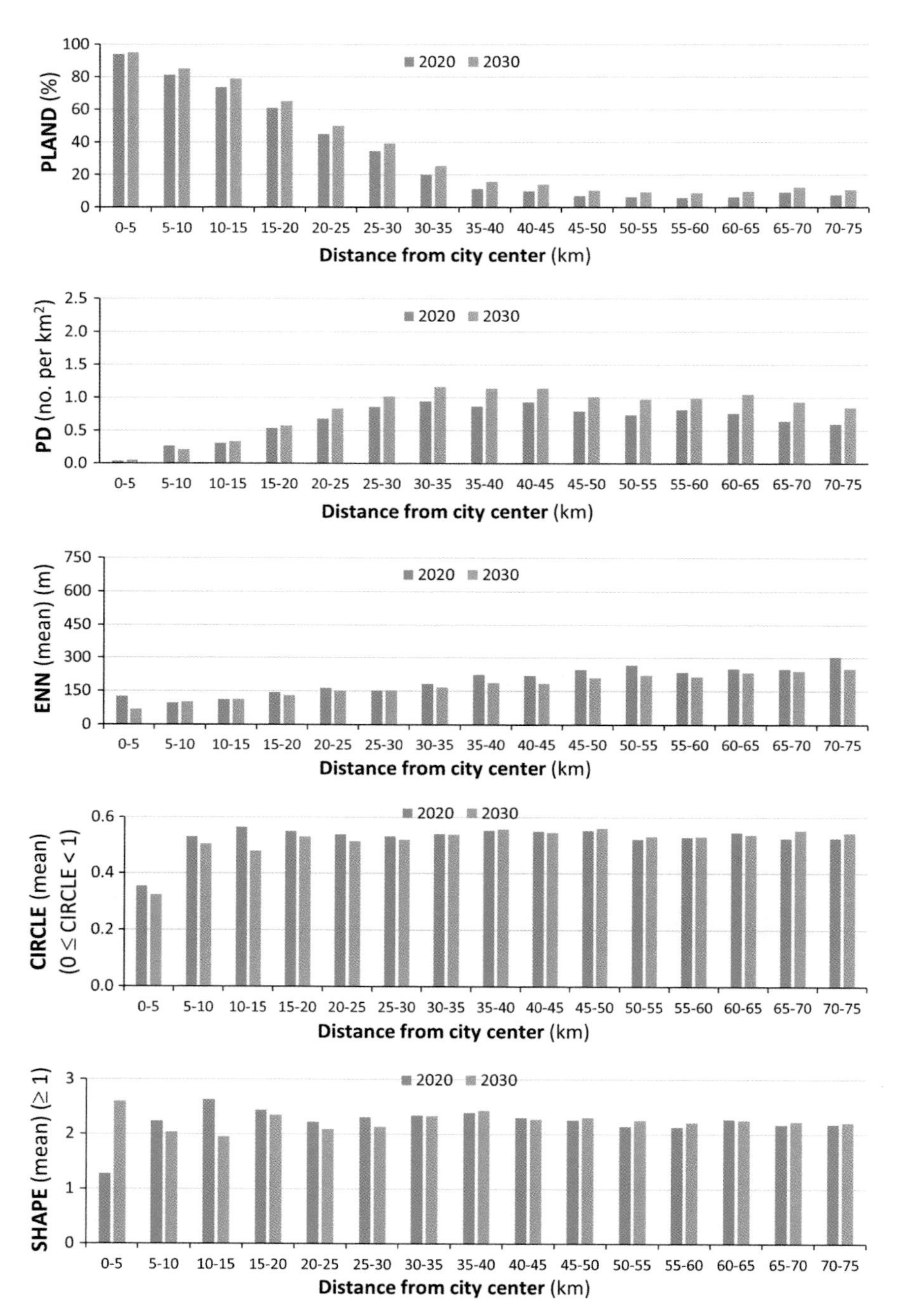

Fig. 8.12 Projected class-level spatial metrics for built-up along the gradient of the distance from city center of Bangkok Metropolitan Area. *Note* The y-axis values are plotted in the same range as those in Fig. 8.10

Table 8.6 Projected landscape pattern of Bangkok Metropolitan Area

Class-level (built-up) spatial metrics	2020	2030
PLAND (%)	18.87	22.57
PD (number per km^2)	0.64	0.82
ENN (mean) (m)	225.56	200.85
CIRCLE (mean) ($0 \leq$ CIRCLE < 1)	0.54	0.54
SHAPE (mean) ($1 \leq$ SHAPE $\leq \infty$)	2.22	2.22

8.5.1 Traffic Problem

Compared with the continuous extension of built-up areas after the 1990s, the provision of public transportation in Bangkok was delayed. As a result, the number of car possession increased and traffic jam in Bangkok was the worst in the world. According to Morisugi and Fukuda (1998), the number of car possession per capita was 0.18 in 1989 but increased to 0.28 in 1993. As the price of fuel was also cheap, car usage rate increased. Also in Bangkok, as shipping was the main traffic means in the past, construction of main road network was delayed. Many streets were like blind alleys, and caused chronic traffic jams. Facing these problems, starting from the construction of the first phase Metropolitan Highway Network (27.1 km) between 1982 and 1987, the second phase Metropolitan Highway Network and inner ring road and outer ring road (National Route No. 9) and main road passing Suvarnabhumi International Airport to Southeast Chon Buri (National Route No. 7) were constructed. However, traffic jam during peak hours of commuting in the morning and evening remains to be a problem. For future sustainable development of Bangkok, as well as further road maintenance including widening of ordinal roads as well as main roads, the strategies are necessary to reduce the amount of traffic into the center. According to land use/cover change analysis, built-up areas in Bangkok expanded with integrating the existing built-up patches. However, the emergence of separated urban cores in the suburb should be expected in the future.

After the 1990s, construction of overhead railways and subways started. By now, two overhead railways (Sukhumvit Line and Si Lom Line) and two subway lines, and ARL (Airport Rail Link) connecting the center of Bangkok and Suvarnabhumi International Airport have been established. The Bangkok Mass Transit System Public Company Limited (BTSC), which manages overhead railways, has a large extension plan of public transportation which will cover the extending urban areas of Bangkok by 2030. The company is also considering introducing Light Rail Transit (LRT). If this extension of public transport network is achieved, reduction of traffic jam and improvement of traffic convenience for residence in extending urban areas is expected.

8.5.2 Water Environmental Problem

Rapid expansion of impermeable built-up areas and population growth in Bangkok caused serious water environmental problems. One problem is water contamination in canals and rivers due to the undeveloped sewage system. The other problem is land subsidence caused by over pumping of groundwater because of the increased water demand. According to Yoshida (1998) and Yoshikoshi (2011), the coverage of sewage system was only 2% in 1996, although main source of water contamination is from industrial and domestic wastewater. As a result, the wastewater flows out into the canals and rivers without cleaning well. Especially in the canals running through central Bangkok, its contamination is so bad that some places have a zero dissolved oxygen record (Fig. 8.13).

On another front, the amount of pumped groundwater increased dramatically after the 1970s. According to Yoshikoshi (2011), the annual groundwater pumping was about 700,000 tons in 1970. It increased to over two million tons in 1999. Due to such increment of groundwater pumping, land subsidence problem came to the surface. Particularly, in the existing built-up areas, the east part of the Chao Phraya River, the total amount of land subsidence can be over 2 m. As an influence of land subsidence, there are cases of leaning and cracking buildings (Fig. 8.14). Also, advancing land subsidence increases flood damage during rainy seasons and prolongs the flood period. In order to address this land subsidence problem, the government introduced regulation and the charging system on the amount of pumping groundwater and the public water supply facilities sourced from rivers were expanded and maintained (Endo 2010).

Fig. 8.13 Seriously polluted canal in central Bangkok. *Source* Author's fieldwork (2006)

Fig. 8.14 A cracked building by land subsidence. *Source* Author's fieldwork (2009)

Even after taking such countermeasures for land subsidence, Bangkok was still suffering from large flood damage in 2011, and it also had a drought in 2015 which caused water shortages. Some measures to improve water quality of canals in urban areas were taken (Fig. 8.15), but are still not enough. Against future expansion of built-up areas, measures for water environmental problems are half way through as well as expansion of public transport. It is necessary to control built-up expansion

Fig. 8.15 A signboard appealing canal water purification. *Source* Author's fieldwork (2006)

and to design a future land use/cover plan which is familiar with water environment for sustainable urban development in Bangkok.

8.6 Conclusion

Bangkok is in an extremely high primary position as the capital city. It has a high density population and viable political and economic prospects. Urban growth process in Bangkok was dominated by extension and densification. The built-up areas will continue to expand in the future. The formation of suburban centers in the outskirts of BMR has progressed. However, the suburban centers can be included into extended urban core and form single and larger urban areas in the future. Countermeasures to stop such extension of BMR and excess concentration should be taken. At the same time, extension of public transport network, road network, and environmental infrastructure development like water supply and sewage systems are necessary to achieve sustainable urban development.

References

Endo T (2010) The roles of government in groundwater management—a case study of land subsidence problem in Bangkok. J Jpn Assoc Hydrol Sci 40:95–108 (Japanese with English abstract)

Estoque RC, Murayama Y, Kamusoko C, Yamashita A (2014) Geospatial analysis of urban landscape patterns in three major cities of Southeast Asia. Tsukuba Geoenviron Sci 10:3–10

Hashimoto T (1998) Urban administration, finance and development policy. In: Tasaka T (ed) Asian large city 1 Bangkok. Nippon Hyoron Sha, Tokyo, pp 281–304 (Japanese)

Ikuta M (2011) Mega-cities in Southeast Asia: widening regional integration including Japan. Kokon-Shoin, Tokyo (Japanese)

Kamusoko C (2017) Methodology. In: Murayama Y, Kamusoko C, Yamashita A, Estoque RC (eds) Urban development in Asia and Africa—geospatial analysis of metropolises. Springer Nature, Singapore, pp 11–46

Morisugi H, Fukuda A (1998) Transportation problems. In: Tasaka T (ed) Asian large city 1 Bangkok. Nippon Hyoron Sha, Tokyo, pp 213–233 (Japanese)

Niwa T (2010) Changing spatial patterns of internal migrations in Thailand: using NESDB data. Q J Geogr 62:83–92 (Japanese with English abstract)

Tamada Y (2003) Change of politics in Thailand. Intriguing ASIA 57:23–33 (Japanese)

Tomosugi T (1998) The formation of urban landscape. In: Tasaka T (ed) Asian large city 1 Bangkok. Nippon Hyoron Sha, Tokyo, pp 45–72 (Japanese)

Tomosugi T (2003) A sketch of Bangkok history. Intriguing ASIA 57:10–22 (Japanese)

Une Y (2009) Process of forming an automotive parts firm agglomeration and spatial characteristics of interfirm linkages in Amata Nakorn Industrial Estate, Thailand. Geogr Rev Jpn 82A:548–570 (Japanese with English abstract)

United Nations (UN) (2015) World urbanization prospects: the 2014 revision. United Nations, New York

Yamashita A (2011) Comparative analysis on land use distributions and their changes in Asian mega cities. In: Taniguchi M (ed) Groundwater and subsurface environments: human impacts in Asian coastal cities. Springer, Tokyo, pp 61–81

Yoshida M (1998) Environmental problems. In: Tasaka T (ed) Asian large city 1 Bangkok. Nippon Hyoron Sha, Tokyo, pp 235–255 (Japanese)

Yoshikoshi A (2011) Urban development and water environment changes in Asian megacities. In: Taniguchi M (ed) Groundwater and subsurface environments: human impacts in Asian coastal cities. Springer, Tokyo, pp 35–59

Chapter 9
Yangon Metropolitan Area

Ronald C. Estoque

Abstract Yangon Metropolitan Area, also known as the Greater Yangon, is Myanmar's largest commercial center and home to the country's former capital, Yangon City. This chapter traces the origin and examines the urban primacy of Yangon Region, the region where this metropolitan area is located. It also examines the recent (1989–2014) and potential future (2014–2030) urban land changes, i.e., changes from non-built-up to built-up lands, in the metropolitan area using geospatial tools and techniques. Finally, it discusses some of the possible key factors influencing the urban development of the area and the potential implications of its rapid population growth and urban land changes to its future sustainable urban development. The analysis showed compelling evidence for Yangon Region's urban primacy over the other regions/states in the country based on population density and proportion of urban population. Over the past 25 years (1989–2014), the area of built-up lands in the study area has increased more than threefold. The geographic location and landscape characteristics of the metropolitan area and its population growth and status as the largest commercial center in the country and the home of the country's former capital city are hypothesized to be among the key factors influencing the spatiotemporal patterns of urban land changes and the overall urban development of the area. The simulated urban land changes indicated that built-up lands would continue to expand in the future (2014–2030) under the influence of infill and sprawl development patterns. The intensifying pressure of urbanization due to rapid population growth and urban land changes poses many challenges that need to be considered in sustainable urban development and landscape planning.

R.C. Estoque (✉)
Faculty of Life and Environmental Sciences, University of Tsukuba, Tsukuba City, Japan
e-mail: estoque.ronald.ga@u.tsukuba.ac.jp; rons2k@yahoo.co.uk

Y. Murayama et al. (eds.), *Urban Development in Asia and Africa*,
The Urban Book Series, DOI 10.1007/978-981-10-3241-7_9

9.1 Origin and Brief History

Yangon Metropolitan Area, also known as the Greater Yangon,[1] is located in Yangon Region, one of the 15 current states/regions of Myanmar. The region's capital is Yangon City, which is also the former capital city of the country. Greater Yangon is geographically located between around 17° 06′ and 16° 35′N latitude and between 95° 58′ and 96° 24′E longitude, east of the Ayeyarwaddy River delta, once known as "the rice bowl of Asia" (Fig. 9.1). Yangon City is located 34 km upstream from the mouth of Yangon River, in the Gulf of Martaban of the Andaman Sea. In 2012, agricultural lands occupied 51% of the Greater Yangon, while urbanized lands occupied 22% (JICA and YCDC 2013).

This section briefly describes the origin and history of Yangon Region, in which the study area is located, based on the *Yangon Region historical context*, a section of the United Nations Development Program's publication, entitled *Local Governance Mapping—The State of Local Governance: Trends in Yangon* (UNDP 2015). The reader is referred to this source for a more detailed account of the origin and history of the region.

According to the *Yangon Region historical context* (UNDP 2015), the Mon people[2] were the first to inhabit the area where the current Yangon Region is located. The region thus became part of various Mon kingdoms, which were also responsible for the establishment of the origins of Shwedagon Pagoda and the foundation for the city of Dagon, which later became Yangon. Nowadays, the Shwedagon Pagoda is Yangon's most famous landmark. After 1057, the place came under the control of Burman kingdoms.

In the early 1850s, following the second Anglo-Burmese War, the entire "Lower Burma," i.e., including today's Ayeyarwady, Yangon, and Bago Regions, came under British colonial rule. During the British period, Yangon Region was integrated into British India as Hanthawaddy district and part of Pegu (i.e., Bago) Division. Yangon, then called Rangoon by the British rulers, was designated as the administrative center. During the 1920s and 1930s, Rangoon emerged not only as an administrative center, but also as a political center. It also became one of Southeast Asia's important cities, prominent in various aspects, such as architecture, culture, transport and logistics, and economic development.

In 1948, Yangon became the center of Burma's independence movement and the capital of the Union of Burma. Following independence, Yangon Division, formerly Hanthawaddy sub-division, became a regular division, centrally administered by the newly independent Burma. After World War II, Yangon continued to grow. However, due to housing problems, slums and squatter settlements also emerged. During the

---

[1]The Greater Yangon (with an area of 1500 km^2) includes Yangon City (with an area of 784 km^2) and parts of the six neighboring townships of Kyauktan, Thanlyin, Hlegu, Hmawbi, Htantabin, and Twantay (JICA and YCDC 2013) (see Fig. 5.15).

[2]The Mon people are an ethnic group from Burma (Myanmar). They were believed to be among the first groups of people to inhabit Southeast Asia (Indochina).

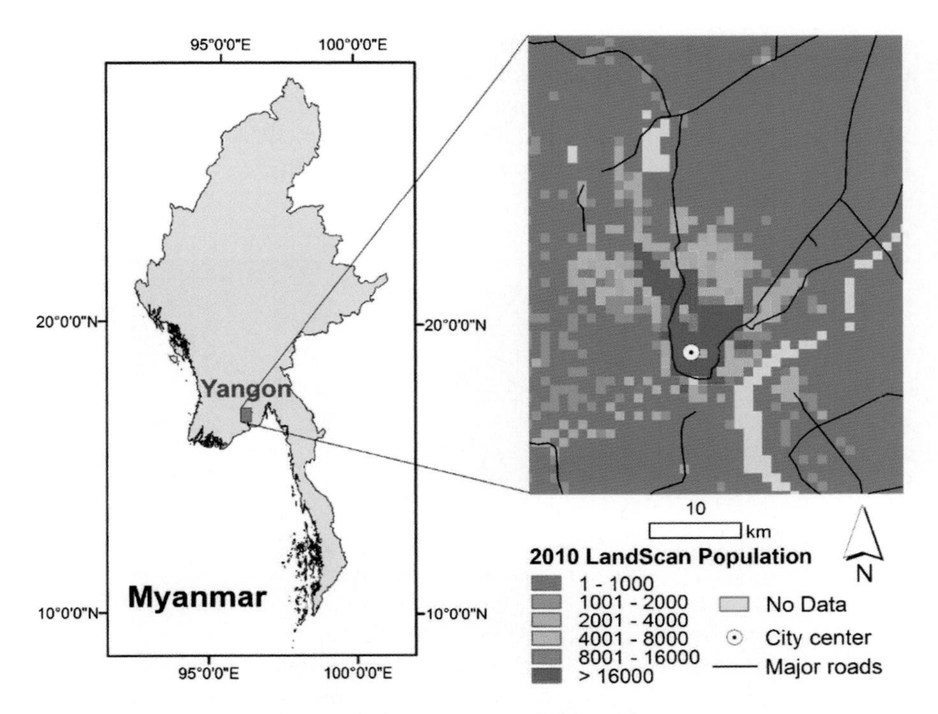

Fig. 9.1 Location and LandScan population of Yangon Metropolitan Area, Myanmar

1950s, while a number of nearby towns were still being built, Yangon City was already among the most cosmopolitan and globalized cities in Asia.

Under the 1974 Constitution of Burma, "states" and "divisions" had the same status. Thus, Yangon Division became a major administrative component of the "Socialist Republic of the Union of Burma," which was then composed of seven states and seven divisions. While the central government had the overall control, people's councils were also introduced. However, the participatory nature of the governing structure was essentially removed with the suspension of the 1974 Constitution in 1988 and when the country was under direct military control.

In 2002, the construction of the new national capital *Naypyitaw* in Pyinmana Township of Mandalay Division was initiated. Between 2005 and 2006, some government offices and staff were transferred from Yangon to the new capital, though a number of government offices were retained in Yangon. In May 2008, a national referendum for the new Constitution was held. Under the new Constitution, Yangon Division became Yangon Region. As a major administrative component of the now known Republic of the Union of Myanmar, Yangon Region has an equal status with the other country's states and regions.

Today, the world-famous architectural heritage of Yangon City, including the colony's administrative buildings, industrial and commercial centers, and main transport infrastructure, is considered an important legacy of the British colonial government. Most of present-day urban structural arrangements and features have their roots during the colonial period. Based on the latest statistics, Yangon Region accounts for more than 14% of the national population (Census Report 2015) and 22% of the gross domestic product (GDP) of the country (JICA and YCDC 2013). Figures 9.2, 9.3, and 9.4 show some snapshots of Yangon.

Fig. 9.2 Inside Yangon City, Myanmar. *Source* Yuji Murayama's fieldwork (2011)

Fig. 9.3 Chinatown in Yangon City, Myanmar. *Source* Yuji Murayama's fieldwork (2011)

Fig. 9.4 Urban development on the outskirt of Yangon City, Myanmar. *Source* Yuji Murayama's fieldwork (2011)

9.2 Primacy in the National Urban System

The Republic of the Union of Myanmar, commonly known as Myanmar and formerly known as Burma, is one of the 11 sovereign states or countries in Southeast Asia. In terms of land area, it is ranked 39th worldwide and second in Southeast Asia (UN 2013). As of 2014 census, Myanmar is ranked 28th worldwide and fifth in Southeast Asia with a population of 50.28 million (Census Report 2015; World Bank 2015a). In terms of population density, Myanmar with 82 people per km^2 is ranked 118th worldwide and eighth in Southeast Asia (World Bank 2015b). It also has an urban population of 14.88 million, i.e., 29.6% of the country's total population (Census Report 2015).

According to the United Nations (UN), the territorial spread of cities of different sizes across the whole territory of one country constitutes a national urban system (UN 2015). Such a system can be linked to the organization of the government at various levels: national, regional, and local levels (Kim and Law 2012; UN 2015). Administratively, Myanmar is officially divided into 15 states/regions, 74 districts, and 412 townships/sub-townships (Census Report 2015). The five largest cities in Myanmar based on total population in 2014 are the following: Yangon with 5.21 million (State/Region: Yangon), Mandalay with 1.23 million (State/Region: Mandalay), Naypyitaw—the national capital—with 1.16 million (State/Region: Naypyitaw), Bago with 0.50 million (State/Region: Bago), and Hpa-An with 0.42 million (State/Region: Kayin) (Census Report 2014).

This section examines the urban primacy of Yangon Region across the whole country of Myanmar. *Urban primacy* indicates the degree of dominance of one urban area based on population, economy, and urban functions and services (Estoque 2017). However, due to data constraints, only population-related indicators were considered, namely total population, population density, and urban population based on the latest population and housing census data (Census Report 2015). More specifically, these indicators were used to compare Yangon with the other regions/states in the country.

Figure 9.5 shows that Yangon is among the smallest regions/states, accounting for only 1.5% of the total land area of the whole country. The regions of Shan, Sagaing, and Kachin are among the largest ones in terms of land area. However, in terms of total population, Yangon has the highest with 7.36 million, followed by Ayeyarwady (6.18 million) and Mandalay (6.17 million) (Fig. 9.6). The regions/states of Kayah, Chin, and Naypyitaw are among the smallest ones in terms of total population.

Due to its small land area and high population, Yangon has the highest population density with 751 people/km^2 (Fig. 9.7). In fact, the population density of Yangon is almost four times and 57 times the density of the second and last in rank regions/states, namely Mandalay and Chin, respectively. It is four and a half times the population density of the country's capital city or region, Naypyitaw (164 people/km^2).

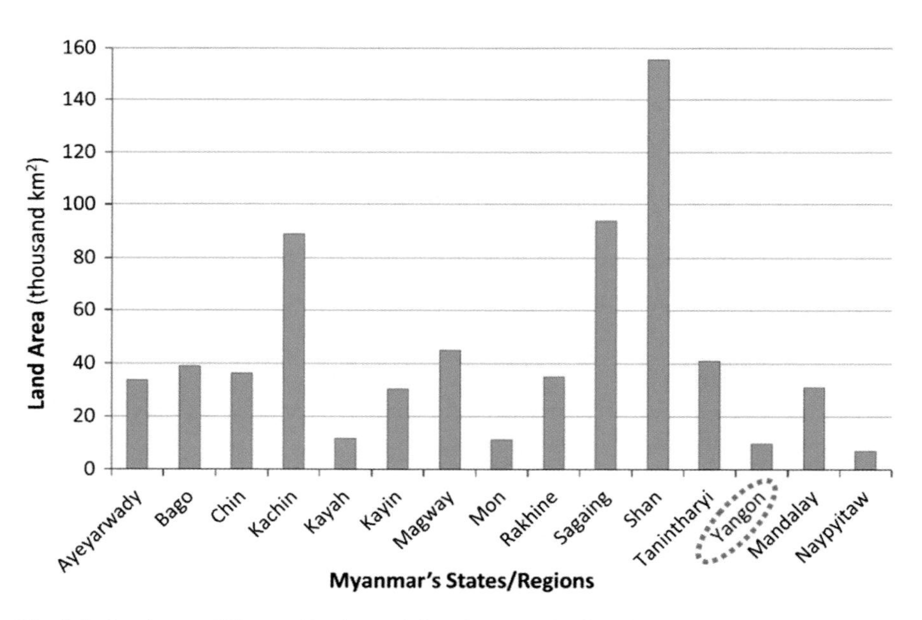

Fig. 9.5 Land area of Yangon Region and the other states/regions in Myanmar. *Note* Area refers to land surface only, not covered by water. *Source* Census Report (2015)

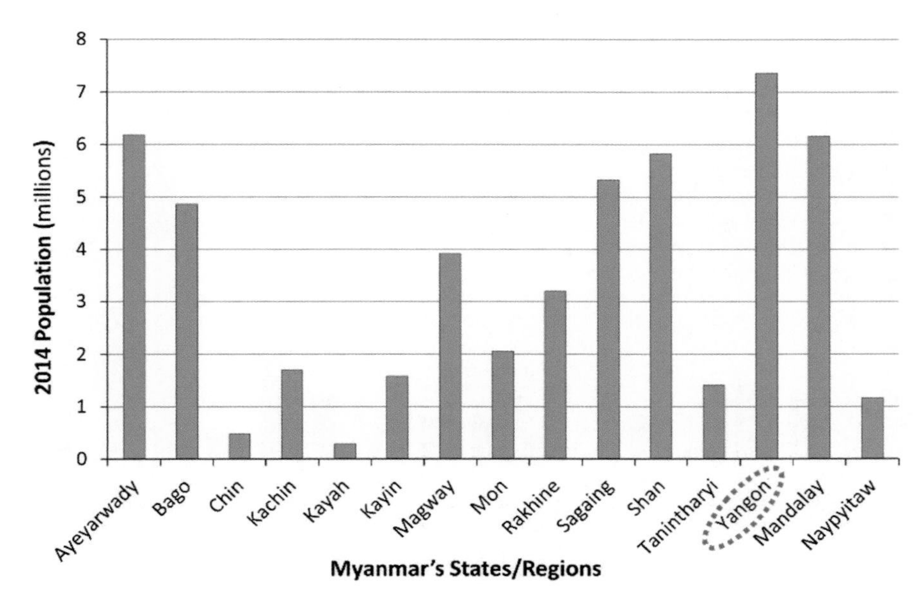

Fig. 9.6 Total population of Yangon Region and the other states/regions in Myanmar. *Data source* Census Report (2015)

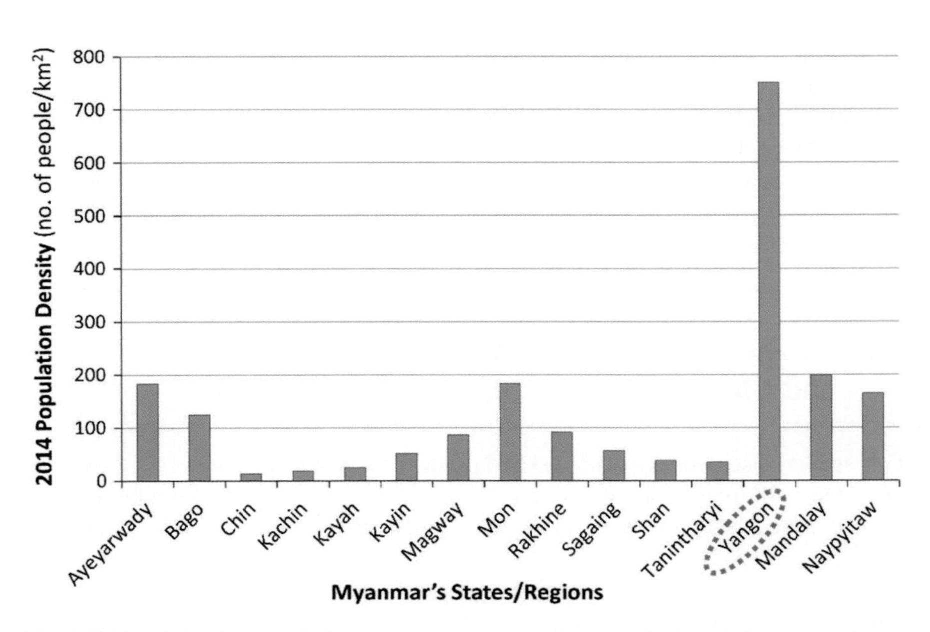

Fig. 9.7 Population density of Yangon Region and the other states/regions in Myanmar. *Source* Census Report (2015)

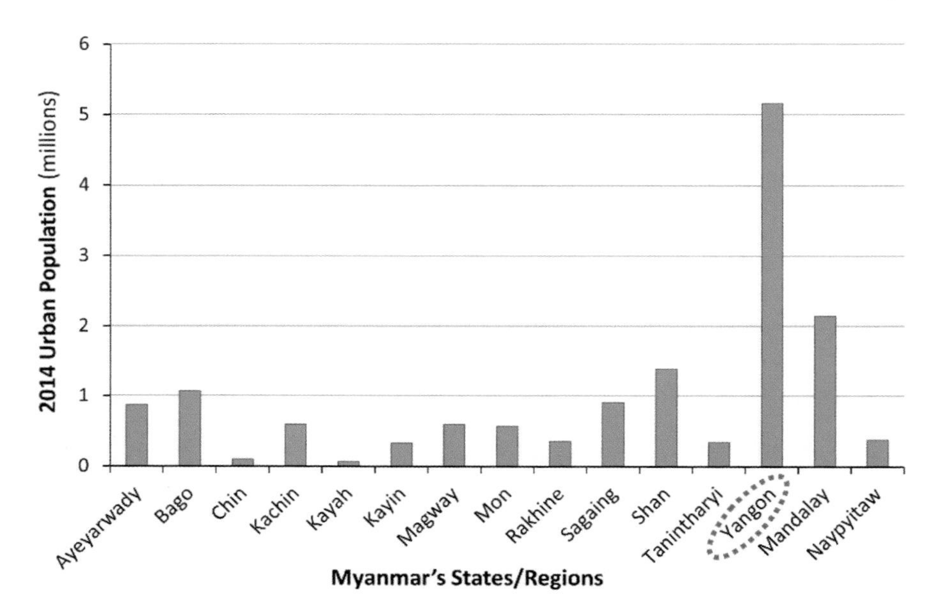

Fig. 9.8 Urban population of Yangon Region and the other states/regions in Myanmar. *Source* Census Report (2015)

In terms of urban population (Fig. 9.8), Yangon also has the highest with 5.16 million urban dwellers, accounting for 34.7% of the country's 2014 total urban population of 14.88 million. Yangon's urban population is about two and a half and seventy times the urban population of Mandalay and Kayah, respectively.

In summary, based on these three population-related indicators, Yangon is clearly the most dominant region in Myanmar today. This is despite of its loss of the status as the home of the country's national capital.

9.3 Urban Land Use/Cover Patterns and Changes (1989–2030)

This section discusses the observed and projected urban land changes, i.e., changes from non-built-up to built-up lands, in Yangon Metropolitan Area. Remote sensing-derived urban land use/cover maps and spatial metrics were used to detect the temporal and spatial patterns of urban land changes. The details of the urban land use/cover mapping, change detection and simulation modeling, and spatial pattern analysis are described in the methodology chapter (Kamusoko 2017). Estoque and Murayama (2017) provide a comparative analysis of the trends and spatial patterns of urbanization in Asia and Africa.

9.3.1 Observed Changes (1989–2014)

Based on the urban land use/cover maps classified from remote sensing satellite imagery (Fig. 9.9), the area of built-up lands has increased more than threefold over the past 25 years (Table 9.1). It increased from 62.69 km^2 in 1989 to 228.89 km^2 in 2014 (Table 9.1). This increase translates to an annual rate of change (increase) of 6.65 km^2/year. The annual rate of change also increased from 6.10 km^2/year during the 1989–1999 period to 6.83 and 7.37 km^2/year during the 1999–2009 and 2009–2014 periods, respectively (Table 9.2). This indicates that the intensity of built-up expansion in the area has been increasing over the past 25 years.

In the context of landscape pattern analysis, the percentage of landscape (PLAND) metric measures the proportion of a particular class at a certain time point relative to the whole landscape. In 1989, the built-up class had a PLAND of 3.22%, which increased to 6.29, 9.81 and 11.87% in 1999, 2009, and 2014, respectively (Table 9.3). The patch density (PD) metric is a measure of fragmentation based on the number of patches per unit area (in this case per 100 ha or 1 km^2), in which a patch is based on an 8-cell neighbor rule. In 1989, the built-up class had a PD of 0.79, which increased to 1.27 and 1.50 in 1999 and 2009, respectively. This indicates that from 1989 to 2009, built-up lands became more fragmented. However, built-up lands in 2014 were more aggregated and less fragmented than in 2009 as indicated by the decrease in PD between these two time points.

The Euclidean nearest neighbor distance (ENN) metric is a measure of dispersion based on the distance of a patch to the nearest neighboring patch of the same class. The mean ENN of the built-up patches in the study area showed an up-down-up pattern, with 189.31 m in 1989, 154.66 m in 1999, 160.76 m in 2009, and 189.96 m in 2014 (Table 9.3). The decreasing trend between 1989 and 1999 can be due to the expansion of the existing built-up patches and the development of new built-up patches in between, but not connected to, the existing ones. The increasing trend between 1999 and 2014 can also be due to the development of new built-up patches, but this time, farther from the existing ones. The four urban land use/cover maps of the area show these two patterns (Fig. 9.9). The increasing trend of PLAND and the overall increase in PD from 1999 to 2014 also support these observations.

The related circumscribing circle (CIRCLE) metric measures the circularity of patches. The value of CIRCLE is 0 for circular or one cell patches and approaches 1 for elongated, linear patches one cell wide. The mean CIRCLE value of the built-up patches in the study area showed a slightly increasing trend during the whole 1989–2014 period (Table 9.3). It can be noted that urban expansion in the northern part of the area, especially during the later periods, followed the road network (Figs. 9.9 and 9.10). Thus, the increasing trend can be due to the development of more elongated built-up patches and the aggregation of much smaller, circular isolated patches.

The shape index (SHAPE) metric is a measure of complexity. This metric has a value of 1 when the patch is square and increases without limit as patch shape becomes more irregular. In the study area, the built-up patches had a decreasing

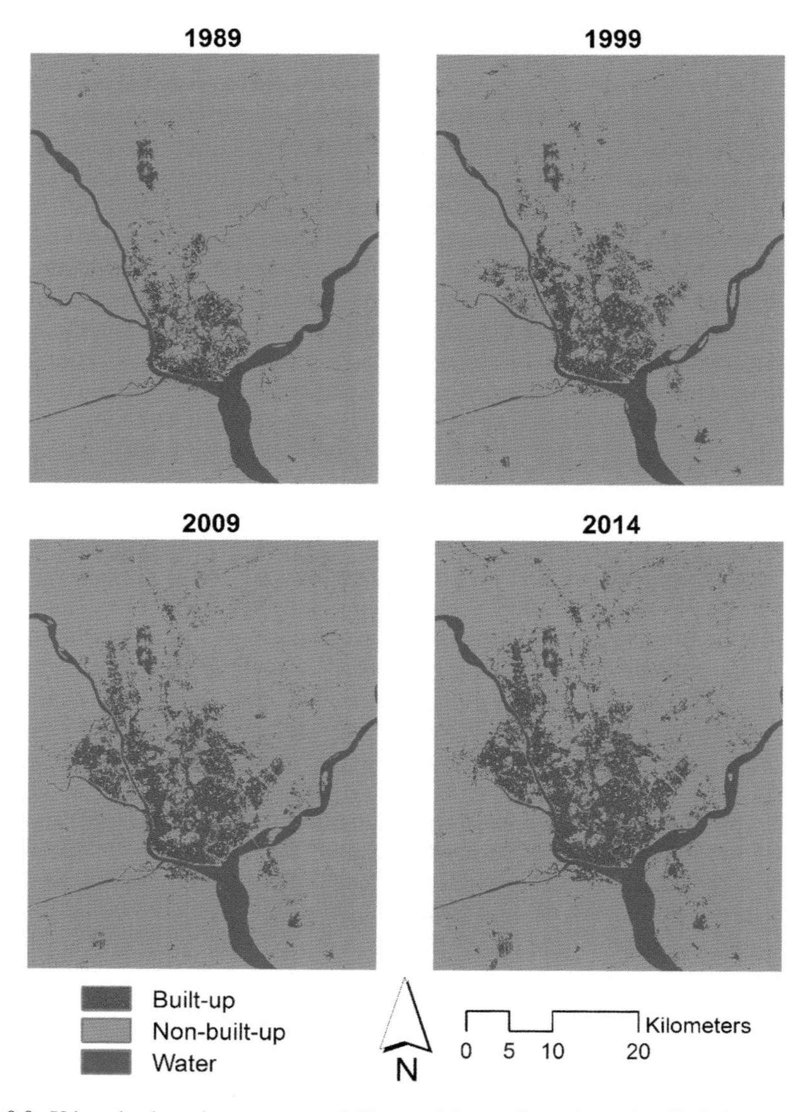

Fig. 9.9 Urban land use/cover maps of Yangon Metropolitan Area classified from Landsat imagery

Table 9.1 Observed urban land use/cover of Yangon Metropolitan Area (km^2)

	1989	1999	2009	2014
Built-up	62.69	123.68	192.02	228.89
Non-built-up	1882.18	1841.98	1765.23	1699.21
Water	142.16	121.38	129.79	158.93
Total	2087.04	2087.04	2087.04	2087.04

Table 9.2 Observed urban land use/cover changes in Yangon Metropolitan Area (km^2)

	1989–1999	1999–2009	2009–2014
Built-up	60.99	68.34	36.87
Annual rate of change (km^2/year)	*6.10*	*6.83*	*7.37*
Non-built-up	−40.20	−76.75	−66.02
Annual rate of change (km^2/year)	*−4.02*	*−7.67*	*−13.20*
Water	−20.79	8.41	29.14
Annual rate of change (km^2/year)	*−2.08*	*0.84*	*5.83*

Table 9.3 Observed landscape pattern of Yangon Metropolitan Area

Class-level (built-up) spatial metrics	1989	1999	2009	2014
PLAND (%)	3.22	6.29	9.81	11.87
PD (number per km^2)	0.79	1.27	1.50	1.43
ENN (mean) (m)	189.31	154.66	160.76	189.96
CIRCLE (mean) ($0 \leq$ CIRCLE < 1)	0.39	0.39	0.40	0.41
SHAPE (mean) ($1 \leq$ SHAPE $\leq \infty$)	1.29	1.29	1.27	1.22

mean SHAPE value during the whole 1989–2014 period (Table 9.3), indicating that the complexity in the shape of the built-up patches had been decreasing.

Figure 9.11 presents all the metrics for the built-up class along the gradient of the distance from the city center across all time periods from 1989 to 2014. The figure shows that PLAND decreases as the distance from the city center increases, indicating that the proportion of built-up lands near the city center was relatively higher. The figure also shows that PD is relatively higher within the 20-km distance, indicating that there were more patches of built-up lands in areas that are in closer proximities to the city center. At farther distances, built-up patches were relatively more dispersed, as indicated by the increasing trend of the mean ENN along the gradient of the distance from the city center. The figure also shows a more or less stable trend of the average values of CIRCLE and SHAPE along the gradient of the distance from the city center. This indicates that the circularity and complexity of built-up patches in the study area were relatively uniform across the gradient of the distance from the city center.

9.3.2 Projected Changes (2014–2030)

The results of the urban land change simulation revealed that the area of built-up lands would increase from 228.89 km^2 in 2014 to 259.54 km^2 in 2020 and 318.79 km^2 in 2030 (Figs. 9.10 and 9.12; Table 9.4). It would increase at the rate of 5.11 km^2/year from 2014 to 2020 and 5.92 km^2/year from 2020 to 2030 (Table 9.5). The simulated patches of built-up lands in 2020 and 2030 would be more aggregated as indicated by the simulated increase in PLAND and decrease in PD

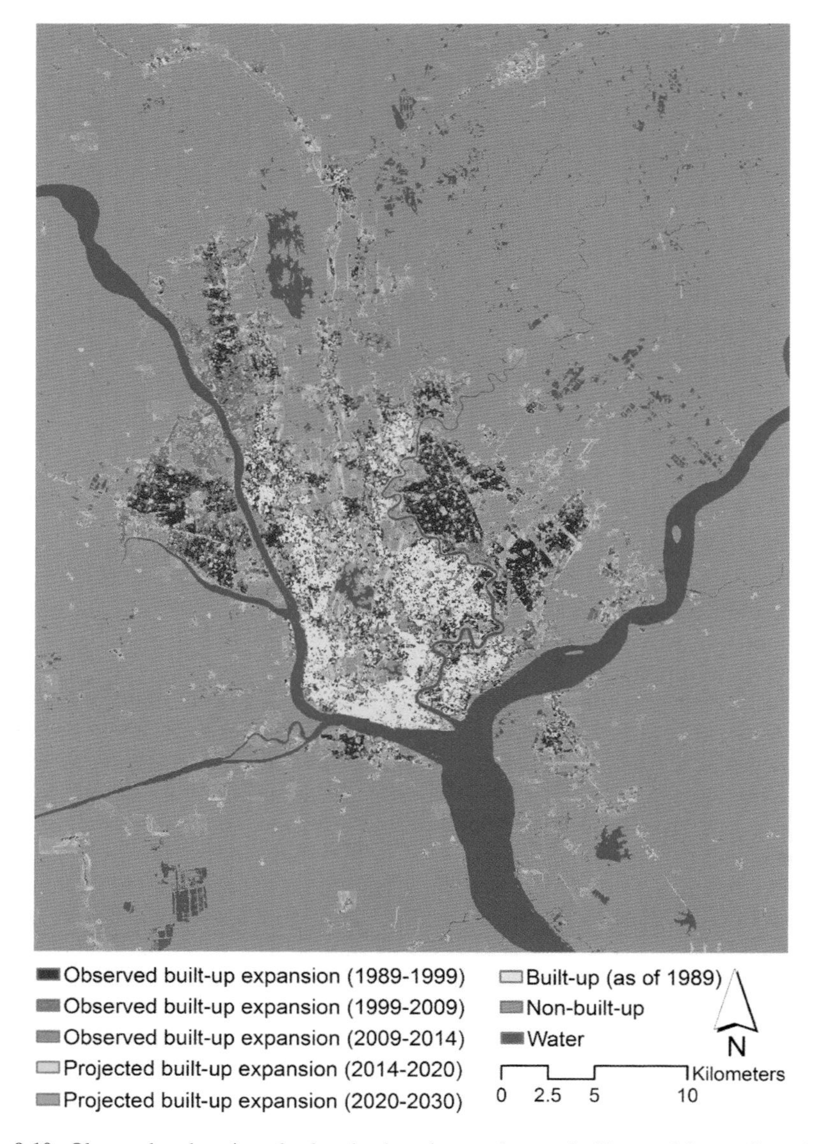

Fig. 9.10 Observed and projected urban land use/cover changes in Yangon Metropolitan Area

(Table 9.6). The simulated increase in mean ENN also indicates that more neighboring (closer) patches would become connected. This simulated aggregation of built-up patches would redefine the average distance between neighboring built-up patches. The simulated increase in the average values of CIRCLE and SHAPE suggests that there would be more elongated/linear and complex patches of built-up lands. It should be noted, however, that between 2020 and 2030, the simulation

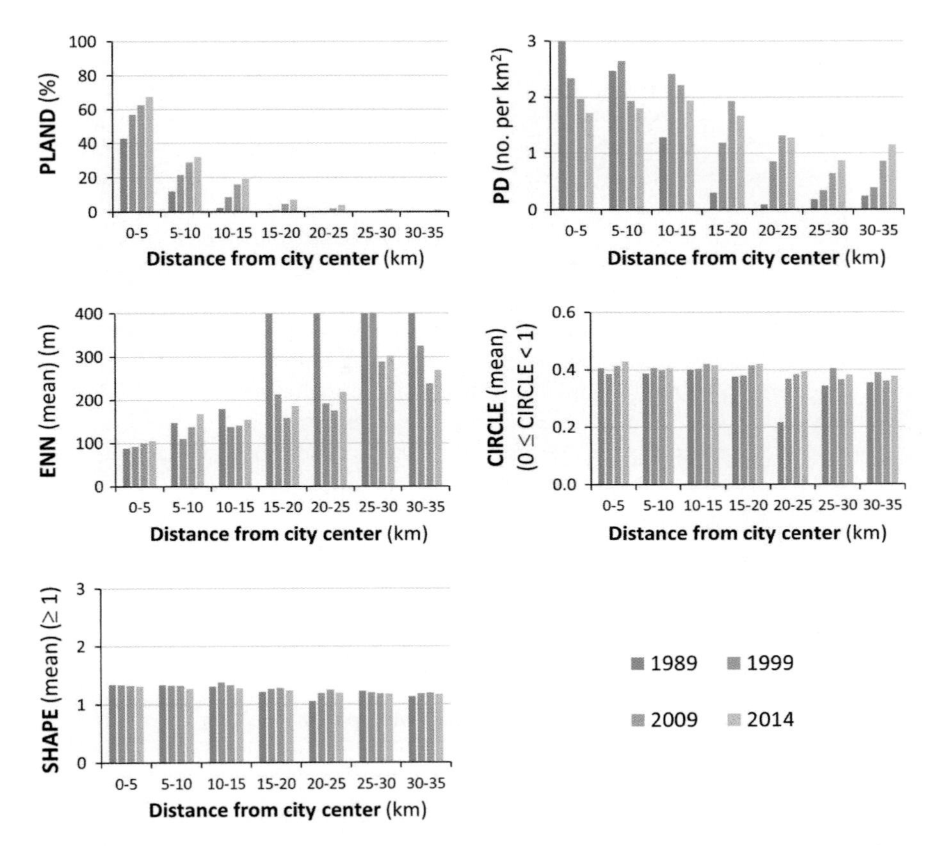

Fig. 9.11 Observed class-level spatial metrics for built-up along the gradient of the distance from city center of Yangon Metropolitan Area. *Note* The y-axis values are plotted in the same range as those in Fig. 9.13

results show that PD would increase, whereas the average values of ENN, CIRCLE, and SHAPE would decrease.

Along the gradient of the distance from the city center (Fig. 9.13), the PLAND of the simulated patches of built-up lands in 2020 and 2030 would also be higher at distances closer to the city center. PD would decrease dramatically in 2020 and 2030, and would be relatively higher in middle distances. On the other hand, the mean ENN would still follow the overall observed pattern during the 1989–2014 period, i.e., mean ENN increases at distances farther from the city center. The mean CIRCLE value would increase in 2020 and 2030, especially at farther distances from the city center, starting from the 5-km distance. Along the gradient of the distance from the city center, however, it would be more or less stable starting from the 10-km distance. The mean SHAPE value would also increase in 2020 and 2030.

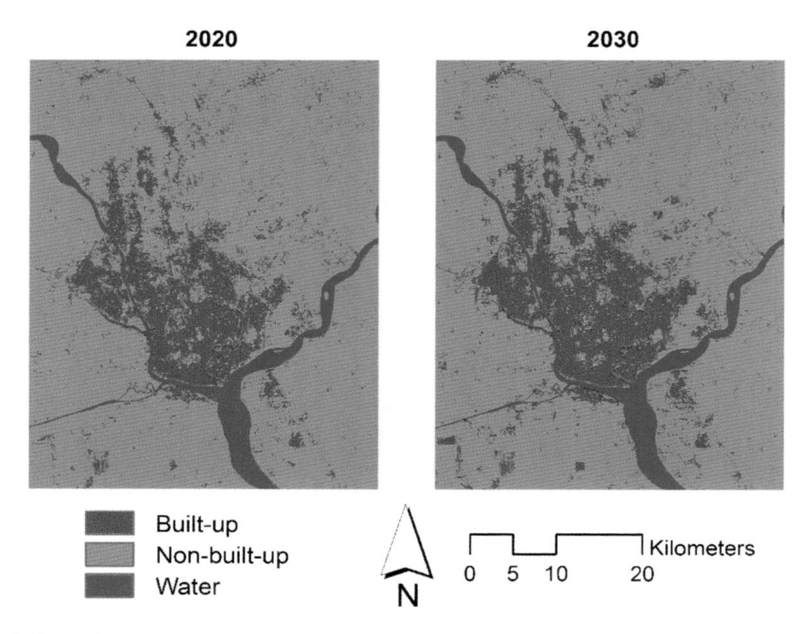

Fig. 9.12 Projected urban land use/cover maps of Yangon Metropolitan Area

Table 9.4 Projected urban land use/cover of Yangon Metropolitan Area (km²)

	2020	2030
Built-up	259.54	318.79
Non-built-up	1668.56	1609.32
Water	158.93	158.93
Total	2087.04	2087.04

Table 9.5 Projected urban land use/cover changes in Yangon Metropolitan Area (km²)

	2014–2020	2020–2030
Built-up	30.65	59.25
Annual rate of change (km²/year)	*5.11*	*5.92*
Non-built-up	−30.65	−59.25
Annual rate of change (km²/year)	*−5.11*	*−5.92*
Water	0.00	0.00
Annual rate of change (km²/year)	*0.00*	*0.00*

It would also have a wavy pattern along the gradient of the distance from the city center. This indicates that the complexity of the simulated built-up patches would greatly vary along the gradient of the distance from the city center.

Table 9.6 Projected landscape pattern of Yangon Metropolitan Area

Class-level (built-up) spatial metrics	2020	2030
PLAND (%)	13.46	16.53
PD (number per km^2)	0.33	0.50
ENN (mean) (m)	281.75	235.52
CIRCLE (mean) ($0 \leq$ CIRCLE < 1)	0.61	0.57
SHAPE (mean) ($1 \leq$ SHAPE $\leq \infty$)	2.75	2.52

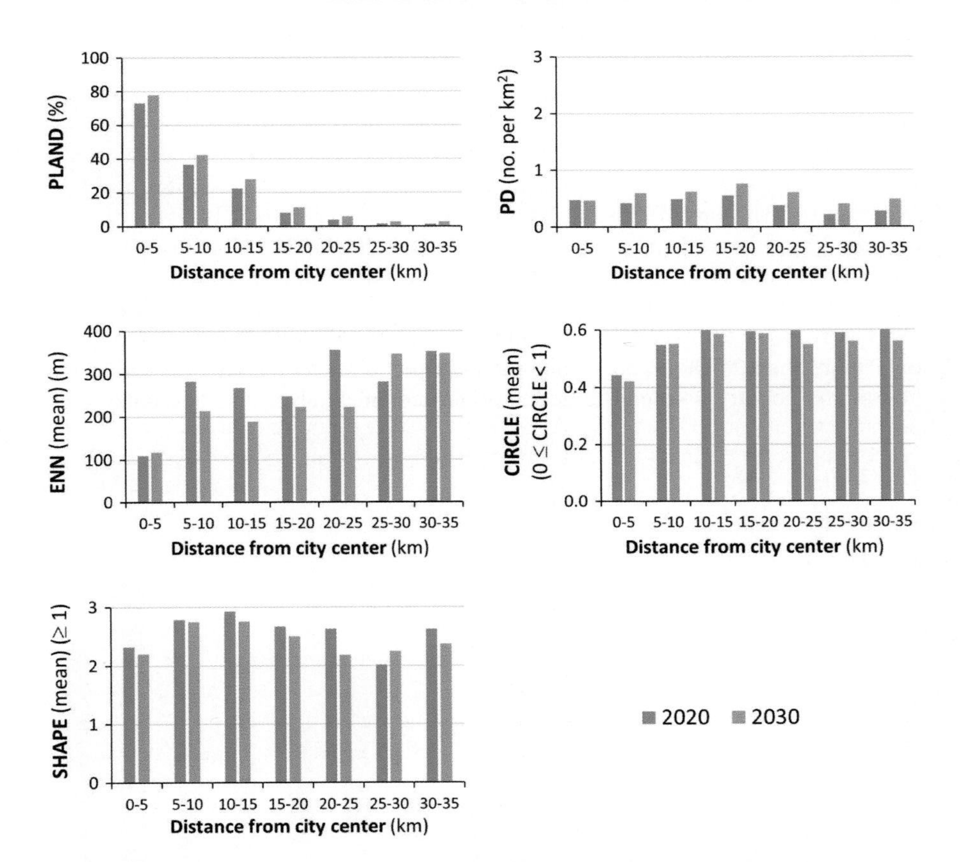

Fig. 9.13 Projected class-level spatial metrics for built-up along the gradient of the distance from city center of Yangon Metropolitan Area. *Note* The y-axis values are plotted in the same range as those in Fig. 9.11

9.4 Driving Forces of Urban Development

As discussed earlier, the urban land changes in the study area over the past 25 years (1989–2014) have been remarkable, with most of the changes occurring north of the main body of the Yangon River and west of one of its tributaries (Sect. 9.3.1; Figs. 9.9 and 9.10). South of the area where the two main tributaries intersect are croplands and low-lying areas. It can be observed that there were not much urban land changes in this part of the area. By looking at the urban land change map (Fig. 9.10), it seems that Yangon River, including its tributaries, has influenced the spatial pattern of urban land changes in the area.

Yangon Region houses the largest commercial center in the country, i.e., the Yangon Metropolitan Area or the Greater Yangon, and the then capital city of the country, Yangon City. Most of the socioeconomic development opportunities are channeled to the region, particularly in Greater Yangon. The socioeconomic dominance of Yangon Region as a whole is an important factor for its rapid population growth, as people continue to flock to the area for better socioeconomic opportunities, especially in the Greater Yangon.

Based on the official census data (Census Report 2015), Yangon Region as a whole has been growing at an annual population growth rate (APGR) of 2.01% over the past 31 years (1983–2014) (Fig. 9.14). From 1983 to 2014, its population increased by 3.39 million, with an annual increase of almost 110 thousand people. Interestingly, Kachin Region had the highest APGR during the same period with 2.03% (Fig. 9.14). But due to its much lower initial population, its increase from 1983 to 2014, i.e., 0.78 million, was much lower than that of Yangon Region, with only more than 25 thousand people increase per year. Mandalay, the region that houses the country's second largest city and commercial center, Mandalay City,

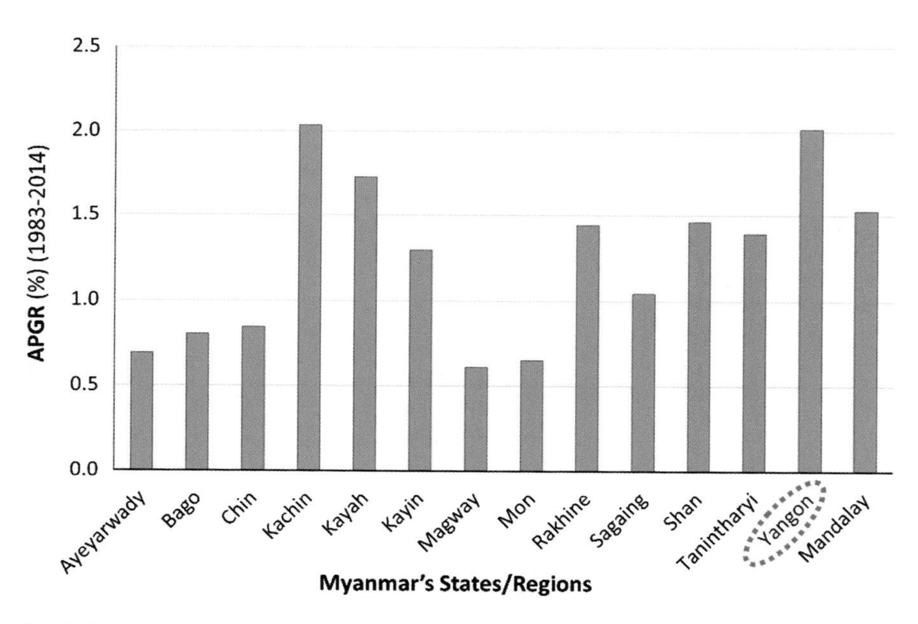

Fig. 9.14 Annual population growth rates (APGRs) of Yangon Region and the other states/regions in Myanmar. *Note* The 2014 population of Naypyitaw, the new capital city which is also considered as one state/region, was included in the 2014 population of Mandalay. *Source* Census Report (2015)

only had an APGR of 1.53% (Fig. 9.14) and a total increase of 2.75 million during the same period.

This population increase seems to have raised the need for various urban services including housing and commercial and business centers, which means that more non-built-up lands had to be converted into built-up. The urban land change analysis from 1993 to 2014 (Sect. 5.3.1) provides some evidence for this proposition. A large and healthy population can provide the manpower needed for the economy: production, distribution, consumption, and resource maintenance. In return, a vibrant and productive economy can create various socioeconomic opportunities for the people. It can also attract more investments and people. Today, Yangon City is considered as Asia's future megacity (Dob 2014). In fact, various major development plans have been crafted to assist the future urban development of the Yangon Metropolitan Area. Some of which are discussed in Sect. 9.5.2.

In general, the geographic location and landscape characteristics of Yangon Metropolitan Area and its population growth and status as the largest commercial center in the country and the home of the country's former capital city are considered to be among the key factors influencing the spatiotemporal patterns of urban land changes (Figs. 9.9, 9.10 and 9.11; Tables 9.1, 9.2 and 9.3) and the overall urban development of the area.

9.5 Implications for Future Sustainable Urban Development

9.5.1 Sustainability, Population Growth, and Urban Land Changes

According to the Sustainable City Agenda of the International Council for Local Environmental Initiatives (ICLEI), "sustainable cities ensure an environmentally, socially, and economically healthy and resilient habitat for existing populations, without compromising the ability of future generations to experience the same" (http://www.iclei.org/activities/our-agendas.html). In other words, a sustainable city must achieve a balance between environmental protection, economic development, and social well-being (Wu 2010; Estoque and Murayama 2014, 2016). In this context, the socioeconomic status of the local people and the environmental quality of Yangon Metropolitan Area, including the extent and quality of its urban green spaces, are important.

Like in many developing and less-developed major cities and metropolitan areas around the world, however, Yangon Metropolitan Area is also faced with various socioeconomic problems, such as poverty (urban poor), poor health care services, unemployment (including gender gap in employment opportunities), lack of support to "persons with disabilities," traffic congestion, and poor waste management, among others (JICA and YCDC 2013). There is also a significant difference in the number of students enrolled at primary schools, middle schools, and high schools in the region, indicating that a large number of students are unable to continue education beyond primary levels (JICA and YCDC 2013).

The region is also vulnerable to natural hazards, such as earthquakes, tropical cyclones, and floods (JICA and YCDC 2013). Deforestation is also a major issue. For example, in 1989, Yangon Region had a total forest cover of 150,000 ha, but in 1998 this figure has decreased to 93,000 ha and further down to 51,000 ha in 2010 (FAO 2014; Raitzer et al. 2015). This shows that during the 1989–2010 period, the region had been losing approximately 4714 ha of forest cover annually. If not addressed, the continuous loss of forest cover in the region will significantly affect the provision of ecosystem services in the area.[3] The Greater Yangon, i.e., the Yangon Metropolitan Area, has also been reported to have three threatened animal species and two threatened plant species (JICA and YCDC 2013).

The implications of the spatiotemporal patterns of built-up expansions in Yangon Metropolitan Area can be further examined using the diffusion-coalescence urban growth theory (Dietzel et al. 2005a,b; Wu et al. 2011; Estoque and Murayama 2015, 2016). The theory suggests that urbanization exhibits a cyclic pattern in time and space driven by two alternating processes: diffusion, in which new urban patches are dispersed from the origin point or seed location, and coalescence, which is the union of individual urban patches, or the growing together of the individual patches into one form or group (Dietzel et al. 2005a, b; Wu et al. 2011; Estoque and Murayama 2015, 2016).

As discussed in Sect. 9.3, the patches of built-up lands in the study area have become more fragmented, as indicated by the overall increase in PLAND and PD between 1989 and 2014. However, built-up patches were less fragmented and more aggregated in 2014 than in 2009. These results provide evidence for the two alternating processes of the urban growth theory mentioned above, i.e., diffusion and coalescence. The increase in PLAND and PD in the earlier periods is an indication that urban development in the area has been undergoing the process of diffusion especially toward the north (Fig. 9.9). The decrease in PD, despite the increase in PLAND, in the later period indicates that urban development has also been undergoing the process of coalescence. The unstable mean ENN values across the 1989–2014 period can be due to the occurrence of the two alternating processes, constantly redefining the average distance between neighboring built-up patches.

The urban land change simulation indicated that built-up lands would continue to expand and undergo the process of aggregation or coalescence in 2020 (Tables 9.3 and 9.6; Figs. 9.10 and 9.12). There would be a slight increase in PD from 2020 to 2030, which suggests that a diffusion process would again follow (Table 9.6). The process of coalescence can result in an infilling growth pattern, whereas the process of continuous diffusion and expansion can result in a sprawl development pattern (Estoque and Murayama 2015). The urban development of

[3]Ecosystem services refer to the benefits that ecosystems (cropland, forest, etc.) generate for and provide to people. Such benefits can be tangible (goods, e.g., cropland for providing food) or intangible (services, e.g., forest for absorbing CO_2), large or small and direct or indirect (MEA 2005; Estoque and Murayama 2013, 2016).

Yangon Metropolitan Area is under the early stage of the two alternating processes, exhibiting some signs of urban sprawl and infilling growth patterns.

While an infilling growth pattern has some potential advantages, e.g., the use of existing infrastructures, the promotion of walkable neighborhoods, and the prevention of the associated external costs of sprawl development, it also has some potential disadvantages, e.g., increased traffic congestion, pollution, limited open space, potential loss of urban green spaces, and crowded services (Estoque and Murayama 2015, 2016). For Yangon Metropolitan Area, infilling pattern poses a threat to its urban green spaces, including forest cover, which are valuable elements for sustainable urban development, especially if such pattern will continue undisrupted. On the other hand, the sprawling pattern detected in the area also has various important implications (e.g., higher urban development costs, greater disturbance or loss of natural habitat, etc.), and thus requires landscape and urban planning in a wider scale.

The loss of forest cover in the region as discussed above can be due to a number of factors, including illegal logging and shifting cultivation in the rural area, as well as land conversion, e.g., from forest to cropland, from forest to residential/commercial land, and from forest to other land uses. But due to a lack of detailed empirical evidence, it was not possible to quantify the direct contribution of the rapid built-up expansion in Yangon Metropolitan Area (Tables 9.1, 9.2 and 9.3; Figs. 9.9, 9.10 and 9.12) to the loss of forest cover in the region. In Yangon City, the more urbanized portion of the region, the observed infilling pattern of urban development may have had an impact on the city's urban green spaces, including the city's urban forests.

Hence, the potential impacts of both sprawl and infilling urban development patterns to the natural environment need to be properly anticipated during landscape and urban planning. There is a need to consider the concept of "sustainable cities," where urban green spaces are kept, improved, restored, or introduced (Estoque 2017).

9.5.2 Current Major Development Plans

Driven by the expected rapid population growth and increasing urbanization pressure (JICA and YCDC 2013; Scobey-Thal 2013; Dob 2014; JICA 2014; MRT and JICA 2014), the government of Myanmar and other concerned bodies have begun planning for the future urban development of the Greater Yangon (i.e., Yangon Metropolitan Area). For instance, in one of the most recent major urban development plans, titled *A Strategic Urban Development Plan of Greater Yangon* (JICA and YCDC 2013), the importance of urban green spaces and the need for landscape and urban development planning in a wider scale have been taken into account (Fig. 9.15).

In the said strategic urban development plan, the slogan "Yangon 2040: The Peaceful and Beloved Yangon—A City of Green and Gold" and its four pillars of development visions are presented (JICA and YCDC 2013). Accordingly, the

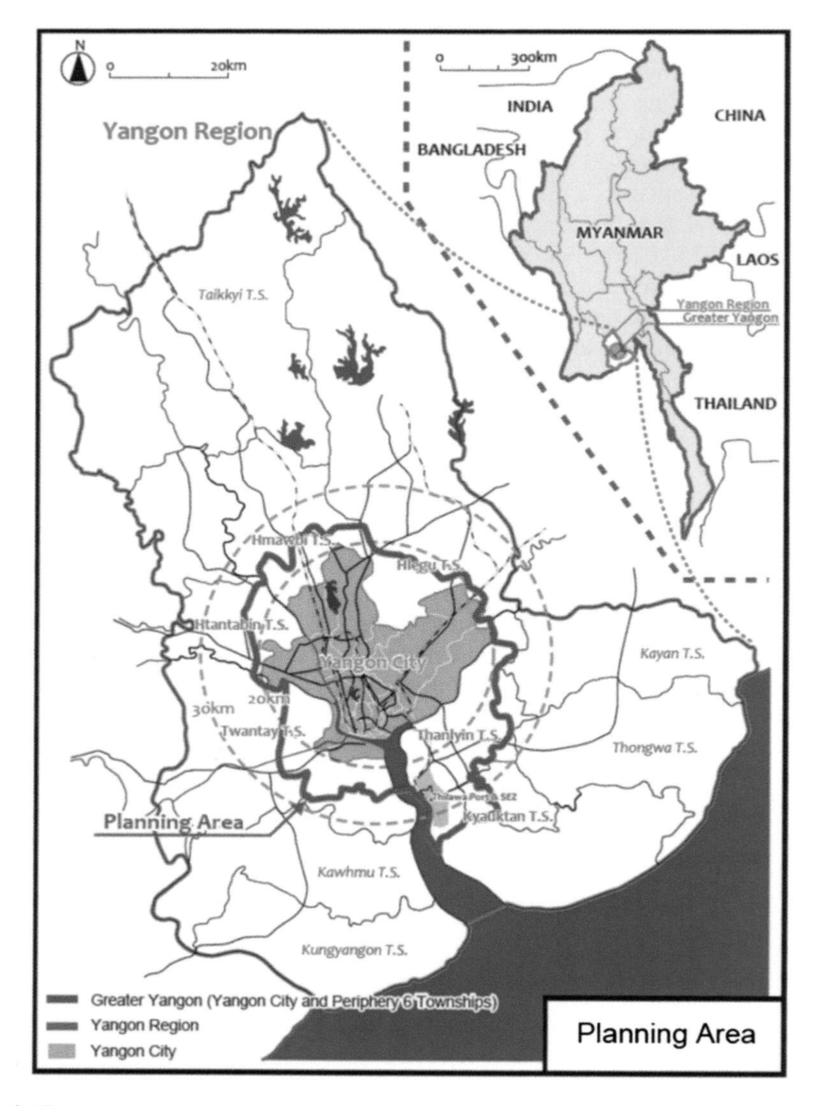

Fig. 9.15 Map showing the planning area for the Greater Yangon. *Source* JICA and YCDC (2013)

slogan expresses the goal of the local people of Yangon toward peace through Myanmar's democratization and their love of Yangon. It also signifies a rich, green natural environment and the gold lighting of Shwedagon Pagoda. The slogan's four pillars of development visions are summarized into four main points: (1) to be an international hub city; (2) to be a comfortable city; (3) to be a well-managed infrastructure city; and (4) to be a city of good-governance (JICA and YCDC 2013).

Another major development plan that is expected to have a major impact on the metropolitan area, socially, economically, and environmentally, is the *Myanmar's National Transport Master Plan: A New Direction* (JICA 2014). The master plan has been formulated based from the national transport vision of Myanmar, i.e., to develop an efficient, modern, safe, and environmentally friendly transportation system in a coordinated and sustainable manner that embraces all transport modes for the benefit of the country and people of Myanmar (JICA 2014). For a more comprehensive urban transport plan for Greater Yangon, a more specific plan called *YUTRA (Yangon Urban Transport Master Plan)*, has also been formulated (MRT and JICA 2014). Specifically, the plan aims to prepare a comprehensive urban transport plan for the Greater Yangon in the short-term (2018), medium-term (2025), and long-term (2035).

All these plans are promising. However, for these plans to be realized, a strong political will among the government leaders, alongside with the full support of the local people, is needed. The local people themselves believe that "there is potential for Yangon again to be one of the most modern, beautiful and liveable cities in Southeast Asia, with modern infrastructure, affordable housing and efficient public transport which conserves its unique cultural heritage" (Oo 2015).

9.6 Conclusions

History shows that Yangon has been among Southeast Asia's important cities, prominent in architecture, culture, and economic development. The analysis showed compelling evidence for the urban primacy of Yangon Region over the other regions/states in the country based on population density and proportion of urban population. Over the past 25 years (1989–2014), the area of built-up lands in the study area has increased more than threefold. The geographic location and landscape characteristics of Yangon Metropolitan Area and its population growth and status as the largest commercial center in the country and the home of the country's former capital city are hypothesized to be among the key factors influencing the spatiotemporal patterns of urban land changes and the overall urban development of the area.

The simulated urban land changes indicated that built-up lands would continue to expand in the future (2014–2030) under the influence of infill and sprawl development patterns. The intensifying pressure of urbanization due to rapid population growth and urban land changes poses many challenges that need to be considered in sustainable urban development and landscape planning. In order to achieve sustainable urbanization and comfortable urban life in Yangon Metropolitan Area, its natural environment, including its valuable urban green spaces such as urban forests and highly productive agricultural areas, needs to be protected and conserved. There is also a need to address the issues on urban poverty, poor healthcare services, unemployment, traffic congestion, and poor waste management. As Yangon has also been dubbed as Asia's future megacity, both the local and national governments, including all sectors of the society, need to push for its sustainable urban development, e.g., by supporting the afore-mentioned current major urban development plans for the area.

References

Census Report (2014) Population and housing census of Myanmar, 2014: Provisional results. Census Report Volume 1. Ministry of Immigration and Population, Myanmar

Census Report (2015) The 2014 Myanmar population and housing census: the Union report. Census Report Volume 2. Ministry of Immigration and Population, Myanmar

Dietzel C, Herold M, Hemphill JJ, Clarke KC (2005a) Spatio-temporal dynamics in California's central valley: empirical links to urban theory. Int J Geogr Inf Sci 19:175–195

Dietzel C, Oguz H, Hemphill JJ, Clarke KC, Gazulis N (2005b) Diffusion and coalescence of the Houston metropolitan area: evidence supporting a new urban theory. Environ Plann B 32:231–236

Dob S (2014) Yangon: challenges and solutions for Asia's future megacity. Accessed on 8 Feb 2016 from http://www.ericsson.com/thinkingahead/the-networked-society-blog/2014/01/17/yangon-challenges-and-solutions-for-asias-future-megacity/

Estoque RC (2017) Manila metropolitan area. In: Murayama Y, Kamusoko C, Yamashita A, Estoque RC (eds) Urban development in Asia and Africa—geospatial analysis of metropolises. Springer Nature, Singapore, pp 85–110

Estoque RC, Murayama Y (2013) Landscape pattern and ecosystem service value changes: Implications for environmental sustainability planning for the rapidly urbanizing summer capital of the Philippines. Landscape Urban Plann 116:60–72

Estoque RC, Murayama Y (2014) Measuring sustainability based upon various perspectives: a case study of a hill station in Southeast Asia. AMBIO: J Human Environ 43:943–956

Estoque RC, Murayama Y (2015) Intensity and spatial pattern of urban land changes in the megacities of Southeast Asia. Land Use Policy 48:213–222

Estoque RC, Murayama Y (2016) Quantifying landscape pattern and ecosystem service value changes in four rapidly urbanizing hill stations of Southeast Asia. Landscape Ecol 31:1481–1507

Estoque RC, Murayama Y (2017) Trends and spatial patterns of urbanization in Asia and Africa: a comparative analysis. In: Murayama Y, Kamusoko C, Yamashita A, Estoque RC (eds) Urban development in Asia and Africa—geospatial analysis of metropolises. Springer Nature, Singapore, pp 393–414

FAO (Food and Agriculture Organization of the United Nations) (2014) Global forest resources assessment 2015. Myanmar. Rome, Country Report

JICA (Japan International Cooperation Agency) (2014) Myanmar's national transport master plan: A new direction. Accessed 8 Feb 2016 from http://www.jica.go.jp/information/seminar/2014/ku57pq00001nep1r-att/kf20140813_01_01.pdf

JICA and YCDC (Japan International Cooperation Agency and Yangon City Development Committee) (2013) A strategic urban development plan of Greater Yangon. Accessed on 8 Feb 2016 from http://www.open/jicareport.jica.go.jp/pdf/12145967.pdf

Kamusoko C (2017) Methodology. In: Murayama Y, Kamusoko C, Yamashita A, Estoque RC (eds) Urban development in Asia and Africa—geospatial analysis of metropolises. Springer Nature, Singapore, pp 11–46

Kim S, Law Marc T (2012) History, institutions, and cities: a view from the Americas. J Reg Sci 52:10–39

MEA (Millennium Ecosystem Assessment) (2005) Ecosystems and human well-being: current state and trends: findings of the condition and trends working group. Island Press, Washington, DC

MRT and JICA (Ministry of Rail Transportation of Myanmar and Japan International Cooperation Agency) (2014) Yangon urban transport master plan of the project for comprehensive urban transport plan of the Greater Yangon (YUTRA). Accessed on 8 Feb 2016 from http://www.jica.go.jp/english/news/field/2014/c8h0vm00008wqgw0-att/YUTRA.pdf

Oo SYM (2015) A vision for the future of Yangon. The Myanmar Times, Myanmar

Raitzer DA, Samson JNG, Nam K-Y (2015) Achieving environmental sustainability in Myanmar. Asian Development Bank, Metro Manila, Philippines

Scobey-Thal (2013) Expecting to boom to 10 million people, Yangon plans for its infrastructural future. Accessed on 8 Feb 2016 from https://nextcity.org/daily/entry/expecting-to-boom-to-10-million-people-yangon-plans-for-its-infrastructural

UN (United Nations) (2013) 2012 Demographic yearbook. United Nations, New York

UN (United Nations) (2015) World urbanization prospects: the 2014 revision. United Nations, New York

UNDP (United Nations Development Programme) (2015) Local governance mapping—the state of local governance: trends in Yangon. Accessed on 8 Feb 2016 from http://www.mm.undp.org/content/dam/myanmar/docs/Publications/PovRedu/Local%20Governance%20Mapping/UNDP_MM_%20State_of_Local_Governance_Yangon.pdf

World Bank (2015a) Population ranking. Accessed on 15 July 2015 from http://data.worldbank.org/data-catalog/Population-ranking-table

World Bank (2015b) Population density (people per sq. km of land area). Accessed on 15 July 2015 from http://data.worldbank.org/indicator/EN.POP.DNST

Wu J (2010) Urban sustainability: an inevitable goal of landscape research. Landscape Ecol 25:1–4

Wu J, Jenerette GD, Buyantuyev A, Redman CL (2011) Quantifying spatiotemporal patterns of urbanization: the case of the two fastest growing metropolitan regions in the United States. Ecol Complex 8:1–8

Chapter 10
Dhaka Metropolitan Area

Syeda Khaleda, Qazi Azizul Mowla and Yuji Murayama

Abstract This chapter traces the origin of Dhaka Metropolis, one of the megacities of Asia, and the capital city of Bangladesh. Urban primacy, urban land use/cover changes as well as the driving forces that influence urbanization were analysed. In addition, the potential implications of urban land use/cover changes on future sustainable urban development were discussed. The urban land use/cover change results revealed that built-up land increased over 10 times (from about 11.6 to 118 km^2) between 1989 and 2014. Rapid urbanization was driven by accessibility (by land, air, and water), the status of Dhaka as the country's capital city, various key legislations, and population and economic growth. The simulated urban land use/cover changes indicated that built-up areas would increase to approximately 169.7 km^2 by 2030. The observed and simulated urban land use/cover changes provide a broad spatial overview, which can be used to understand current and future built-up expansion in Dhaka Metropolis. The results in this chapter can be used to gain useful insights, which can potentially address a number of key urban issues such as high population density, insufficient land for building, the increase in urban poor, traffic congestion, etc. Taking into consideration the current urban growth scenario, both the local and national government need to focus on the sustainable urban development of Dhaka Metropolis.

S. Khaleda (✉)
Department of Disaster Management,
Government of Bangladesh, Dhaka, Bangladesh
e-mail: syedakhaleda@yahoo.com

Q.A. Mowla
Department of Architecture, Bangladesh University of Engineering
and Technology, Dhaka, Bangladesh

Y. Murayama
Faculty of Life and Environmental Sciences,
University of Tsukuba, Tsukuba, Japan

Y. Murayama et al. (eds.), *Urban Development in Asia and Africa*,
The Urban Book Series, DOI 10.1007/978-981-10-3241-7_10

10.1 Origin and Brief History

10.1.1 Context

Dhaka developed on the banks of the River Buriganga. The topography, river courses, and wetlands limited urban development into a north–south orientation (Fig. 10.1). The urban area averages from 5 to 11 km east to west (3–7 miles) and is nearly 50 km north to south (30 miles), totaling to about 400 km^2. The more circular development that would be expected for an inland urban area is precluded by the rivers and wetlands. Dhaka, with 400 years of history as a capital city behind it, is now at a dilemma (Mowla 2013).

In 1765, James Rennel (1792), an English Surveyor, wrote, "the Kingdom of Bengal, particularly the eastern part (Bangladesh), is naturally the most convenient for trade within itself of any country in the world; for the rivers divided into such a number of branches that the people have the convenience of water carriage to and from any principal place." Situated at the center, Dhaka was able to command all these great water routes. This locational advantage gave rise to various urban settlements during various points in history: "there are ruins at Bikrampur, at one time the headquarters of the Sen Dynasty (nine and ten century A.D.), and at Sonargaon, the first capital of the Muhammadens in eastern Bengal; an ancient legend also attached to remains at Rampal, Durduria, Savar and elsewhere" (IGI 1908). Dhaka was the seat of provincial Mughal administration for about one hundred years from 1610 A.D., and later the capital of the newly formed East Bengal-Assam province (1905–11), during the British colonial period. In 1947, Bengal was partitioned between India and Pakistan. Subsequently, Pakistan's portion of Bengal became

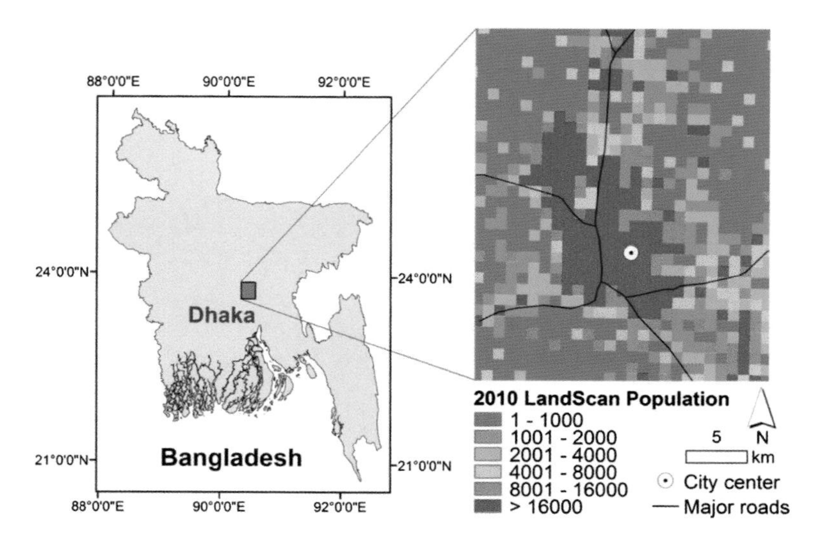

Fig. 10.1 Location and LandScan population of Dhaka Metropolitan Area, Bangladesh

independent in 1971 under the name of Bangladesh. Dhaka, being the major city in the area, was naturally chosen as the capital of the province and then the independent nation. There were political ups and downs, affecting the city's size and morphology. However, Dhaka never ceased to be an important urban center in this region because of its location and geomorphology (Mowla 2012) (Fig. 10.2).

Dhaka started to develop more or less in a fragmented manner after 1947 when it gained regional and political importance. Previously, commercial and residential areas were situated side by side, mostly concentrated along the narrow roads. The old Dhaka still presents this situation with a mixture of commercial, residential, and small industries. After the preparation of a Master Plan of the city in 1958, the commercial center of the city was moved to Motijheel and a high class residential area was developed at Dhanmondi. Housing colonies for government employees, universities, parks, commercial and industrial zones, lakes, and other public facilities were developed gradually to meet the demands of the expanding city. With the development of the city, wide roads and other paved and built-up areas replaced the unpaved areas, natural depressions, and agricultural land. In many cases, natural drainage canals and open water bodies were also filled up for the so-called development works.

10.1.2 Urbanization

The main change that accompanies urbanization is the change in the occupations of the people. People previously engaged in agriculture or cottage industries are seen to shift to modern industrial jobs, commerce, or services. The cities being the hub of these commercial and industrial activities attract more people from the rural areas, causing population to increase. This put pressure on the services that cities provide to its people, causing scarcity which leads to unbalanced distribution, high housing

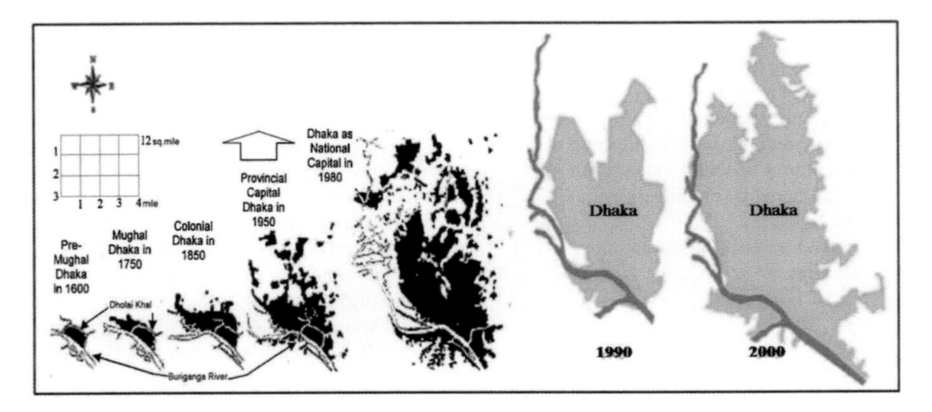

Fig. 10.2 Historic pattern of urban growth in Dhaka. *Source* Mowla (2012)

rents and prices of facilities and finally unhealthy living conditions. Urbanization is defined as a complex process that transforms the rural and natural landscapes into urban and industrial (Antrop 2004). The urbanization of an area has long-term effects on a region and its economy (Mowla 1999).

The most phenomenal urban population growth in Bangladesh occurred during the 1961–1974 inter-census period. Over six million people were living in urban areas consisting roughly 8% of the total population (BBS 1981a). Thus, the percentage increase of the urban population during the 13-year period was striking. The urban population of Bangladesh grew at a much faster rate from 1961–1974 (8.8%) and reached its peak during the period 1974–1981 (10.97%). About 30% of the total increase during 1974–1981 can be explained by the extended definition of urban areas in 1981 (BBS 1981b). That accelerated growth is to a great extent the result of the very recent influx from rural villages. The growth rate of the urban population was 5.4% during the 1981–1991 (BBS 1991). The total urban population increased to 28.6 million by 2001 (BBS 2003).

Migration contributed about 40% to the urban growth in Bangladesh during the 1974–1981 period. The same trend continued in the next 10 years. For some large cities, this share could even be higher, up to 70%, as in the case of Dhaka. Medium term projection shows that urban population will account for 38% of the total population and will exceed 85 million in 2020. This is almost equivalent to the 1981 population of the whole country (ADB 2010). General causes of migration and urbanization are excessive population pressure; lack of productive assets, such as land and other inputs; landlessness and poverty; frequent and severe natural disasters, such as flood, river bank erosion (including climate refugee); absence of law and order; lack of employment and livelihood opportunities; lack of access to basic services such as health, education, water and sanitation; and lack of social and cultural opportunities (applicable for rural rich). The consequences of such migration and urbanization are presented in Table 10.1.

Table 10.1 Consequences of urbanization

Positive	Negative
• Economic benefits: higher productivity, better income etc. • Demographic benefits: lowering of age at marriage, reduction of fertility rate etc. • Sociocultural benefits: modernization. • Political benefits: empowerment, democracy etc. • Improved access to information technology	• Environmental consequences: Water and Air Pollution • Encroachment on productive agricultural land and forest • Extreme pressure on housing, growth of slums and the pressure on urban services • Economic consequences leading to income inequality and poverty, effects of globalization • Social consequences resulting in increased violence and crime, social degradation • Cultural consequences: entry of alien culture, loss of national cultural identity • Habitat of floating people • Political consequences: Criminalization of politics

Source Mowla (2015)

Until 1978, Kolkata was the only megacity in the region where Dhaka was for the first time identified as one of the megacities of the world and second in the region, by the United Nations. It continued to be so, and did not reverse until 2015, as no very radical programs could be implemented for decentralized urbanization and reduction of Dhaka-bound migration flow. Under these conditions, it is expected to outpace Kolkata. Dhaka is on the threshold of launching its structure plan in 2016 for the next 20 years. However, by 2025, the United Nations (2010a) forecasted that Dhaka will reach 23 million, well ahead of Kolkata's projected 19 million (Fig. 10.3).

10.1.3 Planning Endeavors in Dhaka

The first formal planning document for Dhaka was prepared by Patrick Geddes (1917). However, the plan was not implemented. After 1947, several initiatives were taken to plan the provincial capital Dhaka. The most significant planning approaches are as follows: (a) The Master Plan for Dhaka, 1959; (b) Dhaka Metropolitan Area Integrated Urban Development Plan (DMAIUDP), 1981; and (c) Dhaka Metropolitan Development Planning (DMDP), 1995–2015 (also The Task Force Report, 1991). Other major related laws and policies having implications on Dhaka's development are the following: The Building Construction Act, 1952; The Environment Act, 1995; The Wetland Protection Act, 2000; and The Land Use Policies.

(a) The Master Plan for Dhaka, 1959
    The 1959 Dhaka Master plan was prepared for the then Dhaka Improvement Trust (DIT) jurisdiction area. This area had a population of 1,025,000 and covered approximately 220 square miles. The plan assumed a 40% population increase over the next 20-year plan period (1959–1979), with a target figure of 1.47 million. The 1959 Master plan was based on the following physical assumptions:

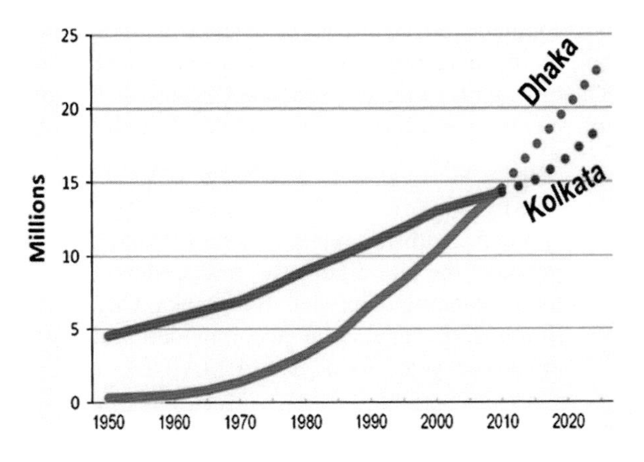

Fig. 10.3 Population comparison 1950–2010 with projection to 2025. *Source* RISE-bd (2013)

- The river Buriganga (a transport artery) will continue to be central to Dhaka's economy.
- No substantial alleviation of the annual flooding will be possible.
- The central part of Dhaka (the old town) will be maintained as the principal business and shopping center.
- The existing population will increase at the rate of 1.75% per annum.
- The south-western loop of the railway through Dhaka will be diverted eastwards, with a new Dhaka station at Kamalapur.
- The part of university will be moved to a new site of about 1000 acres at Faydabad.
- The Dhaka Cantonment will be retained for military use.

Based upon these assumptions, the plan indicated to the higher flood free land for the future expansion toward north. According to the plan, new expansion focused on three locations north of the then urbanized area, namely, Mirpur, Gulshan-Banani, and Tongi-Uttara. These three northern areas together were estimated to accommodate 62% of the projected population within 20 years timeframe. Two assumptions of 1959 Master Plan were subsequently found incorrect. In 1971, the provincial city became the capital of the new nation, Bangladesh and experienced massive migration to the capital.

- Firstly, population growth was projected at 1.75% per annum. Actual growth rate between 1961 and 1974 was approximately 6% per annum.
- Second, it was assumed that the city would continue to grow outward at existing urban densities. In practice it is apparent that the major trend since 1959 was extreme intensification of densities in fewer areas only. The result, in the late 1970s, shows that the development areas proposed in the Master Plan are yet to be developed.
- Above all, the migration factor was not considered in the Master Plan, 1959.

Although the 1959 Master plan was not substantially updated, the plan remained as a framework for development planning by the Dhaka Improvement Trust (DIT). Urban development from the early 1960s to the mid-1970s was loosely regulated by Master plan 1959, which was basically a development control document. In practice, development control, based on the control of building on individual plots, could only be effective in the specific development areas which were planned and leased by DIT. The other parts of the city remained largely uncontrolled, and the lack of updating of the plan left the urban services in the city without a basis for coordinated action.

(b) Dhaka Metropolitan Area Integrated Urban Development Plan (DMAIUDP), 1981

To cope with the huge volume of migrated population in the capital and to manage the city efficiently, besides having sustainable urban development, a new planning approach for Dhaka City was needed to provide an urban framework capable of accommodating industrial and service requirements after the year 2000. The DMAIUDP (1981) did not under estimate the migration factor. Based on the experience of the migration behavior of 1970s and mid-1980s, the DMAIUDP estimated net migration of almost 4 million for the period 1980–2000. The net migration was 61.4% of the total projected population growth. The urban development strategic objectives of the DMAIUDP were the following:

- To ensure that the city is able to perform its essential role efficiently in national development.
- To achieve a balance between urban population and urban employment.
- To provide for the basic needs of the urban population and to secure a more equitable distribution of resources between income groups.

(c) Dhaka Metropolitan Development Planning, 1995–2015 (DMDP 1995–2015) Though the document was based on realistic facts and figures and was quite comprehensive, it was never implemented formally. However, it served as a good resource handbook for the people engaged in the development related fields. Ananya's (2015) study over a period of 13 years from 2000 shows considerable reduction of wetlands and vegetation and excessive increase in built-up area. The Dhaka Metropolitan Development Plan, 1995–2015 was a complete guideline to control and guide the land development of 1528 km^2. Excluding the Dhaka City Corporation North and South (DCC, North and South) area of about 100 km^2, the remaining 1428.9 km^2 (73.3%) was demarcated as nonresidential areas in 1995. About 23.1% was demarcated as agricultural land, 21.1% as flood flow zone, and 7.4% as water body or water retention areas (Table 10.2).

In the last seven years from 2006 to 2013, 21.8% of agricultural land and water bodies have vanished due to the increase in residential area by 32.7% (RDP 2015). Real Estate developers have filled most designated wetlands and flood flow zones of the DMDP 1995. The survey shows that 4.9% of the agricultural land of the RDP 2015 was sand filled—the percentage would be much higher if the DMDP 1995 area is considered (Fig. 10.4).

Table 10.2 Proposed land use of DMDP (1995–2015)

Land use	Area (km^2)	Percentage (%)
Agricultural zone	329.5	23.1
Flood Flow zone	301.8	21.1
Residential zone	320.4	22.4
Government land project	60.5	4.2
Rural settlement zone	145.4	10.2
Water body and water retention areas	106.2	7.4
Others	165.1	11.6
Total	1428.9	100

Source DMDP (1995)

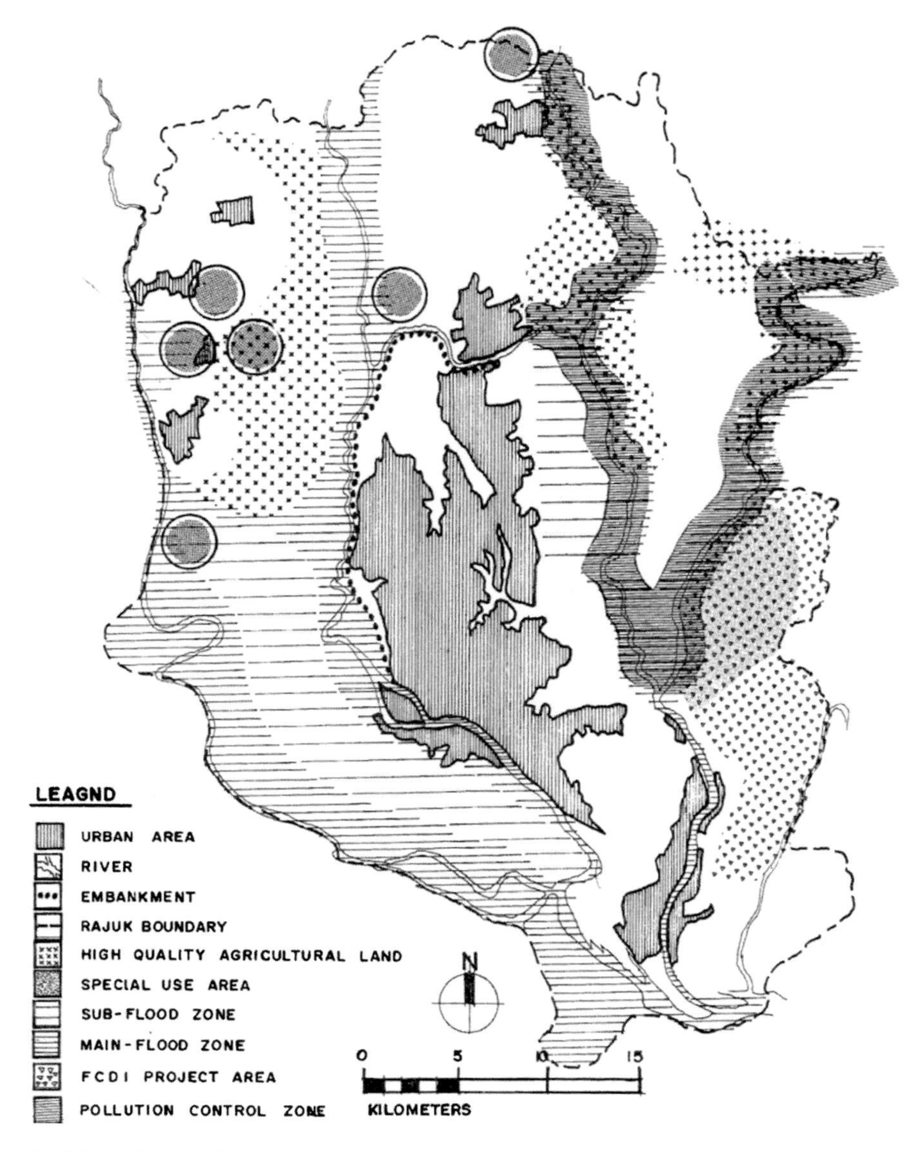

Fig. 10.4 DMDP (1995)—proposed plan policy areas, i.e., built and natural areas for Dhaka. *Source* DMDP (1995), Mowla (2010, 2013)

10.2 Primacy in the National Urban System

A primate city generally dominates over other cities in politics, economy, media, culture, and education. According to Mowla (2013, 2015), unitary governments generally favor the national capital, creating a primate city bias in public services and infrastructure investments, among others. Rent seeking and urban bias by central government bureaucrats result in centralization. This favoritism draws in immigrants. However, according to a World Bank (2007) study, the degree of urban concentration in Bangladesh is higher than many of its comparators with Dhaka's primacy rate being 32%, which is much higher than its optimal (21%) level.

Rapid Urbanization in Dhaka started, not because of its industrialization but due to its primacy, locational advantage, and connectivity with rest of the country. Other factors tried to catch up with increasing population need. However, in the recent years urbanization has been marked by intensified urban pull exerted by the growth of industries and services, particularly export-oriented industries in the large cities, construction, and foreign remittances. The local entrepreneurs have performed an impressive job within a short span of time since their take-off in the early 1990s. The garment industries took off in Dhaka (accommodating more than 75% of such units). Accessibility to buyers, ability to communicate with overseas firms, financial and banking opportunities, and other location advantages drew industrial and other public sector investments to the country's urban areas, particularly the bigger cities, Dhaka being the most privileged. The sectoral approach to economic development has created a regional imbalance in the field of employment and investment, favoring more urbanized regions.

The heart of the problem, be it city governance or overall governance of the country, lies in the country's de facto centralized governance system. Excessive concentration of political and economic powers to a handful of ministries and bureaucratic offices, primarily located in the capital, is making the country a highly centralized state. The development budgets of the Dhaka City Corporations (North and South), for example, are 13 folds larger than that of Chittagong City Corporation, although the latter hosts about one-third population of the former. A number of studies show that the nexus between over expansion of primate cities and economic growth is negative. Economic models suggest that centralization increases urban concentration. Research also shows that when the primate city is the national capital, it is 25% or more, suggesting that resource centralization goes with political centralization. Excessive urban concentration leads to degradation of the quality of life owing to congestion, contamination of ground water and poor air quality, among others.

In 1961, slightly more than 5% of the country's population lived in the urban areas. According to the 1991 census estimate, 22% of the country's population is urban. Correspondingly, the level of urbanization in Bangladesh stands to over 25% by now. Nearly 50% of the national urban population happens to be concentrated in the four metropolitan cities: Dhaka, Chittagong, Khulna, and Rajshahi. This four-city-primacy situation has sharpened in every census. However, Dhaka alone enjoys a clear primate status in its urban structure. In 1961, Dhaka's share of the urban population of Bangladesh was 20%. The share rose to 24% in 1974, 26% in 1981, and 30% in 1991. Dhaka in 2004 with a population of over 10 million people contained 33% of the country's urban population (of about over 30 million). Dhaka's urban population had been growing at an average rate of 4% since independence in 1971, when the national population growth rate was about 2.2%

(World Bank 2007), which means that physically Dhaka had been expanding partly by transforming agricultural land into urban land. Urbanization and urban growth take place through a combination of three components, such as (1) natural increase of the native urban population; (2) area redefinition or reclassification or annexation; and (3) rural–urban (or other forms of internal) migration.

Dhaka is the capital city of Bangladesh, located centrally and accommodating about 10% of the country's population. The current (in 2014) population of Dhaka is 16,982,000 (United Nations 2015). The United Nations projection for Dhaka's population for 2030 is 27 million. According to its estimation, the annual average rate of change has been 3.6% between the year 2010 and 2015. The capital hosts about 33% of the country's urban population and 25% of economic activities, employing 35% of the total urban labor force.

10.3 Urban Land Use/Cover Patterns and Changes (1989–2030)

10.3.1 Observed Changes (1989–2014)

The built-up areas increased by over 10 times between 1989 and 2014. The built up area increased from 11.6 km^2 in 1989 to 118 km^2 in 2014 (Table 10.3) at an annual rate of 4.3 km^2/year. The annual rate of change during the "1989–2000" period was slightly higher (3.5 km^2/year) than during the "2000–2010" period (3.3 km^2/year) (Table 10.4). However, the annual rate of change increased substantially to about 8.8 km^2/year during the "2010–2014." The built-up change rate is similar to the urban changes envisioned by the 1995 DMDP (Fig. 10.4).

Table 10.3 Observed urban land use/cover of Dhaka Metropolitan Area (km^2)

	1989	2000	2010	2014
Built-up	11.59	49.87	82.63	117.99
Non-built-up	666.96	577.43	583.12	563.67
Water	40.27	91.53	53.07	37.17
Total	718.84	718.84	718.84	718.84

Table 10.4 Observed urban land use/cover changes in Dhaka Metropolitan Area (km^2)

	1989–2000	2000–2010	2010–2014
Built-up	38.27	32.76	35.36
Annual rate of change (km^2/year)	*3.48*	*3.28*	*8.84*
Non-built-up	−89.53	5.69	−19.45
Annual rate of change (km^2/year)	*−8.13*	*0.56*	*−4.86*
Water	51.26	−38.45	−15.90
Annual rate of change (km^2/year)	*4.66*	*−3.84*	*−3.97*

Table 10.5 shows all built-up landscape metrics between 1989 and 2014. The percentage of landscape (PLAND) metric measures the proportion of a particular class relative to the whole landscape. The PLAND metric increased from 1.7 to 17.3% in 2014 (Table 10.5). This indicates that built-up areas increased rapidly between 1989 and 2014. The patch density (PD) metric measures fragmentation based on the number of patches of built-up area per unit area (in this case per 100 ha or 1 km^2), in which a patch is based on an 8-cell neighbor rule. The PD increased from 1.1 in 1989 to 1.96 in 2014 (Table 10.5). This indicates that over the years the built-up areas in Dhaka became less fragmented and more aggregated.

The Euclidean nearest neighbor distance (ENN) metric measures dispersion based on the distance of a patch to the nearest neighboring patch of the same class. The mean ENN increased from 134.1 m in 1989 to 151.5 m in 2000 due to the diffusion of new built-up patches. However, the mean ENN decreased to 133.7 m in 2010, and to 120.6 m in 2014, which is attributed to new built-up patches between the old patches (that is, infill development). The same pattern is also observed, particularly the increase in PLAND and PD during the same period (Table 10.5). The decrease in mean ENN from 2000 to 2014 on one hand, and the increase in PLAND and PD on the other suggest that Dhaka Metropolitan's urban form became more compact due to infilling development.

The related circumscribing circle (CIRCLE) metric measures the circularity of patches. The CIRCLE value ranges from 0 for circular or one cell patches to 1 for elongated and linear patches one cell wide. While the mean CIRCLE value slightly increased from 0.39 to 0.41 during the "1989–2014" period (Table 10.5), urban expansion followed rail and road networks and water bodies in the outer portions of Dhaka Metropolitan Area (Fig. 10.5). The shape index (SHAPE) metric, a measure of complexity, has a value of 1 when the patch is square and increases without limit as patch shape becomes more irregular. The mean SHAPE value was relatively stable between 1989 and 2014 (Table 10.5), indicating simple forms or low complexity of the built-up patches.

Figure 10.6 presents all the metrics for the built-up class of Dhaka Metropolis along the gradient of the distance from the city center from 1989 to 2014. The PLAND metric decreases as the distance from the city center increases, indicating that the proportion of built-up land near the city center is higher. Similarly, PD decreases drastically as it approaches 5 km distance from the city center in 1989. However, PD increases slightly as it approaches to 5 km distance in 2010 and 2014, and then decreases with distances gradually. This indicates that there were

Table 10.5 Observed landscape pattern of Dhaka Metropolitan Area

Class-level (built-up) spatial metrics	1989	2000	2010	2014
PLAND (%)	1.71	7.95	12.41	17.31
PD (number per km^2)	1.05	1.62	1.61	1.96
ENN (mean) (m)	134.07	151.45	133.69	120.57
CIRCLE (mean) ($0 \leq$ CIRCLE < 1)	0.39	0.41	0.41	0.41
SHAPE (mean) ($1 \leq$ SHAPE $\leq \infty$)	1.26	1.28	1.28	1.27

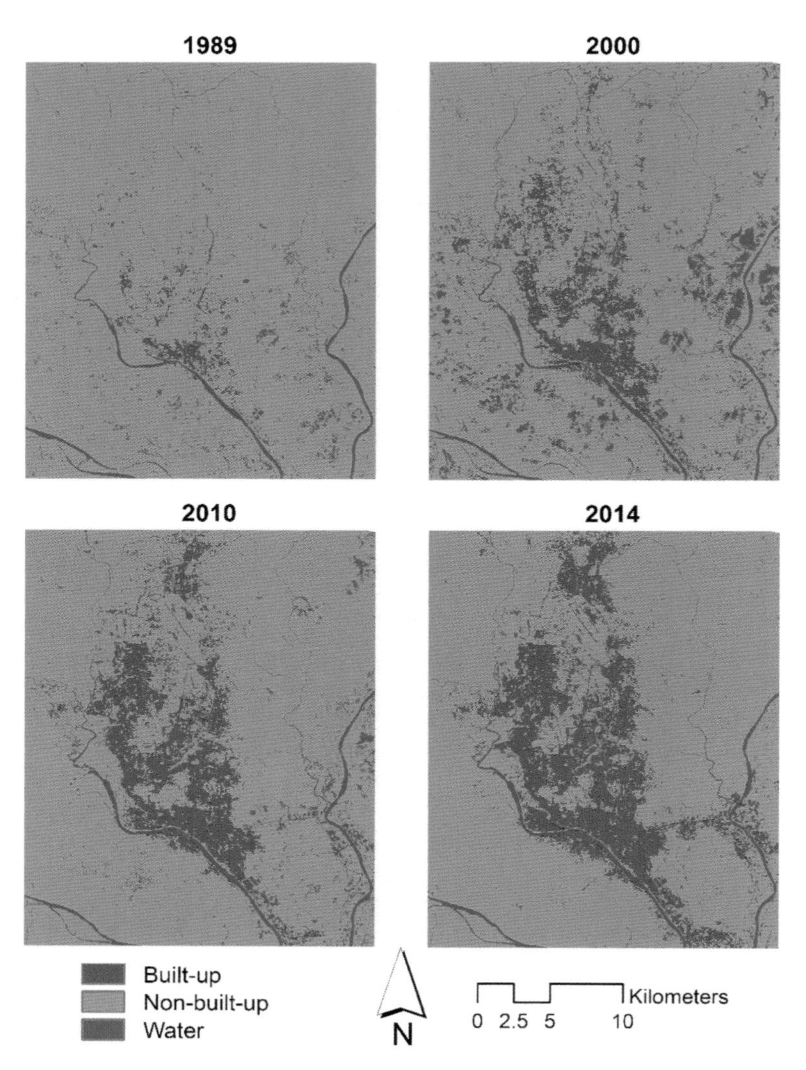

Fig. 10.5 Urban land use/cover maps of Dhaka Metropolitan Area classified from Landsat imagery

more patches of built-up land near the city center. The built-up patches were relatively more dispersed with further distances, as indicated by the increasing trend of mean ENN. The mean CIRCLE value did not vary much along the gradient of the distance from the city center. Similarly, the mean SHAPE value did not vary a lot. This suggests a low complexity of the built-up areas in Dhaka Metropolitan Area. Therefore, the urban form was more uniform and stable.

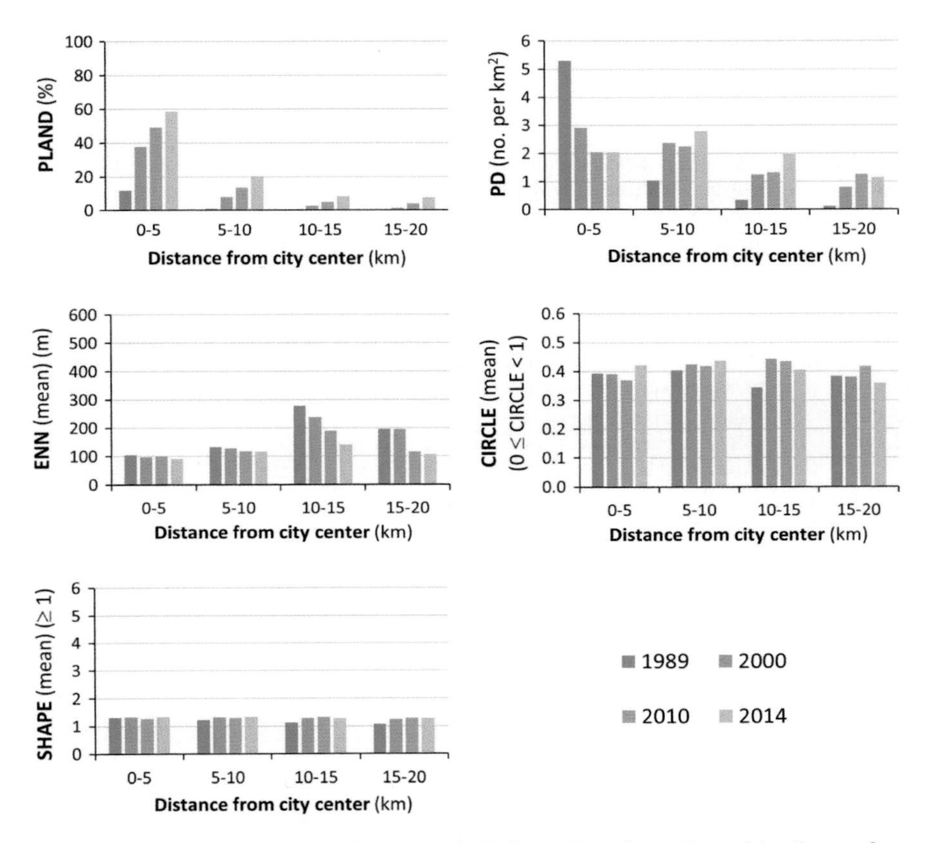

Fig. 10.6 Observed class-level spatial metrics for built-up along the gradient of the distance from city center of Dhaka Metropolitan Area. *Note* The y-axis values are plotted in the same range as those in Fig. 10.10

A study by Dewan and Yamaguchi (2009) for a period of 1960–2005 and another study by Byomkesh, Dewan, and Yamaguchi (2012) for a period between 1975 and 2005 found urban expansion, which resulted in considerable reduction of cultivated land, vegetation, and diminishing water bodies. It is clear that over the years built-up area has increased distinctly. It is also noteworthy that there is an evident decrease in vegetation and low land (Tables 10.6 and 10.7).

Being surrounded by a network of river, flood plains, and wetlands, the topography of Dhaka offers a natural relief to the congested metropolis of Dhaka. Filling up of the water bodies poses two prong problems of unsuitability for construction and reduced water retention capacity besides diminishing public places and increasing the urban water logging. Existing man-made drainage and sewerage system in Dhaka is not adequate for a growing metropolis. Dhaka's topography

Table 10.6 Trend of land cover typology and change (%)

Land Cover Type	Description of Land Cover	1999	2009	Change
Built-up Area	All residential, commercial and industrial areas, villages, settlements, and transportation infrastructure	24.35	37.44	58.32
Water body	River, permanent open water, lakes, ponds, canals, and reservoirs	11.46	13.54	9.26
Vegetation	Trees, shrub lands and semi natural vegetation: including gardens, parks, playgrounds, etc.	29.19	22.10	−31.60
Low Land	Permanent or seasonal wetlands, low-lying areas, marshy land, and all cultivated areas	14.12	9.13	−22.21
Fallow Land	Fallow land, abandoned sites, earth and sand in fills, construction sites solid waste landfills and exposed soils	20.88	17.97	−13.77

Source BBS (2009), Mowla (2015)

Table 10.7 Change of major land use pattern in Dhaka (%)

Source	Agriculture	Residential	Water body	Circulation
DAP survey 2006	54.80	28.10	9.57	1.82
RDP survey 2013	42.49	37.36	7.84	2.44

Source RAJUK (2015), Mowla (2015), DAP (2006)

provided a natural drainage system toward its peripheral river and wetland systems, which was never brought into the planning framework (Mowla 2010). Many natural ponds which served as the runoff reservoir and meeting the domestic water needs were filled up to create areas for housing and road. Being one of the megacities of the world and without any hydrological planning, remaining surface water is extensively being polluted and threatened for destruction (Mowla 2013). The water bodies and green spaces preserved for city's breathing place are fast diminishing due to the encroachments, introduction of new roads, business activities, and housing development (Fig. 10.7).

10.3.2 Projected Changes (2014–2030)

The simulated urban land use/cover changes show that built-up areas would increase from 118 km^2 in 2014 to 169.7 km^2 by 2030 (Tables 10.8 and 10.9; Figs. 10.8 and 10.9). The landscape pattern metrics show that built-up patches in 2020 and 2030 would be disaggregated, as indicated by the increase in PLAND and decrease in PD (Table 10.10). The increase in mean ENN also indicates that more neighborhood patches would become isolated. The increase in mean CIRCLE and

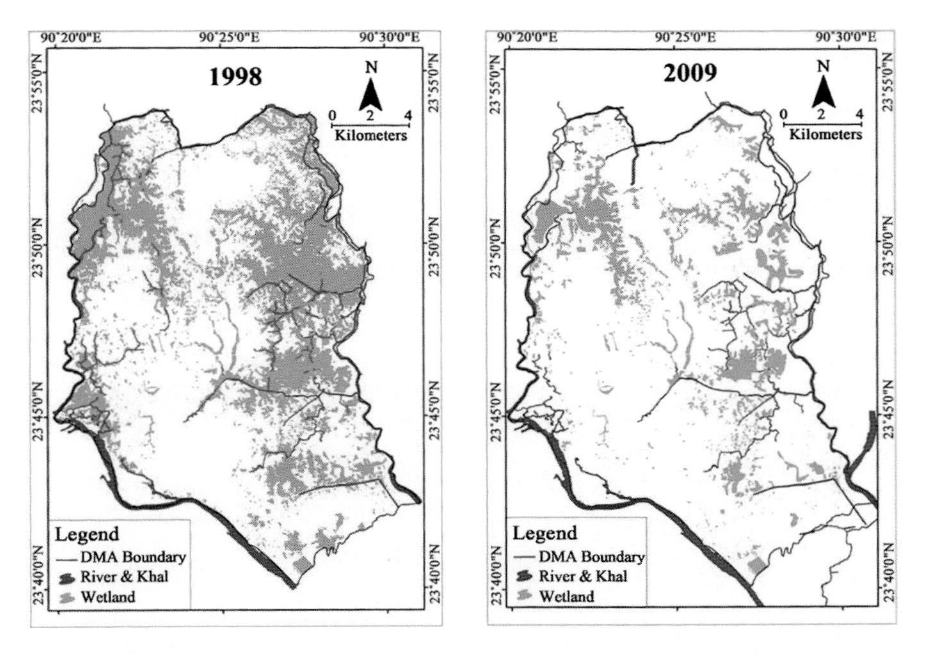

Fig. 10.7 Wet land scenario of Dhaka Metropolis from 1998 to 2009. *Source* Mowla (2010, 2013)

Table 10.8 Projected urban land use/cover of Dhaka Metropolitan Area (km^2)

	2020	2030
Built-up	141.15	169.69
Non-built-up	540.51	511.97
Water	37.17	37.17
Total	718.84	718.84

SHAPE values indicates more connected, elongated/linear, and relatively more complex patches of built-up land, respectively.

Figure 10.10 indicates that PLAND (i.e., the percentage of patches of built-up areas) would also be higher in closer proximities to the city center in 2020 and 2030. The PD would decrease dramatically by 2020 and 2030. However, PD would still be relatively higher in middle distances. The mean ENN would increase highly, especially in distances of 10–15 km by 2030. The mean CIRCLE value would also increase by 2020 and 2030, especially within the 5–10 km distance buffer zones. The mean SHAPE value would also increase and be more complex. However, the SHAPE metric would be relatively more uniform or stable along the gradient of the distance to city center (Fig. 10.10).

Table 10.9 Projected urban land use/cover changes in Dhaka Metropolitan Area (km^2)

	2014–2020	2020–2030
Built-up	23.15	28.54
Annual rate of change (km^2/year)	*3.85*	*2.85*
Non-built-up	−23.15	−28.54
Annual rate of change (km^2/year)	*−3.85*	*−2.85*
Water	0.00	0.00
Annual rate of change (km^2/year)	*0.00*	*0.00*

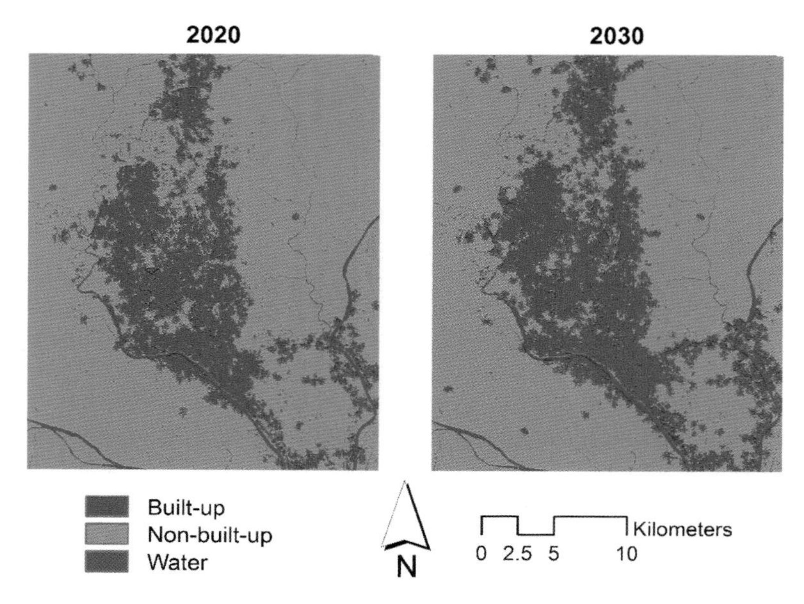

Fig. 10.8 Projected urban land use/cover maps of Dhaka Metropolitan Area

10.4 Driving Forces of Urban Development

Dhaka has grown from a small settlement within the confines of River Buriganga and the Dholai *Khal* (canal) to a sprawling metropolis of about 16 million people. Dhaka is encircled by Buriganga River on its south and south west, Turag River on its west-northwest and Balu River on its east connected to Turag River by Tongi to the North. The spatial development followed the prong of flood free terrace originating from the old nucleus along Buriganga River towards north as a part of Madhupur terrace (Dhaka Terrace) of pre-ostacian age. The Dhaka Terrace sloped toward eastern and western flood plains, marshes, and rivers. Water bodies and rivers have historically played an important role in the spatial development, life and liveability of Dhaka. Population growth due to natural increase and rural–urban

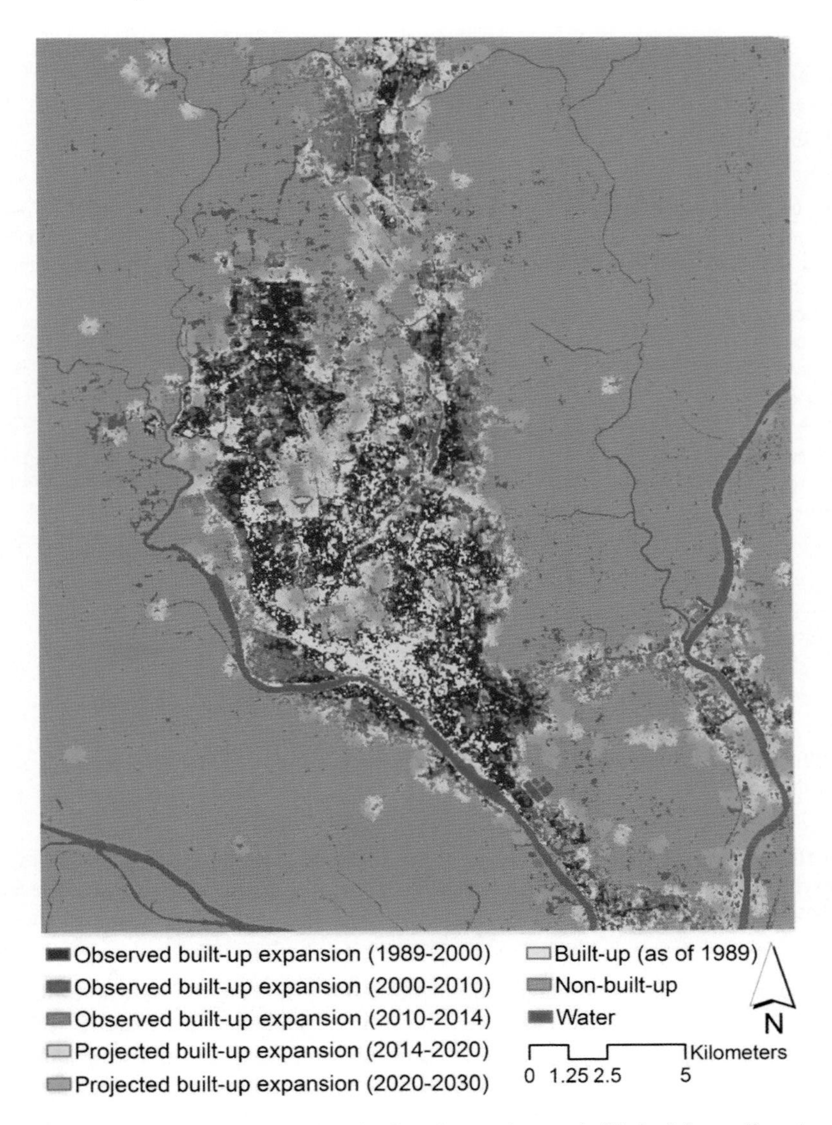

Fig. 10.9 Observed and projected urban land use/cover changes in Dhaka Metropolitan Area

Table 10.10 Projected landscape pattern of Dhaka Metropolitan Area

Class-level (built-up) spatial metrics	2020	2030
PLAND (%)	20.71	24.89
PD (number per km^2)	0.07	0.08
ENN (mean) (m)	386.35	533.19
CIRCLE (mean) ($0 \leq$ CIRCLE < 1)	0.67	0.66
SHAPE (mean) ($1 \leq$ SHAPE $\leq \infty$)	6.39	5.99

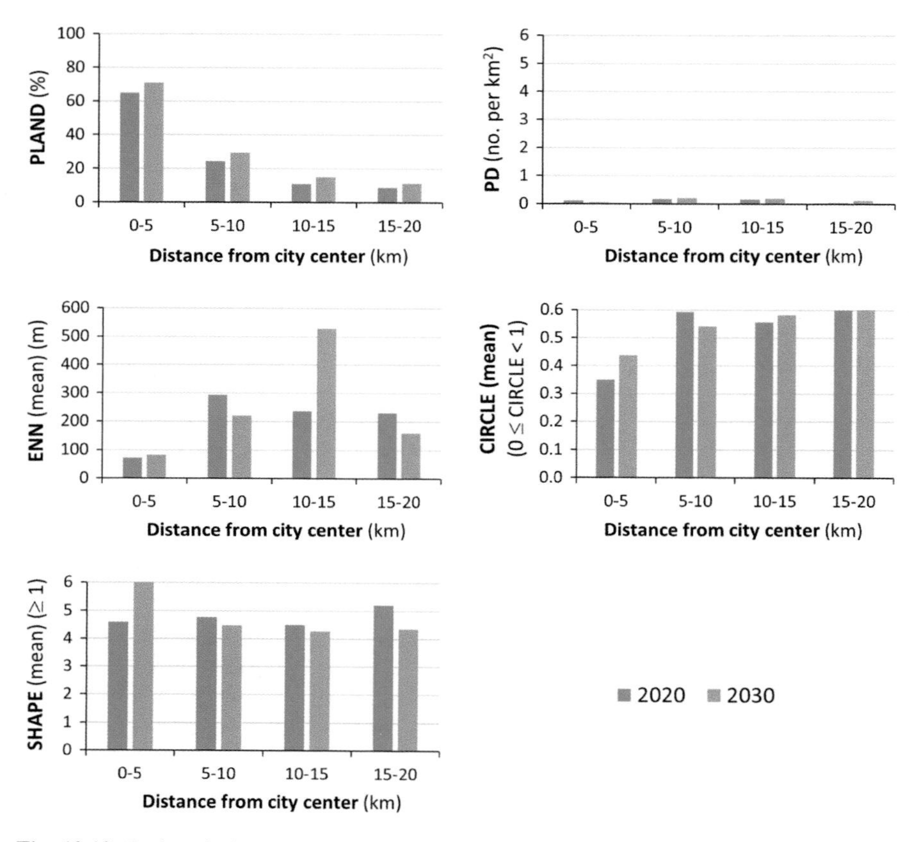

Fig. 10.10 Projected class-level spatial metrics for built-up along the gradient of the distance from city center of Dhaka Metropolitan Area. *Note* The y-axis values are plotted in the same range as those in Fig. 10.6

migration caused rapid urbanization. Dhaka attracts rural migrants where the vast majority of productive activities are concentrated and the vast majority of paid employment opportunities are located. Recent economic and industrial development, particularly the garments sector development (since 1980s), increased the demand for labor, especially females who get higher social status and freedom.

Urbanization, without considering the geomorphology of Dhaka during recent times, has left a deep scar in the city's environment. It needs some strategic decisions and quick actions to remain liveable. Water logging, pollution, changes in hydro-geological system, land subsidence, and building collapse are some of the severe consequences of these environmental interventions or changes. Easy, cheap, and fast intercity transportation and communication relieve pressure on the center. Investment in modern intercity transport and communication together with decentralization of uses and development of satellite towns may prompt urban de-concentration as manufacturers or other users could locate their establishments in the hinterlands or satellites. To reduce in-migration, externalities such as congestion or pollution arising from higher population density should be priced (through tax and other measures). This could lead to de-congestion. The lack of pricing or ineffective regulation means that immigration into the capital city is underpriced or subsidized or in fact encouraged.

10.5 Implications for Future Sustainable Urban Development

Dhaka City is one of the fastest growing megacities in the world. This rapid increase has occurred since the independence in 1971. The urban land use/cover change analysis shows that built-up area increased by over 10 times between 1989 and 2014. Future urban land use/cover changes are projected to increase by over 1.4 times between 2015 and 2030. The built-up areas would be highly concentrated near the city center, i.e. in Motijheel, Dhanmondi, and Shantinagar. In the future, built-up area is expected to expand toward the north following the water bodies, national railway, and Dhaka-Mymensingh highway. Urban development would mainly be concentrated in Mirpur and Uttara occupying areas within the 10–15 km and 15–20 km from the city center. This excludes the in-between area where the cantonment and international airport are situated. The simulated landscape metric results indicated that built-up areas in Mirpur would become isolated in the future. However, the simulated landscape metrics indicated that built-up areas in Uttara would become more connected and relatively more complex.

Studies show that urban populations in Bangladesh have grown from 5% in 1971 to 27% in 2008, suggesting that approximately 41 million people are currently living in urban areas (UNICEF Bangladesh 2010). The huge population growth occurred due to rapid and uncontrolled urbanization, which in turn gave rise to poverty. Dhaka has been burdened by many challenges such as unplanned urbanization, extensive urban poverty, water logging, growth of urban slums and squatters, traffic jam, environmental pollution, and other socioeconomic problems.

More than 60% of the urban population is concentrated in four major metropolitan cities. According to United Nations (2010b) Dhaka itself absorbs more than 40% of the entire urban population of Bangladesh. It is estimated that more than 50% of the population would be living in urban areas by 2025, which means that about 58 million would be living in Dhaka. In Bangladesh, urban population increased from 19.8% in 1990 to 28.1% in 2010 and projected to 33.9% in 2020 (United Nations 2010b). The UN-ESCAP (1993) however, projected a base line of 16.4% from 1990 to 38.2% in 2020, which has come to be more realistic. It is assumed that the 2010 study considered the Dhaka City Corporation area, while the 1993 study was comprised of Dhaka Metropolitan areas. It is estimated that the trend of urbanization in its present form will continue, as such, the development is envisioned to be a guided one. The objective of the next structure plan is to accelerate the primacy of Dhaka and Chittagong in a proportionate manner. Each city would have well knitted satellites in close proximity around them. Dhaka would be primate city as before.

10.6 Concluding Remarks

The urbanization of Bangladesh is interlinked with the intense development of Dhaka City. In particular, geomorphology guided the physical expansion of Dhaka. The historical process of urban development in Dhaka City presents different trends based on political development. Dhaka developed as a political and administrative city. Furthermore, the concentration of economic and commercial activities resulted in Dhaka becoming the primate city of the country. The urbanization activities in Dhaka City have been attracting tremendous growth, catering for the needs of the country's capital. Overall, Dhaka City has experienced its highest rate of physical and population growth in recent decades, transforming it into a megacity.

While the urban growth of Dhaka Metropolis has been remarkable, there are a number of key urban issues that need to be taken into consideration for future urban development. Examples of these key urban issues are the increase in urban poor, congestion, disaster preparedness, etc. The geospatial analysis results in this chapter show that the built-up areas increased rapidly, and are projected to further increase in the future. Guided decentralization at the regional level as well as compact development should be the urban policy for Dhaka. In order to overcome the key issues outlined above, local and national government as well as other sectors of the society, need to implement sustainable urban development plans for the Dhaka Metropolis. If such sustainable urban development plans are implemented effectively, this primate city will have better prospects of becoming a world-class city.

References

ADB (2010) City cluster economic development—Bangladesh Case Study, report prepared by ADB; AusAID and CUS. Dhaka for Asian Development Bank, Asia

Ananya TH (2015) Land cover change detection of dhaka metropolitan during 2000–2013. In CUS bulletin on urbanization and development, no. 68, Jan–June 2015, Dhaka

Antrop M (2004) Landscape change and the urbanization process in Europe. Landscape Urban Plan 67:9–26

BBS (1981a) Bangladesh population census 1981, report on urban area: national series. Bangladesh Bureau of Statistics, Ministry of Planning, Dhaka

BBS (1981b) Bangladesh population census 1981: analytical findings and national tables. Bangladesh Bureau of Statistics, Ministry of Planning, Dhaka

BBS (1991) Bangladesh population census 1991 urban area report. Bangladesh Bureau of Statistics, Ministry of Planning, Dhaka

BBS (2003) Population census 2001, national report (Provisional). Bangladesh Bureau of Statistics, Ministry of Planning, Dhaka

BBS (2009) Population census 2001, national report. Bangladesh Bureau of Statistics, Ministry of Planning, Dhaka
Byomkesh T, Dewan A, Yamaguchi N (2012) Urbanization and green space dynamics in greater Dhaka, Bangladesh. Landscape Ecol Eng 8:45–58
DAP (2006) Detail area plan. RAJUK, Dhaka
Dewan A, Yamaguchi N (2009) Using remote sensing and GIS to detect and monitor land use and land cover change in Dhaka metropolitan of Bangladesh during 1960–2005. Environ Monit Assess 150:237–249
DMDP (1995) Dhaka metropolitan development plan, 1995–2015. RAJUK, Dhaka
DMAIUDP (1981) Dhaka metropolitan area integrated urban development plan (DMAIUDP-81). RAJUK, Dhaka
Geddes Patrick (1917) Report on town planning. Bengal Secretariat Book Depot, Dhaka, Calcutta
IGI (1908) Imperial Gazetteer of India. XI Oxford at Clarendon press
Rennel J (1792) Memoir of a map of Hindustan. 2nd ed, London
Mowla QA (1999) Contemporary morphology of Dhaka: lessons from the context. Oriental Geogr 43:51–66
Mowla QA (2010) Role of water bodies in Dhaka for sustainable urban design. Jahangirnagar Plan Rev 8:1–9
Mowla QA (2012) Dhaka: a mega city of persistence and change. In: Misra RP (ed) Urbanization in South Asia—focus on mega cities. Cambridge University Press, New Delhi
Mowla QA (2013) Dwindling urban water-bodies of Dhaka and the city fabric: a post-mortem. In: Jahan S, Kalam AKMA (eds) Dhaka metropolitan development area and its planning problems, issues and policies. BIP Publication, Dhaka
Mowla QA (2015) Review of Dhaka structure plan 2016–2035. Official Report Submitted to RAJUK
RAJUK (2015) Dhaka structure plan, 2016–2035 (Draft-Consultation Copy), Dhaka
RDP (2015) Regional development planning—Dhaka structure plan, 2016–2036. RAJUK, Dhaka
RISE-bd (2013) Minimum wage—an insight, research initiative for social equity. Accessed on 20 Sept 2015 from http://risebd.com/2013/
Bangladesh UNICEF (2010) Understanding urban inequalities: a prerequisite for achieving vision 2011. UNICEF Bangladesh, Dhaka
UN-ESCAP (1993) The state of urbanization in Asia and the pacific. New York
United Nations (2010a) World urbanization prospects 2010. The Economic and Social Affairs, United Nations, New York
United Nations (2010b) The state of Asian cities 2010/11. UN-HABITAT, UN-ESCAP, UNEP and UCLG-ASPAC
United Nations (2015) World urbanization prospects 2014. The Economic and Social Affairs, United Nations, New York
World Bank (2007) Development and the next generation. World Bank, Washington, DC

Chapter 11
Kathmandu Metropolitan Area

Rajesh Bahadur Thapa

Abstract This chapter analyzes the origin and brief history of the Kathmandu Metropolitan Area (KMA), the national capital of Nepal, which forms the core of the nation's most populous metropolitan area. It examines the urban primacy, urban land use/cover change patterns, and driving forces that influence the rapid urbanization of the KMA. In addition, it discusses the prospective implications of these elements for the future sustainable urban development of the metropolitan area. The KMA has been important economically, administratively, and politically for hundreds of years. During the past 25 years (1989–2014), the KMA has experienced tremendous growth that is expected to continue through 2030. These results suggest that the current urban development process is in a critical stage in which urban and fringe frontier areas will create unprecedented stress on land resources that will be manifested in river and forest ecosystems and other environmentally sensitive areas. These changes are driven by various interrelated physiographic as well as socioeconomic factors. Similar to many developing cities, the KMA has issues of poor management of urban expansion and infrastructure as well as disaster preparedness, resulting in environmental and socioeconomic consequences. However, possibilities are available for improving the urban environment and managing the potential land demands in the metropolitan area through the strict enforcement of sustainable urban development policies and changes in the current urbanization trend. The Gorkha earthquake on April 25, 2015, in Kathmandu has afforded an opportunity to revitalize the city. If such improvement measures are implemented, living conditions will be improved, enabling the KMA to become a world-class city.

R.B. Thapa (✉)
Geospatial Solutions, International Centre for Integrated Mountain Development, Khumaltar, Lalitpur, Nepal
e-mail: thaparb@gmail.com

R.B. Thapa
Earth Observation Research Center, Japan Aerospace Exploration Agency (JAXA), 2-1-1 Sengen, Tsukuba City, Ibaraki 305-8505, Japan

Y. Murayama et al. (eds.), *Urban Development in Asia and Africa*,
The Urban Book Series, DOI 10.1007/978-981-10-3241-7_11

11.1 Origin and Brief History

The Kathmandu Metropolitan Area (KMA) is the capital and main political center of Nepal (Fig. 11.1). The area is situated on a bowl-shaped valley floor and includes one of the oldest human settlements in the central Himalayas. With a history and culture dating back 2000 years, the city has evolved from lakebed to paddy agriculture to urban society today. The agricultural landscape has transformed dramatically since the 1960s, into an urban form stretching across the valley, driven by the vehicular arteries and migration into the capital (Thapa et al. 2008). As of census 2011, the KMA is composed of five prominent urban centers including Kathmandu Metropolitan City, Lalitpur Sub-Metropolitan City, Bhaktapur Municipality, Kirtipur Municipality, and Madhyapur Thimi Municipality.

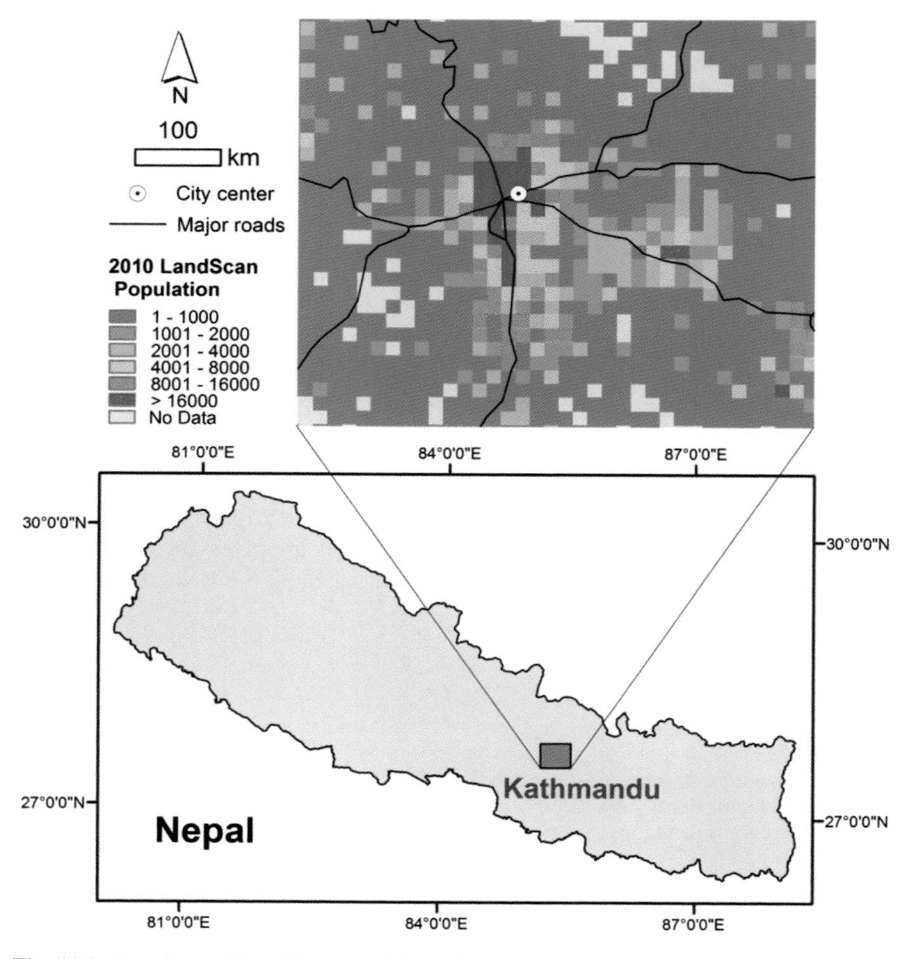

Fig. 11.1 Location and LandScan population of Kathmandu Metropolitan Area, Nepal

The historical development of the KMA is closely associated with the origin of Kathmandu valley. Historical chronicles reveal that this valley was originally a huge lake that was drained by Manjushree, a Chinese saint, who cut a passage through the ridge at the Chobar Gorge, south of Kathmandu city. The valley was inhabited after the water was drained. Gradually, settlements were agglomerated into a town between the Swayambhu and Gujeswari areas (Ranjitkar 1983; Jha 1996). The town was governed by a cow herder named Gopalbansi from 900 to 700 B.C., by a buffalo herder named Mahisapalas from 700 to 625 B.C., and by the Kirati, the indigenous people of the valley, from 625 B.C. to A.D. 100 (Regmi 1999). Subsequently, the Lichavi migrated from the Gangetic plains of Northern Bihar, developed the valley, and ruled until the twelfth century. An Indo-Tibet trade route was opened in the Lichavi era in the seventh century.

Mallas from far western Nepal invaded the Lichavi kingdom and ruled the valley from A.D. 1257 to 1768. King Jayasthiti Malla (1380–1395), the most popular king of this period, was known as a great politician, reformer, and judicious administrator. He introduced the Hindu-based caste system and classified society into the various castes, a system still in use today. He classified the land into four categories: Abbal (highly productive), Doyam (medium productive), Sim (productive), and Chahar (less productive). In later years of the Malla era, the Kathmandu valley was divided into three small kingdoms (i.e., Kathmandu, Patan, and Bhaktapur), which led to the end of the Malla rule. The Malla period, which lasted until the eighteenth century, is marked by extraordinary achievements in urban planning, architecture, arts/crafts, infrastructure, and development of sociocultural institutions for urban management. The urban structure of the Kathmandu valley consists of the three principal capital cities: Durbar Squares of Kathmandu, Patan, and Bhaktapur (Fig. 11.2). Secondary cities, including Thimi and Kirtipur, and satellite settlements of Deopatan, Chabahil, Naxal, Bungamati, Harisiddhi, and Panauti show evidence of urban development during the Malla period (Shah 2003).

Prithivi Narayan Shah, the king of Gorkha, conquered the three cities of Kathmandu, Patan, and Bhaktapur in 1768 and declared Kathmandu as the capital of unified Nepal. Afterward, the city was developed as the main political and administrative center of the country. In 1846, Jung Bahadur Kunwar came into power and founded the Rana reign (1846–1950). Rana rulers built large palaces on prime agriculture land at the city's periphery. New settlements began to develop around the palaces, which eventually led to the suburbanization process in Kathmandu. This was known as the starting era of productive agriculture land encroachment in the valley. A democratic movement emerged in the later years of the Rana reign, which eventually ended the rule in 1950.

Construction of the Tribhuvan highway linking the region to India in the 1950s and Araniko highway linking the region to China in the 1960s (DoR 2004) widened the commercialization and external influences in the valley. These roads provided access for the citizens residing in outer valley regions to migrate into Kathmandu. An international airport in the country constructed in 1949, which began operation in some Indian cities in the late 1950s, helped to attract wider communities around the world into Kathmandu. Along with the establishment of infrastructure and easy

Fig. 11.2 Durbar squares that developed in the Malla period including Kathmandu, Patan, and Bhaktapur. *Source* Patan (S. Ale's fieldwork, 2015) and all others (Author's fieldwork, 2014)

access to the valley, the earlier settlement area changed from paddy agricultural land to a modern urban region. The agricultural landscape transformed intensely since the 1960s into an urban form stretching across the valley that was driven by migration into the capital and the development of vehicular arteries (Thapa and Murayama 2010).

Reflecting the long history of the KMA, the region has a great variety of cultural heritage sites including historic settlements, palaces, monuments, religious sites such as temples and monasteries, historic ponds, taps, and finely embellished public wells. The historic arts and crafts of six of the ten United Nations Educational, Scientific and Cultural Organization (UNESCO) World Heritage Sites in Nepal, including Pashupatinath, Boudhanath, and Swayambhunath temples, and the Durbar Squares of Basantapur, Patan, and Bhaktapur hold particular significance. Along with numerous other historical and religious monuments and unique festivals of national and international importance, these sites account for the influx of tourists in the city and eventually contribute a significant part of the national gross domestic product (GDP). However, the devastating "Gorkha" earthquake on April 25, 2015, ruined many heritage sites, the recovery from which may take many years.

11.2 Primacy in the National Urban System

11.2.1 Nepal, Urban System, and Urbanization

Nepal is in South Asia and shares border with China in the North and India in the other directions and covers a total area of 147,181 km^2. The land begins at 70 m above sea level from Kanchan Kalan in Tarai in the southern belt and passes through a hilly region of the Himalayas in the northern belt including Mount Everest, the world's highest point, at 8848 m. As of 2014, Nepal is divided administratively into 5 development regions, 14 zones, 75 districts, 189 urban centers (including metropolitan cities, submetropolitan cities, and municipalities), and more than 3276 villages.

The first official census in 1952/54 reported that only ten settlements in Nepal met the criteria to be considered urban centers, five of which were in Kathmandu valley. Data on urbanization have become available since that time. Frequent changes in the definitions of urban political territories and the incorporation of new ones in Nepal often complicate the study of urban areas. In the 1970s, a regional development strategy was adopted, and a series of north–south growth corridors and growth centers were identified to concentrate development efforts for achieving full economic development and to encourage agglomeration of the economy. These measures resulted in important contributions to urbanization (Thapa 2009).

According to the census in 2011, 17.1% of the 26.5 million population of Nepal lives in urban areas (CBS 2013). The urban population growth accelerated with an annual growth rate of 5.9% in the 1980s. In the 1990s, the 6.6% annual rate of urbanization was among the highest in the Asia Pacific Region (ADB/ICIMOD 2006). This rate is higher than that in Sri Lanka (2.2%), India (2.9%), Pakistan (4.4%), Bangladesh (5.3%), and Cambodia (6.2%) (Portnov et al. 2007). The number of urban places in 1981 in Nepal was 23 and increased to 58 in 2001. The high growth rate of the urban population is attributed to natural growth, migration, the designation of additional municipal towns, and the expansion of municipal towns amalgamating rural areas (Thapa and Murayama 2009). However, the annual urbanization rate fell 3.4% during 2001–2011 (MUD 2015). During that period, the country experienced political instability and little attention was given to the ongoing rural–urban agglomeration, causing a number of urban areas and their boundaries to remain unchanged. Nevertheless, 131 additional urban municipalities were added in 2014, accounting for 38.26% of the urban population in Nepal. More than half of the urban population in Nepal resides in the hills; the remaining population resides in the Tarai region. Physiographic and connectivity characteristics are also important determinants of urbanization in Nepal, resulting in wide variation in the regional levels of urbanization. The level of urbanization in Kathmandu valley and Pokhara valley accounts for 96.97 and 79.52%, respectively; the level is only 18.28% for the rest of the hills. Similarly, the inner Tarai valleys have 41.97% urbanization compared with 38.94% for the rest of Tarai (MUD 2015).

11.2.2 Primacy of the Kathmandu Metropolitan Area

The KMA is the most populous urban area in Nepal (Fig. 11.3), followed by Pokhara and Biratnagar urban areas in second and third places, respectively. A large difference in population size between the KMA and the other large urban areas can be observed. For example, the KMA is 5.4, 7.0, 9.7, and 10.3 times larger than Pokhara, Biratnagar, Bharatpur and Birgunj, respectively. During the last decade, the KMA experienced an average annual growth rate of 4.5% while Pokhara and Bharatpur were more than 5% (MUD 2015).

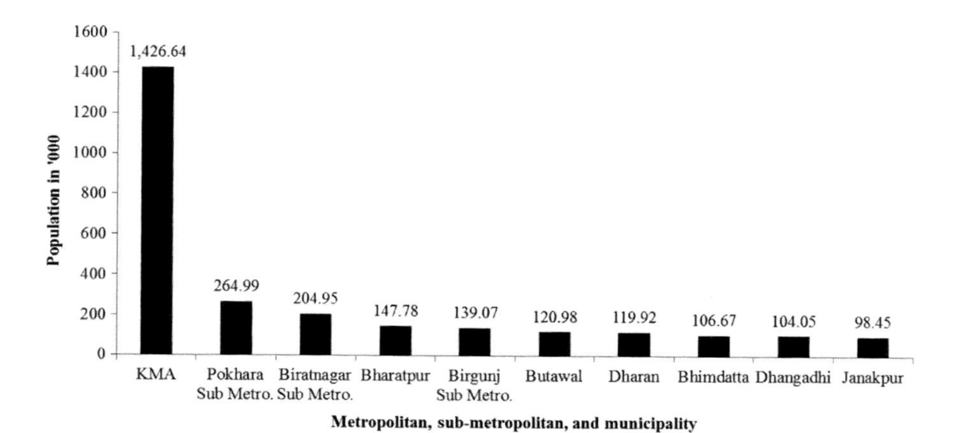

Fig. 11.3 The 10 most populated urban areas in Nepal. *Note* The KMA includes Kathmandu Metropolitan City, Lalitpur Sub-Metropolitan City, Bhaktapur Municipality, Kirtipur Municipality, and Madhyapur Thimi Municipality. *Source* CBS (2013)

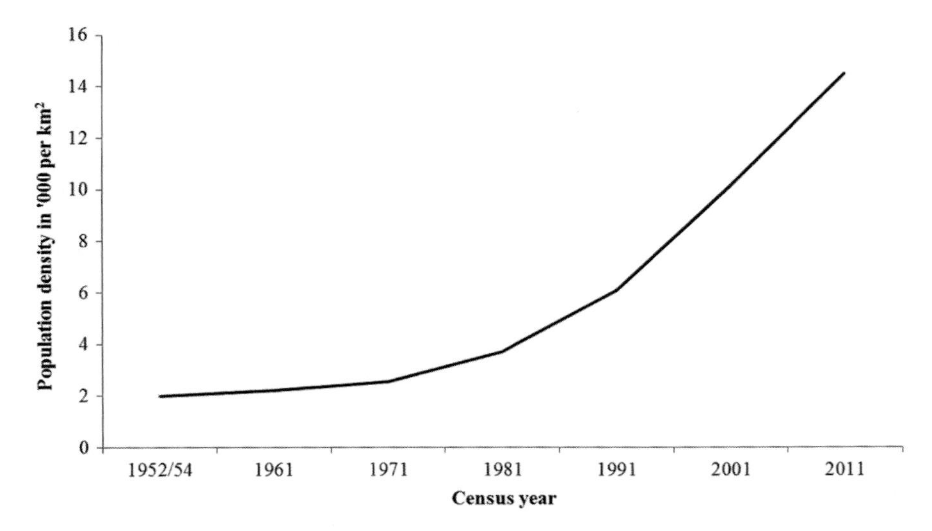

Fig. 11.4 Population density in Kathmandu Metropolitan Area. *Source* CBS, Government of Nepal (http://www.cbs.gov.np/)

The KMA has experienced population influxes in various periods with an increase in urban population from 196,777 in 1952/54 to 1,426,641 in 2011 (Thapa and Murayama 2009; CBS 2013). The KMA covers only 98.33 km^2; the population density, an indicator of urbanization, has increased from 2000 to 14,500 people per square kilometer in the same period (Fig. 11.4). A very high increase has been observed since the 1980s.

Currently, KMA houses 32% of the country's total urban population. The population increase in the metropolitan area led to decreases in highly productive agricultural land as well as to changes in farming practices. Rapid urbanization and the introduction of new agricultural technology have encouraged the KMA's farmers to change their farming patterns from traditional low value crops to new high value crops (Thapa et al. 2008). Until 1990, the urban area was limited to that within the ring road (Thapa and Murayama 2009), which is almost equal to the Yamanote Rail Line of Tokyo in length (Kobayashi 2006), owing to accessibility to commercial areas and the movement of residents from the crowded urban core.

During the 1980s and 1990s, urban growth of the KMA generally occurred in the north–south direction. It was mainly due to the land bordering the west which was undulating and difficult to develop, and the international airport that impeded expansion to the east (ICIMOD 2007). The area between Kathmandu and Bhaktapur was gradually occupied by low-density ribbon development (Thapa and Murayama 2009) that currently remains. The challenges and opportunities in Kathmandu have attracted people from different parts of the country. The migration of people of prominent status has made Kathmandu the most competitive city in the country and has helped its prosperity (Thapa et al. 2008). Significant increases in movement to the metropolitan area from other parts of Nepal during the last several years have occurred as a result of political turmoil and unemployment in rural areas. This migration, coupled with natural population increases, has caused tremendous population growth in the valley (Thapa and Murayama 2010).

A planning approach to urban development in Nepal began in 1956 with the formulation of the First Plan (1956–1961). Several periodic plans have been executed since then. In 1961, the government defined an urban setting as a settlement having a population of 5000 or more that includes urban facilities such as markets, industry, and service centers. Population size, revenue generation, and availability of facilities and services appear to form the basis for urban or municipal area. Some areas have been classified, de-classified, and re-classified as municipalities during the past 60 years, and the territorial boundaries of many settlements have been re-drawn to include surrounding rural areas to meet the population size criteria. Kirtipur and Madhyapur Thimi municipalities in the valley failed to meet such criteria and were removed from the urban centers listed in the 1971 census. However, they were included in the 2001 census.

The Town Development Committee Act was promulgated in 1963 to form a judicial body for reinforcing urban development projects. This act was further amended in 1973, 1988, and 1996 (Gyawali 1997). The first physical development plan of Kathmandu valley was finalized in 1969 to accelerate the city development process of Kathmandu (HMGN 1969). This plan aimed to preserve historical and cultural heritage, guide urban development through land use planning, and enhance population densification in fringe areas. The plan adopted a multi-nucleated regional growth strategy linked to the dispersed settlement in the valley, continuation of existing growth tendencies of the Kathmandu–Patan complex, and bidirectional development of Bhaktapur by reinforcing transportation links and expanding settlements. The government promulgated a Town Development Implementation Act in 1972 to implement the plan. The Kathmandu Valley Town Development Committee (KVTDC) was formed under this act in 1976 to assign the overall responsibility of planning and regulation of urban growth in Kathmandu valley.

In 1976, the Kathmandu Valley Town Development Plan (KVTDP) formulated three broad zoning concepts: Zone A as city core (Kathmandu and Lalitpur), Zone B as city fringe, and Zone C as planned settlements in rural villages of the valley. The plan, prepared by KVTDC, led to the development of a ring road (27.8 km) around Kathmandu and Lalitpur municipalities in the mid-1970s that significantly revitalized the urbanization process in the rural areas of the city's periphery, particularly in Bhaisepati, Dallu, Galfutar, and Kuleshwor. Later, the KVTDC launched two programs: guided land development and land-pooling schemes. To address the rising issues and problems resulting from the rapid urbanization trend in the valley, the KVTDP was revised in 1984 and renamed as the Kathmandu Valley Physical Development Concept (KVPDC). In this plan, major consideration was given to transportation facilities, protection of arable lands, and improvement of urban environments (HMGN 1991).

The Government of Nepal prepared the Structural Plan of Kathmandu Valley in 1987 to provide guidelines for the physical development of metropolitan Kathmandu for the year 2010. Zoning was proposed to preserve agricultural lands and environmentally sensitive areas; however, this plan was not implemented due to political changes in 1990 (ICIMOD 2007). In 1991, Kathmandu Valley Urban Development Plans and Programs was prepared to make various strategic recommendations related to land use, the environment, the infrastructure, financial investments, and institutional aspects for urban development in the valley (HMG 1991). Urban expansion focused on greater Kathmandu (i.e., Kathmandu and Lalitpur) surpassing the ring road and the present municipal boundaries. The government published Environment Planning and Management of Kathmandu Valley in 1999. It analyzes the existing ecological and environmental problems of the valley and the possibilities of limiting the growth of Kathmandu through migration to secondary adjoining towns (IUCN 1999).

The Kathmandu Valley Mapping Program was launched in 1999 to improve the capability of the cities in the valley urban development in a geographically sustainable manner by introducing a geographic information system-based Urban Management Information System (KVMP 2002). The KVTDC in 2002 prepared the Long Term Development Concept for Kathmandu Valley 2020, the latest 20-year planning document. This plan focuses on developing Kathmandu city as central city core; Lalitpur and Bhaktapur as subcity cores; Gokarna, Thimi, Kirtipur, and Harisiddhi as towns; Thankot, Tokha, Sankhu, Lubhu, Chapagaon, and Pharping as traditional settlements; and Nagarkot and Saibu Bhaisepati as nucleated centers (KVTDC 2002). Kathmandu Metropolitan City, Lalitpur Sub-Metropolitan City, Bhaktapur Municipality, Kirtipur Municipality, and Madhyapur Thimi Municipality are also included in the design and implementation of urban development plans at the local level. Furthermore, National Urban Policy 2007 has provided guidance for urban development nationwide. This policy developed the National Urban Development Strategy in 2015 which will guide urban development (Fig. 11.5) in the valley and throughout the nation for the next 15 years.

11.3 Urban Land Use/Cover Patterns and Changes (1989–2030)

11.3.1 Observed Changes (1989–2014)

Figure 11.6 shows that the built-up areas increased almost fivefold during the past 25 years (1989–2014). Built-up areas increased from 14.2 km^2 in 1989 to 75.8 km^2 in 2014 (Table 11.1), at an annual rate of 2.5 km^2/year. The annual rate of change during the 1989–1999 period was a slightly higher, at 1.95 km^2/year than that during the 1999–2010 period, at 1.70 km^2/year (Table 11.2). However, the annual

Fig. 11.5 From the *top*: Swayambhunath and Boudhanath cultural world heritage sites and cityscape and infrastructures in the foothills of Kathmandu Valley. *Source* Cityscape in the *bottom left* (S. Ale, fieldwork, 2015) and all others (Author's fieldwork, 2014)

rate of change increased substantially to 5.81 km^2/year during the 2010–2014 period. The current results do not match the previous study from Thapa and Murayama (2009, 2012) due to the different land use/cover classification approaches and the remote sensing data used. In addition, the land use/cover classification accuracies between the previous and current studies are different. Furthermore, the mapping approach in the present study is designed to be applicable to all cities in the study project, which might not agree with the urban landscape of the KMA owing to physiographic complexity. Lack of sufficient training data for land use/cover classification could be another reason.

Table 11.3 shows the built-up class landscape metrics for KMA. The percentage of landscape (PLAND) metric measures the proportion of a particular class at a certain time relative to the entire landscape. In 1989, the PLAND metric was 2.08% and increased to 4.96, 7.71, and 11.13% in 1999, 2010, and 2014, respectively (Table 11.3). The patch density (PD) metric is a measure of fragmentation based on the number of patches per unit area, which in this case is per 100 ha or 1 km^2, and in which a patch is based on an 8-cell neighbor rule. The PD metric increased from 0.6 to 2.4 in 2014 (Table 11.3). This indicates that the built-up areas became more fragmented over the years. On the contrary, the infill development process in the city core and immediate fringe areas decrease the neighborhood distances between land use/cover patches and increase physical connectivity. This implies a higher probability of homogeneous landscape development in the upcoming decades

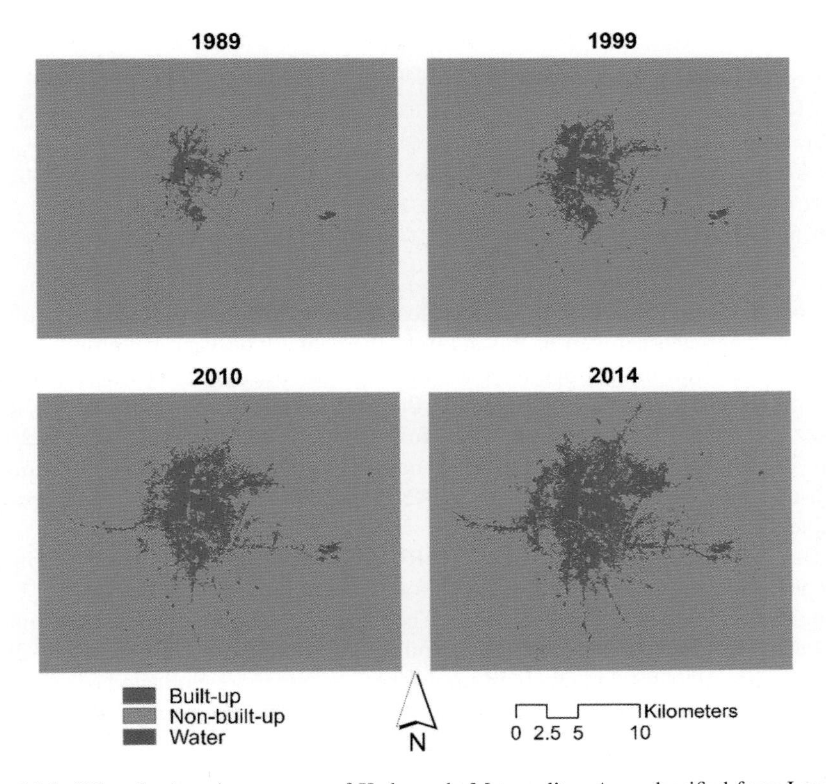

Fig. 11.6 Urban land use/cover maps of Kathmandu Metropolitan Area classified from Landsat imagery

Table 11.1 Observed urban land use/cover of Kathmandu Metropolitan Area (km^2)

	1989	1999	2010	2014
Built-up	14.17	33.74	52.51	75.78
Non-built-up	666.10	646.81	628.16	604.79
Water	1.36	1.07	0.95	1.06
Total	681.63	681.63	681.63	681.63

Table 11.2 Observed urban land use/cover changes in Kathmandu Metropolitan Area (km^2)

	1989–1999	1999–2010	2010–2014
Built-up	19.57	18.76	23.26
Annual rate of change (km^2/year)	*1.95*	*1.70*	*5.81*
Non-built-up	−19.28	−18.65	−23.37
Annual rate of change (km^2/year)	*−1.92*	*−1.69*	*−5.84*
Water	−0.28	−0.11	0.10
Annual rate of change (km^2/year)	*−0.02*	*−0.01*	*0.02*

Table 11.3 Observed landscape pattern of Kathmandu Metropolitan Area

Class-level (built-up) spatial metrics	1989	1999	2010	2014
PLAND (%)	2.08	4.96	7.71	11.13
PD (number per km^2)	0.64	1.01	1.48	2.37
ENN (mean) (m)	168.12	150.12	137.17	125.00
CIRCLE (mean) ($0 \leq$ CIRCLE < 1)	0.37	0.38	0.38	0.28
SHAPE (mean) ($1 \leq$ SHAPE $\leq \infty$)	1.24	1.19	1.23	1.18

(Thapa and Murayama 2009). The landscape results illustrate the weakness of the mapping approach adopted in this study to the actual landscape patterns.

The Euclidean nearest-neighbor distance (ENN) metric is a measure of dispersion based on the distance of a patch to the nearest neighboring patch of the same class. The mean ENN metric decreased from 168.1 m in 1989 to 125.0 m in 2014. This is attributed to disaggregation of non-built-up patches or fragmentation of built-up areas with increases in PLAND and PD during the same period (Table 11.3).

The related circumscribing circle (CIRCLE) metric measures the circularity of patches, and its value ranges from 0 for circular or one cell patches to 1 for elongated, linear patches one cell wide. The mean CIRCLE value increased slightly between 1989 and 1999 (Table 11.3) and remained stable during the 1999–2010 period. This indicates that few built-up patches became more elongated or exhibited a linear pattern in the later decades (Figs. 11.6 and 11.7). The mean CIRCLE value decreased sharply during 2010–2014 period, indicating the formation of more circular patches.

The shape index (SHAPE) metric is a measure of complexity. This metric has a value of 1 when the patch is square and increases without limit as the patch shape becomes more irregular. The mean SHAPE values varied between 1989 and 2014 (Table 11.3), indicating complex dynamics in the shapes of the built-up patches during the different periods.

Figure 11.8 presents all of the metrics for the built-up class of KMA along the gradient of the distance from the city center across all time periods from 1989 to 2014. The figure shows that PLAND decreased as the distance from the city center increased, indicating that the proportion of built-up land near the city center was relatively higher. In contrast, PD increased until it approached a distance of 10 km from the city center and then decreased at farther distances. This suggests the presence of more patches of built-up land in middle distances and that the patches were more dispersed at farther distances, as shown by the increasing trend of mean ENN across the distance from the city center. The figure also shows a slightly increasing trend of the mean CIRCLE value along the gradient of the distance from the city center, indicating that the patches of built-up land were slightly more elongated or linear at farther distances. However, despite the variability of these metrics along the gradient of the distance from the city center, the complexity of the built-up areas was much more uniform or stable, as implied by the mean SHAPE value.

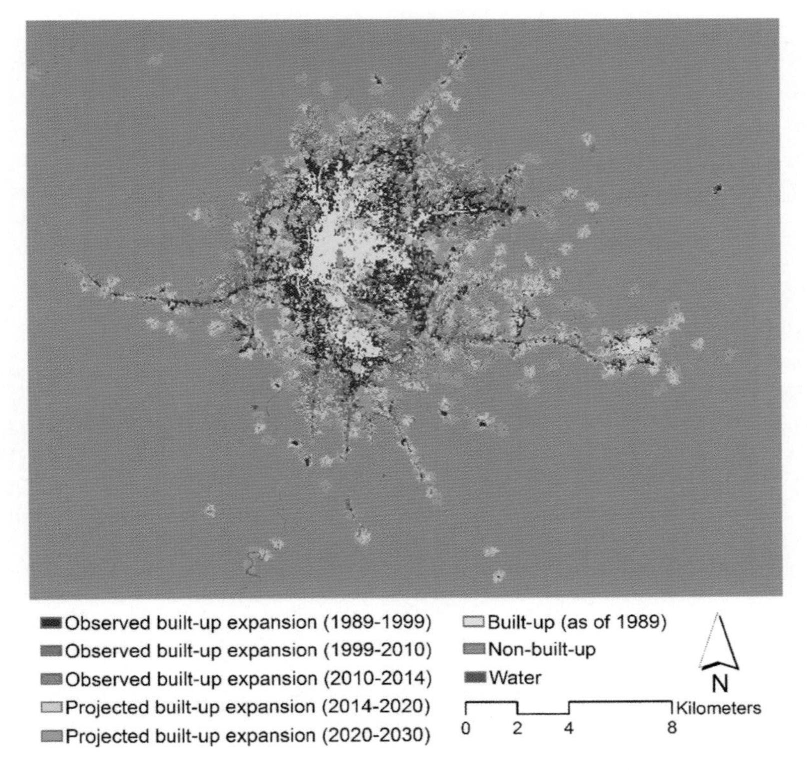

Fig. 11.7 Observed and projected urban land use/cover changes in Kathmandu Metropolitan Area

11.3.2 Projected Changes (2014–2030)

The projected land use/cover maps (Fig. 11.9; see also Fig. 11.7) indicated that built-up land would increase from 75.78 km^2 in 2014 to 85.93 km^2 by 2020 and 102.33 km^2 by 2030, with 1.69 and 1.64 km^2/year annual rates of change, respectively (Tables 11.4 and 11.5). The results of the landscape pattern analysis show that the predicted patches of built-up land in 2020 and 2030 will be more aggregated, as indicated by the increase in PLAND and decrease in PD (Table 11.6). The increase in mean ENN also suggests that more neighborhood (closer) patches will become connected. The increases in the mean CIRCLE and SHAPE values indicate more increases in connected elongated/linear and complex patches of built-up land, respectively.

Along the gradient of the distance from the city center (Fig. 11.10), the PLAND of the predicted 2020 and 2030 patches of built-up land will also be higher for distances closer to the city center. The PD will decrease dramatically by 2020 and 2030 but will still be relatively high in middle distances. In contrast, the mean ENN will increase, although it will still follow the pattern of increases at distances farther

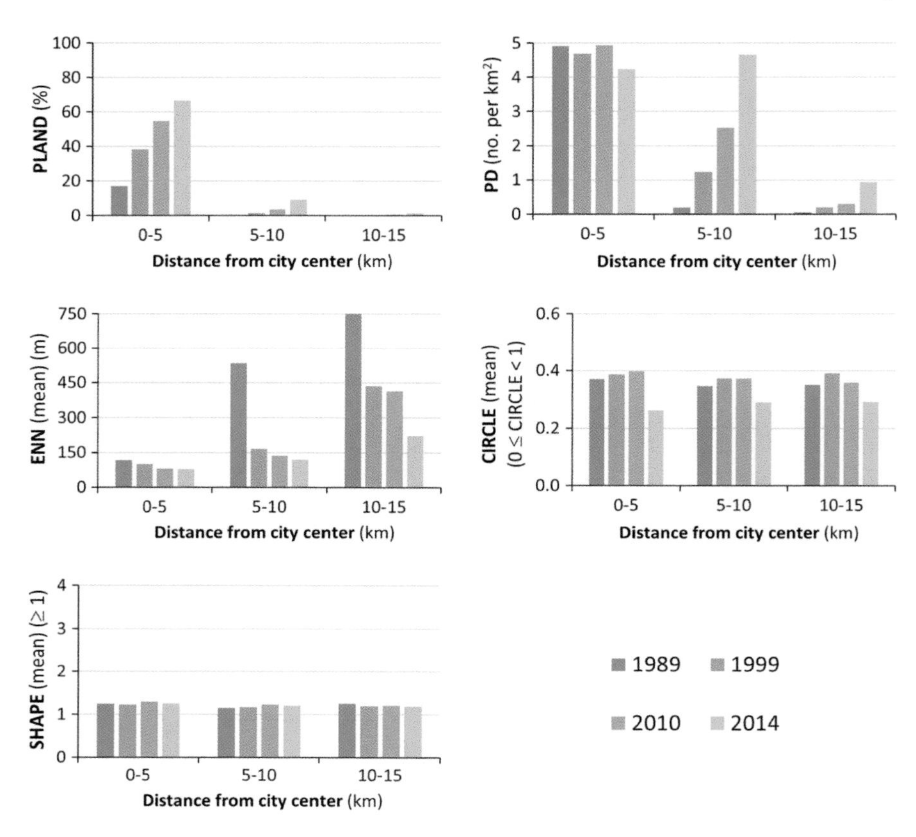

Fig. 11.8 Observed class-level spatial metrics for built-up areas along the gradient of the distance from city center of Kathmandu Metropolitan Area. *Note* The y-axis values are plotted in the same range as those in Fig. 11.10

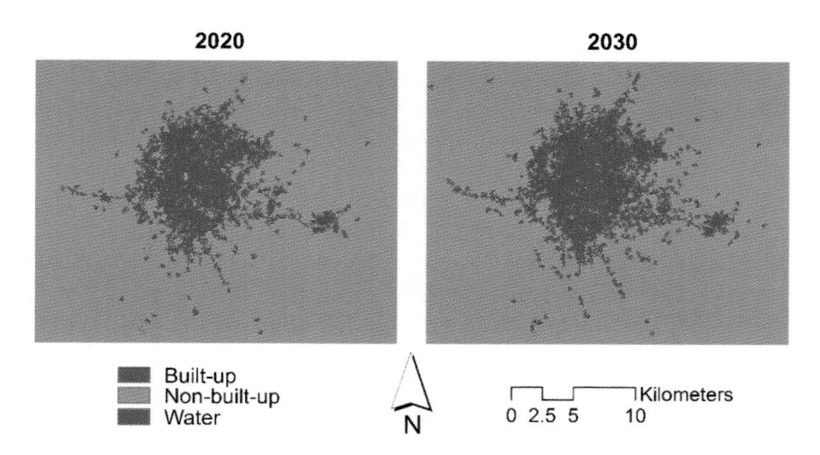

Fig. 11.9 Projected urban land use/cover maps of Kathmandu Metropolitan Area

Table 11.4 Projected urban land use/cover of Kathmandu Metropolitan Area (km^2)

	2020	2030
Built-up	85.93	102.33
Non-built-up	594.63	578.23
Water	1.06	1.06
Total	681.63	681.63

Table 11.5 Projected urban land use/cover changes in Kathmandu Metropolitan Area (km^2)

	1999–2010	2010–2014
Built-up	10.15	16.40
Annual rate of change (km^2/year)	*1.69*	*1.64*
Non-built-up	−10.15	−16.40
Annual rate of change (km^2/year)	*−1.69*	*−1.64*
Water	0.00	0.00
Annual rate of change (km^2/year)	*0.00*	*0.00*

Table 11.6 Projected landscape pattern of Kathmandu Metropolitan Area

Class-level (built-up) spatial metrics	2020	2030
PLAND (%)	12.63	15.04
PD (number per km^2)	0.10	0.14
ENN (mean) (m)	349.10	283.09
CIRCLE (mean) ($0 \leq$ CIRCLE < 1)	0.61	0.57
SHAPE (mean) ($1 \leq$ SHAPE $\leq \infty$)	4.24	4.09

from the city center, particularly for the 2020 predicted patches. The mean CIRCLE value will also increase by 2020 and 2030, particularly at farther distances. The mean SHAPE value will also increase but will be relatively more uniform or stable along the gradient of the distance to the city center (Fig. 11.10).

11.4 Driving Forces of Urban Development

Socioeconomic processes such as migration, urban sprawl, agriculture, and forest patterns often contribute to urbanization. As urbanization proceeds, more land is used for the production of goods and services, and more residential land is necessary to accommodate those who migrate to the city. Urbanization is typically driven by a variety of related forces based on different spatial and temporal settings. In general, the urban development causes land use changes according to the complex interaction of behavioral and structural factors associated with the demand, technological capacity, and social relations affecting demand and capacity, which

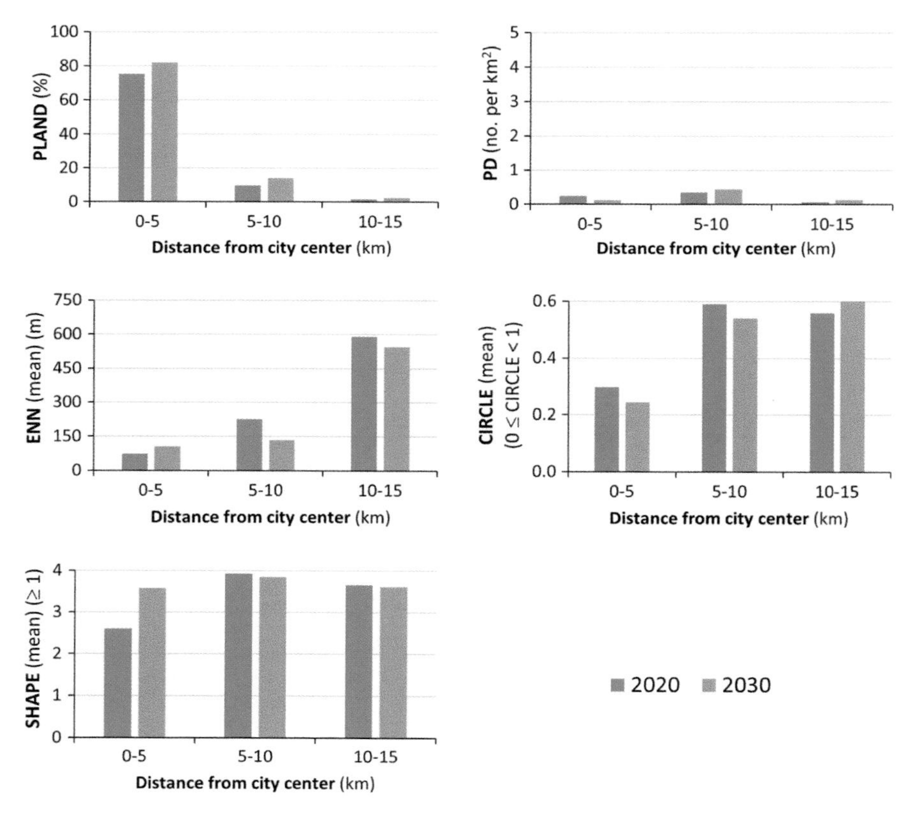

Fig. 11.10 Projected class-level spatial metrics for built-up along the gradient of the distance from city center of Kathmandu Metropolitan Area. *Note* The y-axis values are plotted in the same range as those in Fig. 11.8

ultimately strains the environment. People, government plans and programs, landforms, landscape change processes, and available resources often cause differences in the importance of various factors. In the KMA, the similarities among these factors are apparent. The changes in the spatial pattern of the urban area in the KMA in the 1970s and 1980s were driven by transportation accessibility and the implementation of government plans and policies (Thapa et al. 2008). However, with the increase in environmental and socioeconomic awareness of the residents of the metropolitan area, several emerging factors have resulted in changes to the landscape in recent decades. Thapa and Murayama (2010) synthesized various data and evidence and identified seven representative driving forces in the KMA's urban development ranging from biophysical to socioeconomic factors. These include physical conditions, public service accessibility, economic opportunities, land market, population growth, political situation in the country, and government plans and policies. The impacts of these forces in urban area development differ with urban spatial characteristics. For example, these forces have varying importance among the city core, fringe, and rural areas of a metropolitan area.

The same study quantified these seven driving forces, ranked them according to importance from the first for the lowest impact force to the seventh for the highest, and analyzed their impact for shaping the KMA's urban development based on three urban geographic characteristics: core, fringe, and rural (Thapa and Murayama 2010). The political situation factor was the highest in rural areas but was only the fifth and fourth in the fringe and core areas, respectively. Political turmoil was present during the past decade in Nepal when many residents migrated to rural areas in the valley, which was considered to be the safest place in the country. In the 6th position, the impact of the population growth factor was recognized as the 2nd highest in both the core and the rural areas but ranked 7th, the highest impact factor, in the fringe areas. This shows that population growth has a higher impact on landscape change in the fringe areas in the metropolitan area. The land market factor appears to be modest, ranking 6th, 5th, and 3rd in the fringe, rural, and core areas, respectively. The physical condition factor had the lowest impact in the city core and the fringe areas but ranked 4th in the rural areas. The economic opportunities factor had the highest rank in the city core area but was the 4th and 3rd in the fringe and rural landscapes, respectively. This is attributed to more opportunities in the city core area ranging from employment to business ownerships. Several traditional residential areas in the core are being transformed into business complexes and multi-story housing. However, this factor plays a smaller role in the rural landscape owing to the lack of opportunities for business ownership and employment. The role of plans and policies was relatively weak in the metropolitan area, ranking as the 2nd lowest impact factor in all of the landscape types in the valley. The residents of Kathmandu have indicated that the government plans and policies performed poorly in the changing landscape. Owing to the minimal accessibility to public services in rural areas, this factor ranked the lowest in these areas. With increasing proximity to the city core, its impact order gradually improved, from rank 3rd in the fringe to the 5th in the core areas. To sum it up, the economic opportunities, population growth, and political situation factors are the highest ranking factors affecting change in the core, fringe, and rural areas, respectively. Two factors, population growth and land market, ranked 6th. The physical condition factor appears to be the lowest impact factor in the core and fringe areas, whereas the public service accessibility factor was the lowest in rural areas.

Antrop (2005) reported that natural disasters sometime play important roles in shaping urban development. Indeed, the KMA was severely affected by the April 25, 2015, Gorkha earthquake and subsequent aftershocks that left about 9000 people dead or missing in the country, many injured, and several thousands of houses and many infrastructures destroyed, including world heritage sites. In the upcoming decade, this earthquake event may become the major driving force of urban development in the KMA based on the many plans under discussion and financial investment available for rebuilding the city. In addition, although political instability remains, the turmoil of the past may no longer be a driving force in the future.

11.5 Implications for Future Sustainable Urban Development

Considering the spatial patterns of land use/cover change in the past, the rate of conversion from non-built-up to built-up areas has been quite rapid, with scattered patches of urban development in peri-urban and rural areas characterizing the urban sprawl in the valley. Biophysical characteristics, socioeconomic conditions, neighborhood interactions, and transportation accessibility in the KMA have played important roles in producing these spatial patterns. The spatial diffusion of the built-up areas has spread outward from the city core and along the major roadways. This result may be attributed to agglomeration of built-up areas in the urban frontiers, which forced the farmers to migrate to rural areas. In addition, road expansion and market accessibility to rural areas have encouraged farmers to develop agricultural activities, which gradually converted the landscape into built-up areas. Growing business and economic opportunities led to a population influx in the valley, accelerating housing demands that eventually increased the built-up surfaces. The villages experienced higher population growth, receiving larger proportions of migration. The higher proportions of economically active residents are strongly associated with the urban development patterns in the KMA (Thapa 2009). The form of urban sprawl and regional imbalance of its spatial distribution in the past years could have been caused by weaknesses in the local government regulations (Thapa et al. 2008).

In recent years, the city has become home to 1.5 million people. Although rapid urbanization is a sign of economic prosperity for a city, it also brings biophysical changes that are often problematic to the environment. The KMA, surrounded by complex mountainous terrain, has very limited land resources for new development. The burgeoning population, unguided urban development and daunting urban environmental problems (Thapa et al. 2008) are serious concerns in Kathmandu. Haphazard and unguided processes of land use change create diverse consequences in the city such as inadequacies in housing and urban services in addition to air pollution. Although several new economic reform policies were initiated after the restoration of democracy in the country, the decade-long political turmoil, public insecurity, and lack of an elected body led to the collapse of a significant number of industrial firms and businesses in the KMA. Environmental protection regulations and market globalization were also pertinent factors in this decline. The weaknesses in transportation and other infrastructures are also responsible for the sluggishness of economic changes in the area. A period that appeared to promise great benefits from the new policies of economic liberalization and economic transformation of the KMA has proved, to a large extent, to be a wasted opportunity. However, the decline of manufacturing industries and emergence of new services in the city may offer hope for revitalizing the cityscape environment in the future. The current horizontal trend of urban growth in the metropolitan region is considerable. More vertical growth through the development of multi-story buildings and additional multi-family residential areas are necessary for conserving the limited land resources in the metropolitan area. Effective urban development strategies and land use planning need to be finalized at the local and regional levels. The success of such plans will depend on better communication between the different line agencies and stakeholders.

For sustainable urban development, a national urban development strategy was formulated in 2015 with a time horizon of 15 years (MUD 2015). Strategies have been conceived for achieving desirable conditions in each major theme—such as infrastructure, environment, economy and finance, in addition to the social, economic and cultural visions of urban areas, reflecting the highest values of society. Each strategy is backed by a number of activities recommended for each lead and supportive agency within the different levels of the government, nongovernmental organizations (NGOs), and the private sector. Furthermore, strengthening urban–rural linkages is envisioned, as well as upgrading inter- and intra-regional road connectivity standards; facilitating higher level functions in major regional urban centers; improving connectivity infrastructure in key urban centers; facilitating small towns in realizing their comparative advantages; creating infrastructure for "smart" towns in priority locations; promoting green environment, heritage, and tourism friendly economic functions in the KMA; and integrating future provincial capitals in regional and national urban systems.

Balanced integration of natural and sociocultural environments would be beneficial for sustainable urbanization in the KMA. Issues of urban safety, culture, agriculture, forest, and land and environmental pollution should be addressed. The promotion of a multi-hazard approach in handling disasters and climate change; internalization of a resilient perspective in land use regulations and building codes and bylaws; and enhanced awareness and preparedness in dealing with disaster risk and vulnerability at levels of both the government and local communities will be helpful. Promoting peri-urban agriculture (Thapa and Murayama 2008) for fresh food, vegetables, and horticultural products; adding more vegetation cover in urban areas; maintaining open space and parks; preserving heritage sites and museums; and promoting innovative art, architecture, and culture in areas of potential urban development would provide a better urban environment and improve the KMA. In addition, the effects of the Gorkha earthquake may change the cityscape of the metropolitan area, as indicated by the numerous ongoing discussions for rebuilding the area; such changes will also affect the projected landscape of the KMA in 2020 and 2030. Furthermore, Nepal decided in September 2015 to adopt the federal state model of governance, which would also affect urban development. However, many years may be required for the impacts on urban areas to be apparent.

Acknowledgements I wish to thank Ms. Shailja Ale for helping with field data collection in Kathmandu.

References

ADB/ICIMOD (2006) Environment assessment of Nepal: emerging issues and challenges. Kathmandu

Antrop M (2005) Why landscapes of the past are important for the future. Landscape Urban Planning 70:21–34

CBS (Central Bureau of Statistics) (2013) Statistical year book of Nepal. His Majesty's Government of Nepal, Kathmandu

DoR (Department of Road) (2004) List of important roads and status. Road Statistics, Government of Nepal

Gyawali H (1997) A case study on municipal development fund in Nepal. Town Development Fund Board, Kathmandu

HMGN (His Majesty's Government of Nepal) (1969) Physical development plan for the Kathmandu Valley. Department of Housing and Physical Planning, Kathmandu

HMGN (His Majesty's Government of Nepal) (1991) Kathmandu Valley urban development plans and programmes. Department of Housing and Physical Planning, Kathmandu

ICIMOD (2007) Kathmandu Valley environment outlook. Kathmandu

IUCN (The World Conservation Union) (1999) Environmental planning and management of the Kathmandu Valley. His Majesty's Government of Nepal, Ministry of Population and Environment, Kathmandu

Jha HK (1996) Hindu—Buddhist festivals of Nepal. Nirala Publications, New Delhi

Kobayashi M (2006) Social change in Kathmandu related with globalization and liberalization. Bulletin 44:23–38

KVMP (Kathmandu Valley Mapping Program) (2002) Technical note: integrated action planning. Kathmandu Valley Mapping Program, KVMP—TN10, Kathmandu

KVTDC (Kathmandu Valley Town Development Committee) (2002) Long term development concept of Kathmandu valley. Kathmandu Valley Urban Development Committee, Kathmandu

MUD (2015) National urban development strategy. Government of Nepal, Ministry of Urban Development

Portnov BA, Adhikari M, Schwartz M (2007) Urban growth in Nepal: does location matter? Urban Stud 44:915–937

Ranjitkar NG (1983) Change in agricultural land use and land value in urban fringe of Kathmandu City. Ph.D. Dissertation, Institute of Humanities and Social Sciences, Tribhuvan University, Nepal

Regmi RR (1999) Dimension of Nepali society and culture. SAAN Research Institute, Kathmandu

Shah B (2003) Heritage conservation and planning new development in Bhaktapur, Nepal. In 6th US/ICOMOS international symposium on managing conflict & conservation in historic cities. Annapolis, Maryland, Apr 24–27

Thapa RB (2009) Spatial process of urbanization in Kathmandu Valley, Nepal. Ph.D. Dissertation. Graduate School of Life and Environmental Sciences, University of Tsukuba, Ibaraki, Japan

Thapa RB, Murayama Y (2008) Land evaluation for peri-urban agriculture using analytical hierarchical process and geographic information system techniques: a case study of Hanoi. Land Use Policy 25:225–239

Thapa RB, Murayama Y (2009) Examining spatiotemporal urbanization patterns in Kathmandu valley, Nepal: remote sensing and spatial metrics approaches. Remote Sens 1:534–556
Thapa RB, Murayama Y (2010) Drivers of urban growth in the Kathmandu valley, Nepal: examining the efficacy of the analytic hierarchy process. Appl Geogr 30:70–83
Thapa RB, Murayama Y (2012) Scenario based urban growth allocation in Kathmandu Valley, Nepal. Landscape Urban Planning 105:140–148
Thapa RB, Murayama Y, Ale S (2008) Kathmandu. Cities 25:45–57

Chapter 12
Tehran Metropolitan Area

Niloofar Haji Mirza Aghasi and Ronald C. Estoque

Abstract Tehran Metropolitan Area is the primary urban center of the Islamic Republic of Iran and the home of the country's capital city, Tehran City. This chapter examines the spatiotemporal patterns of urban land changes, i.e., changes from non-built-up to built-up lands, in Tehran Metropolitan Area in recent decades, including the potential future urban land changes. The analysis revealed that the area of built-up lands in the study area has increased more than threefold over the past 26 years (1988–2014). The patches of built-up lands in the area have also become more fragmented. The geographic location, landscape characteristics and road network of Tehran Metropolitan Area, including its population growth and status as the primary urban center in the country and home of the country's capital city, are hypothesized to be among the key factors influencing the spatiotemporal patterns of urban land changes and the overall urban development of the area. The simulated urban land changes indicated that the area of built-up lands would continue to increase in the future (2014–2030) under the influence of continuous expansion and an infill urban development pattern. The intensifying pressure of urbanization due to continuous population growth and urban land changes poses many challenges that need to be considered in sustainable urban development and landscape planning.

12.1 Origin and Brief History

Tehran Metropolitan Area is located in Tehran Province, one of the 31 provinces of the Islamic Republic of Iran. Tehran City is the capital city of both Tehran Province and the country. Geographically, Tehran Province is located in the north of the

N. Haji Mirza Aghasi (✉)
Graduate School of Life and Environmental Sciences,
University of Tsukuba, Tsukuba, Japan
e-mail: nia922@yahoo.com

R.C. Estoque
Faculty of Life and Environmental Sciences,
University of Tsukuba, Tsukuba, Japan

Y. Murayama et al. (eds.), *Urban Development in Asia and Africa*,
The Urban Book Series, DOI 10.1007/978-981-10-3241-7_12

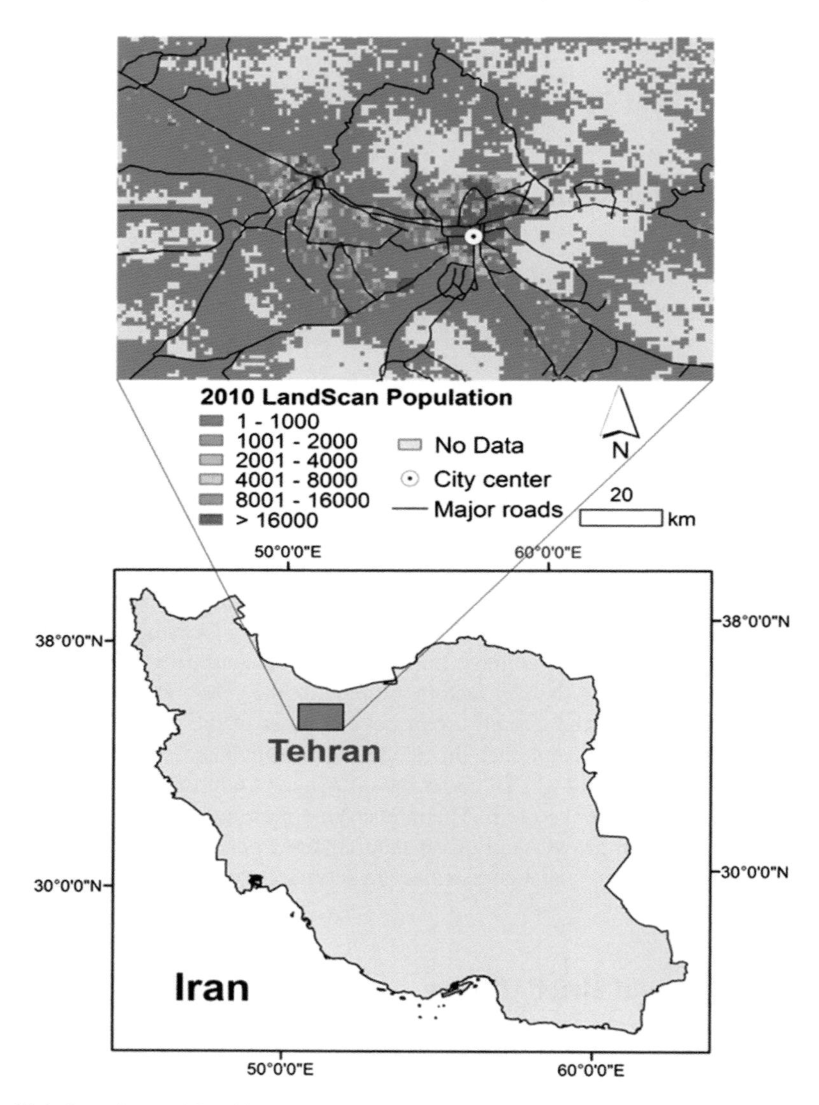

Fig. 12.1 Location and LandScan population of Tehran Metropolitan Area, Iran

central plateau of Iran (Fig. 12.1). The elevation within the extent of the study area (Fig. 12.1) ranges from 851 m in the south to 4343 m in the north, while the city center is situated at about 1150 m above sea level. The metropolitan area is located between a rugged terrain in the north and a desert area in the south. The climate in the area varies considerably from the north to the south. The northern regions are cold and dry, while southern parts are warm and dry.

In the 1920s and 1930s, the city needed to expand and change. These changes began at the time of Reza Shah Pahlavi. Tehran Bazaar was also bisected with this policy, and many historic buildings were demolished to improve accessibility inside the capital city. During World War II, Great Britain and Soviet troops entered the city of Tehran.

Tehran was the venue of the Tehran Conference in 1943. This conference was a strategy meeting of Joseph Stalin, Franklin D. Roosevelt, and Winston Churchill from November 28 to December 1, 1943. It was held in the Soviet Union's embassy in Tehran. It was the first of the World War II conferences of the "Big Three" Allied leaders (the Soviet Union, the United States, and the United Kingdom).

In the 1960s and 1970s (the period of Mohammad Reza Shah Pahlavi), Tehran quickly developed and many construction projects were completed (www.tehran.ir/). In 1973, Tehran Metropolitan Area, also known as Greater Tehran, was formed, covering the contiguous cities of Tehran, Ray, Shemiranat, and other areas. Since then, the landscape of the area has been transformed gradually. Today, Tehran Metropolitan Area is the primary urban center in the country. However, today the city is also faced with some urbanization-related issues, such as traffic-related problems (Figs. 12.2 and 12.3).

12.2 Primacy in the National Urban System

The Islamic Republic of Iran is divided into 31 provinces, 429 counties, 1057 districts, 1245 cities, and 2589 rural districts (Presidency of Islamic Republic of Iran, Management and Planning Organization, Statistical Center of Iran 2016). Tehran Province houses Tehran Metropolitan Area or the Greater Tehran. Tehran City is the capital of not only the Tehran Metropolitan Area and Tehran Province, but also the country. Tehran City has a land area of 730 km^2, while Tehran Metropolitan Area covers 13,689 km^2 (Presidency of Islamic Republic of Iran, Management and Planning Organization, Statistical Center of Iran 2011).

Urban primacy indicates the degree of dominance of one urban area (e.g., city or region) based on population, economy, and urban functions and services (Estoque 2017). The urban primacy of Tehran Province is highlighted in this chapter by comparing its land area, population, and population density with those of the other provinces in the country.

With a land area of 13.69 thousand km^2 (www.citypopulation.de), Tehran Province is the 3rd smallest province in the country in terms of area (Fig. 12.4a). However, in terms of total population and population density, it is the number one province in the country. In 2011, Tehran Province had a total population of 12.18 million (Fig. 12.4b) and a population density of 890 people/km^2 (Fig. 12.4c).

In addition, Tehran Province contributes approximately 26% of the country's gross domestic product (Ministry of Industry and Mining 2015). This makes it the richest province in the country. As of 2011 census data, 16.21% of the country's total population live in the province (www.citypopulation.de). Furthermore, Tehran

Province is the most industrialized province of Iran. With the presence of Tehran City and Tehran Metropolitan Area, Tehran Province is the overall most dominant province of Iran in terms socioeconomics, commanding urban primacy across the country.

12.3 Urban Land Use/Cover Patterns and Changes (1988–2030)

This section discusses the observed and projected urban land changes, i.e., changes from non-built-up to built-up lands, in Tehran Metropolitan Area. Remote sensing-derived urban land use/cover maps, and spatial metrics were used to detect the

Fig. 12.2 A view of Valiasr St. (Hashemifar), near Tehran City Theater, Tehran City, Iran. The street is also used as a parking space. *Source* First author's fieldwork (2015)

Fig. 12.3 A traffic accident along Azadi St., Tehran City, Iran. *Source* First author's fieldwork (2015)

temporal and spatial patterns of urban land changes. The details of the urban land use/cover mapping, change detection and simulation modeling, and spatial pattern analysis are described in the methodology chapter (Kamusoko 2017). Estoque and Murayama (2017) provide a comparative analysis of the trends and spatial patterns of urbanization in Asia and Africa.

12.3.1 Observed Changes (1988–2014)

The urban land change analysis revealed that the area of built-up lands has increased more than threefold over the past 26 years (1988–2014) (Figs. 12.5 and 12.6; Table 12.1). It increased from 301.83 km^2 in 1988 to 923.35 km^2 in 2014 (Table 12.1). This increase translates to an annual rate of change (increase) of 23.90 km^2/year. The annual rate of change during the 2000–2010 period with 21.06 km^2/year was higher than during the 1988–2000 period with 11.94 km^2/year (Table 12.2). It was even higher during the 2010–2014 period with 66.91 km^2/year. This indicates that the rate of urban land changes in the area has been accelerating over the past two and a half decades.

The percentage of landscape (PLAND) metric measures the proportion of a particular class relative to the whole landscape. In 1988, the built-up class had a PLAND of 2.24%, which increased to 3.30, 4.87, and 6.85% in 2000, 2010, and 2014, respectively (Table 12.3). The patch density (PD) metric is a measure of fragmentation based on the number of patches per unit area (in this case per 100 ha

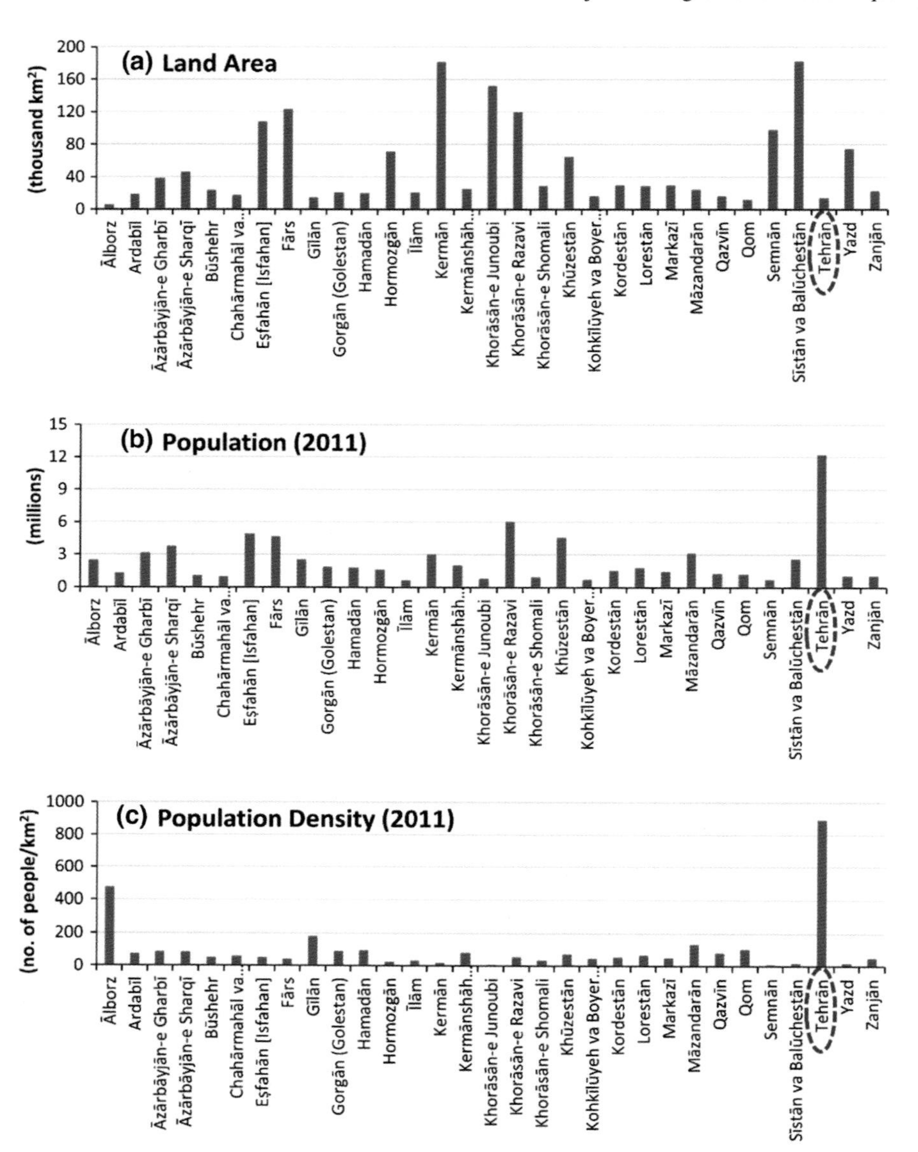

Fig. 12.4 **a** Land area, **b** population, and **c** population density of the 31 provinces of Iran (2011). *Source* www.citypopulation.de/Iran-MajorCities.html

or 1 km^2), in which a patch is based on an 8-cell neighbor rule. In 1988, the built-up class had a PD of 1.85, which also increased to 2.11, 2.92, and 4.99 in 2000, 2010, and 2014, respectively. This indicates that built-up lands have become more fragmented over the years.

The Euclidean nearest-neighbor distance (ENN) metric is a measure of dispersion based on the distance of a patch to the nearest neighboring patch of the same class. The mean ENN of the built-up patches in the area decreased from 127.43 m in 1988 to 123.59, 120.53, and 109.61 m in 2000, 2010, and 2014, respectively (Table 12.3). This decrease can be due to the development of new built-up patches in between neighboring built-up patches (but not necessarily connected patches). These new built-up patches redefined the average distance between neighboring built-up patches. This is supported by the observed increase in PLAND and PD over the years.

The related circumscribing circle (CIRCLE) metric measures the circularity of patches. The value of CIRCLE is 0 for circular or one cell patches and approaches 1 for elongated, linear patches one cell wide. The mean CIRCLE value of the built-up patches in the area was in general stable over the 1988–2014 period (Table 12.3). The shape index (SHAPE) metric is a measure of complexity. This metric has a value of 1 when the patch is square and increases without limit as patch shape becomes more irregular. The mean SHAPE value of the built-up patches in the area was also stable over the 1988–2010 period. During the 2010–2014 period, it decreased, indicating that the shape of built-up patches in the area has become less complex (Table 12.3).

Figure 12.7 presents all the metrics for the built-up class along the gradient of the distance from the city center across all time periods from 1988 to 2014. The

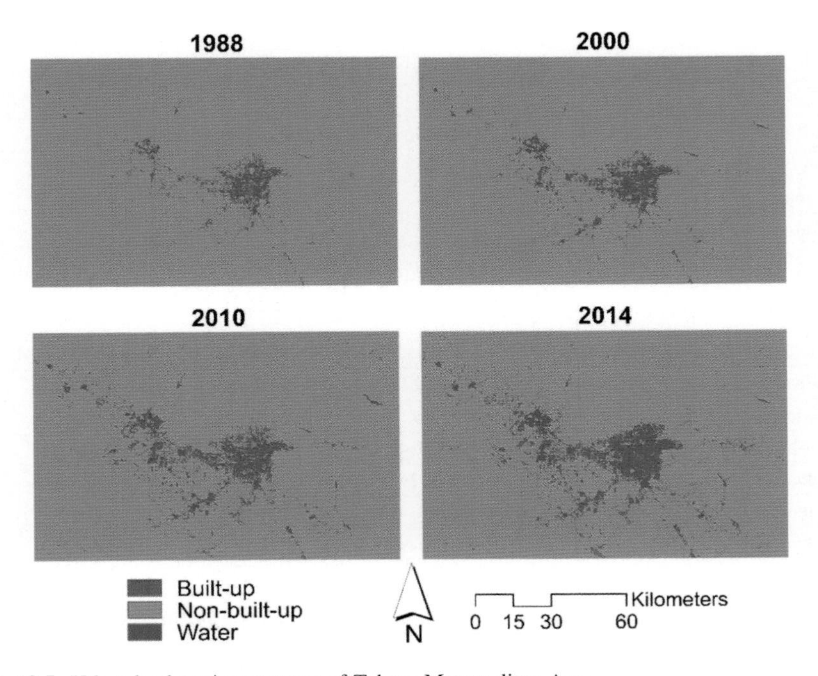

Fig. 12.5 Urban land use/cover maps of Tehran Metropolitan Area

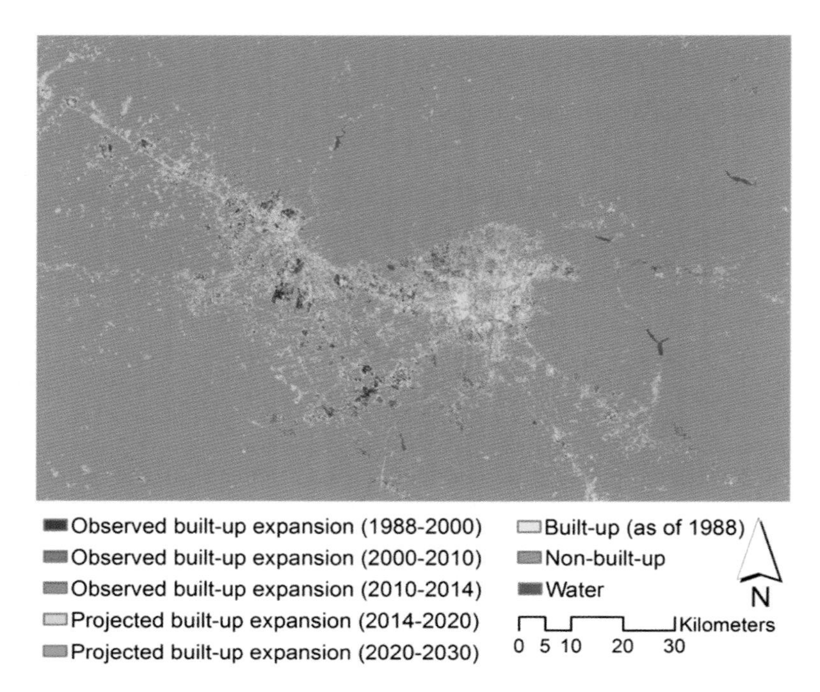

Fig. 12.6 Observed and projected urban land use/cover changes in Tehran Metropolitan Area

Table 12.1 Observed urban land use/cover of Tehran Metropolitan Area (km^2)

	1988	2000	2010	2014
Built-up	301.83	445.06	655.69	923.35
Non-built-up	13,186.21	13,038.72	12,807.70	12,558.02
Water	11.20	15.46	35.85	17.87
Total	13,499.24	13,499.24	13,499.24	13,499.24

Table 12.2 Observed urban land use/cover changes in Tehran Metropolitan Area (km^2)

	1988–2000	2000–2010	2010–2014
Built-up	143.24	210.63	267.65
Annual rate of change (km^2/year)	*11.94*	*21.06*	*66.91*
Non-built-up	−147.49	−231.02	−249.68
Annual rate of change (km^2/year)	*−12.29*	*−23.10*	*−62.42*
Water	4.26	20.39	−17.98
Annual rate of change (km^2/year)	*0.35*	*2.04*	*−4.49*

figure shows that PLAND decreases as the distance from the city center increases, indicating that the proportion of built-up lands near the city center was relatively higher. The figure also shows that PD increases first as it approaches the 15-km

Table 12.3 Observed landscape pattern of Tehran Metropolitan Area

Class-level (built-up) spatial metrics	1988	2000	2010	2014
PLAND (%)	2.24	3.30	4.87	6.85
PD (number per km^2)	1.85	2.11	2.92	4.99
ENN (mean) (m)	127.43	123.59	120.53	109.61
CIRCLE (mean)($0 \leq$ CIRCLE < 1)	0.27	0.28	0.28	0.26
SHAPE (mean) ($1 \leq$ SHAPE $\leq \infty$)	1.20	1.22	1.21	1.16

distance from the city center and then decreases in farther distances. Built-up patches were relatively more dispersed in farther distances, as indicated by the increasing trend of the mean ENN across the distance from the city center. The figure also shows that the mean CIRCLE and SHAPE values were almost uniform across the gradient of the distance from the city center. This indicates that the circularity and complexity of the patches of built-up lands along the gradient of the distance from the city center were more or less stable.

12.3.2 *Projected Changes (2014–2030)*

The results of the urban land change simulation revealed that the area of built-up lands would increase from 923.35 km^2 in 2014 to 1089.62 km^2 in 2020 and 1289.90 km^2 in 2030 (Fig. 12.8; Table 12.4). It would increase at the rate of 27.71 km^2 per year from 2014 to 2020 and 20.03 km^2 per year from 2020 to 2030 (Table 12.5). The simulated patches of built-up lands in 2020 and 2030 would be more aggregated as indicated by the simulated increase in PLAND and decrease in PD (Table 12.6). The simulated increase in mean ENN also indicates that more neighboring (closer) patches would become connected. This simulated aggregation of built-up patches would redefine the average distance between neighboring built-up patches. The simulated increase in the average values of CIRCLE and SHAPE indicates more connected, elongated/linear, and complex patches of built-up lands, respectively.

Along the gradient of the distance from the city center (Fig. 12.9), the PLAND of the simulated built-up patches in 2020 and 2030 would also be higher in areas closer to the city center, especially in the first 20 km distance. PD would decrease dramatically and would also be higher in areas closer to the city center, especially in the first 15 km distance. By contrast, the mean ENN would increase dramatically, though it would still follow the pattern, i.e., mean ENN increases in areas farther from the city center. The mean CIRCLE and SHAPE values would also increase and would also be higher in areas farther from the city center (Fig. 12.9).

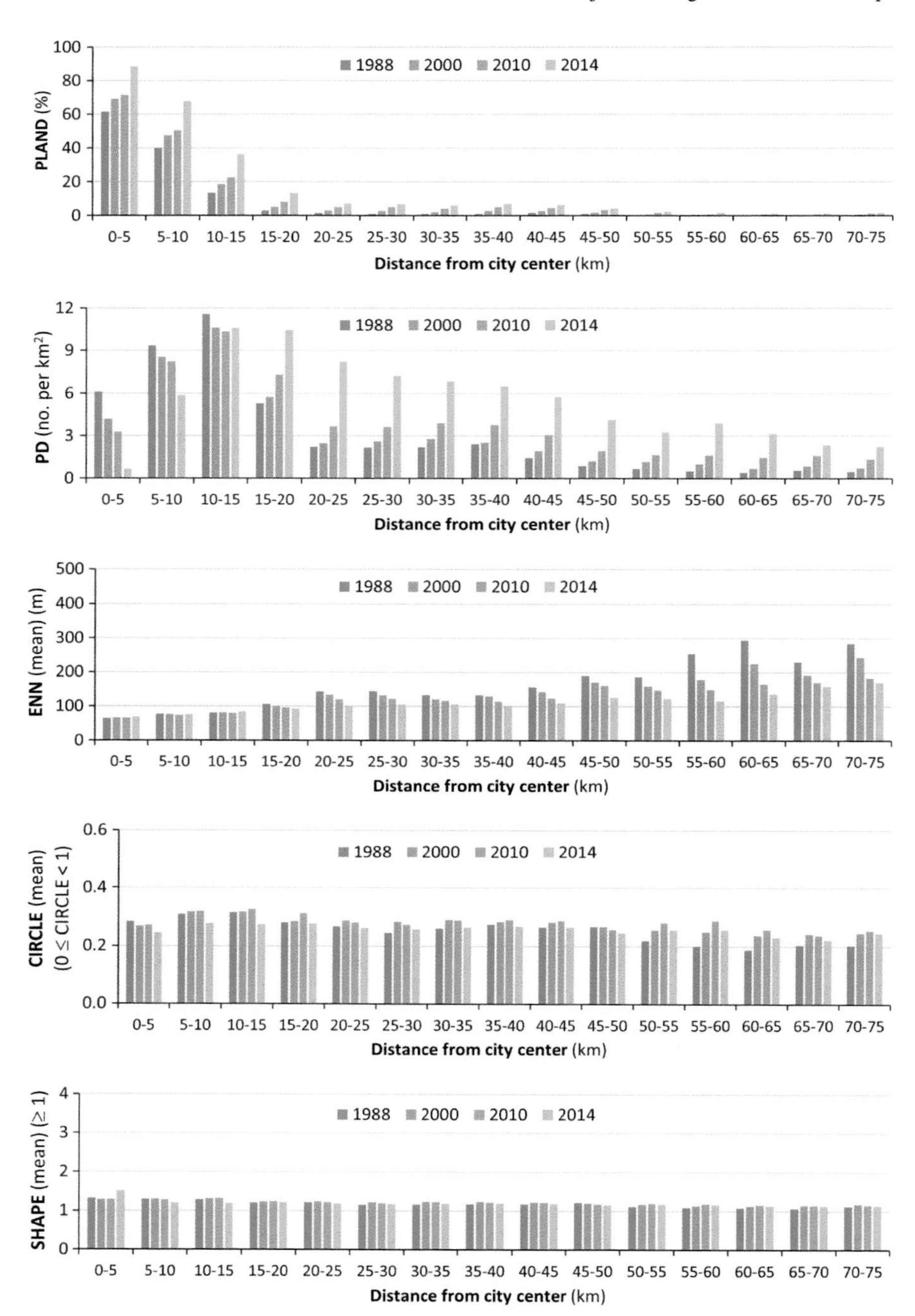

Fig. 12.7 Observed class-level spatial metrics for built-up along the gradient of the distance from city center of Tehran Metropolitan Area. *Note* The y-axis values are plotted in the same range as those in Fig. 12.9

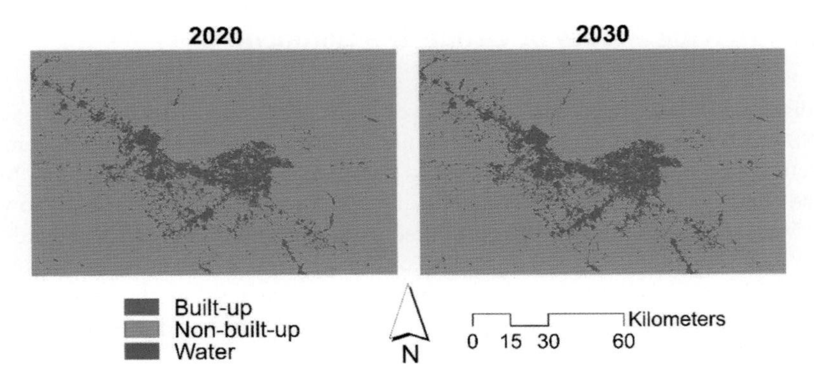

Fig. 12.8 Projected urban land use/cover maps of Tehran Metropolitan Area

Table 12.4 Projected urban land use/cover of Tehran Metropolitan Area (km^2)

	2020	2030
Built-up	1089.62	1289.90
Non-built-up	12,391.75	12,191.47
Water	17.87	17.87
Total	13,499.24	13,499.24

Table 12.5 Projected urban land use/cover changes in Tehran Metropolitan Area (km^2)

	2014–2020	2020–2030
Built-up	166.27	200.28
Annual rate of change (km^2/year)	*27.71*	*20.03*
Non-built-up	−166.27	−200.28
Annual rate of change (km^2/year)	*−27.71*	*−20.03*
Water	0.00	0.00
Annual rate of change (km^2/year)	*0.00*	*0.00*

Table 12.6 Projected landscape pattern of Tehran Metropolitan Area

Class-level (built-up) spatial metrics	2020	2030
PLAND (%)	8.08	9.57
PD (number per km^2)	0.46	0.37
ENN (mean) (m)	228.76	251.26
CIRCLE (mean)($0 \leq$ CIRCLE < 1)	0.39	0.45
SHAPE (mean) ($1 \leq$ SHAPE $\leq \infty$)	2.30	2.68

12.4 Driving Forces of Urban Development

As discussed above, the urban land changes in the study area over the past 26 years (1988–2014) have been remarkable. Built-up lands have been expanding in all directions, except toward the north due to the rugged terrain in this direction. Most of the changes seem to have followed the pattern of road network, especially toward the west, southwest, and southeast directions, but more especially toward the northwest direction (Figs. 12.5 and 12.6). In general, it seems that the spatial pattern of urban land changes in the area has been influenced by the spatial pattern of road network and the topographic characteristics (slope and elevation) of the area.

Tehran Metropolitan Area houses the capital city of the country, i.e., Tehran City. Most of the socioeconomic development opportunities have been, and still are, channeled to the metropolitan area and the city. The socioeconomic dominance of Tehran Province in general, in which Tehran Metropolitan Area and Tehran City are located, is a major factor for its rapid population growth. People continue to flock to the area to seek for better socioeconomic opportunities.

Based on the official census data (www.citypopulation.de), the population of Tehran Province grew from 8.1 million in 1986 to 12.2 million in 2011. This shows that the population of the province has been growing at annual growth rate of 1.65%, equivalent to an annual increase of 164 thousand people over the past 25 years. This growth of population in the area has been fueled by rural–urban migration as reported in the Atlas of Tehran Metropolis (http://atlas.tehran.ir).

This population increase also raised the need for various urban services including housing, roads, and commercial and business centers, which means that more non-built-up lands had to be converted into built-up. The urban land change analysis from 1988 to 2014 (Sect. 12.3.1) provides some evidence for this proposition.

In general, the geographic location, landscape characteristics, and road network of Tehran Metropolitan Area, including its population growth and status as the primary urban center in the country and home of the country's capital city, are considered to be among the key factors influencing the spatiotemporal patterns of urban land changes (Figs. 12.5, 12.6 and 12.7; Tables 9.1–9.3) and the overall urban development of the area.

12.5 Implications for Future Sustainable Urban Development

A sustainable city must achieve a balance between environmental protection, economic development, and social well-being (Wu 2010; Estoque and Murayama 2016). In this context, the socioeconomic status of the local people and the

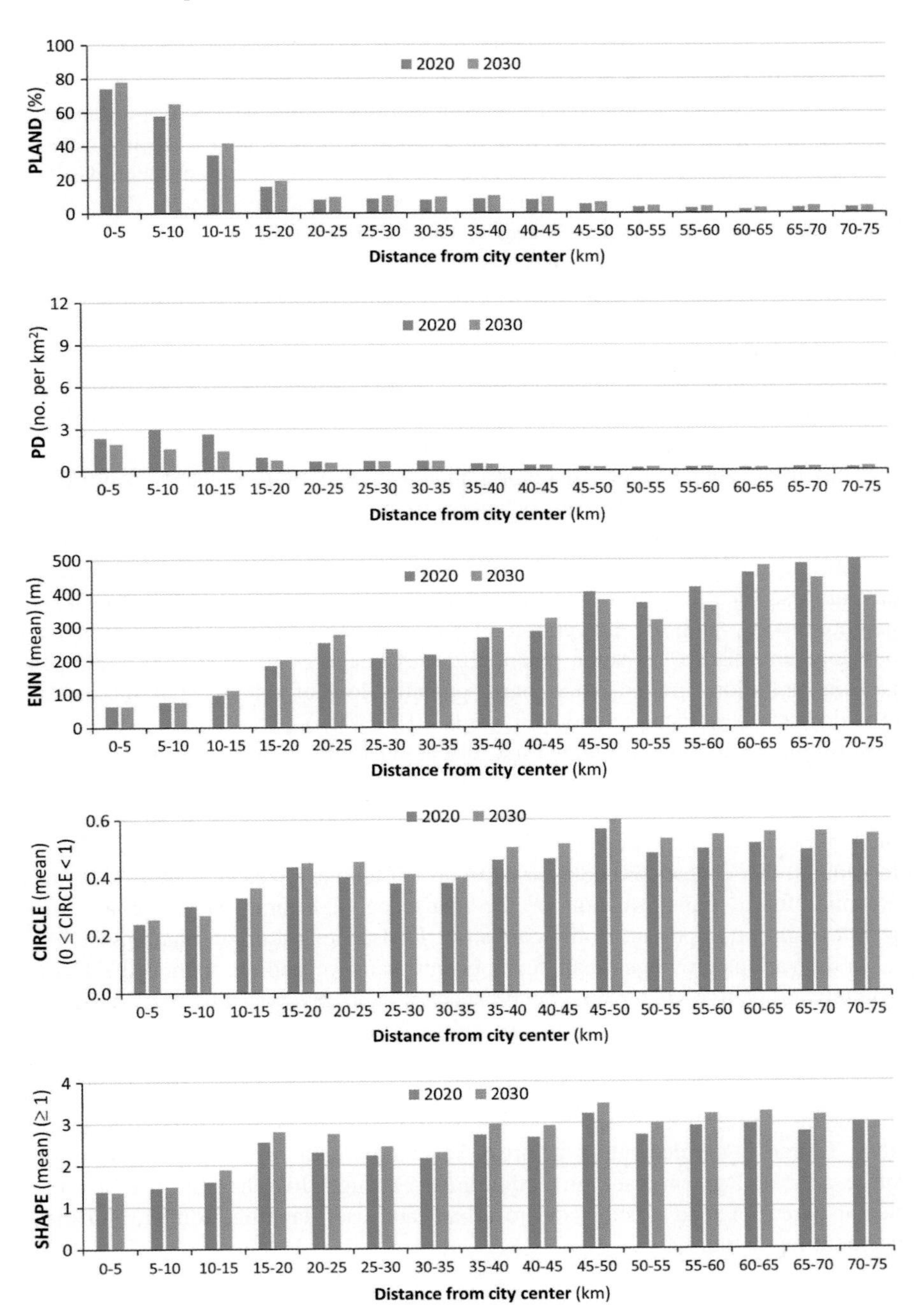

Fig. 12.9 Projected class-level spatial metrics for built-up along the gradient of the distance from city center of Tehran Metropolitan Area. *Note* The y-axis values are plotted in the same range as those in Fig. 12.7

environmental quality of Tehran Metropolitan Area, including the extent and quality of its urban green spaces, are important.

Like in many developing and less-developed major cities and metropolitan areas around the world, however, Tehran Metropolitan Area is also facing various socioeconomic and environment-related problems, such as unemployment, working children, high economy class differences, climate change, insufficient urban transport system, limited urban green spaces, water supply problems, noise, and varying land prices, among others. As discussed in Sect. 12.3, the patches of built-up lands in the area have become more fragmented or dispersed as indicated by the overall increase in PLAND and PD between 1989 and 2014. The process of continuous diffusion and expansion of built-up patches can result in a sprawl development pattern (Estoque and Murayama 2015). Sprawl urban development has some potential disadvantages, such as the consequent external costs, loss of wildlife habitat, and the disturbance of wider natural landscapes.

The urban land change simulation indicated that built-up lands continue to expand and undergo the process of aggregation or coalescence in the near future (Tables 9.3 and 9.6; Figs. 12.7 and 12.9). The process of coalescence can result in an infilling growth pattern. While an infilling growth pattern has some potential advantages, e.g., the use of existing infrastructures, the promotion of walkable neighborhoods, and the prevention of the associated external costs of sprawl development, it also has some potential disadvantages, e.g., increased traffic congestion, pollution, limited open space, potential loss of urban green spaces, and crowded services (Estoque and Murayama 2015, 2016).

Overall, the findings presented and discussed in this chapter on the spatiotemporal urban land changes in Tehran Metropolitan Area supplement the findings of other related studies that have been conducted in the area (e.g., Javadian et al. 2011; Shahraki et al. 2011; Arsanjani et al. 2013; Amini et al. 2014; Bokaie et al. 2016). In general, this chapter provides an overview of the past–present (1988–2014) and potential future transformation of the landscape of Tehran Metropolitan Area, which might be important in the context of landscape and urban planning.

In the various previous urban development plans for Tehran, such as the Tehran Municipality's five-year plan, the prospect was to preserve the Iranian–Islamic originality of Tehran and push for the cultural and economic progress to improve the quality of life in the region. One of the current urban development plans for Tehran is the so-called Detailed Plan. This plan is a strategic document for urban management in Tehran. It serves as the basis for all the plans for the city. In this plan, Tehran is divided into four areas: residential area, work space area, mixed space area, and green space protection area. Through this plan, the use of a particular parcel of land is subjected to urban land use planning, zoning, and management regulations.

12.6 Concluding Remarks

This chapter has examined the spatiotemporal pattern of urban land changes, i.e., changes from non-built-up to built-up lands, in Tehran Metropolitan Area. The analysis revealed that the area of built-up lands in the study area has increased more than threefold over the past 26 years (1988–2014). The patches of built-up lands in the area have also become more fragmented. The geographic location, landscape characteristics and road network of Tehran Metropolitan Area, including its population growth and status as the primary urban center in the country and home of the country's capital city, are hypothesized to be among the key factors influencing the spatiotemporal patterns of urban land changes and the overall urban development of the area.

The simulated urban land changes indicated that built-up lands would continue to expand in the future (2014–2030) under the influence of continuous expansion and an infill urban development pattern. The intensifying pressure of urbanization due to continuous population growth and urban land changes poses many challenges (as discussed above), which need to be considered in sustainable urban development and landscape planning.

References

Amini H, Taghavi-Shahri SM, Henderson SB, Naddafi K, Nabizadeh R, Yunesian M (2014) Land use regression models to estimate the annual and seasonal spatial variability of sulfur dioxide and particulate matter in Tehran, Iran. Sci Total Environ 488–489:343–353

Arsanjani JJ, Helbich M, Kainz W, Boloorani AD (2013) Integration of logistic regression, Markov chain and cellular automata models to simulate urban expansion. Int J Appl Earth Obs Geoinf 21:265–275

Bokaie M, Zarkesh MK, Arasteh PD, Hosseini (2016) Assessment of urban heat island based on the relationship between land surface temperature and land use/land cover in Tehran. Sustain Cities Soc 23:94–104

Estoque RC (2017) Manila metropolitan area. In: Murayama Y, Kamusoko C, Yamashita A, Estoque RC (eds) Urban development in Asia and Africa—geospatial analysis of metropolises. Springer Nature, Singapore, pp 85–110

Estoque RC, Murayama Y (2015) Intensity and spatial pattern of urban land changes in the megacities of Southeast Asia. Land Use Policy 48:213–222

Estoque RC, Murayama Y (2016) Quantifying landscape pattern and ecosystem service value changes in four rapidly urbanizing hill stations of Southeast Asia. Landscape Ecol 31:1481–1507

Estoque RC, Murayama Y (2017) Trends and spatial patterns of urbanization in Asia and Africa: a comparative analysis. In: Murayama Y, Kamusoko C, Yamashita A, Estoque RC (eds) Urban development in Asia and Africa—geospatial analysis of metropolises. Springer Nature, Singapore, pp 393–414

Javadian M, Shamskooshki H, Momeni M (2011) Application of sustainable urban development in environmental suitability analysis of educational land use by using AHP and GIS in Tehran. Procedia Eng 21:72–80

Kamusoko C (2017) Methodology. In: Murayama Y, Kamusoko C, Yamashita A, Estoque RC (eds) Urban development in Asia and Africa—geospatial analysis of metropolises. Springer Nature, Singapore, pp 11–46
Ministry of Industry and Mining (2015) Province's share in GDP. Tehran, Iran
Presidency of Islamic Republic of Iran, Management and Planning Organization, Statistical Center of Iran (2011) Statistical yearbook. Tehran, Iran
Presidency of Islamic Republic of Iran, Management and Planning Organization, Statistical Center of Iran (2016) Statistical yearbook. Tehran, Iran
Shahraki AZ, Sauri D, Serra P, Modugno S, Seifolddini F, Pourahmad A (2011) Urban sprawl pattern and land-use change detection in Yazd, Iran. Habitat Int 35:521–528
Wu J (2010) Urban sustainability: an inevitable goal of landscape research. Landscape Ecol 25:1–4

Part III
Urbanization in Africa

Chapter 13
Dakar Metropolitan Area

Courage Kamusoko

Abstract Globally, the demand for timely land use/cover change information has increased over the past decades given the rapid pace of urbanization. The objective of this chapter was to analyze observed and simulated land use/cover changes between 1989 and 2030 in Dakar Metropolitan Area. The land use/cover maps for 1989, 1999, 2010, and 2014 indicated that built-up areas increased substantially over the study period. Generally, built-up expanded during the "1989–1999" and "2010–2014" epochs, while built-up expansion slowed down during the "1999–2010" epoch. Built-up growth in Dakar Metropolitan Area was characterized by a combination of sprawl and densification. Future land use/cover simulations (up to 2030) indicated that the current land use/cover change trends such as the increase in built-up areas and decrease in non-built-up areas as well as urban sprawl would continue to persist. The observed and simulated land use/cover changes provide a panoramic view of built-up expansion as well as a simulated urban growth scenario for Dakar Metropolitan Area. These results convey important insights about urban expansion, which could potentially be used to implement the "2035 Dakar Urban Masterplan."

13.1 Origin and Brief History

The Dakar Metropolitan Area analyzed in this chapter comprises Dakar region (or Greater Dakar area). The Dakar region (Fig. 13.1a), which covers approximately 550 km^2 of the national territory, is located in the Cap Vert peninsula between 17° and 17° 10′ west longitude and 32° 14′ and 14° 35′ north latitude (Agence Nationale de la Statistique et de la Démographie 2016a). The Dakar region is limited to the east by Thies region and the Atlantic Ocean to the north, south, and west. The Dakar region has undergone several administrative changes since the colonial era (Agence Nationale de la Statistique et de la Démographie 2016a).

C. Kamusoko (✉)
Asia Air Survey Co. Ltd, Kawasaki, Japan
e-mail: kamas72@gmail.com

Y. Murayama et al. (eds.), *Urban Development in Asia and Africa*,
The Urban Book Series, DOI 10.1007/978-981-10-3241-7_13

Currently, it is divided into four departments, namely Dakar, Pikine, Rufisque, and Guédiawaye. These four departments (Sy 2013) are subdivided into ten districts (Fig. 13.1b). Dakar department comprises Almadies, Dakar Plateau, Grand-Dakar, Parcelles Assainies districts, while Pikine department consists of Dagoudane,

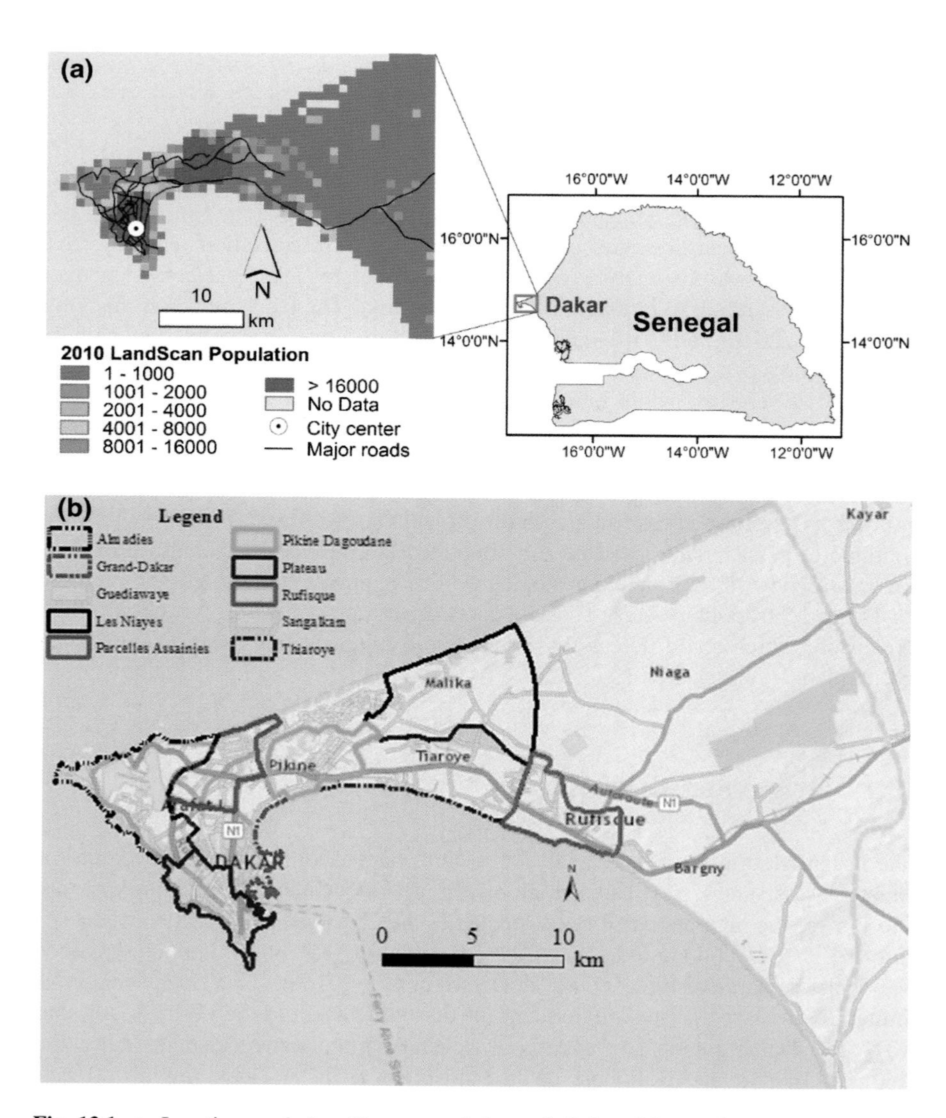

Fig. 13.1 a Location and LandScan population of Dakar Metropolitan Area, Senegal. **b** Departments (Arrondissements) of Dakar region overlaid on ESRI map data. *Note* Boundaries were downloaded from United Nations Office for the Coordination of Humanitarian Affairs OCHA Regional Office for West and Central Africa (https://data.hdx.rwlabs.org/organization/ocharowca?sort=metadata_modified+desc&res_format=zipped+shapefile)

Niayes, Thiaroye. Rufisque department is composed of Rufisque and Sangalkam districts, while Guédiawaye consists of the Guédiawaye district (Agence Nationale de la Statistique et de la Démographie 2016a). The Dakar region has 43 communes, of which 19 are in the department of Dakar, 5 in the Guédiawaye, 16 in Pikine, and 3 in the Rufisque (Agence Nationale de la Statistique et de la Démographie 2016a). In total, there are four cities (Dakar, Pikine, Rufisque, and Guédiawaye) and three rural areas, which are located in the department of Rufisque (Bambilor, Tivaouane Peulh-Niagha, and Yène) (Agence Nationale de la Statistique et de la Démographie 2016a).

Generally, the relief of the Dakar region is low. In terms of geomorphology, it is divided into three main parts from west to east, namely the western part, peninsula of Cape Verde, and the eastern part (Agence Nationale de la Statistique et de la Démographie 2016). The western part is comprised of the southern, central, and northwest zones. The altitude of the southeast zone varies between 15 and 40 m (Agence Nationale de la Statistique et de la Démographie 2016). This zone corresponds to neighborhoods of Plateau in the department of Dakar, which are underlain by volcanic flows and rock outcrops (Agence Nationale de la Statistique et de la Démographie 2016). The central area, which has an altitude of less than 10 m, consists of sand areas underlain by clay and limestone (Agence Nationale de la Statistique et de la Démographie 2016). This area includes the popular residential area of the Medina as well as the industrial zone. The northwest portion corresponds to the volcanic rock areas that are characterized by an average altitude of over 60 m. This area is home to traditional villages of Ouakam, Ngor, and Yoff, and the Leopold Sedar Senghor international airport (Agence Nationale de la Statistique et de la Démographie 2016). The second major part of the peninsula of Cape Verde includes a series of sand dunes, where Pikine and Guediawaye are located. A number of dry lakes are also found in this area, which is mainly used for market gardening and horticulture. The eastern part of the region includes hills and elevations below 50-m plates, where Rufisque department is located. The area is underlain by limestone as well as sandstone (Agence Nationale de la Statistique et de la Démographie 2016). The temperature in the Dakar region varies between 17 and 25 °C from December to April, and between 27 and 30 °C from May to November (Agence Nationale de la Statistique et de la Démographie 2016). The rainy season is generally three or four months from July to October (Agence Nationale de la Statistique et de la Démographie 2016).

According to Bigon (2012), before the official French occupation of the Cape Vert peninsula—over which Dakar extends—in 1857, Dakar consisted of several Lebu villages numbering approximately 10,000 residents. The first city plan of Dakar was prepared by Pinet Laprade, the French Governor of Senegal in 1862 (Arecchi 1985). The orthogonal plan in which a port, administrative headquarters, military and commercial facilities were marked was known as Dakar-ville (Bigon 2012). This area became the heart of the city and grew following the Senegal–Niger railway line in 1885 (Bigon 2012). The second city plan was designed in 1901 to extend the limit of the city area in response to population growth. According to Bigon (2012), the Dakar colonial administration followed two residential schemes in the early twentieth century. The first extended northwest of Dakar-ville and involved forcing the Lebu people to resettle in Medina, where no basic infrastructure was provided (Bigon 2012). Médina was created after a plague epidemic in order to separate the residential area for the indigenous population from the French cordon area (Arecchi 1985; Bigon 2012). Dakar was designated the capital of French West Africa Federation in 1902 after Saint Louis (Bigon 2012). The second scheme inaugurated by the colonial administration from the mid-1900s was the creation of the new urban quarters designated to house its employees (Bigon 2012). This was established on the high plateau at the tip of the peninsula, south of Dakar-ville, and is known simply as the Plateau (Bigon 2012).

The population of Dakar increased to 92,600 persons in 1936, strengthening the administrative, military, and commercial activities (Creevey 1980; Arecchi 1985; Bigon 2012). Given the population growth, a master plan was prepared by Michel Ecochard in 1967 (Arecchi 1985). The master plan was prepared for a population of 500,000 habitants, which was more than three times the population in 1945. To date, Dakar continues to play an important role in West Africa in general and Senegal in particular. For example, the Autonomous Port of Dakar, Leopold Sedar Senghor international airport, and the Dakar–Bamako Railways Company are located in the metropolitan area (Agence Nationale de la Statistique et de la Démographie 2016).

The spatial structure of Dakar is largely based on the French colonial era development model, which is characterized by five major zones of built-up areas that are interlinked by several axes (Ministry of Urban Renewal, Housing and Living Environment and JICA 2016). In general, the Dakar region plays a significant role as the economic center producing about 80% of the national economic outputs. Within Dakar region, the Plateau plays an important role in administration, manufacturing, and logistics (Ministry of Urban Renewal, Housing and Living Environment and JICA 2016). In the north, Almadies provide an international airport, an international exhibition center, five-star hotels, and offices (Ministry of Urban Renewal, Housing and Living Environment and JICA 2016). Pikine, Guédiawaye, and Rufisque provide residential areas in the Dakar region. Pikine and Guédiawaye are already densely built-up areas, while Rufisque still has vacant space for new development (Ministry of Urban Renewal, Housing and Living Environment and JICA 2016).

13.2 Primacy in the National Urban System

According to the Agence Nationale de la Statistique et de la Démographie (2016), the population of Senegal was estimated at 13,508,715 inhabitants in 2013. Approximately 45% (6,102,798 inhabitants) of the population reside in urban areas, while 55% (7,405,911 inhabitants) live in rural areas. The population is mostly concentrated to the west and center of the country, while the east and north are sparsely populated (Agence Nationale de la Statistique et de la Démographie 2016).

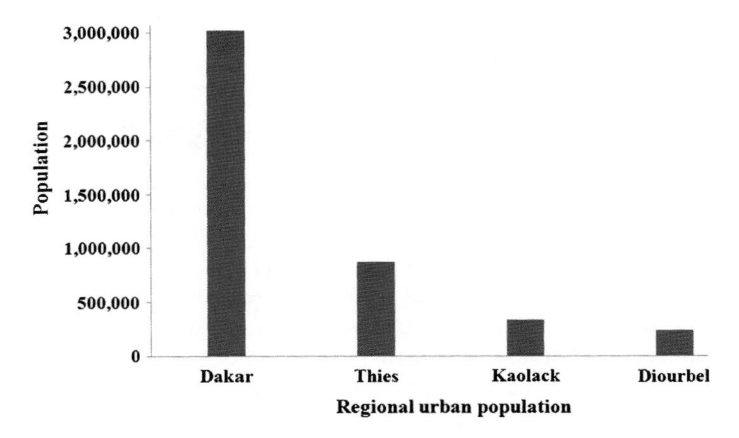

Fig. 13.2 Urban population distribution (2013) for Dakar, Thies, Diourbel, and Kaolack regions. *Source* Agence Nationale de la Statistique et de la Démographie (ANSD), Senegal

The population of Dakar region was approximately 3,137,196 people in 2013 which represents 23% of the population of Senegal (Agence Nationale de la Statistique et de la Démographie 2016).

Figure 13.2 shows the urban population in the most populous regions of Senegal. A high spatial imbalance in the urban population distribution of Senegal is conspicuous, which shows the urban primacy of Dakar Metropolitan Area. While the Dakar region occupies only 0.3% of the total national territory, it is the most populated region of Senegal with a population density of 5704 persons/km^2 (Agence Nationale de la Statistique et de la Démographie 2016). The Dakar region has an urbanization rate of 96%, which accounts for about 50% of the urban population in Senegal. The demographic profile of the region shows that its population is very young with 44.5% of people below 20 years of age (Agence Nationale de la Statistique et de la Démographie 2016a). Over 75% of the urban population in the Dakar region is concentrated in the urban areas of Dakar, Guédiawaye, and Pikine departments (Agence Nationale de la Statistique et de la Démographie 2016a).

Thies region is the second most urbanized region after Dakar, with an urbanization rate of 48.8% (Agence Nationale de la Statistique et de la Démographie 2016b). However, the region has only 14% of the urban population of Senegal (about 872,111 inhabitants). While Kaolack region is the fourth most populous region (after Dakar, Thies, and Diourbel), it is the third most urbanized region in Senegal (Agence Nationale de la Statistique et de la Démographie 2016d). Approximately 6% of the urban population in Senegal lives in Kaolack region (Agence Nationale de la Statistique et de la Démographie 2016d). The Diourbel region is the fourth most urbanized region hosting approximately 4% of the urban population in Senegal (Agence Nationale de la Statistique et de la Démographie 2016c).

13.3 Urban Land Use/Cover Change and Landscape Analysis

13.3.1 Land Use/Cover Change Analysis (1989–2014)

Land use/cover maps and statistics for 1989, 1999, 2010, and 2014 are shown in Fig. 13.3 and Table 13.1. The results indicate that in 1989, built-up areas occupied 30.6 km^2, while non-built-up areas occupied 611.3 km^2. In 1999, built-up areas increased to 47.9 km^2, while non-built-up areas decreased to 587.8 km^2. However, substantial spatial expansion in built-up areas and subsequent decreases in non-built-up areas was observed in 2010. Analysis of the 2010 land use/cover map revealed that built-up areas occupied 97.5 km^2, while non-built-up areas decreased to 513.9 km^2. For the 2014 epoch, built-up and non-built-up areas occupied 118.6 and 504.4 km^2, respectively. In general, built-up areas increased substantially from

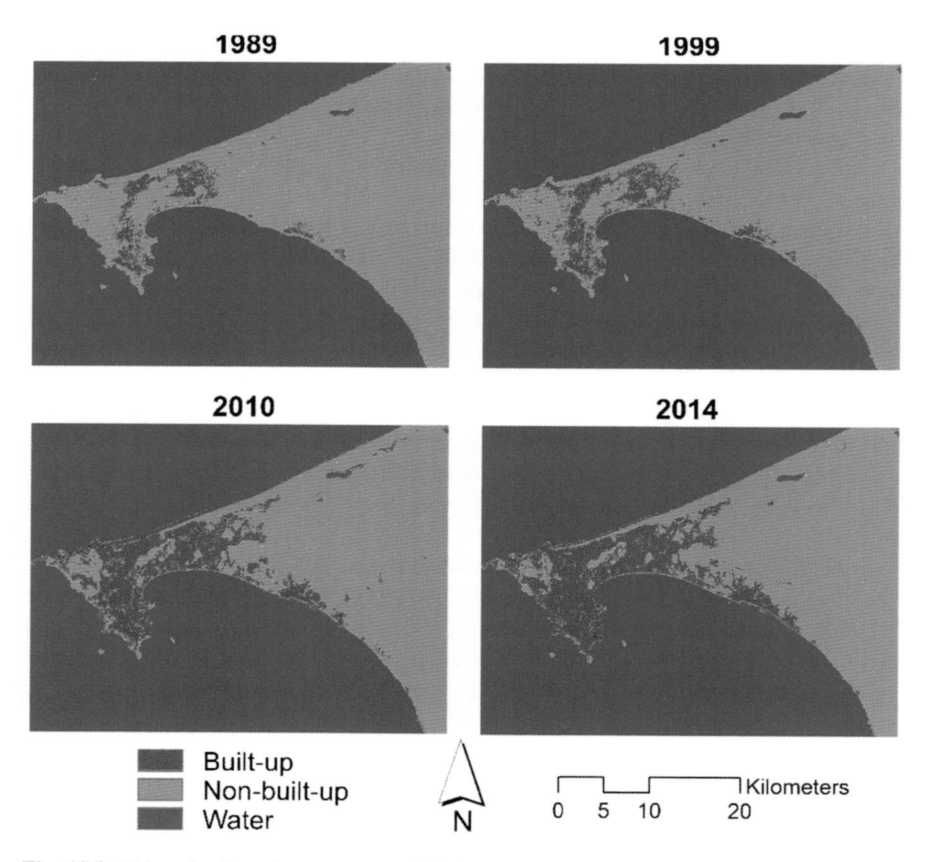

Fig. 13.3 Urban land use/cover maps of Dakar Metropolitan Area classified from Landsat imagery

Table 13.1 Observed urban land use/cover of Dakar Metropolitan Area (km^2)

	1989	1999	2010	2014
Built-up	30.6	47.9	97.5	118.6
Non-built-up	611.3	587.3	513.9	504.4
Water	957.9	964.2	988.5	976.9

30.6 to 118.6 km^2 between 1989 and 2014, whereas non-built-up areas decreased by 106.9 km^2. The water area varied slightly over the study period due to seasonal changes (Fig. 13.3).

Figure 13.4a shows that rates of land use/cover changes varied during the "1989–1999," "1999–2010," and "2010–2014" epochs. For example, "non-built-up to built-up" change was 17.3 km^2 at an annual rate of 1.7 km^2 between 1989 and 1999. A zonal analysis, which involved overlaying land use/cover changes with the Dakar district boundary, was performed in order to understand areas where the major changes occurred. The zonal analysis revealed that the major "non-built-up to built-up" changes occurred in Grand-Dakar, Almadies, Parcelles Assainies, and Thiaroye during the "1989–1999" epoch (Fig. 13.4a and b). Upward urban growth continued during the "1999–2010" epoch. The "non-built-up to built-up" changes increased significantly to 46.9 km^2 at an annual rate of 4.5 km^2. The major "non-built-up to built-up" changes occurred in Almadies, Les Niayes, Grand-Dakar, and Thiaroye (Fig. 13.4a and b). However, there was a decrease in "non-built-up to built-up" changes during the 2010–2014 epoch. The "non-built-up to built-up" changes decreased to 21.1 km^2 at an annual rate of 5.3 km^2. The major "non-built-up to built-up" changes also occurred in Almadies, Plateau, Les Niayes, and Thiaroye. Generally, the land use/cover change analysis revealed high rates of built-up expansion for the "1989–1999" and "1999–2010" epochs, while built-up growth slowed down during the "2010–2014" epoch.

13.3.2 Landscape Change Analysis (1989–2014)

The percentage of landscape (PLAND), patch density (PD), Euclidean nearest-neighbor distance (ENN), related circumscribing circle (CIRCLE), and shape index (SHAPE) spatial metrics for the built-up class were computed using FRAGSTATS 4.2 (McGarigal et al. 2012). PLAND is a landscape composition metric, which measures the proportion of a particular class relative to the whole landscape (McGarigal et al. 2012). The PLAND metric is 0% when the corresponding class is rare in the landscape, while PLAND is 100% when the entire landscape is dominated by a single class (McGarigal et al. 2012). The PD metric is a measure of fragmentation based on the number of patches per unit area. Low PD signifies that patches are compact and less fragmented, while high PD implies high fragmentation. The ENN metric is a patch isolation metric that measures dispersion based on the average distance to the nearest-neighboring patch of the same class (McGarigal et al. 2012). Generally, isolation measures the degree to which patches are spatially

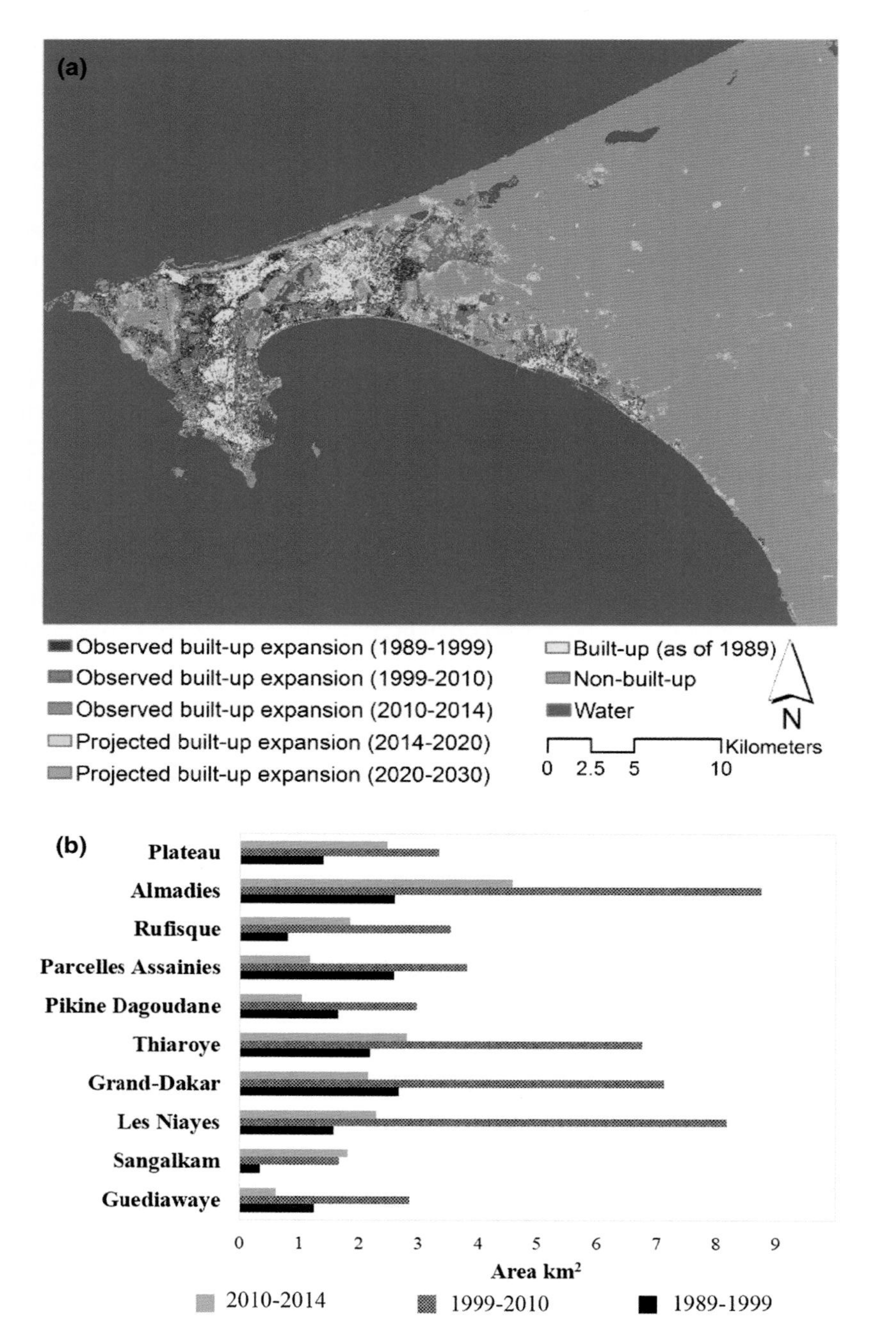

Fig. 13.4 **a** Observed and projected urban land use/cover changes in Dakar Metropolitan Area **b** observed urban land use/cover changes (built-up expansion) in Dakar Metropolitan Area (km^2)

Table 13.2 Observed landscape pattern of Dakar Metropolitan Area

Class-level (built-up) spatial metrics	1989	1999	2010	2014
PLAND (%)	4.8	7.5	15.9	19.0
PD (number per km^2)	1.9	2.4	0.1	0.1
ENN (mean) (m)	97.9	78.6	478.7	338.3
CIRCLE (mean) ($0 \leq$ CIRCLE < 1)	0.3	0.3	0.6	0.5
SHAPE (mean) ($1 \leq$ SHAPE $\leq \infty$)	1.2	1.3	3.4	2.2

isolated. ENN approaches 0 as the distance to the nearest neighbor decreases. CIRCLE is a shape metric that focuses on geometric complexity (McGarigal et al. 2012). Note that CIRCLE provides a measure of overall patch elongation. For example, CIRCLE is equivalent to 0 for circular or one cell patches and approaches 1 for elongated and linear patches one cell wide. The SHAPE metric measures the complexity of patch shape compared to a standard shape (e.g., square or circle) of the same size (McGarigal et al. 2012).

The observed class-level spatial/landscape metrics for the built-up class between 1989 and 2014 are shown in Table 13.2. The PLAND metric increased from 4.8 to 19% over the study period, which indicates an increase in built-up class in the metropolitan area. However, the low PLAND metric shows that most of the study area is dominated by non-built-up areas. The PD increased slightly from 1.9 per 100 km^2 to 2.4 per 100 km^2 during the "1989–1999" epoch, suggesting that built-up class became less compact due to rapid built-expansion. However, the PD decreased from 2.4 per 100 km^2 to 0.1 per 100 km^2 during the "1999–2014" epoch, which implies that built-up class became compact due to densification. The mean ENN decreased substantially from 97.9 to 78.6 m between 1989 and 1999, and increased significantly from 78.6 to 478.7 m between 1999 and 2010. However, the mean ENN decreased from 478.7 to 338.3 m between 2010 and 2014. The mean ENN indicated high levels of dispersion over the study period. The mean CIRCLE value varied slightly from 0.3 to 0.6 over the study period suggesting that the built-up pattern is less complex. Lastly, the mean SHAPE value also varied slightly indicating that the built-up area is regular in most parts of the study area.

Figure 13.5 shows spatial metrics for the built-up areas of Dakar Metropolitan Area (from 1989 to 2014) along the gradient of the distance from city center. In general, PLAND increased between 1989 and 2014 within all the distance buffer zones. Nevertheless, the magnitude of change was greater within the 0–15 km buffer zone, which indicates that the proportion of the built-up class near the city center was relatively high. The PD metric varied within the different buffer zones over the study period. For example, PD increased within the 0–5 km and 5–10 km buffer zones during 1989 and 1999, which implies fragmentation of the built-up class. However, PD decreased significantly within the 0–5 and 5–10 km buffer zone between 2010 and 2014 implying that built-up patches were becoming less fragmented. While the mean ENN varied over the study period, the most significant increase was observed in 2014, particularly within the 25–30 km distance buffer zone. The variation in mean ENN suggests a complex and mixed urban development processes characterized by sprawl

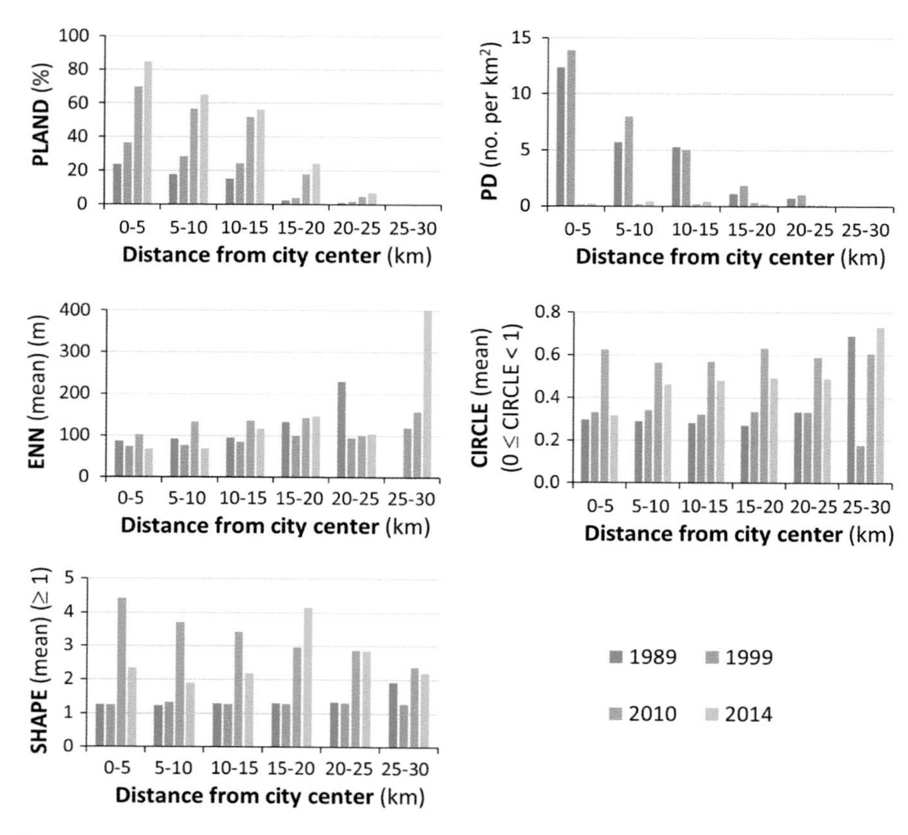

Fig. 13.5 Observed class-level spatial metrics for built-up along the gradient of the distance from city center of Dakar Metropolitan Area. *Note* The y-axis values are plotted in the same range as those in Fig. 13.7

and densification between 1989 and 2014. The mean CIRCLE and SHAPE values did not show a clear trend over the study period. However, the mean CIRCLE and SHAPE values are relatively high in 2010, which imply that the built-up class was complex and irregular. This attributed to the rapid built-up expansion that occurred in 2010. Generally, built-up expansion between 1989 and 1999 was relatively slow while built-up areas expanded rapidly between 1999 and 2014.

13.3.3 Driving Forces of Urban Development

Urban development in Dakar Metropolitan Area was driven by various and complex socioeconomic, political, and natural driving factors. According to the

UN-Habitat (2014), Dakar Metropolitan Area is increasing by more than 100,000 inhabitants annually in the current decade. Literature review shows that Dakar Metropolitan Area grew from approximately 1.3 million people in 1985 to about 3.1 million people in 2013 (Arrechi 1985; Agence Nationale de la Statistique et de la Démographie 2016). Urban growth of Dakar Metropolitan Area, like in the other cities in West Africa, was driven by natural increase in city populations rather than by rural–urban migration (UN-Habitat 2014). However, lack of appropriate urban development policies coupled with a dearth of formal real estate market mechanisms has resulted in failure to meet the housing needs of the expanding population (Scott et al. 2013). Consequently, there is a proliferation of unplanned settlements, which is one of the key drivers of urban development (UN-Habitat 2012; Scott et al. 2013). This situation is observed across most of the cities in sub-Saharan Africa. In general, the majority of housing developments in Dakar Metropolitan Area were spontaneous and self-built (Scott et al. 2013). Furthermore, as in most Sub-Sahara African countries, the colonial administration policy of "dual cities" (Bigon 2012) fueled the growth of informal settlements in the outskirts, which is also one of the major causes of the current urban development pattern. According to Arrechi (1985), many people were moved to Dagoudane–Pikene area—which was about 13 km from central Dakar—in 1952. To date, approximately 50% of the residents of Dakar Metropolitan Area live in Pikine, which has a population density of 10,166 inhabitants/km^2 (Scott et al. 2013). Figure 13.3 shows that the urban expansion of the Dakar Metropolitan area has been dominant eastwards. The rate of growth was especially rapid after Senegal became an independent state in 1960 (Arrechi 1985).

Economic growth and availability of economic opportunities in Dakar also increased urban development. While the Dakar Metropolitan Area represents only 0.3% of Senegal, more than 80% of the national economic activities occur in the metropolitan area (Sy 2013). As a result, Dakar Metropolitan Area attracts migrants. In order to understand the economic factors behind the rapid urban development in Dakar Metropolitan Area, the World Bank's gross national income (GNI) index was used as a proxy. According to the World Bank (2016), "GNI is the sum of value added by all resident producers plus any product taxes (less subsidies) not included in the valuation of output plus net receipts of primary income (compensation of employees and property income) from abroad." The GNI is converted to international dollars, which has the same purchasing power as the US dollar (World Bank 2016). The World Bank statistics (2016) indicated that GNI increased from about $9 billion to about $15 billion between 1990 and 2000. However, between 2000 and 2010, GNI increased significantly from about $15 billion to about $27 billion, which corresponds to high rate of built-expansion between 1999 and 2010. Furthermore, the slowdown in built-up expansion between 2010 and 2014 is also related to the decrease of the economy in Senegal over the same period. According to the World Bank statistics (2016), GNI increased slightly from about $27 billion to about $34 billion between 2010 and 2014. Given that many Senegalese live in the diaspora, remittances from abroad have impact in the urban development in Dakar Metropolitan Area as people living abroad invest in properties. According to

Newland and Plaza (2013), Senegal established an investment fund for Senegalese living abroad, which has financed 804 projects worth 20 billion CFA francs ($40 million).

13.4 Projected Future Land Use/Cover Changes

13.4.1 Projected Land Use/Cover Changes

Future land use/cover changes for 2020 and 2030 (Fig. 13.6) were simulated based on the calibration scenarios between 1989 and 2010 (see Chap. 2). Based on the 1989 land use/cover base map, the multiple transition probabilities and the "1989–2010" transition potential maps, the land change model projected that built-up areas would increase substantially from 118.6 km^2 in 2014 to 146.7 km^2 in 2030, while non-built-up areas would decrease from 504.5 to 476.3 km^2 over the same study period (Table 13.3). The annual rate of built-up change was projected to be 7.6 km^2 for the "2014–2020" epoch, and 20.6 km^2 for the "2020–2030" epoch (Table 13.3). The simulated built-up areas in 2020 and 2030, and the observed built-up areas in 1990 indicated that urban expansion will continue in the future (Fig. 13.6).

The spatial metric results (Table 13.4) for the built-up class indicate that PLAND would likely increase from 20.2 to 23.5% between 2020 and 2030, which indicates an increase of built-up areas in the future. Furthermore, PD would increase slightly from 0.4 per 100 km^2 to 0.7 per 100 km^2, which implies that most of the built-up class will be slightly fragmented. The mean ENN would decrease slightly from 239.6 to 219.4 m between 2020 and 2030, which implies high levels of dispersion for the built-up areas. The mean CIRCLE value would likely remain

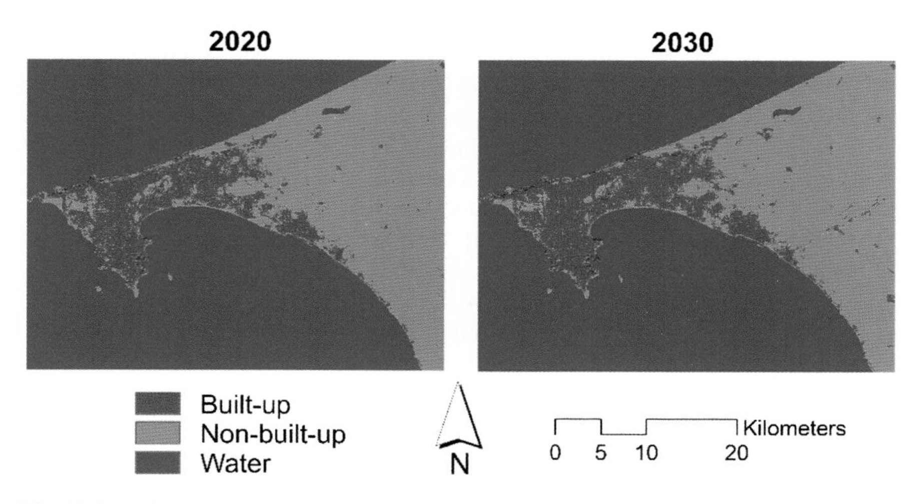

Fig. 13.6 Projected urban land use/cover maps of Dakar Metropolitan Area

Table 13.3 Projected urban land use/cover of Dakar Metropolitan Area (km^2)

	2020	2030
Built-up	126.1	146.7
Non-built-up	496.9	476.3
Water	976.9	976.9

Table 13.4 Projected landscape pattern of Dakar Metropolitan Area

Class-level (built-up) spatial metrics	2020	2030
PLAND (%)	20.2	23.5
PD (number per km^2)	0.4	0.7
ENN (mean) (m)	239.6	219.4
CIRCLE (mean) ($0 \leq$ CIRCLE < 1)	0.4	0.4
SHAPE (mean) ($1 \leq$ SHAPE $\leq \infty$)	2.1	1.8

constant over the study period, while the mean SHAPE value would decrease slightly (Table 13.4).

Figure 13.7 shows spatial metrics for the built-up areas of Dakar Metropolitan Area (2020 and 2030) within distance buffer zones from the city center. Results show that PLAND would decrease with increasing distance in 2020 and 2030. This implies that built-up patches would be continuous near the city center and scattered further away from the city center. The simulated PD results for Dakar are quite interesting given that the PD metric varied within the distance buffer zones. For example, the PD was low within the 0–5 km distance buffer zone in 2020 and 2030. In contrast, the PD increased within the 5–10 km distance buffer zone in 2020 and 2030. This suggests that the built-up patches are compact and less fragmented within the 0–5 km distance buffer zone, but more fragmented within the 5–10 km distance buffer zones. Nonetheless, the PD decreased again within the 10–15 km distance buffer in 2020 and 2030, indicating less fragmentation, and then increased again within the 15–20-km buffer zones implying fragmentation. Thereafter, the PD would decrease within the 20–25-km and 25–30-km distance buffer zones. The variation in PD suggests that the Dakar Metropolitan Area has two or more urban cores. As a result, densification and sprawl processes would be active in many different urban cores of Dakar Metropolitan Area (polycentric urban development) in the future. Generally, the mean ENN would likely increase with distance for 2020 and 2030, which implies that built-up patches would likely become scattered over the 20–25-km buffer zones. The mean CIRCLE value indicated high variations in 2020, which suggests that the built-up patch would be less stable. In contrast, the mean CIRCLE value indicated low variations in 2030 implying that the built-up patch would be more stable in the future. Nonetheless, the mean SHAPE value would vary in 2020 implying that the built-up patches will be more irregular. Contrarily, the mean SHAPE value would vary less in 2030 implying that the built-up patches will be less irregular. The spatial metrics suggest that future urban development characterized by a combination of densification and sprawl processes would be centered on two or more urban cores within Dakar Metropolitan Area.

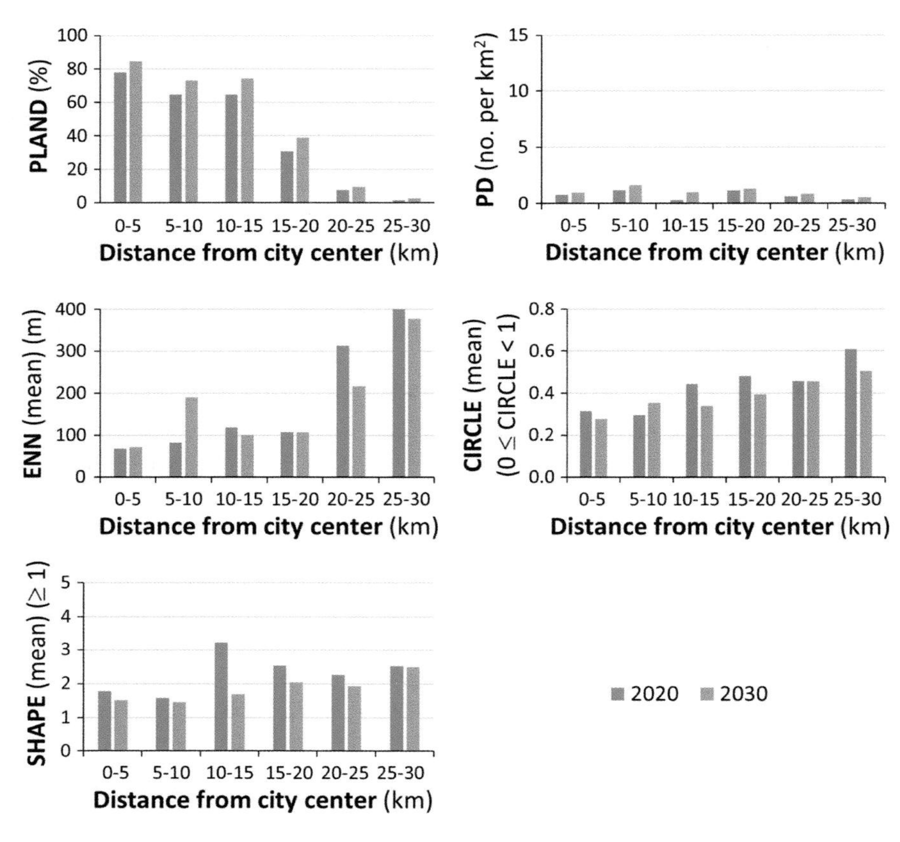

Fig. 13.7 Projected class-level spatial metrics for built-up along the gradient of the distance from city center of Dakar Metropolitan Area. *Note* The y-axis values are plotted in the same range as those in Fig. 13.5

13.4.2 Implications for Future Sustainable Urban Development

The rapid urban expansion in Dakar Metropolitan Area has major implications on sustainable urban development. Generally, population increase coupled with historical and current housing policies as well as the economic changes had significant influence on the spatial expansion of built-up areas in Dakar. According to the UN Nations (2012), over 50% of the urban population in Dakar currently lives in informal settlements. As in other counties in sub-Saharan Africa, rapid expansion of built-up areas creates major difficulties for urban planning, particularly the provision of basic social services and infrastructure development. As mentioned before, a poor urban policy framework coupled with weak real estate market mechanism among other factors pose major obstacles for sustainable urban development. Therefore, an in-depth understanding of spatial patterns and underlying processes of built-up expansion in Dakar Metropolitan Area is required.

Within this context, built-up change analysis and simulations in this chapter provide an analytical framework in order to assess urban growth as well as analyze its underlying driving forces. For example, the current built-up changes and spatial metrics suggest that the Dakar Metropolitan Area has two or more urban cores, which are characterized by a combination of densification and sprawl processes. Therefore, policy and decision makers need to identify first the major urban cores where urban development is likely to occur, and then focus attention on reducing discontinuous and scattered built-up expansion. This chapter has shown that built-up expansion was rapid in Les Niayes, Almadies, Grand-Dakar, and Thiaroye districts between 1989 and 2014. Furthermore, the simulated land use/cover changes indicated that built-up expansion would continue under the current urban development scenario. Therefore, to implement effective sustainable urban development in Dakar Metropolitan Area, more priority should be given to Les Niayes, Almadies, Grand-Dakar, and Thiaroye districts. This can be accomplished by integrating mixed land use patterns, which would minimize transport and infrastructure development costs.

Government and local authorities in Senegal have intensified efforts to improve sustainable urban development based on the “2035 Dakar Urban Masterplan” in Dakar Metropolitan Area (Ministry of Urban Renewal, Housing and Living Environment and JICA 2016). However, implementation of the “2035 Dakar Urban Masterplan” requires timely monitoring of urban development. The observed and simulated land use/cover maps in this chapter can be used as a base to monitor built-up expansion in Dakar Metropolitan Area. In addition, the observed and simulated land use/cover changes can also be used as an initial scientific or diagnostic tool for guiding sustainable urban development. This can also possibly assist researchers, policy makers, and other stakeholders in assessing the implications of future built-up expansion.

13.5 Summary and Conclusions

The objective of this chapter was to analyze observed and simulated land use/cover changes in Dakar Metropolitan Area. The land use/cover maps for 1989, 1999, 2010, and 2014 along with change analyses and spatial metrics indicated significant built-up expansion over the study period. Built-up growth in Dakar Metropolitan Area was attributed to complex socioeconomic, political, and natural driving factors. The evolution of Dakar Metropolitan Area reflects the multiple and complex nature of a polycentric urban development. The analysis of the spatial metrics with distance buffer zones revealed that built-up expansion was characterized by a combination of densification and sprawl from different urban cores. Although urban growth dynamics are too complex to capture in a short period, this chapter provided an overview of built-up expansion in Dakar Metropolitan Area. Future land use/cover simulations (up to 2030) indicated that the land use/cover change trends such as the increase in built-up areas and decrease in non-built-up areas would continue under the current growth scenario. This chapter has also demonstrated that an integrated approach incorporating remote sensing, spatial metrics, and simulation tools help to understand land use/cover changes in Dakar Metropolitan Area. Therefore, land use/cover products could potentially assist decision makers and other stakeholders with built-up extent in Dakar Metropolitan Area.

Acknowledgements The comments and suggestions by Dr. Souleye Wade (African Institute of Geomatics Group) were most helpful and are appreciated.

References

Agence Nationale de la Statistique et de la Démographie (ANSD) (2016) Situation Economique et Sociale du Senegal en 2013. Ministere de L'Economie, des Finances et du Plan. Republique du Senegal. Accessed on 22 Feb 2016 from http://www.ansd.sn/ressources/publications/1-demographie-SESN2013.pdf

Agence Nationale de la Statistique et de la Démographie (ANSD) (2016a) Service Régional de la Statistique et de la Démographie de Dakar. Ministere de L'Economie, des Finances et du Plan. Republique du Senegal. Accessed on 22 Feb 2016 from http://www.ansd.sn/index.php?option=com_ansd&view=titrepublication&id=23&Itemid=294

Agence Nationale de la Statistique et de la Démographie (ANSD) (2016b) Service Régional de la Statistique et de la Démographie de Thiès. Ministere de L'Economie, des Finances et du Plan. Republique du Senegal. Accessed on 22 Feb 2016 from http://www.ansd.sn/index.php?option=com_ansd&view=titrepublication&id=23&Itemid=294

Agence Nationale de la Statistique et de la Démographie (ANSD) (2016c) Service Régional de la Statistique et de la Démographie de Diourbel. Ministere de L'Economie, des Finances et du Plan. Republique du Senegal. Accessed on 22 Feb 2016 from http://www.ansd.sn/index.php?option=com_ansd&view=titrepublication&id=23&Itemid=294

Agence Nationale de la Statistique et de la Démographie (ANSD) (2016d) Service Régional de la Statistique et de la Démographie de Kaolack. Ministere de L'Economie, des Finances et du Plan. Republique du Senegal. Accessed on 22 Feb 2016 from http://www.ansd.sn/index.php?option=com_ansd&view=titrepublication&id=23&Itemid=294

Arecchi A (1985) Dakar. Cities 2:198–211

Bigon L (2012) Garden City' in the tropics? French Dakar in comparative perspective. J Hist Geogr 38:35–44

Creevey L (1980) Religious attitudes and development, Senegal. World Dev 8:503–512

McGarigal K, Cushman SA, Ene E (2012) FRAGSTATS v4: spatial pattern analysis program for categorical and continuous maps. Computer software program produced by the authors at the University of Massachusetts Amherst. Accessed 1 July 2015 from http://www.umass.edu/landeco/research/fragstats/fragstats.html

Ministry of Urban Renewal, Housing and Living Environment and JICA (2016) Project for urban master plan of Dakar and neighboring area for 2035. Final Report Summary

Newland K, Plaza S (2013) What we know about diasporas and economic development. Migration Policy Institute, Policy Brief No. 5

Scott P, Cotton A, Khan MS (2013) Tenure security and household investment decisions for urban sanitation: the case of Dakar, Senegal. Habitat Int 40:58–64

Sy AB (2013) Dakar: a fragmented agglomeration. Accessed on 20 Jan 2016 from http://www.urbanafrica.net/news/dakar-a-fragmented-agglomeration/

United Nations Human Settlements Programme (UN-HABITAT) (2012) The State of world's Cities 2012/2013. Nairobi, Kenya

United Nations Human Settlements Programme (UN-Habitat) (2014) The state of African cities 2014: re-imagining sustainable urban transitions. Nairobi, Kenya

World Bank, International Comparison Program database (2016) Accessed on 21 Mar 2016 from http://databank.worldbank.org/data/reports.aspx?source=2&country=SEN&series=&period=#

Chapter 14
Bamako Metropolitan Area

Courage Kamusoko

Abstract The objective of this chapter was to analyze observed and projected land use/cover changes between 1990 and 2030 in Bamako Metropolitan Area. The land use/cover change analysis revealed significant built-up expansion for the "1990–2000" and "2010–2014" epochs, while built-up expansion slowed down during the "2000–2010" epoch. Built-up growth in Bamako Metropolitan Area was characterized by a low-density urban sprawl moving outward from the urban core into the surrounding rural areas. Generally, vacant lands in the surrounding rural areas were converted to residential and urban land uses. Future land use/cover simulations (up to 2030) indicated that the current land use/cover change trends, such as the increase in built-up areas and decrease in non-built-up areas as well as low-density urban sprawl, would continue to persist. The observed and simulated land use/cover changes provide an overview of built-up expansion as well as a simulated urban growth scenario for Bamako Metropolitan Area. This could potentially assist decision-makers with general built-up change information that can be used to guide sustainable urban development in Bamako Metropolitan Area.

14.1 Origin and Brief History

Bamako Metropolitan Area is located in the southwest of Mali (Fig. 14.1a). The metropolitan area is separated into the northern and southern parts by the Niger River. Bamako Metropolitan Area has a mean annual temperature of 28.2 °C and a mean annual precipitation of 82.6 mm. High temperatures are recorded in March, April, and May, while temperatures vary from 26.7 °C in November to 25.2 °C in February. The rainy season occurs in summer, with highest rainfall recorded in July, August, and September.

According to Dembele (2004), pre-colonial Bamako, which was located along the Niger River, was a secondary commercial center of West Africa. Pre-colonial

C. Kamusoko (✉)
Asia Air Survey Co. Ltd, Kawasaki, Japan
e-mail: kamas72@gmail.com

Y. Murayama et al. (eds.), *Urban Development in Asia and Africa*,
The Urban Book Series, DOI 10.1007/978-981-10-3241-7_14

Fig. 14.1 **a** Location and LandScan population of Bamako Metropolitan Area, Mali. ► **b** Communes of Bamako District overlaid on OpenStreetMap and hillshade. *Note* Commune boundaries were downloaded from United Nations Office for the Coordination of Humanitarian Affairs Mali country office (https://data.hdx.rwlabs.org/organization/ocha-mali?page=2#datasets-section). **c** Aerial view of residential areas under development in Commune VI. *Source* Author's fieldwork (2016). **d** Government buildings in Commune IV. *Source* Author's fieldwork (2016). **e** The tall orange building is the BCEAO Tower (Malian headquarters of the Central Bank of West African), which is located to the north of Niger River. *Source* Author's fieldwork (2016) **f** Residential and shops in Commune V. *Source* Author's fieldwork (2016)

Bamako was inhabited by the Bambara, Bozo, Maure, and Arab ethnic groups (Dembele 2004). The French occupied Bamako—as a strategic point in order to invade West Sudan—at the end of the nineteenth century. In 1904, the former capital of West Sudan was relocated to Bamako (Vaa 2000). During the pre-independence era, the development of the city was moderate and gradual. In 1960, Bamako became the capital of Mali when the country gained independence. Although the original site of the city was located on the north bank, construction of the Martyrs Bridge and the King Fahd Bridge promoted the development of Bamako toward the south (Vaa 2000). Bamako has continued to sprawl particularly to the south, which does not have major physical constraints as the north (Fig. 14.1b). According to estimations, the population of Bamako grew from approximately 757,051 in 1987 to 2,156,177 in 2009 (World Bank 2015). While population estimates vary according to sources, Bamako was estimated to be increasing at an annual urban growth rate of 5.8% (Farvacque-Vitkovic et al. 2007).

The current spatial structure of Bamako is largely based on the French colonial era development model. The district of Bamako is divided into six communes (Fig. 14.1b). Each commune is administered by a municipal council and an elected mayor. Commune I is bordered to the north by the rural commune of Djalakorodji and Sangarebougou, to the northeast by Moribabougou, to the east by the rural commune of Gabakourou and Baguineda Camp, to the south by Commune VI, and to the west by Commune II. Commune II is limited to the north by the rural commune of Djalakorodji, to the east by Commune I, to the south by Commune V, and to the west by Commune III. The industrial sector of the city is located in Commune II. Commune III is bordered to the north by the rural commune of Dogodouman, to the east by Commune II, to the south by the portion of the Niger River and Commune V, and to the west by Commune IV. Commune III is the administrative and commercial center of Bamako. Commune IV is limited to the north and east by Commune III, to the west by Dogodouman, and to the south by Mande and the left bank of the Niger River. Commune V is bordered to the north by Communes II, III, and IV as well as parts of the Niger River, to the south by the rural commune of Kalaban Coro, and to the east by the Commune VI. Commune VI, where the airport is located, is the largest in Bamako District. This commune is bordered to the north by Commune I, to the east by Baguineda Camp, to the south and west by the rural commune of Kalaban Coro. Figure 14.1c–f shows selected sites in Bamako Metropolitan Area.

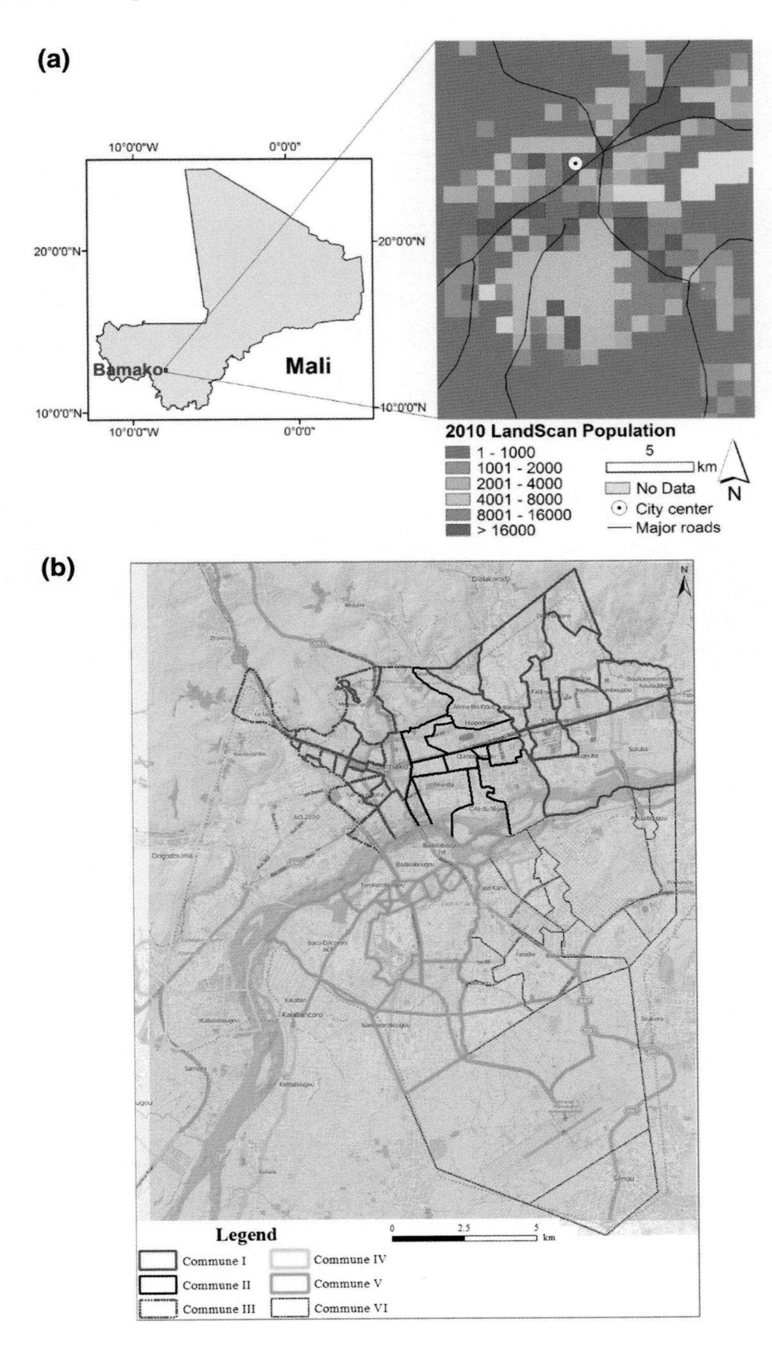
(a)
10°0'0"W
0°0'0"
20°0'0"N
10°0'0"N
Bamako
Mali
2010 LandScan Population
1 - 1000
1001 - 2000
2001 - 4000
4001 - 8000
8001 - 16000
> 16000
5
km
No Data
City center
Major roads
N
(b)
Legend
0
2.5
5
km
Commune I
Commune II
Commune III
Commune IV
Commune V
Commune VI

Fig. 14.1 (continued)

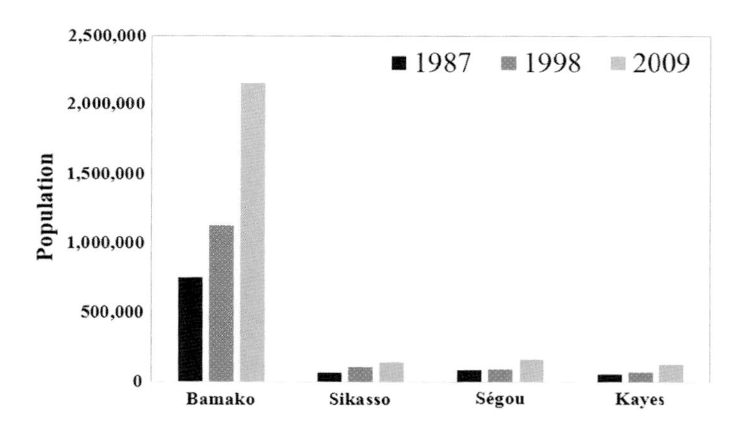

Fig. 14.2 Population distribution in major cities of Mali. *Source* World Bank (2015)

14.2 Primacy in the National Urban System

In Mali, urban population grew from 2,595,596 in 1998 to 4,766,170 in 2010 (World Bank 2007). To date, about 35% of the population in Mali is estimated to live in urban areas (World Bank 2007). Figure 14.2 summarizes the details of the four most populous cities and urban centers in Mali. Approximately 50% of the urban dwellers who live in urban areas in Mali are residents of Bamako. According to the World Bank report (2015), Bamako has a population of more than 2.1 million, representing 15% of the total population in Mali. Segou, Sikasso, and Kayes, the three largest cities after Bamako, have populations of 166,000, 143,000, and 127,000, respectively. Figure 14.2 shows that urban population is highly concentrated in Bamako.

According to the rank-size rule, the largest city should only be twice as large as the second largest city, while the second largest city should be three times as the third largest city. Bamako is more than six times larger than Segou, the second largest city. This clearly demonstrates the status of Bamako as a primate city. In addition, the population Bamako is approximately two times larger than the total population of Segou, Sikasso, and Kayes. This clearly shows the spatial imbalance of the cities in Mali, which is largely attributed to the concentration of economic opportunities and political power in the capital city. The dominance of Bamako is also attributed to the long distances between cities, poor intercity transport, and communication network (World Bank 2015). According to the World Bank report (2015), most urban places are located along the major rivers in the south of Mali, where land is more suitable for agriculture, and the transport network is denser. Furthermore, the primacy of Bamako is exacerbated by refugees who come from northern parts of the country and from other neighboring countries (World Bank 2015). This concentration of inhabitants has led to increased poor living and sanitation conditions in the city.

14.3 Urban Land Use/Cover Change and Landscape Analysis (1990–2014)

14.3.1 Land Use/Cover Change Analysis (1990–2014)

Land use/cover maps and statistics of Bamako Metropolitan Area for 1990, 2000, 2010, and 2014 are shown in Fig. 14.3 and Table 14.1. The results indicate that built-up areas occupied 6.2 km^2, while non-built-up areas occupied 386.7 km^2 in 1990. Note that reference data were not available for the 1990 land use/cover map. Therefore, there is uncertainty associated with the 1990 land use/cover map, which is reflected by the low quantity of built-up areas. However, significant spatial expansion in built-up areas and subsequent decreases in non-built-up areas were observed in 2000. Built-up areas increased to 46.3 km^2, while non-built-up areas

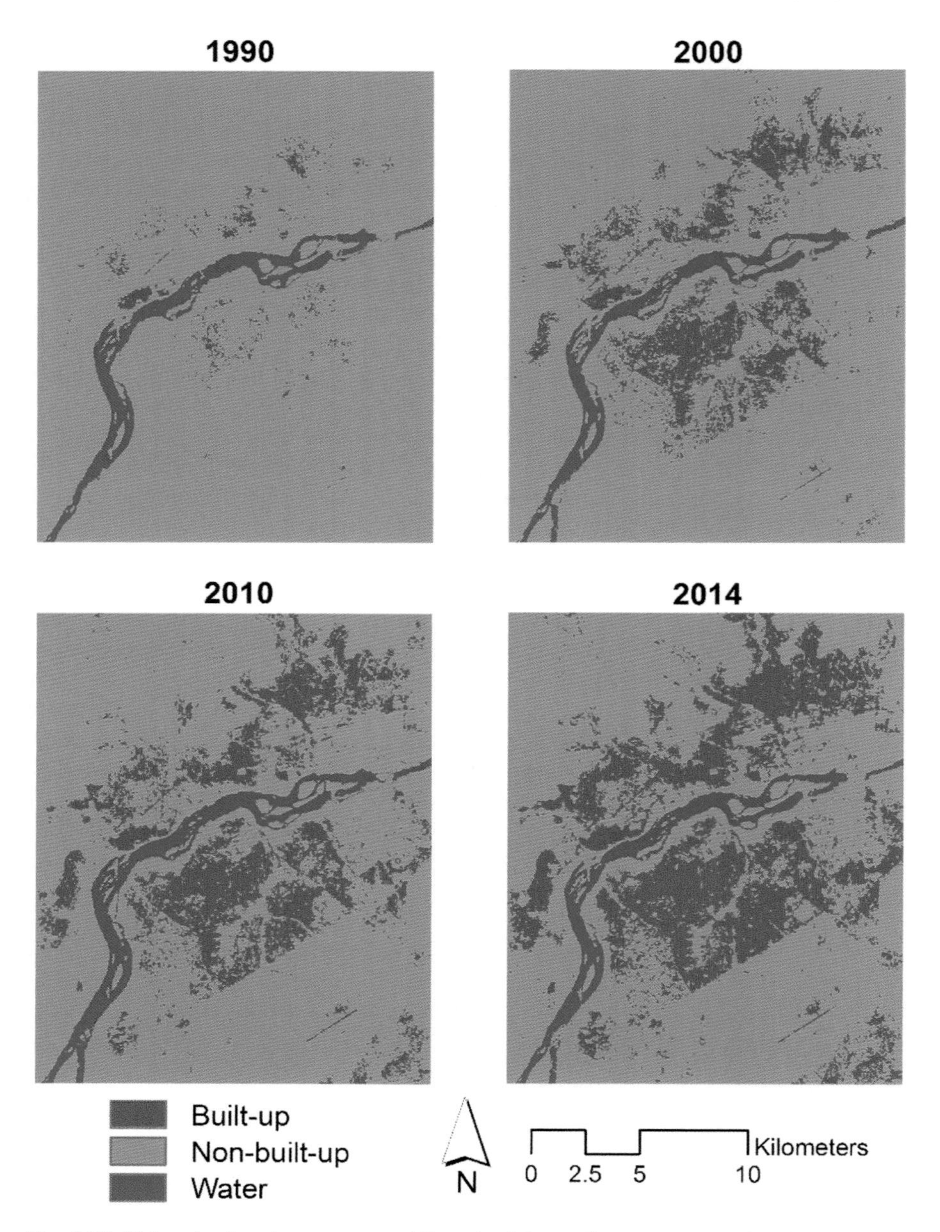

Fig. 14.3 Urban land-use/cover maps of Bamako Metropolitan Area classified from Landsat imagery

Table 14.1 Observed urban land use/cover of Bamako Metropolitan Area (km^2)

	1990	2000	2010	2014
Built-up	6.2	46.3	72.8	103.9
Non-built-up	386.7	347.3	317.4	289.0
Water	18.1	17.3	20.8	18.0

decreased to 347.3 km^2. Furthermore, analysis of the 2010 land use/cover map revealed additional increases in built-up areas which occupied 72.8 km^2, while non-built-up areas decreased to 317.4 km^2. For the 2014 land use/cover map, built-up and non-built-up areas occupied 103.9 and 289 km^2, respectively. In general, built-up areas increased substantially by 97.7 km^2 between 1990 and 2014, whereas non-built-up areas decreased. The water area varied slightly over the study period due to seasonal changes (Fig. 14.3).

Figure 14.4a shows that rates of land use/cover changes varied during the "1990–2000," "2000–2010," and "2010–2014" epochs. For example, "non-built-up to built-up" change was approximately 40.1 km^2 at annual rate of 4 km^2 between 1990 and 2000. We also performed a zonal analysis by overlaying land use/cover changes with the Bamako District boundary in order to understand areas where the major changes occurred. The major "non-built-up to built-up" changes occurred in communes V and VI in the south, and communes I and IV in the north of Niger River (Fig. 14.4a and b). However, during the "2000–2010" epoch, "non-built-up to built-up" changes slowed down to approximately 26.4 km^2 at an annual rate of 2.6 km^2. The major "non-built-up to built-up" changes also occurred in communes V and VI in the south and communes I and IV in the north (Fig. 14.4a and b). The "2010–2014" epoch saw a slight increase in "non-built-up to built-up" changes. The "non-built-up to built-up" changes increased to 31.1 km^2 at an annual rate of 7.8 km^2. The major "non-built-up to built-up" changes also occurred in the same communes as the previous epochs. Generally, the land use/cover change analysis revealed high rates of built-up expansion for the "1990–2000" and "2010–2014" epochs. However, built-up growth slowed down during the "2000–2010" epoch.

Land use/cover change analysis shows that built-up expansion between 1990 and 2000 was largely dominated by extension and infill developments (Fig. 14.4a). For example, the expansion of built-up areas in communes V and VI in the south and communes I and IV in the north shows infill and extension developments. This is because most of the growth of newly developed areas in 2000 occurred in areas that were already urbanized in 1990. Extension also occurred with the expansion of the existing built-up areas. However, the growth of built-up areas during the "2000–2010" period was dominated by the extension and leapfrog developments. Leapfrog development is characterized by the conversion of non-developed areas into newly developed areas outside of existing urban built-up areas. This development is conspicuous in the southern parts of the Niger River, especially in Commune VI and the rural commune of Kalaban Coro (Fig. 14.1b and Fig. 14.4a). The "non-built-up to built-up" change patterns between 2010 and 2014 also indicated the continuation of extension and leapfrog developments in Bamako Metropolitan Area.

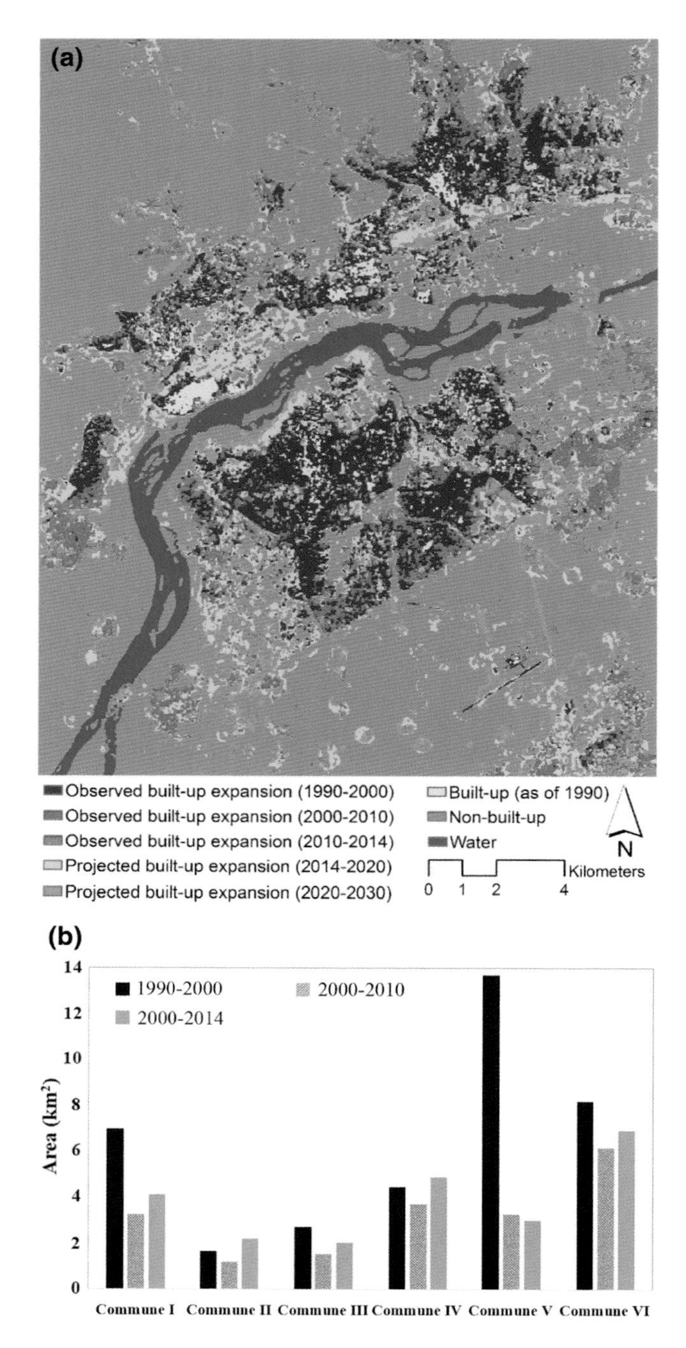

Fig. 14.4 **a** Observed and projected urban land use/cover changes in Bamako Metropolitan Area and **b** observed gains of built-up at the commune level for "1990–2000," "2000–2010," and "2010–2014" epochs

14.3.2 Landscape Change Analysis (1990–2014)

The percentage of landscape (PLAND), patch density (PD), Euclidean nearest-neighbor distance (ENN), related circumscribing circle (CIRCLE), and shape index (SHAPE) spatial metrics for the built-up class were computed using FRAGSTATS 4.2 (McGarigal et al. 2012). PLAND is a landscape composition metric, which measures the proportion of a particular class relative to the whole landscape (McGarigal et al. 2012). The PLAND metric is 0% when the corresponding class is rare in the landscape, while PLAND is 100% when the entire landscape is dominated by a single class (McGarigal et al. 2012). The PD metric is a measure of fragmentation based on the number of patches per unit area. Low PD signifies that patches are compact and less fragmented, while high PD implies high fragmentation. The ENN metric is a patch isolation metric that measures dispersion based on the average distance to the nearest neighboring patch of the same class (McGarigal et al. 2012). Generally, isolation measures the degree to which patches are spatially isolated. ENN approaches 0 as the distance to the nearest neighbor decreases. CIRCLE is a shape metric that focuses on geometric complexity (McGarigal et al. 2012). Note that CIRCLE provides a measure of overall patch elongation. For example, CIRCLE is equivalent to 0 for circular or one cell patches and approaches 1 for elongated and linear patches one cell wide. The SHAPE metric measures the complexity of patch shape compared to a standard shape (e.g., square or circle) of the same size (McGarigal et al. 2012).

The observed class-level spatial metrics for the built-up class between 1990 and 2014 are shown in Table 14.2. The PLAND metric increased from 1.6 to 26.4% over the study period, which indicates an increase in built-up class in the metropolitan area. However, the low PLAND metric shows that most of the study area is dominated by non-built-up areas. The PD increased slightly from 1.6 to 3.2 per km^2 during the "1990–2000" epoch, suggesting that built-up class became less compact due to rapid built-up expansion. The mean ENN decreased substantially from 125.6 to 112.3 m between 1990 and 2010. However, the mean ENN increased slightly between 2010 and 2014. While the mean ENN shows some variations, the values indicate low levels of dispersion between the built-up areas. The mean CIRCLE and SHAPE values did not change significantly over the study period,

Table 14.2 Observed landscape pattern of Bamako Metropolitan Area

Class-level (built-up) spatial metrics	1990	2000	2010	2014
PLAND (%)	1.6	11.8	18.7	26.4
PD (number per km^2)	1.6	3.0	3.6	3.2
ENN (mean) (m)	125.6	122.0	112.3	117.6
CIRCLE (mean) ($0 \leq$ CIRCLE < 1)	0.4	0.4	0.4	0.4
SHAPE (mean) ($1 \leq$ SHAPE $\leq \infty$)	1.2	1.3	1.3	1.3

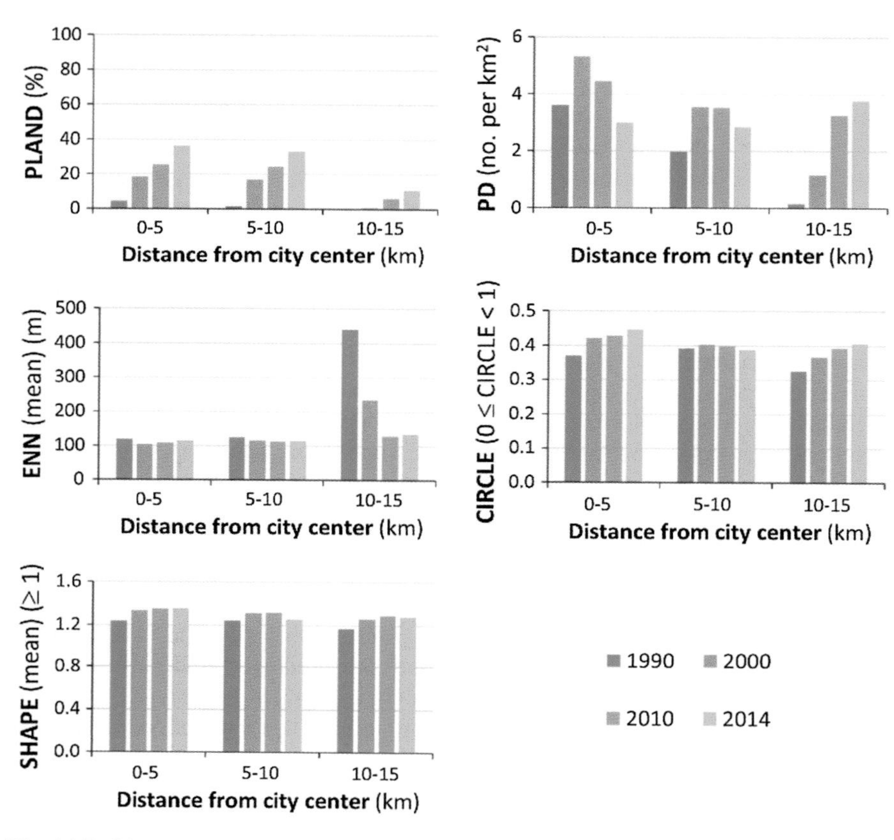

Fig. 14.5 Observed class-level spatial metrics for built-up along the gradient of the distance from city center of Bamako Metropolitan Area. *Note* The y-axis values are plotted in the same range as those in Fig. 14.8

suggesting that the built-up pattern is less complex and regular in most parts of the study area.

Figure 14.5 shows spatial metrics for the built-up areas of Bamako Metropolitan Area (from 1990 to 2014) along the gradient of the distance from the city center. Generally, PLAND increased between 1990 and 2014 within all the distance buffer zones. However, the magnitude of change was greater within the 0–15-km buffer zone. This indicates that the proportion of the built-up class near the city center is relatively high. The PD metric varied within the different buffer zones over the study period. For instance, PD increased within the 0–5 and 5–10-km buffer zones between 1990 and 2000, which suggests that the built-up patches were fragmented. Nevertheless, PD decreased within the 0–5-km and 5–10-km buffer zone between 2010 and 2014, suggesting that built-up patches were becoming less fragmented. Yet, PD increased substantially within the 10–15 km distance buffer zone over the study period, indicating that built-up patches became less compact with distance. This is mainly attributed to outward and leapfrog developments. The mean ENN metric was relatively constant within the 0–5 and 5–10-km buffer zones over the study period. However, the mean ENN substantially decreased within the 10–15-km buffer zones, which indicates low dispersion level. This implies that built-up patches within the 10–15-km zones were more connected. Generally, the mean CIRCLE and SHAPE values did not show a clear trend over the study period. Nevertheless, the relatively low mean CIRCLE and SHAPE values suggest that the built-up class is less complex and more regular. In general, the analysis of the spatial metrics with distance buffer zones indicates that 1990 built-up areas were scattered given that the initial built-up area in 1990 was substantially small. Built-up expansion between 2000 and 2014 suggests that the built-up patches were coalescing. On the other hand, the built-up areas became discontinuous and scattered further away from the city center, which is partly attributed to outward and leapfrog developments.

14.3.3 Driving Forces of Urban Development

Urban development is influenced by many complex underlying and proximate driving forces, which vary over space and time (Turner et al. 1995; Geist and Lambin 2001). Qualitative analysis of the driving forces of urban development for the "1990–2000," "2000–2010," and "2010–2014" epochs was based on the literature review. Generally, the land use/cover changes and the subsequent built-up expansion in Bamako Metropolitan Area were driven by a number of socioeconomic, political, and natural driving factors.

According to the UN-Habitat (2012), Bamako Metropolitan Area is one of the ten fastest growing cities in the world. The metropolitan area (i.e., Bamako District and the surrounding rural communes) has experienced high population growth due to rural–urban migration and high natural population growth. The UN-Habitat (2012) estimated that approximately 2,350,000 people were living in Bamako Metropolitan Area in 2011. While the population in Bamako District was reported to be increasing at approximately 5.8% per year, the annual growth rate of the population in the surrounding rural communes was higher—ranging from 6.2 to 17.2%. As a result, the demand for land for housing in Bamako Metropolitan Area has increased due to massive population increase.

Economic growth and availability of economic opportunities in Bamako also increased urban development. According to Durand-Lasserve et al. (2013), the income of urban households as well as the emergence of an urban middle class, albeit fairly small combined with remittances from Malians living in the diaspora, also resulted in housing developments in Bamako Metropolitan Area. In addition, land in the city and the surrounding rural communes were also acquired for speculative investments given lack of other investment opportunities in Mali

(Durand-Lasserve et al. 2013). This led to a high demand for land since investors speculated that property prices would increase in the future if the area became urbanized (Durand-Lasserve et al. 2013). Equally important is the influence of political factors. For instance, land was usually allocated before elections as part of the election campaign strategy and patronage (Durand-Lasserve et al. 2013). However, after elections, some of the land was sold in order to develop residential areas (Durand-Lasserve et al. 2013).

While the proportion of urban population living in slums has decreased from 94.2 to 65.9% in Bamako, unplanned settlement is one of the key drivers of urban development (UN-Habitat 2012). This is partly attributed to complex land tenure system whereby land in urban and peri-urban areas is governed by a combination of state and customary laws (Durand-Lasserve et al. 2013). As a result, land in Bamako Metropolitan Area is generally expensive. This forces the majority of urban dwellers to illegally squat on public land or to enter into informal land use agreements with traditional authorities on the urban periphery. Furthermore, as in most sub-Saharan African countries, poor local governance combined with corruption in land allocation as well as expensive land registration processes has fueled the growth of informal settlements in the outskirts, which led to increased built-up expansion (Durand-Lasserve et al. 2013). According to Feracque-Vitkovic et al. (2007), the growth of Bamako is uncontrolled since traditional land owners, local government, and national authorities have the right to sell or distribute land. Last but not least, people have been forced to migrate to urban areas during periods of recurring droughts (Diallo and Zhengyu 2010). For example, the farmers and nomads have migrated to urban areas in search of better economic opportunities.

14.4 Projected Future Land Use/Cover Changes

14.4.1 Projected Land Use/Cover Changes

Future land use/cover changes for 2020 and 2030 (Fig. 14.6) were simulated based on the calibration scenarios between 1990 and 2010 (see Chap. 2). Based on the 1990 land use/cover base map, the multiple transition probabilities, and the "1990–2010" transition potential maps, the land change model projected that built-up areas would increase substantially from 103.9 km^2 in 2014 to 151.8 km^2 in 2030, while non-built-up areas would decrease from 289.6 to 241.2 km^2 over the same study period (Fig. 14.7). The annual rate of built-up change was projected to be 3.8 km^2 for the "2014–2020" epoch and 2.5 km^2 for the "2020–2030" epoch.

The simulated built-up areas in 2020 and 2030, and the observed built-up areas in 1990 indicated that urban expansion will continue with infill and leapfrog developments particularly to the north and south of Niger River (Fig. 14.4a). For

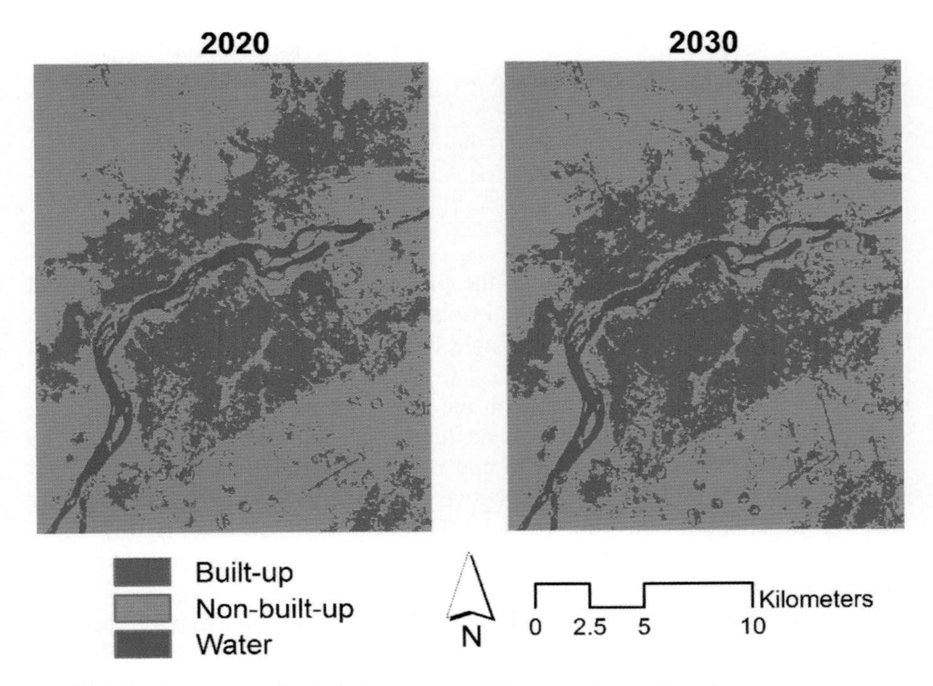

Fig. 14.6 Projected urban land use/cover maps of Bamako Metropolitan Area

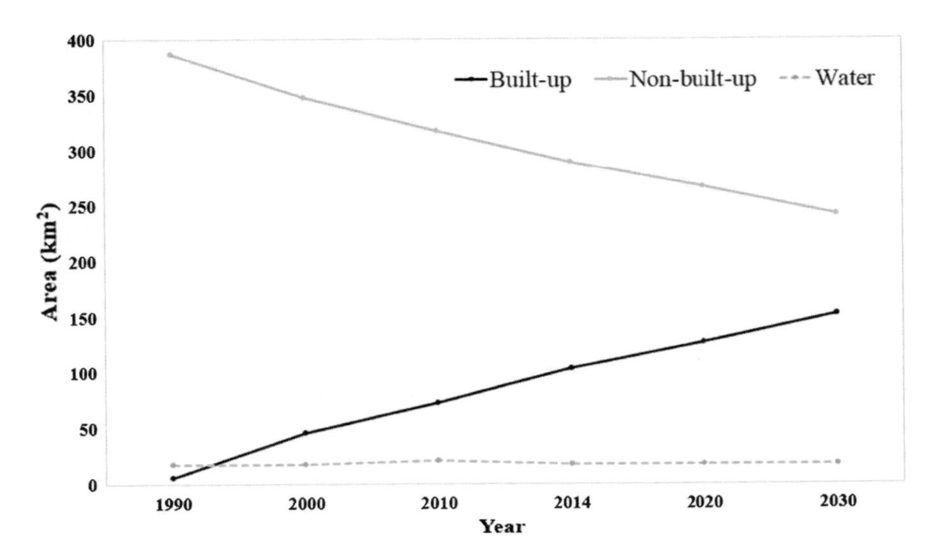

Fig. 14.7 Simulated future urban land use/cover trend under the current scenario

Table 14.3 Projected landscape pattern of Bamako Metropolitan Area

Class-level (built-up) spatial metrics	2020	2030
PLAND (%)	32.2	38.6
PD (number per km^2)	3.1	2.9
ENN (mean) (m)	96.8	100.0
CIRCLE (mean) ($0 \leq$ CIRCLE < 1)	0.4	0.4
SHAPE (mean) ($1 \leq$ SHAPE $\leq \infty$)	1.2	1.3

example, built-up areas to the south of the airport in Commune VI (Fig. 14.1c) and parts of the rural communes of Kalaban Coro are expected to grow substantially in the future. The spatial metric results (Table 14.3) for the built-up class indicate that PLAND would likely increase from 32.2 to 38.6% between 2020 and 2030. This implies an increase in built-up areas in the future. However, PD would decrease slightly from 3.1 to 2.9 per km^2 in the future, which implies that most of the built-up class will be less fragmented and more compact. The mean ENN would increase slightly from 96.8 to 100 m between 2020 and 2030, which implies low levels of dispersion for the built-up areas. The mean CIRCLE and SHAPE values would likely remain constant over the study period (Table 14.3), indicating that built-up would become more regular in the future.

Figure 14.8 shows spatial metrics for the built-up areas of Bamako Metropolitan Area (2020 and 2030) within the distance buffer zones from the city center. Results show that PLAND would decrease with distance in 2020 and 2030, which implies that built-up patches would be continuous near the city center, and scattered further away from the city center. The PD metric would remain constant within the 0–5-km and 5–10-km buffer zones and then increase over the 5–10-km buffer zone. Therefore, built-up expansion will be more compact near the city center and more fragmented further from the city center. This implies that extension and infill developments would continue near the city center, and leapfrog development further from the city center. The results show that the mean ENN would likely decrease with distance for 2020, which indicates low dispersion levels near the city center. However, for 2030, the mean ENN would likely decrease within the 0–5-km and 5–10-km buffer zones and then increase further away from the city center. This implies that built-up patches would likely become scattered within the 10–15-km buffer zones. The mean CIRCLE value shows slight variations with distance over the study period, which suggests that the built-up patch would be more stable. Nonetheless, the mean SHAPE value would vary in 2020 and 2030, implying that the built-up patches will be more irregular with distance. In summary, the results indicate that built-up areas would likely be continuous near the city center in the future, which implies continuation of infill and extension developments. However, the built-up would likely be discontinuous and scattered with further distance from the city center in 2020 and 2030.

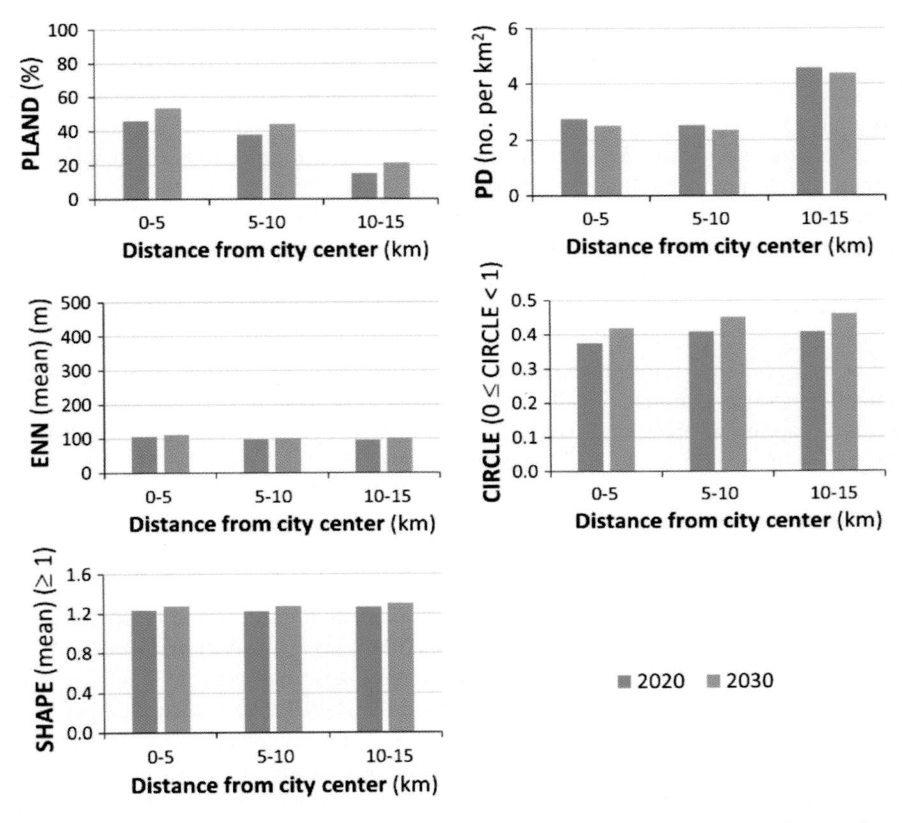

Fig. 14.8 Projected class-level spatial metrics for built-up along the gradient of the distance from city center of Bamako Metropolitan Area. *Note* The y-axis values are plotted in the same range as those in Fig. 14.5

14.4.2 Implications for Future Sustainable Urban Development

The rapid pace of urban expansion in Bamako Metropolitan Area has major implications on sustainable development given that the spatial expansion of built-up areas is being driven by informal access to land for housing. For example, over 65.9% of the urban population in Bamako currently lives in slums and informal settlements (UN-Habitat 2012). As a result, rapid expansion of built-up areas creates major difficulties for urban planning, particularly the provision of basic services and infrastructure development. As mentioned before, poor governance within local authorities coupled with a complex land tenure system and a malfunctioning land sector also pose major obstacles for sustainable development (Farvacque-Vitkovic et al. 2007). Consequently, vacant land available for housing in the metropolitan area is costly. Within this context, an in-depth understanding of spatial patterns and underlying processes of built-up expansion in Bamako Metropolitan Area is required. Therefore, the land use/cover change analysis and simulations in this chapter provide an analytical framework in order to assess urban growth as well as to analyze its underlying driving forces.

The simulated land use/cover changes have several implications for sustainable urban development for Bamako Metropolitan Area. For example, the results indicated that simulated future built-up expansion under the current scenario would be dense near the city center, and discontinuous and scattered with farther distance from the city center. According to Farvacque-Vitkovic et al. (2007), urban sprawl in the surrounding areas has resulted in high infrastructure and commuting distance costs. As a result, local authorities need to focus attention on reducing discontinuous and scattered built-up expansion. This can be accomplished by integrating mixed land use patterns, which would minimize transport and service delivery costs. In addition, open land would be used in a cost-effective manner, which will improve sustainable urban development in the long term.

While the local authorities in Bamako have intensified efforts to improve sustainable urban development (Farvacque-Vitkovic et al. 2007), urban sprawl will likely continue unless proper urban policies based on clear scientific evidence are formulated. The local authorities and central government have prioritized the integration of population in the surrounding areas as a cornerstone of sustainable urban development in Bamako Metropolitan Area (Farvacque-Vitkovic et al. 2007). This study produced simulated land use/cover maps that provide a visual and quantitative representation of built-up expansion in Bamako Metropolitan Area. For example, the observed and simulated land use/cover changes under the current scenario provide built-up change information, which can be used as an initial scientific or diagnostic tool for guiding sustainable urban development. This can possibly assist researchers, policy makers, and other stakeholders in assessing the implications of future built-up expansion and urban development policy alternatives.

14.5 Summary and Conclusions

The land use/cover maps for 1990, 2000, 2010, and 2014 along with change analyses and spatial metrics revealed that significant built-up expansion occurred during the study period. Built-up growth in Bamako Metropolitan Area was characterized by low-density urban sprawl moving outward from the urban core into the surrounding rural areas. The built-up expansion was dominated by infill, extension, and leapfrog developments, which are attributed to socioeconomic, political, and natural driving factors. The analysis of the spatial metrics with distance buffer zones revealed that built-up areas were expanding in a continuous and high-density form near the city center. In contrast, the built-up expansion was discontinuous and scattered as housing developments occurred further from the city center due to outward and leapfrog developments. While urban growth dynamics are too complex to capture in a short period, this chapter provided an overview of built-up expansion in Bamako Metropolitan Area.

Future land use/cover simulations (up to 2030) indicated that the land use/cover change trends, such as the increase in built-up areas and decrease in non-built-up areas, would continue under the current growth scenario. The spatial metric results indicated that future built-up areas would likely expand in a continuous and homogenous form near the city center. However, built-up areas would likely be discontinuous and scattered with farther distance from the city center in 2020 and 2030. Since built-up expansion affects open and agriculture land, the current land use/cover changes will have impact on the environment in Bamako Metropolitan Area. This chapter has demonstrated that an integrated approach incorporating remote sensing, spatial metrics, and simulation tools help to understand land use/cover changes in Bamako Metropolitan Area. Therefore, land use/cover products could potentially assist decision-makers and other stakeholders with built-up extent in the metropolitan area. This can be used to guide sustainable urban land use planning and development in Bamako Metropolitan Area, in particular, and West Africa in general.

References

Dembele M (2004) French colonization and urban evolution in Djenne and Bamako. In: Falola T, Salm SJ (eds) Globalization and urbanization in Africa. Africa World Press, Eritrea

Diallo BA, Zhengyu B (2010) Land cover change assessment using remote sensing: case study of Bamako, Mali. Researcher 2:7–14

Durand-Lasserve A, Durand-Lasserve M, Selod H (2013) Systemic analysis of land markets and land institutions in West African cities. Rules and practices: the case of Bamako, Mali. The World Bank Development Research Group, Environment and Energy Team

Farvacque-Vitkovic C, Csalis A, Diop M, Eghoff C (2007) Development of the cities of Mali-challenges and priorities. World Bank, Africa Region Working Paper Series Number 104a

Geist HJ, Lambin EF (2001) What drives tropical deforestation? A meta-analysis of proximate and underlying causes of deforestation based on subnational case study evidence. LUCC Report Series 4. Louvain-la-Neuve

McGarigal K, Cushman SA, Ene E (2012) FRAGSTATS v4: spatial pattern analysis program for categorical and continuous maps. Computer software program produced by the authors at the University of Massachusetts Amherst. Accessed 1 July 2015 from http://www.umass.edu/landeco/research/fragstats/fragstats.html

Turner II BL, Skole D, Sanderson S, Fischer G, Fresco L, Leemans R (1995) Land-use and land-cover change. Science/Research Plan. IGBP Report No.35, HDP Report No.7. IGBP and HDP, Stockholm and Geneva

United Nations Human Settlements Programme (UN-HABITAT) (2012) The State of world's cities 2012/2013. Nairobi, Kenya

Vaa M (2000) Housing policy after political transition: the case of Bamako. In Environment and Urbanization-Poverty Reduction and Urban Governance

World Bank (2015) Mali: geography of poverty in Mali. PREM 4 Africa Region

Chapter 15
Nairobi Metropolitan Area

Charles N. Mundia

Abstract This chapter examines the origin and brief history of Nairobi Metropolitan Area, which is Kenya's principal economic and cultural center and one of the largest and fastest growing cities in Africa. This chapter looks at the urban primacy, urban land use/cover change patterns, and the factors that have influenced the urbanization of Nairobi, as well as the potential implications of these factors to the future urban development of this Metropolitan. Nairobi is an example of an African colonial city, with colonial origins, which shaped its structure and management at the time of Kenya's transition to independence. Nairobi was born of the European colonial project and was first established as a transportation center, before it grew to become an administrative center. Like other African cities, after independence Nairobi was characterized by a rapid increase in rural to urban migration, accompanied by the proliferation of small-scale trade and commodity production. Nairobi has grown remarkably, with its urbanization driven by various interrelated factors. Land use/cover change analysis for Nairobi Metropolitan Area shows that the built-up areas would continue to increase at an average annual rate of change of 1.49 km^2/year. The results of the landscape pattern analysis show that built-up land would be more aggregated but with disconnected, nonlinear, and complex patches of built-up land as Nairobi continues to expand. Nairobi's accessibility as the regional hub, its relative position as the gateway to eastern African region, its status as the country's capital, the adoption of various urban development strategies, and its population and economic growth, together, have combined to drive its urban development. Nairobi is, however, faced by a myriad of urban challenges that need to be taken into consideration in its future development including traffic congestion, inadequate urban housing, mushrooming slums, urban poor, unemployment, delinquency, crime, unavailability of clean water, inadequate drainage and sanitation, lack of adequate public transport, environmental degradation, and disaster unpreparedness. The Government of Kenya has embarked on an ambitious Nairobi Metro 2030 vision to spatially redefine the Nairobi

C.N. Mundia (✉)
Institute of Geomatics, GIS and Remote Sensing,
Dedan Kimathi University of Technology, Nyeri, Kenya
e-mail: ndegwa.mundia@gmail.com

Y. Murayama et al. (eds.), *Urban Development in Asia and Africa*,
The Urban Book Series, DOI 10.1007/978-981-10-3241-7_15

Metropolitan Area and create a world-class city region which is envisaged to generate sustainable wealth and quality of life for its residents, investors, and visitors. It is hoped that this will be realized so that Nairobi can become a world-class metropolitan.

15.1 Origin and Brief History

Nairobi is the capital and largest city in Kenya (Fig. 15.1) and is main economic, administrative, and cultural center. It is the most populous city in eastern Africa and one of the most prominent and fastest growing cities in Africa (Rakodi 1995; Mundia and Murayama 2010). The city and its surrounding areas (Fig. 15.2) also form the Nairobi County, one of the 47 counties in Kenya under a new devolved system of governance. The area where Nairobi is currently located was an uninhabited swamp until a supply depot of the Kenya–Uganda railway was built around 1900 (Boedecker 1936). The site was chosen because it offered a number of favorable factors including a suitable stopping place via railway line enroute to Uganda, adequate water supply from nearby rivers, ample land for rail development, and elevated cooler ground to the west suitable for residential purposes (Foran 1950; Walmsley 1957). The supply depot soon became the railway headquarters and was named Nairobi, after the Masai name "Enkare Nairobi," which means "a place of cool waters" (Hirst and Lamba 1994). The place was a grazing land and livestock watering point for the local Masai people and had no permanent African settlements.

Nairobi was totally rebuilt in the early 1900s after an outbreak of plague and the burning of the original town (White et al. 1948). Soon after, the spatial patterns

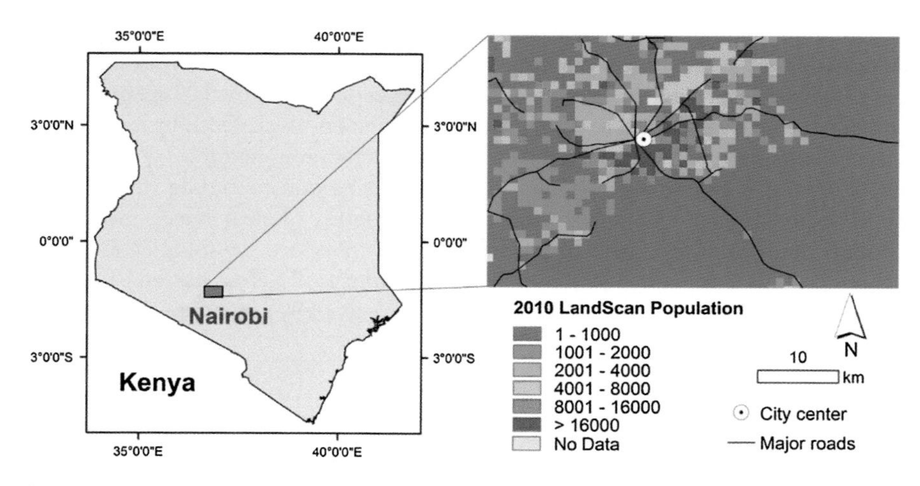

Fig. 15.1 Location and LandScan population of Nairobi Metropolitan Area, Kenya

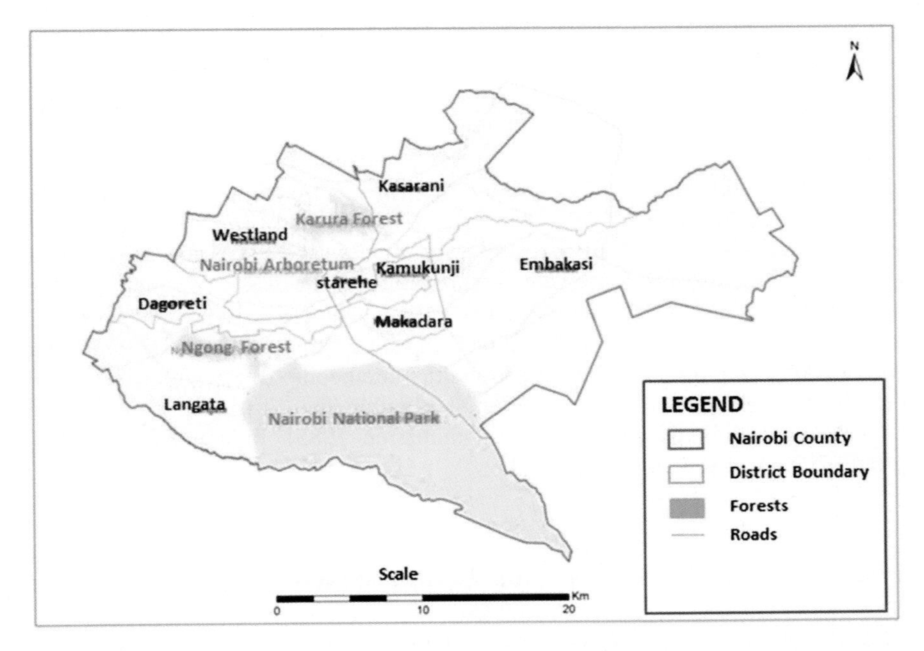

Fig. 15.2 Post-independence administrative boundary of Nairobi showing 8 constituencies

around the depot and the railway station emerged. The Europeans established their homes on the elevated area to the west, away from Asians and Africans leading to exclusive European residential settlements. Asian employees, formerly working for the railway, established shops, later known as Indian Bazaars, not far from the railway station. African workers lived in employee housing and shanty villages to the east (White et al. 1948). By 1905, the original railway depot and camp had grown into an urban center of over 10,000 people and definite land use zones had appeared. The Europeans mainly occupied the cooler Westland, the Indians in the north, and the African workers were mainly concentrated on the periphery (White et al. 1948). In the same year, Nairobi replaced Mombasa as the capital of the British East Africa protectorate. It continued to grow under British rule, and many British people settled within the city's suburbs. With the influx of more non-African settlers, it expanded rapidly. By 1909, much of its internal structure, especially the roads, was already established (Boedecker 1936). The continuous expansion of the city began to anger the neighboring local Masai people because the city was taking up their grazing land to the south which they needed for their livestock. In the west, it also angered the Kikuyu people who wanted their land returned to them (Foran 1950). In 1919, Nairobi was declared a municipality under Nairobi city council, and at the same time, the boundary was extended to include the peri-urban settlements (Croix 1950). The boundary was extended again in 1927 to cover 30 miles^2 (White et al. 1948). From 1928 to the time of independence in 1963, the boundary remained the same with only minor additions and excisions taking place. Upon independence, Nairobi became the capital of the new republic and the city was expanded further (Foran 1950; Walmsley 1957) (Fig. 15.2). It has grown rapidly since independence, and this growth has put pressure on the city's infrastructure (Karuga 1993; Mundia and Aniya 2005).

Nairobi is currently a characteristic blend of modernism and traditionalism. The city is a brawling, dynamic maelstrom of cultures and enterprises that reflects its diverse heritage (Nairobi City Commission 1985; Hirst and Lamba 1994). Its many contrasts are reflected in the variety of tribes, races, as well as the geographical juxtaposition of the city lying in close proximity to the African wilderness. Just outside the city is the Nairobi National Park which has a rich variety of wild animals and the only national park within a city anywhere in the world (Mundia and Aniya 2005). Nairobi has a diverse and multicultural composition with a number of prominent churches, mosques, and temples within the city serving the various cultures and religions (Nairobi City Commission 1985). Figures 15.3 and 15.4 provide views of the city of Nairobi.

15.2 Primacy in the National Urban System

15.2.1 Kenya and Its National Urban System

Kenya is located on the equator with the Indian Ocean lying to the southeast and is bordered by five East African countries of Tanzania to the south, Uganda to the west, South Sudan to the northwest, Ethiopia to the north, and Somalia to the north east. At 580,367 km^2, Kenya with a population of about 45 million people is the world's forty-seventh largest country by size and lies between latitudes 5°N and 5°S, and longitude 34° and 42°E (Stren and White 1989). From the coast on the Indian Ocean, the low plains rise to central highlands. The highlands are bisected by the Great Rift Valley, with a fertile plateau lying to the east (Walmsley 1957; Syagga et al. 2001).

Kenya adopted a new constitution in 2010 which marked a major milestone in the way the country is governed. Political power and economic resources were dispersed from a centralized government in Nairobi to the grassroots in a devolved form of governance. As a result, 47 counties were established (Fig. 15.5) based on cultural, ethnological, and geographical characteristics. Under a democratic–presidential form of government, Kenya is politically and administratively divided into these 47 counties which are the primary administrative and political divisions in the country (Omolo 2010).

Urbanization in Kenya has a long history, going back to as early as the ninth century when urban trading centers started along the Kenyan coast. Many urban centers, however, started during the pre-independence period, when they were used as centers for administrative and political control by the colonial government (Foran

Fig. 15.3 Nairobi Metropolitan Area as seen from upper hill and from the rooftop of KICC building. *Source* Author's fieldwork (2015)

1950). The urbanization patterns in Kenya have tended to reflect the development of British colonization rather than traditional African settlement patterns. The rate of urbanization in the country is one of the highest in the world with an estimated annual rate of growth of the urban population at 7.0% (Cohen 2004). The growth of urban population which has resulted from both natural population growth and rural–urban migration has led to an increased demand for resources needed to meet the demand for infrastructural services (Olima 1997). Population statistics show that the proportion of Kenyans living in urban centers increased from 5.4% in 1948 to 15.3% in 1979, to 21.4% in 1989, to 28.2% in 1999, and to 38.4% in 2009.

Fig. 15.4 Sections of Nairobi's CBD taken along major city roads. *Lower left* Traffic congestion along Moi Avenue; *lower right* slum area, showing informal settlement in Kibera, Nairobi. *Source* Author's fieldwork (2015)

There are currently about 230 urban centers in Kenya, with 45% of the urban population living in the capital city, Nairobi (Government of Kenya 2007). The Nairobi Metropolitan Area is the most populous in East and Central Africa with a

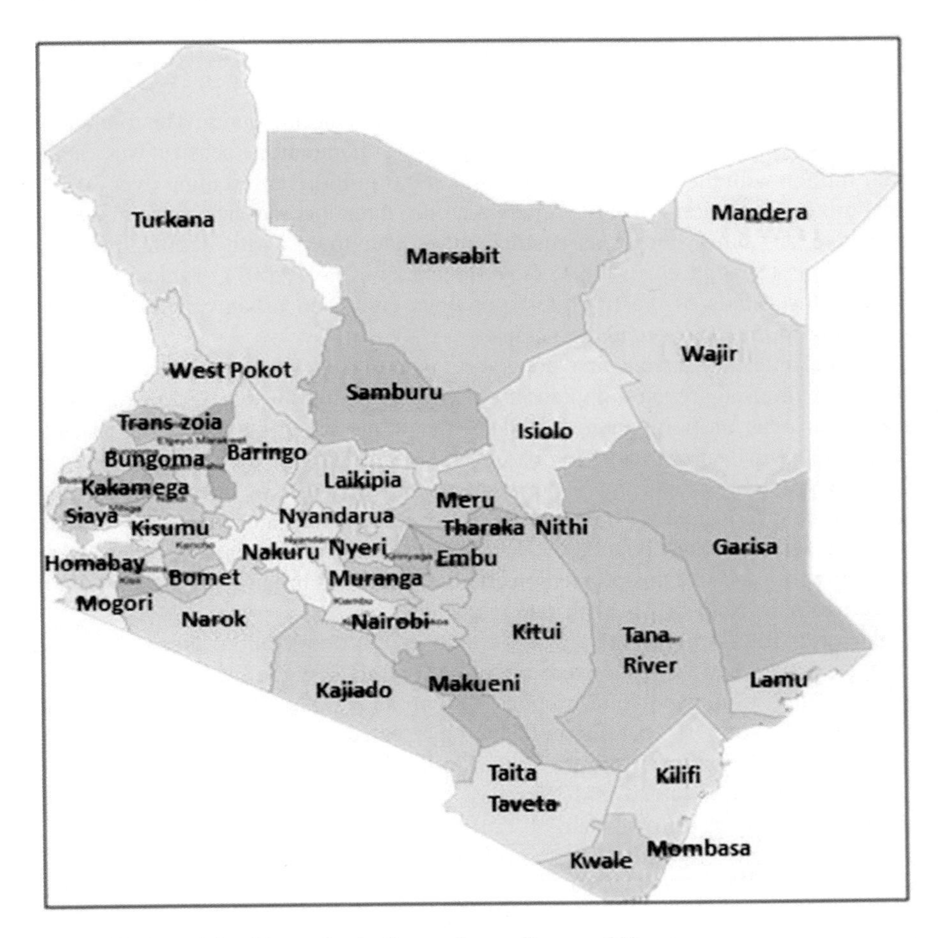

Fig. 15.5 Map of the 47 counties in Kenya. *Source* Survey of Kenya

population of 4 million people spread over 685 km^2. Kenya is the largest economy in East Africa and the ninth largest in Africa.

In the national framework for spatial planning, the 47 county headquarters spread throughout the country have been identified as the country's leading industrial, financial, and technological centers that serve as the main national growth hubs. The territorial spread of urban centers of different sizes across the whole country constitutes the national urban system. The spatial distribution of these urban centers provides an overview of Kenya's national urban system and the regional social and economic agglomerations in the country (Government of Kenya 2008).

15.2.2 Primacy of Nairobi Metropolitan Area

Nairobi city and its surrounding area also form the Nairobi County. The total urban population in Kenya according to the 2009 population census was about 12.0 million with the population of Nairobi at 4.0 million representing over 30% of total urban population in Kenya. There are only three incorporated cities in Kenya, but there are other numerous municipalities and towns with significant urban populations (Syagga et al. 2001; Government of Kenya 2007). Table 15.1 summarizes the details of the five most populous cities and urban centers in Kenya according to the 2009 population census.

At the time of Kenya's first population census in 1948, there were 17 urban centers with an aggregate population of 285,000 people (Table 15.2). The urban population was proportionately small (5.0% of the total) but disproportionately concentrated in Nairobi and Mombasa (41.0 and 32.0% of the total urban population, respectively) with the majority of the urban dwellers being non-Africans (Government of Kenya 2007). By 1962, the number of urban centers had doubled to 34 and the urban population increased to 671,000 people, with Nairobi accounting for 34% of this population. While the overall urban growth rate stood at 6.3% per year, Nairobi's growth rate was 4.6%.

The growth of urban centers, both in numbers and population, accelerated after independence when Africans were allowed to migrate to the urban areas without any legal and administrative restrictions. The urban population grew to 1 million in 1969, at a rate of 7.1% per annum. This represented about 10% of the total population, with Nairobi and Mombasa accounting for about 67% of the total urban population: Nairobi (48%) and Mombasa (19%). This period also saw Nairobi recording the highest growth rate of 12.0%.

By 1979, the overall level of urbanization had risen to about 15.0% with over 90 urban centers and an urban population of 2.3 million (Obudho and Owuor 1991). Nairobi and Mombasa accounted for over 50% of the total urban population: Nairobi (36%) and Mombasa (15.2%). Although the urban population increased from 2.3 million in 1979 to close to 4.0 million in 1989, the growth rate was only

Table 15.1 List of the five populous cities and municipalities in Kenya

	City/town	Status	Population	County
1.	Nairobi	Metropolitan	3,375,000	Nairobi
2.	Mombasa	City	1,200,000	Mombasa
3.	Kisumu	City	409,928	Kisumu
4.	Nakuru	Municipality	307,990	Nakuru
5.	Eldoret	Municipality	289,380	Uasin Gishu

Source Kenya Bureau of Statistics, 2009 population census

Table 15.2 Distribution of urban population in Kenya by size of urban center, 1948–2009

Size of urban pop. ('000)	1948		1962		1969		1979		1989		1999		2009	
	#	%	#	%	#	%	#	%	#	%	#	%	#	%
Over 100,000	1	6	2	6	2	4	3	3	6	4	20	10	19	8
20,000–99,999	1	6	2	6	2	4	13	15	21	15	82	42	90	39
10,000–19,999	2	12	3	9	7	15	11	12	19	14	18	9	28	12
5000–9999	3	18	11	32	11	24	22	24	32	23	23	12	30	13
2000–4999	10	58	16	47	25	53	42	46	61	44	51	27	64	28
Total no. of urban centers	17	100	34	100	47	100	91	100	139	100	194	100	230	100
Total urban pop. (millions)	0.28		0.75		1.06		2.31		3.88		5.4		12.0	
Total country pop. (millions)	5.4		8.6		10.9		15.3		21.4		28.2		38.4	
Urban pop. as % of country pop.	5.1		7.8		9.9		15.1		18.1		19.3		31.3	

Source Compiled from Kenya Population Censuses, 1948–2009

slightly above 5.0% compared to over 7.0% in the previous decade. With about 140 urban centers, the 1989 population results indicated that about 19% of the population resided in the urban areas. Nairobi and Mombasa accounted for about 45% of the total urban population (34 and 12%, respectively). In 2009, about 31.3% of the population lived in urban areas, of which more than half were in the five big urban centers in Kenya—Nairobi, Mombasa, Nakuru, Kisumu, and Eldoret. The urban growth rate reduced to about 3.5%, but the number of urban centers increased to 230 (Obudho and Owuor 1991; Kenya National Bureau of Statistics 2013).

As a consequence, the urban primacy index has shown an upward trend between 1979 and 2009 indicating that most of the Kenyan urban population lives in Nairobi. In other words, Nairobi continues to be a major urban center for socioeconomic and political activities in Kenya (Kenya National Bureau of Statistics 2013). Overall, based on population and GDP, the urban primacy of Nairobi Metropolitan Area relative to the other urban centers in Kenya is evident. The primacy of Nairobi City creates an imbalance in the urban hierarchy and development processes in the region. It is hoped that the new devolved system of governance in Kenya together with the current national framework for physical planning will promote country-side developments and a more balanced national urban system. At the moment, however, unlike Nairobi, there is no functional governing structure for the other urban areas which will be a great challenge in their respective landscape and urban development planning and implementation. There is, however, an ongoing national physical planning policy initiative that will ensure proper governance structure for the other urban areas to guide their growth and development. This is expected to ensure a more balanced national urban system that can spur nationwide development (Government of Kenya 2007).

15.3 Urban Land Use/Cover Patterns and Changes (1988–2030)

15.3.1 Observed Changes (1988–2014)

The results of the land use/cover mapping and change detection for Nairobi Metropolitan Area (Figs. 15.6 and 15.7) show that the built-up area increased from 43.32 km^2 in 1988 to 182.18 km^2 in 2014 (Table 15.3), a 320% increase in a period of 27 years, with an annual rate of change (increase) of 5.14 km^2/year. The annual rate of change during 1988–2000 was 2.91 km^2/year which increased during the 2000–2010 period to 4.15 km^2/year (Table 15.4). During the period 2010–2014, however, the annual rate of change increased substantially to 15.63 km^2/year.

Table 15.5 summarizes the results of the landscape pattern analysis using spatial metrics for the built-up class. The percentage of landscape (PLAND) metric

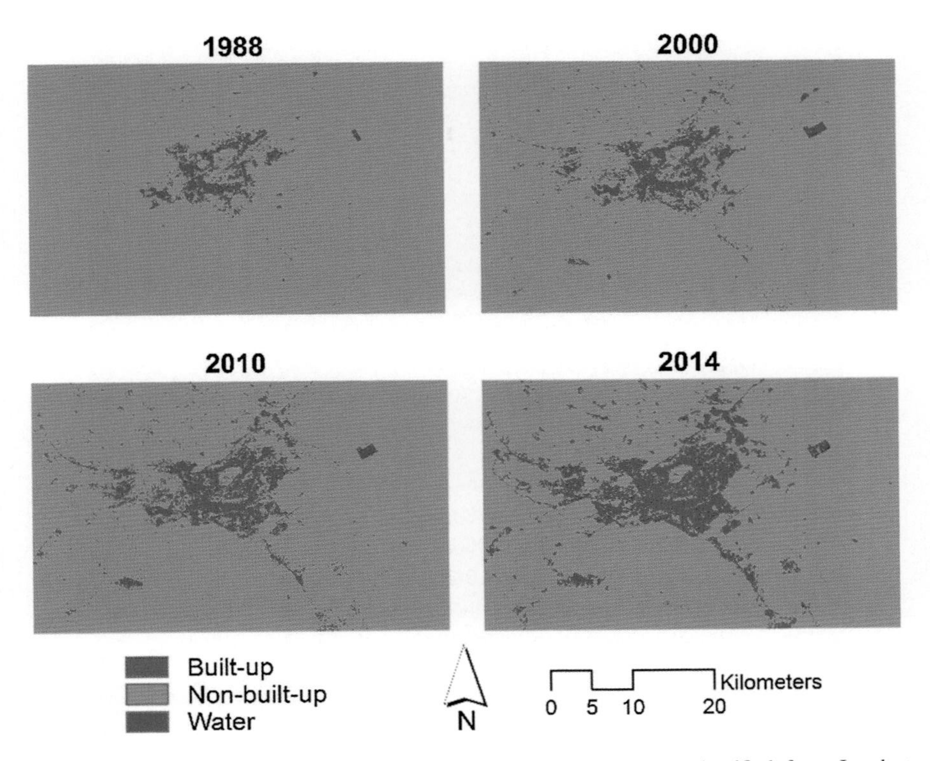

Fig. 15.6 Urban land use/cover maps of Nairobi Metropolitan Area classified from Landsat imagery

measures the proportion of a particular class at a certain time point relative to the whole landscape. In 1988, the PLAND of Nairobi Metropolitan Area's built-up land was 2.66%, and this increased to 4.82, 7.36, and 11.22% in 2000, 2010, and 2014, respectively (Table 15.5). The patch density (PD) metric measures fragmentation based on the number of patches per km^2, in which a patch is based on an 8-cell neighbor rule. For Nairobi, the PD of its built-up land was 0.43 in 1988, which changed to 1.13 and 1.57 for 2000 and 2010, respectively. The PD for Nairobi then decreased to 0.12 in 2014 suggesting that the built-up land for Nairobi Metropolitan Area became less fragmented and more aggregated.

The Euclidean nearest-neighbor distance (ENN) metric is a measure of dispersion based on the average distance to the nearest neighboring patch of the same class. The mean ENN value of Nairobi's built-up land increased from 138.91 m in 1988 to 164.43 m in 2000 and then declined to 157.95 m in 2010. The mean ENN then increased to 353.50 in 2014. The increase during the 1988–2000 period could have been due to the aggregation of neighboring patches of built-up land, as also indicated by the increase in PLAND and decrease in PD during the same period (Table 15.5). The decrease of mean ENN from 2000 to 2010 could be due to the

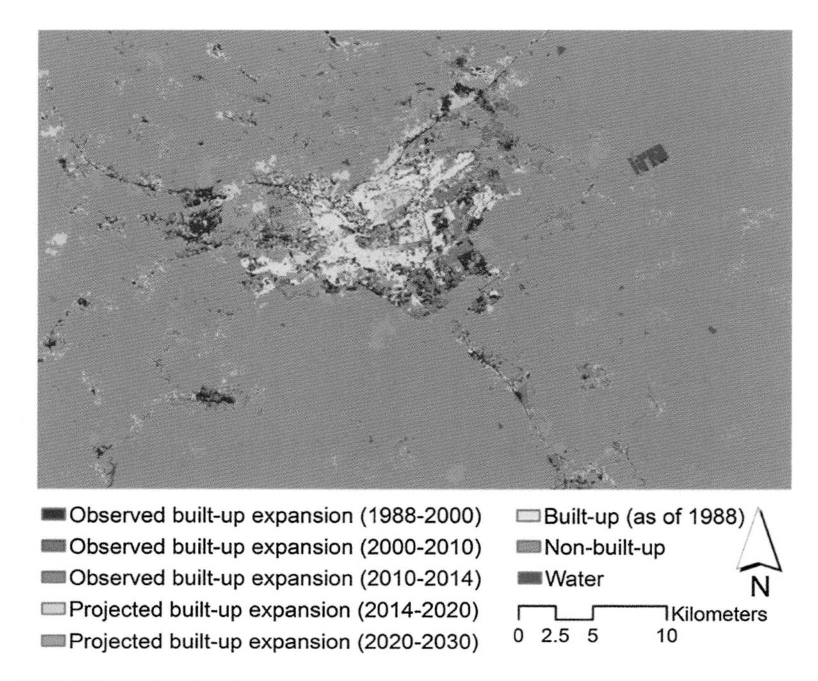

Fig. 15.7 Observed and projected urban land use/cover changes in Nairobi Metropolitan Area

Table 15.3 Observed urban land use/cover of Nairobi Metropolitan Area (km^2)

	1988	2000	2010	2014
Built-up	43.32	78.19	119.67	182.18
Non-built-up	1583.56	1543.56	1505.23	1441.03
Water	4.28	9.41	6.26	7.94
Total	1631.16	1631.16	1631.16	1631.16

expansion of the old patches, as indicated by the increase in PLAND and PD during the same period.

The related circumscribing circle (CIRCLE) metric measures the circularity of patches. The CIRCLE value ranges from 0 for circular or one cell patches to 1 for elongated, linear patches one cell wide. The mean CIRCLE value of Nairobi Metropolitan Area's built-up land generally showed an increasing trend during the 1988–2014 period (Table 15.5), indicating that substantial patches of built-up land became more elongated exhibiting a linear pattern. It is notable that Nairobi's urban expansion seems to follow the road network (Figs. 15.6 and 15.7). The increasing trend could therefore be due to the aggregation of smaller, circular isolated patches.

The shape index (SHAPE) metric is a measure of complexity. This metric has a value of 1 when the patch is square and increases without limit as patch shape becomes more irregular. Nairobi's built-up land had 1.25 mean SHAPE value for

Table 15.4 Observed urban land use/cover changes in Nairobi Metropolitan Area (km^2)

	1988–2000	2000–2010	2010–2014
Built-up	34.87	41.48	62.51
Annual rate of change (km^2/year)	*2.91*	*4.15*	*15.63*
Non-built-up	−40.00	−38.33	−64.20
Annual rate of change (km^2/year)	*−3.33*	*−3.83*	*−16.05*
Water	5.13	−3.15	1.69
Annual rate of change (km^2/year)	*0.43*	*−0.32*	*0.42*

Table 15.5 Observed landscape pattern of Nairobi Metropolitan Area

Class-level (built-up) spatial metrics	1988	2000	2010	2014
PLAND (%)	2.66	4.82	7.36	11.22
PD (number per km^2)	0.43	1.13	1.57	0.12
ENN (mean) (m)	138.91	164.43	157.95	353.50
CIRCLE (mean) ($0 \leq$ CIRCLE < 1)	0.39	0.41	0.39	0.62
SHAPE (mean) ($1 \leq$ SHAPE $\leq \infty$)	1.25	1.25	1.25	2.75

the period 1988–2010, which increased from 1.25 to 2.75 between 2010 and 2014 (Table 15.5), indicating complexity in the shape of the built-up patches. The observed landscape metrics correspond to the observed build-up expansion for Nairobi Metropolitan Area for the different time epochs as shown in Tables 15.3 and 15.4.

Analyses of the metrics for the built-up class of Nairobi Metropolitan Area along the gradient of the distance from the city center across all time periods from 1988 to 2014 (Fig. 15.8) show that PLAND decreases as the distance from the city center increases. This indicates that the proportion of built-up land near the city center is relatively higher. By contrast, PD increases first until it approaches 20 km distance from the city center and then starts to decrease.

This indicates that there were more patches of built-up land in middle distances. These were relatively more dispersed in farther distances, as indicated by the increasing trend of mean ENN across the distance from the city center. Analyses of the metrics also show a slight increase of the mean CIRCLE value along the gradient of the distance from the city center, indicating that the patches of built-up land were slightly more elongated or linear in farther distances. However, despite the variability of these metrics along the gradient of the distance from the city center, the complexity of the built-up land of Nairobi Metropolitan Area was much more uniform or stable as indicated by the mean SHAPE value.

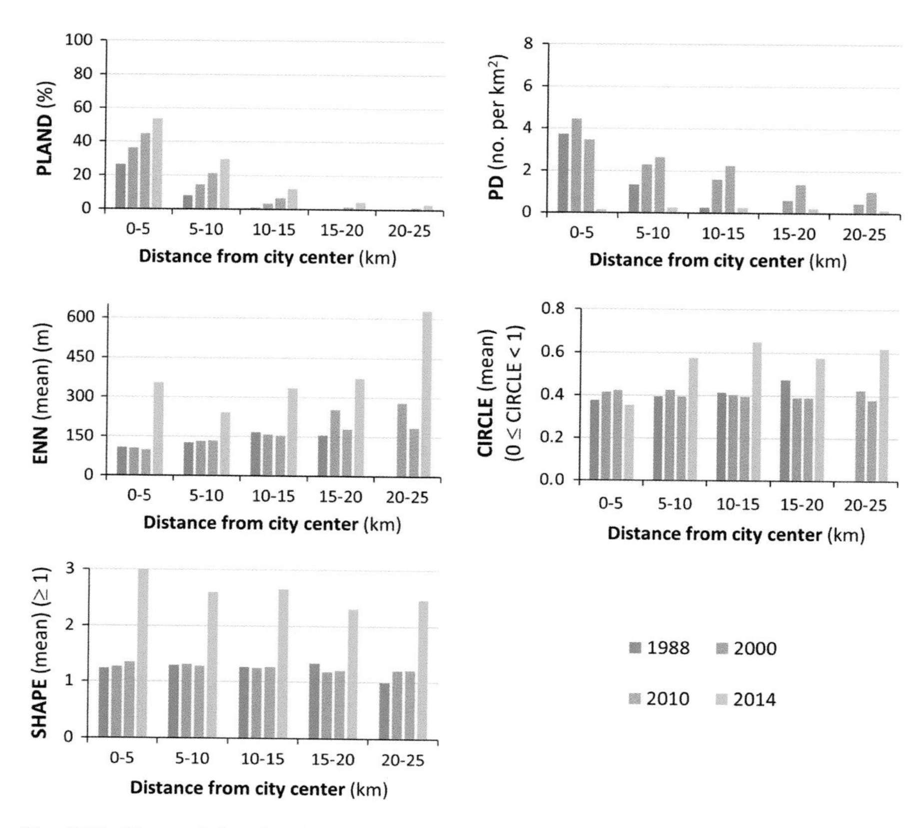

Fig. 15.8 Observed class-level spatial metrics for builtup along the gradient of the distance from city center of Nairobi Metropolitan Area. *Note* The y-axis values are plotted in the same range as those in Fig. 15.10

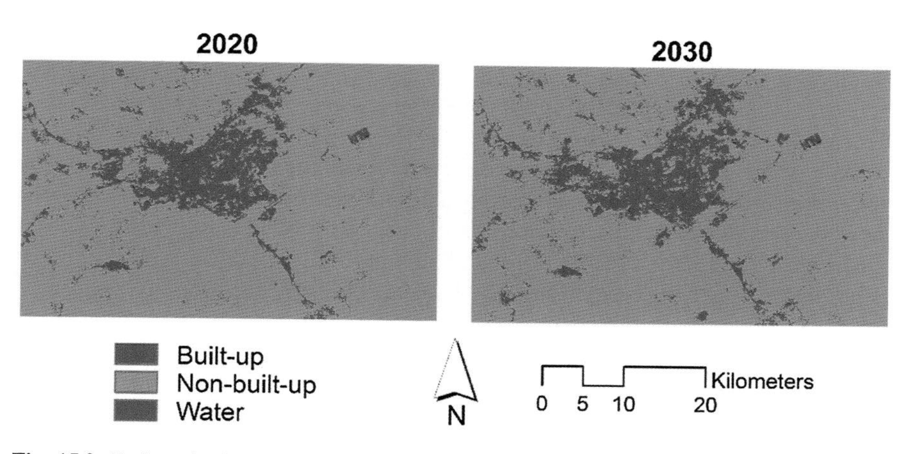

Fig. 15.9 Projected urban land use/cover maps of Nairobi Metropolitan Area

Table 15.6 Projected urban land use/cover of Nairobi Metropolitan Area (km^2)

	2020	2030
Built-up	184.70	210.38
Non-built-up	1438.51	1412.83
Water	7.94	7.94
Total	1631.16	1631.16

Table 15.7 Projected urban land use/cover changes in Nairobi Metropolitan Area (km^2)

	2014–2020	2020–2030
Built-up	2.52	25.68
Annual rate of change (km^2/year)	*0.42*	*2.57*
Non-built-up	−2.52	−25.68
Annual rate of change (km^2/year)	*−0.42*	*−2.57*
Water	0.00	0.00
Annual rate of change (km^2/year)	*0.00*	*0.00*

Table 15.8 Projected landscape pattern of Nairobi Metropolitan Area

Class-level (built-up) spatial metrics	2020	2030
PLAND (%)	11.38	12.96
PD (number per km^2)	4.14	4.84
ENN (mean) (m)	78.25	78.44
CIRCLE (mean) ($0 \leq$ CIRCLE < 1)	0.23	0.22
SHAPE (mean) ($1 \leq$ SHAPE $\leq \infty$)	1.07	1.07

15.3.2 Projected Land Use/Cover Changes (2014–2030)

The projected results are shown in Fig. 15.9 (see also Fig. 15.7) and Tables 15.6, 15.7, and 15.8. The projected land use/cover for Nairobi Metropolitan Area shows that the built-up land would increase from 182.18 km^2 in 2014 to 184.70 km^2 by 2020 and 210.38 km^2 by 2030, with an average annual rate of change of 1.49 km^2/year (Tables 15.6 and 15.7).

The results of the landscape pattern analysis show that the predicted patches of built-up land in 2020 and 2030 would be more aggregated as indicated by the increase in PLAND (Table 15.8). The decrease in mean ENN also indicates that there would be more patches that would be disconnected as Nairobi expands. The increase in the mean CIRCLE and SHAPE values indicates more disconnected patches and nonlinear and complex patches of built-up land.

Along the gradient of the distance from the city center (Fig. 15.10), the PLAND of the predicted 2020 and 2030 patches of built-up land would also be higher at distances closer to the city center. The PD would still be relatively higher in middle distances. The mean ENN would increase at distances farther from the city center, especially for the 2020 predicted patches. The mean CIRCLE value would decrease

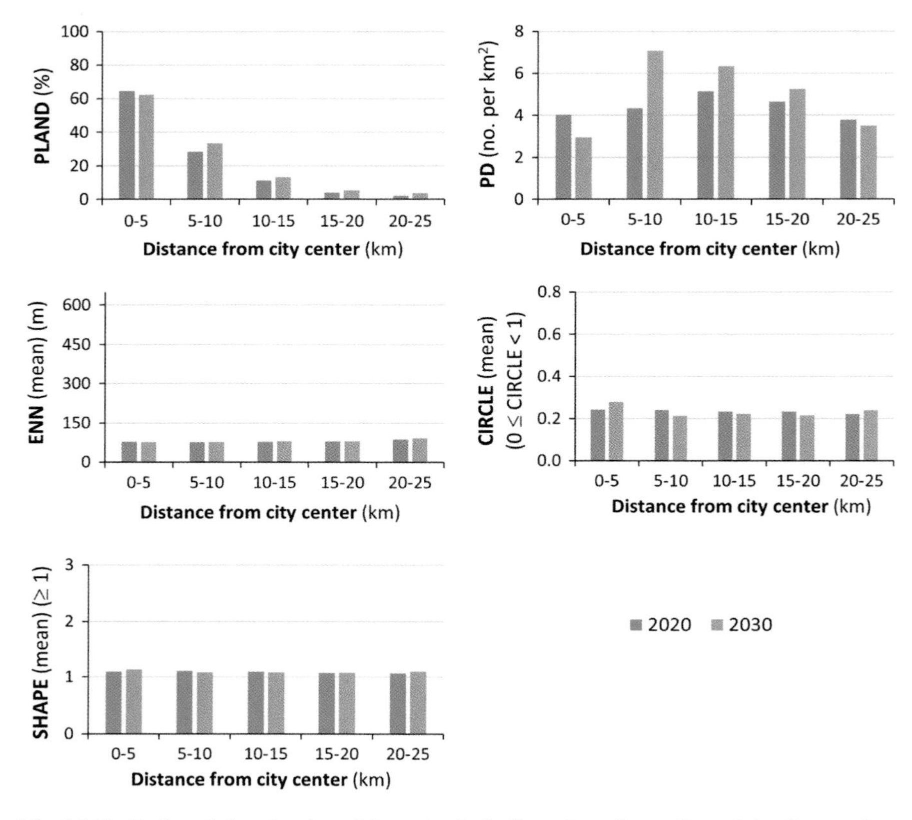

Fig. 15.10 Projected class-level spatial metrics for builtup along the gradient of the distance from city center of Nairobi Metropolitan Area. *Note* The y-axis values are plotted in the same range as those in Fig. 15.8

by 2020 and 2030, especially in distances 0–20 km. The mean SHAPE value would be relatively more uniform (Table 15.8).

15.4 Driving Forces of Urban Development

The land use/cover changes revealed for Nairobi City have occurred as a result of interactions of a number of environmental as well as demographic and socioeconomic forces. Some factors that have influenced this urban expansion include the following.

15.4.1 Some Key Urban Development Initiatives for Nairobi Metropolitan Area

1. The 1898 Plan of Nairobi. This was the first plan for Nairobi which was a town layout for the railway depot, with the main feature being the Nairobi railway station and the railway line. The plan strikingly resembles the town layout of Nairobi CBD today.
2. The 1926 Plan for a Settler Capital. In this plan, the city area was expanded to 77 km^2 to accommodate the growing population.
3. The 1948 Master Plan for a Colonial Capital. This master plan introduced the neighborhood units and zoning scheme for different users in the city (White et al. 1948).
4. The 1973 Nairobi Metropolitan Growth Strategy. This strategy was a long-term structure planning policy guide with broad long-term policy directions, strategies, and possibilities for the development of Nairobi City (Nairobi Urban Study Group 1973). The plan was meant to guide the development of the city up to the year 2000. The 1973 Nairobi Metropolitan Growth Strategy recommended decentralization and development of alternative service centers; modification, upgrade, and extension of the road network; formulation of realistic housing programs; and extension of the city boundary to the west and northeast, as well as the growth of satellite towns surrounding the city (Nairobi Urban Study Group 1973). As much as the 1973 Nairobi Metropolitan Growth Strategy was a tool for state intervention, it supported the interests of a few (Emig and Ismail 1980; Anyamba 2004). The interests of the urban majority were neglected as segregation was enhanced based on economic and class lines as opposed to racial and class lines. In the process, the urban majority were marginalized further and informalization thrived (Anyamba 2004).
5. The 1984–1988 Nairobi City Commission Development Plan, which outlined the development needs of all sectors: housing, health and environment, sewerage, social services, transport and public works, manpower development, and financial management (Nairobi City Commission 1985). This plan remained only on paper, and nothing much was achieved as regards its implementation.
6. "The Nairobi we want convention of 1993." The Nairobi City Convention was organized to map out strategies and practical actions toward a better Nairobi. The recommendations of this convention were broadly organized around four areas, namely (1) issues dealing with the use of space and the physical environment; (2) problems pertaining to the provision of services; (3) issues relating to the social sector; and (4) administrative, legal, and political issues (Karuga 1993). Unfortunately, much of these ideas were not taken into account in the planning of Nairobi.
7. The Kenya Local Government Reform Program (KLGRP) of 1999. This initiative embarked on policy and legal framework changes aimed at decentralization and local authority empowerment. This initiative introduced the Local Authorities Transfer Fund (LATF), a grant from the central government to equip

local authorities with means to provide their citizens with basic services. The key objectives of LATF were to (a) enable local authorities to improve and extend service delivery; (b) resolve municipal debts; and (c) improve local revenue mobilization, accountability, and financial management (Owuor et al. 2006). Within the same reform framework, the government in 2001 introduced the Local Authority Service Delivery Action Plan (LASDAP), which is a participatory planning and budgeting tool for identifying local priority needs in the local authorities.

8. Creation of the Ministry of Nairobi Metropolitan Development in 2008. This ministry was charged with the development issues of the Nairobi Metropolitan Area, aiming at area-wide governance interventions. Specifically, the ministry was in charge of roads, bus, and rail infrastructure—creating an efficient transport system; replacing slums with affordable low-cost and rental housing; enforcing planning and zoning regulations; facilitating efficient water supply and waste management infrastructures; and promoting, developing, and investing in sufficient public utilities, public services, and infrastructure (Government of Kenya 2008).
9. Nairobi Metro 2030 vision of 2010. Responding to urban growth projections and in an attempt to address current and future Nairobi Metropolitan Area challenges, the Government of Kenya embarked in an ambitious Nairobi Metro 2030 vision to spatially redefine the Nairobi Metropolitan Area and create a world-class city region expected to generate sustainable wealth and quality of life for its residents, investors, and visitors. The plan's elaboration and implementation fell under the responsibilities of the then Ministry of Nairobi Metropolitan Development. The vision of the Nairobi Metropolitan Area was to create a best managed metropolis in Africa, providing a dynamic and internationally competitive and inclusive economy supported by world-class infrastructure and a skilled labor force (Government of Kenya 2008). Based on the core values of innovation, enterprise, sustainability, coresponsibility, self-help, and excellence, the strategy was to optimize the role of the Nairobi Metropolitan Area in national development by building on existing strengths, including Nairobi's hub function in air transportation, the large number of regional and international bodies already present in the city, and its educational and research institutions. Nairobi Metro 2030 sought to brand and promote Nairobi as East Africa's key gateway city by creating a framework for comprehensively addressing a broad range of policy areas, including the economy, trunk and social infrastructures, transportation, slums and housing, safety and security, and financing. The Nairobi Metropolitan Area covers the 3000 km^2 that depend on Nairobi's regional core functions for employment and social facilities. Planning would initially involve a 40 km radius, despite Nairobi's functional outreach covering about 100 km. Apart from Nairobi Municipality itself, the Nairobi Metropolitan Area vision affected 14 other adjacent independent local authorities (Government of Kenya 2008). The implementation of this strategy started in 2010, and some progress has been noted.

15.4.2 Rapid Development

The economic development has been one of the dominant driving forces. Nairobi's gross domestic product (GDP) was about US $254 million in 1975, $970 million in 1985, and $1.65 billion in 1995 (Republic of Kenya 2002). The national economic survey of the year 2000 put Nairobi's GDP at $2.25 billion (Republic of Kenya 2002). The economic development has led to the establishment of more industries, the boom of real estate and subsequently to the expansion of the built-up areas. The unregulated small-scale businesses have expanded rapidly, and the employment in this sector is estimated at 500,000 people (Republic of Kenya 2002). The increase in economic development as measured by the changes in the GDP values reflects the change in urban expansion. The economic development which grew much faster in the period 1975–1985 (153% growth) led to a higher rate of urban expansion. The period 1988–2000 had a lower rate of urban expansion, which can be explained by the slow economic development (70% growth) during the period 1985–1995.

15.4.3 Urban Population Growth

The 1969 population census put Nairobi's population at slightly over half a million. The population rose to 1.35 million by 1989 against a national total population of 23 million (Development Solutions for Africa 1992). The current population is estimated at 4.2 million. This rapid urban population growth reflects a natural population increase among the urban residents (52%) as well as migration of people from rural areas to the city (48%). As a consequence, Nairobi's urban primacy index has shown an upward trend between 1979 and 2010 indicating that most of the Kenyan urban population lives in Nairobi (Mundia and Aniya 2005). The substantial population growth in Nairobi is responsible for the land use/cover changes shown in Fig. 15.6. Nairobi's economy, public services, and infrastructure have not managed to keep up with the increasing population. The city management has been unable to cope with the increasing demand for efficient city services since the rapid urban growth has outpaced the capacity of local authorities to provide and maintain infrastructure and basic services (Stren and White 1989; Mundia and Murayama 2010). The population, which has been growing at a rate of 4% annually, has contributed to the urban sprawl as well as the mushrooming slums, and the increased land use/cover changes. Poor planning in addition to the population increase has made worse the already existing physical, social, economic, and environmental problems (Mundia and Murayama 2010).

15.4.4 Physical Factors

The physical setting of Nairobi City has also influenced the expansion directions. From Figs. 15.6 and 15.7, the northeast and westward expansions have tended to follow the flat areas. In the areas to the east, where flat land and the general topography offer lower land and residential building costs, the poor road network, the greater prevalence of clay soils, and the drainage problems have reduced these advantages. In the western part of Nairobi, where the ground is higher with rugged topography, expansion is constrained by the existence of steep slopes. Major constraints to the expansion of Nairobi City include the national park to the south of the built-up area, and the safety zone and noise corridor around the Nairobi international airport. The national park, a protected area within the city right next to the built-up area, makes it a unique and valuable resource, not only as a tourist attraction, but also as an ecological counterpoint to noise, traffic, pollution, and stress of the urban environment, with all the inherent benefits that an unspoiled natural environment provides to its surrounding area. It is therefore a resource that must be preserved, and this has checked the southward expansion of the city (Mundia and Aniya 2005).

15.5 Implications for Future Sustainable Urban Development

Urbanization is one of the many human activities that have a serious impact on the natural environment, both locally and globally (Grimm et al. 2008; Wu 2010; Seto et al. 2011; Dahiya 2012; Estoque and Murayama 2014). It is also "the most drastic form of land transformation that result in irreversible landscape changes" (Estoque and Murayama 2014, p. 943). Sustainable urban development, characterized by a well-balanced relationship between environmental, social, and economic aspects of society, is an important component and an indispensable part of the sustainability goal of human kind (Estoque and Murayama 2014). For the case of Nairobi Metropolitan Area, though its socioeconomic conditions improved over the years, its overall sustainability remains a critical issue. This section examines the potential implications of its population growth and built-up expansion pattern to its future sustainable urban development.

15.5.1 Population Growth

The Population of Nairobi Metropolitan Area has increased from 500,000 in 1970 to the current 4.2 million. Nairobi has experienced rapid growth in terms of population compared to other major cities in Africa. The large population of Nairobi has been a key factor to its primacy over all urban centers in Kenya. However, its increasing population density and the rate of urbanization have overstretched the capacity of infrastructure and services and have caused various socioeconomic problems including uncontrolled growth and spreading of slum settlements, unemployment, delinquency, crime, unavailability of clean water, inadequate drainage and sanitation, lack of adequate public transport and environmental degradation, and urban poverty among other urban challenges. Much of Nairobi's urban footprint is unplanned settlement driven by rapid population growth and urban poverty, among other things. Population increase coupled with the acute shortage of urban housing and the problem of inadequate shelter have manifested themselves in the rapid formation and growth of informal settlements and tenement structures matched by deficiencies in the supply of the most basic infrastructure and public facilities required for human habitation. The sprawling informal settlements handicap the city's delivery of social services and negatively impact the quality of life. There is an urgent need for the Government of Kenya to ensure that there is corresponding increase in delivery of urban services to cater for the expanding population of its primate city.

As noted earlier, Nairobi's gross domestic product has been on the increase. The economic development has led to the establishment of more industries, boom in real estate development and subsequently to the expansion of the built-up areas. The increase in economic development as measured by the changes in the GDP values reflects in the change in urban expansion. Nairobi's gross domestic product, however, requires to consistently grow at a much higher level, if the identified urban challenges are to be managed. The national and county governments have indicated their intention to ensure continued economic growth and per capita socioeconomic development in addition to sorting out Nairobi's urban growth challenges to ensure sustainable growth trajectory.

15.5.2 Built-Up Expansion Pattern

Built-up expansion pattern of Nairobi Metropolitan Area exhibits aspects of concentric and sector urban growth models. Urban expansion has not taken place evenly in all directions but more along certain directions. Expansion has taken place around the periphery and also through infilling and sprawl development. Nairobi shows a star-shaped urban sprawl emanating from the city center and centered on the main roads. Some of the challenges that have been noted and which are related to this growth pattern include traffic congestion, inadequate urban housing, mushrooming slums, urban poor, unemployment, delinquency, crime, unavailability of clean water, inadequate drainage and sanitation, lack of adequate public transport, environmental degradation, and disaster unpreparedness (Mundia and Aniya 2005, 2006, 2007; Mundia and Murayama 2010; Kamusoko et al. 2011).

The sprawling pattern of Nairobi Metropolitan Area also has an important implication to urban planning. Analysis of urban expansion indicates that deliberate planning is lacking in Nairobi's built-up expansion pattern and that the principles of urban planning and regulations have not been adhered to. Some key urban development strategies initiated previously have not been followed through. Comprehensive planning is therefore needed to help Nairobi manage its resources better and deal with various challenges. The main challenge at present is learning how to cope with rapid urban growth which requires concerted effort by all stakeholders to redirect their collective energies and available resources in devising viable urban management strategies.

15.5.3 Current Major Development Plans

Two key urban development strategies are notable. The first is the Nairobi Metropolitan Development Plan of 2008. Under this development plan, the boundaries of the metropolitan were expanded to include adjoining towns and municipalities.

The plan's goal included the following:

1. Developing integrated road, bus, and rail infrastructure for the metropolitan area to provide an efficient mass transport system;
2. Replacing informal settlements with affordable low-cost housing;
3. Developing and enforcing planning and zoning regulations;
4. Preparing a spatial plan for the metropolitan area;
5. Developing efficient water supply and waste management infrastructure;
6. Promoting, developing, and investing in sufficient public utilities, public services, and world-class infrastructure for transforming Nairobi into a global competitive city for investment and tourism;
7. Identifying and implementing strategic projects and programs requiring support by the government;
8. Promoting the Nairobi Metropolitan Area as a regional and global services center for financial, information, and communication technology, health, education, business, tourism and other services; and
9. Developing a sustainable funding framework for the development of identified urban and metropolitan areas.

The implementation of this development plan started in 2010, but the results are yet to be seen. Given the many challenges facing Nairobi and the effects these have on a large number of people as well as on the country's economy and its international reputation, Kenya needs to move quickly and fully to implement this development strategy.

The second and the newly launched key development plan for Nairobi is the integrated urban development master plan for the city of Nairobi. The purpose of this integrated plan is to provide a guiding framework to manage urban development in Nairobi City County from 2014 to 2030, integrate all urban development sectors, and realize the goals of Kenya Vision 2030 for the city county of Nairobi. This master plan has integrated all the existing master plans of various infrastructures within the city of Nairobi and its surrounding. Infrastructures integrated in this plan include urban transport, railway, airport, power, water supply, sewerage, telecommunication, and solid waste management. The plan's goals include the following:

1. To provide spatial order of physical investments;
2. To enhance quality of life for inhabitants;
3. To guide investments by providing location criteria; and
4. To embrace the evolving urban policy regime in integrating social, economic, environmental, and political issues under one unitary framework.

It is hoped that with the new devolved system of governance, this integrated urban development master plan for the city of Nairobi will be implemented as envisaged.

15.6 Concluding Remarks

Nairobi Metropolitan Area has experienced rapid urban expansion driven by various factors including its position as the regional hub and gateway to the East African region, status as Kenya's capital city, its population and economic growth, and the adoption of a number of urban development strategies to guide its urban development. Land use change analysis for Nairobi Metropolitan Area shows that the built-up land would continue to increase at an average annual rate of change of 1.49 km^2/year. The results of the landscape pattern analysis show that built-up land would be more aggregated but with more disconnected, nonlinear, and complex patches of built-up land. Key drivers of urban expansion include environmental, demographic, and socioeconomic forces as well as some urban development policy initiatives that have been adopted.

Though Nairobi's growth has been phenomenal, there are a number of key urban challenges that need to be taken into consideration in its future development. The key challenges include traffic congestion, inadequate urban housing, mushrooming slums, urban poor, unemployment, delinquency, crime, shortage of clean water, inadequate drainage and sanitation, lack of adequate public transport, environmental degradation, and disaster unpreparedness.

The county government of Nairobi and the National Government of Kenya have embarked on an ambitious Nairobi Metro 2030 vision to spatially redefine the Nairobi Metropolitan Area and create a world-class city region which is envisaged to generate sustainable wealth and quality of life for its residents, investors, and visitors. The county government of Nairobi has also unveiled a new integrated master plan for Nairobi to provide a guiding framework to manage urban development in Nairobi Metropolitan Area and Nairobi County for the period

2014–2030, integrate all urban development sectors, and realize the goals of Kenya Vision 2030 for the city county of Nairobi. It is hoped that these initiatives will finally be realized so that Nairobi can grow into a world-class metropolitan.

References

Anyamba TJC (2004) In-formalization of a planned neighborhood in Nairobi. A paper presented at the international conference on "Future vision and development for urban development", Cairo, Egypt, 20–22 Dec 2004

Boedecker C (1936) Early history of Nairobi township. MacMillan Library, Nairobi

Cohen B (2004) Urban growth in developing countries: a review of current trends and a caution regarding existing forecasts. World Dev 32:23–51

Croix L (1950) City status for Nairobi. Commonwealth Surv 43:23–24

Dahiya B (2012) Cities in Asia 2012: demographics, economics, poverty, environment and governance. Cities 29:S44–S61

Development solutions for Africa (1992) Strategic health plans for the Nairobi area. Development Solutions for Africa, Nairobi, pp 174–192

Emig S, Ismail Z (1980) Notes on the urban planning of Nairobi. School of Architecture, Copenhagen

Estoque RC, Murayama Y (2014) Measuring sustainability based upon various perspectives: a case study of a hill station in Southeast Asia. AMBIO 43:943–956

Foran R (1950) Rise of Nairobi: from campsite to city. Phase in the history of Kenya's capital which is soon to receive a royal charter. Crown Colonist 20:161–165

Government of Kenya (2007) Kenya vision 2030. Government Printers, Nairobi

Government of Kenya (2008) Nairobi Metro 2030: a world class African metropolis. Government Printers, Nairobi

Grimm NB, Faeth SH, Golubiewski NE, Redman CL, Wu J, Bai XM, Briggs JM (2008) Global change and the ecology of cities. Science 319:756–760

Hirst T, Lamba D (1994) The struggle for Nairobi. Mazingira Institute, pp 64–83

Kamusoko C, Mundia CN, Murayama Y (2011) Recent advances in GIS and remote sensing analysis in sub-Sahara Africa. Nova Science Publishers, USA

Karuga JG (1993) Actions towards a better Nairobi: report and recommendations of the Nairobi city convention, July 1993, City Hall, Nairobi, 2002, National development Plan. Government printer, Nairobi, pp 3–15

Kenya National Bureau of Statistics (2013) Economic survey. Government Printers, Nairobi

Mundia CN, Aniya M (2005) Analysis of land use changes and urban expansion of Nairobi City using remote sensing and geographical information systems. Int J Remote Sens 26:2831–2849

Mundia CN, Aniya M (2006) Dynamics of land use/cover changes and environmental degradation of Nairobi City using GIS. Land Degrad Dev 17:97–108

Mundia CN, Aniya M (2007) Modeling urban growth of Nairobi city using cellular automata and geographical information systems. Geogr Rev Jpn 80:777–788

Mundia CN, Murayama Y (2010) Modeling spatial processes of urban growth in African cities: a case study of Nairobi City. Urban Geogr 31:259–272

Nairobi City Commission (1985) Development plan 1984–1988. Nairobi City Commission, Nairobi

Nairobi Urban Study Group (1973) Nairobi metropolitan growth strategy report, vol I: Main report. Nairobi City Commission, Nairobi

Obudho RA, Owuor SO (1991) Urbanization and regional planning of metropolitan Nairobi, Kenya. A paper presented at the international conference on big cities of Africa and Latin America: Urban facilities and cultural practices, Toulouse, France, 27–29 Nov 1991

Olima WHA (1997) The conflicts, shortcomings, and implications of the urban land management system in Kenya. Habitat Int 21:319–331

Omolo A (2010) Devolution in Kenya: A critical review of past and present frameworks. In: IEA devolution in Kenya: prospects, challenges and the future. IEA research paper series No. 24, Institute of Economic Affairs, Nairobi

Owuor SO, Charler B, Chretin M, Schaffner B (2006) Urban planning and management in small and medium-size towns. Les Cahiers d'Afrique de l'Est, pp 23–48

Rakodi C (1995) Harare. Inheriting a settler colonial city: change or continuity? Wiley, London

Republic of Kenya (2002) National development plan. Government Printers, Nairobi, pp 7–19

Seto KC, Fragkias M, Guneralp B, Reilly MK (2011) A meta-analysis of global urban land expansion. PLoS ONE 6:e23777

Stren RE, White RR (1989) African cities in crises: managing rapid urban growth. West Views Press, London, pp 103–162

Syagga PM, Mittulah WV, Gitau SK (2001) Nairobi situational analysis—a consultative report. Government of Kenya and United Nations Center for Human Settlements (UN-HABITAT), Nairobi

Walmsley RW (1957) Nairobi: the geography of a new city. East African Literature Bureau, Nairobi

White LWT, Silberman L, Anderson PR (1948) Nairobi: master plan for a colonial capital. A Report Prepared for the Municipality of Nairobi, HMSO, London

Wu J (2010) Urban sustainability: an inevitable goal of landscape research. Landscape Ecol 25:1–4

Chapter 16
Lilongwe Metropolitan Area

Kondwani Godwin Munthali

Abstract This chapter traces the origin and brief history of Lilongwe, the capital of Malawi in southern Africa. Its primacy in the urban landscape of the country is assessed by examining the urban land use/cover change patterns and the driving forces that are influencing its rapid urbanization. Lilongwe started off as a dusty colonial town which became an administrative center by 1904. Its growth had been slow until it became the capital in 1975 when built-up area grew sevenfold from 5.08 km^2 in 1990 to 34.73 km^2 in 2013. Furthermore, it is projected that its built-up area would have increased from 34.73 km^2 in 2013 to 41.24 km^2 by 2020 and 53.08 km^2 by 2030, respectively. Much of this growth is due to rural–urban migration, natural increase, and reclassification of settlements apart from being physically central, making it easily accessible. However, a significant portion (16–21%) of this growth has been occurring in informal settlements. As such, the urbanization of Lilongwe has been, to a large extent, unsustainable as the proportion of the urban poor is relatively high not just in monetary terms but also along the lines of access to basic urban services. All is not lost, however, for the primate city as the growth of informal settlements continues to decrease over the years. While this improvement is observed despite having several loopholes in the legal framework to support proper urban planning, it is only hoped that the strategic plans outlined for the city get implemented.

16.1 Origin and Brief History

Lilongwe is the capital city of Malawi, and it is the economic and political hub, having served as the seat of government since 1975 when the capital relocated from Zomba (see Fig. 16.1). The establishment of Lilongwe as the new capital after independence has been documented as one of Hastings Kamuzu Banda's

K.G. Munthali (✉)
Computer Science Department, Chancellor College,
University of Malawi, Zomba, Malawi
e-mail: kmunthali@gmail.com

Y. Murayama et al. (eds.), *Urban Development in Asia and Africa*,
The Urban Book Series, DOI 10.1007/978-981-10-3241-7_16

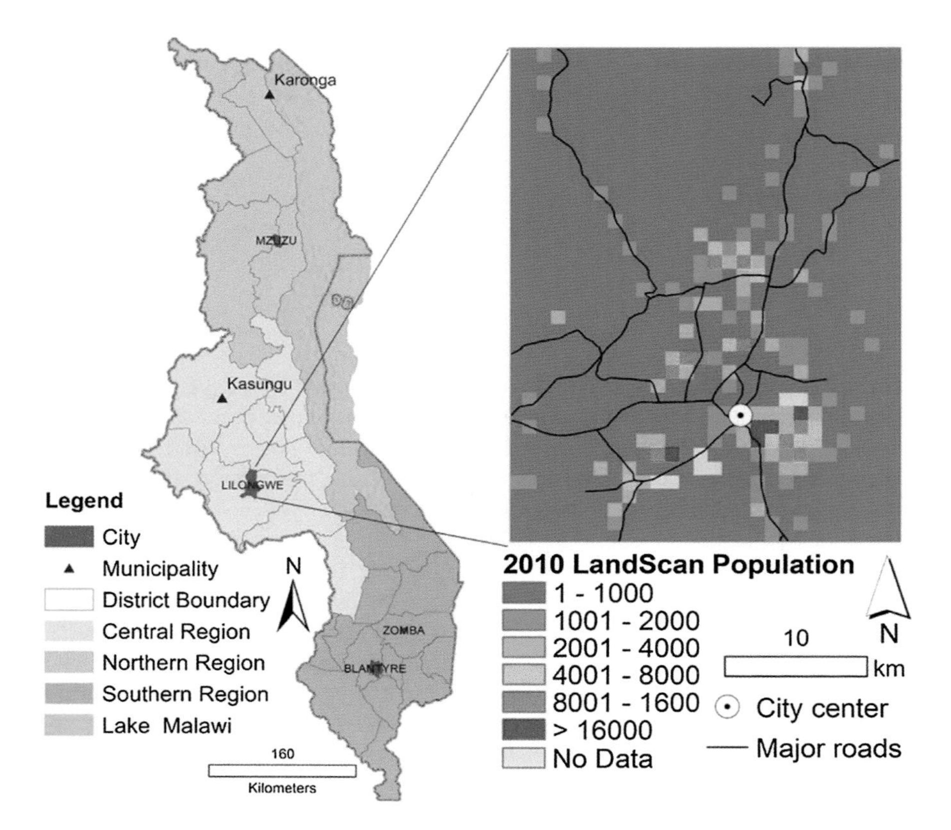

Fig. 16.1 Location and LandScan population of Lilongwe Metropolitan Area, Malawi

(Malawi's first president after independence) dreams in prison in Southern Rhodesia (Zimbabwe) during the struggle for independence. Others have then argued whether the relocation project was a matter of personal gratification, with underlying ethnic politics, or an effort to balance regional development (Englund 2002). The former argument has been supported by scholars who emphasized that Banda's primary vision of national building was to create a Chewa ethnicity in Central Region upon attaining independence from the British (Fig. 16.1) (Vail and White 1989). Others pushed the argument further that the choice was not only because Lilongwe was in his Chewa speaking central region but more importantly that it was nearby his home town of Kasungu (Connell 1972) (Fig. 16.1).

However, the choice of Lilongwe as the new capital could not have been as arbitrary an issue to be underpinned to personal gratification and ethnic politics alone. The colonial capital, Zomba, and its neighboring city of Blantyre as the commercial center had already concentrated much of the little urban growth in the southern region (see Table 16.1, 1966 urban population distribution) (Englund 2002). These two urban centers were distributed, largely, from the need for administrative and trading centers in the colonial period (Potts 1985). This saw urban and agricultural development concentrated in the southern region. As a result, Blantyre attracted a major share of industrial and commercial urban-based investment (Potts 1985). Though a capital, Zomba's role was limited to administrative functions which, in themselves, generated very little growth (Potts 1985).

Despite colonial concentration of development in the south, Lilongwe was already an administrative center by 1904, and the provincial headquarters for the Central Province, then known as Central Angoniland, shortly thereafter (Cole-King 1971). The emergence of tobacco as the main cash crop in the 1920s and improving communication and road services further improved its standing as a growth center in the colony (Englund 2002). It was of virtually the same size as Zomba by 1966 and thus joint second to Blantyre in the urban hierarchy (see Table 16.1) (Potts 1986). In any case, Zomba and Malawi had a peripheral economic position within the British Federation of Rhodesia and Nyasaland (that included Zimbabwe, Malawi, and Zambia) and in the southern African region as a whole, thereby mitigating against its significant urban development (Potts 1985). For instance, most of the local or national headquarters of many commercial and industrial firms operating in Malawi (Nyasaland then) were not in Zomba or Blantyre but in Harare (Salisbury then) in Zimbabwe (Southern Rhodesia then) (Potts 1985).

Capital cities are centers of administration and urban symbols of power and nationalism within a national framework. First, they ought to serve a more central position and provide a renewed focus for national pride apart from promoting ethnic accord by choosing a "neutral" site for a capital (Best 1970; Salau 1977; Moore 1982). These were emphasized in the relocation programs for Abuja in Nigeria and Gaborone in Botswana (Potts 1985). Second, they are used as "growth poles" to alleviate regional disparities and stimulate simultaneous further development (Friedman 1973). Lilongwe relates to all these general common features despite the official reasons given at the time of relocation, which were that Zomba had two

Table 16.1 Population of Malawi and its cities (million)

City	1966	1977	1987	1998	2008	2015[a]	2018[a]	2028[a]
Blantyre	0.11	0.23	0.33	0.50	0.66	0.88	0.99	1.43
Zomba	0.02	0.02	0.04	0.07	0.09	0.14	0.17	0.28
Mzuzu	0.01	0.02	0.04	0.09	0.13	0.22	0.27	0.47
Lilongwe	0.02	0.10	0.22	0.44	0.67	1.04	1.23	2.01
Total urban population	0.16	0.37	0.65	1.10	1.55	2.28	2.66	4.19
Total population	4.04	5.55	7.99	9.93	13.08	15.32	17.93	24.54
% of urban population[b]	4	7	8	11	12	15	15	17

Sources NSO (1978, 2008a, b)

[a]Projected population data

[b]Actual urban population percentages are slightly higher as this table excludes data from municipalities and town assemblies

major shortcomings: terrain limitations and that it was not suitably placed for future international airport (Connell 1972).

The British colonial government rejected to fund the establishment of the new capital (Potts 1986) arguing that Banda simply had to expand Zomba and concentrate on developing the agricultural sector. Banda announced the relocation of the capital in parliament in 1965 with a loan funding from the apartheid government of South Africa (Richards 1974). The South African influence extended to Lilongwe's town planning as evidenced by the sharp zoning of land use types where vast unbuilt-upon land separates residential areas (Englund 2002) characteristic of the apartheid regime. However, founded on the western bank of the Lilongwe River in 1924, segregation had characterized Lilongwe even before the South African government's involvement. All natives, for instance, were ordered to live on the eastern bank, while Asians were allocated the southeastern portion on the same side of the river as the natives (Kalipeni 1999).

Work began in July 1970 on Lilongwe, as a new capital city of Malawi, and it was a third post-independence capital in Africa after Nouakchott and Gaborone in Mauritania and Botswana, respectively (Connell 1972). By 1974, nearly 160 km of road had been built and over 80 km of water supply mains installed, while the supply of electricity witnessed an annual increase of over 20% (Richards 1974). The relocation of central administration had been virtually completed, with all government ministries moving to Lilongwe by 1978, together with most diplomatic missions (Potts 1985).

16.2 Primacy in the National Urban System

16.2.1 The Urban System in Malawi

Malawi is a landlocked country in Southern Africa bordered to the south by Mozambique, Zambia to the west, and Tanzania to the northeast. The country lies between latitude 9° and 18°S and longitudes 32° and 36°E. Malawi is known as the "Warm Heart of Africa" because of its friendly people with a beaming smile and a genuine disarming warmth. The country covers 118,484 km^2 out of which 94,276 km^2 (79.4%) is land, while the rest is water, particularly Lake Malawi which is the third largest fresh water lake in Africa (see Fig. 16.1).

Traditionally, the country is divided into 208 local traditional authority areas headed by local cultural leaders. Geographically and administratively, Malawi is divided into three main regions: north, center, and south with regional government headquarters in Mzuzu, Lilongwe and Blantyre cities, respectively. Zomba city, as the former capital, serves as headquarters for the "political and technical" eastern region. Malawi is further divided into districts (28): 10 in the central, 6 in the north, and 12 in the south. The southern region has the largest population share (44.8%) followed by the center (42.1%) and the north (13.1%) though the center has more land (35,592 km^2) followed by the south (31,753 km^2) and the north (26,931 km^2) (NSO 2008a). Consequently, the south has a high regional population density (185 persons per km^2) with six (Blantyre, Chiradzulu, Thyolo, Mulanje, Phalombe, and Zomba) out of the eight high-density districts found in this region (NSO 2008a). Politically, Malawi has 193 constituencies constituting a 193-member parliament of central governments legislative arm and 462 wards that supply membership to the district councils under decentralization's local government authorities. The wards are the smallest administrative unit, though for census purposes Malawi has enumeration areas which are smaller than the wards.

During the independence (July 6, 1964), the urbanization level was 4% with the majority living in the four main centers of Blantyre, Zomba, Mzuzu, and Lilongwe (see Table 16.1) (Potts 1985). The urbanization has grown to 10.7, 14.4, and 15.3% as of 1987, 1998, and 2008, respectively (NSO 2008a). Malawi is the third smallest country (in terms of surface area) of the nine southern African countries, larger only to Lesotho and Swaziland (World Bank 2015a). However, Malawi is fourth in terms of population size (16.7 million) exceeded only by South Africa (54 million), Mozambique (27.22), and Angola (24.22 million) (World Bank 2015b). Consequently, Malawi is the most densely populated country in southern Africa (177 persons per km^2) ranked at number 66 in the world with Swaziland coming second (74 persons per km^2) in southern Africa and ranked 133 in the world (World Bank 2015c).

The United Nations defines a national urban system as the spread of cities of different sizes across the whole territory of a country (United Nations 2015). It determines the need for specific urban development policies consistent with the size, growth, and function of each of the cities (United Nations 2015). As a result, it is linked to the organization and function of the government at various levels: national, regional, and local levels (Kim and Law 2012; United Nations 2015). Malawi's national urban system is comprised of four main cities: one located in each region except for the southern region which has two; and two municipalities of Kasungu and Karonga. Figure 16.1 shows the spatial distribution of these centers that form the regional social and economic framework of Malawi. Figure 16.2 shows parts of Lilongwe and Blantyre cities.

16.2.2 Primacy of Lilongwe

There is an interplay of forces toward population agglomeration and distribution that makes the empirical relationships between a country's existing urban system and its future development both multidimensional and nonlinear (United Nations 2015). To this extent, in most countries, the largest city tends to be home to a relatively high proportion of the urban population, making such a city a "primate city" (United Nations 2015). The United Nations defines a city as "primate" in a country if it accounts for at least 40% of the urban population in a particular year and that it has at least 1 million inhabitants with its degree measured by the

Fig. 16.2 **a**, **b** Views of parts of Lilongwe City Center; **c** Victoria Avenue, Blantyre city; **d** makeshift bridge over the Lilongwe river connecting two parts of Lilongwe market; **e** semipermanent houses in an informal settlement in Lilongwe; **f** waste disposal in Lilongwe city. *Source* **a** https://ayileche.wordpress.com/; **b**, **c** Author's fieldwork; **d** kelvininstrangeland.wordpress.com; **e**, **f** Nora Lindstrom (2014)

proportion of the country's urban population living in it (United Nations 2015). Lilongwe is estimated to have a population of 1,037,294 which represents 45% of the projected total urban population as of 2015 (Blantyre 884,497; 39%) (see Table 16.1; Fig. 16.3) (NSO 2008a, b).

Similar to all primate cities, Lilongwe's primacy is affected by urbanization. However, it is to remain a primate city accounting for close to half the total urban population in Malawi by the year 2028 (NSO 2008b). By then, it is projected that Lilongwe will have a population of 2,009,841 followed by Blantyre at 1,429,650

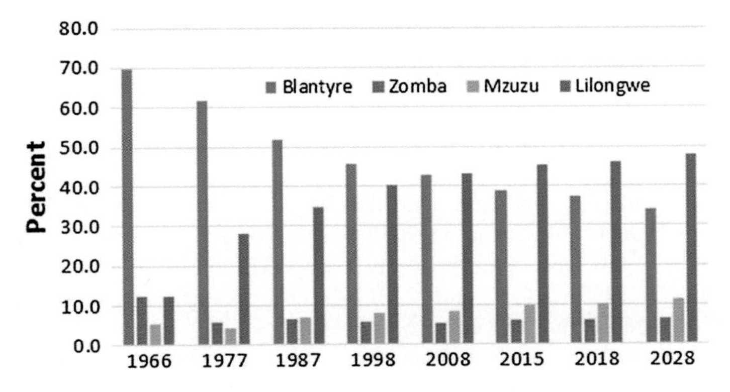

Fig. 16.3 City population against total urban population (%)

representing 48 and 34% of the total urban population and 8 and 6% of the total population, respectively (NSO 2008b) (see Table 16.1; Fig. 16.3). Figure 16.3 shows the declining dominance of Blantyre in the urban population and being overtaken by Lilongwe. Lilongwe has a land surface area of 392 km^2, which is bigger than all the other three combined. Blantyre comes second at 225 km^2 making it more densely populated than Lilongwe at 2940 and 1707 persons per km^2, respectively (NSO 2008a, b).

The city's population has been growing at an annual rate of 15.16% between 1966 and 1977, 7.75% from 1977 to 1987, 6.17% from 1987 to 1998, and 4.18% between 1998 and 2008 (NSO 2008b). The dramatic population changes in the early years are mostly due to boundary changes that occurred between the different early post-independence census exercises (Englund 2002). For instance, the drastic population changes between 1966 and 1977 for Lilongwe coincided with the capital relocation project, which involved an addition of 350 km^2 and thousands of people within the city boundary (Potts 1986). Otherwise, between 1977 and 2008 Lilongwe has grown at an annual growth rate of 6.04%, with Blantyre (second largest city) growing at 3.46% within the same period (NSO 2008b).

Lilongwe has therefore grown, both in terms of the population contained within its boundaries and also the extent of the built-up area that the population occupies. A list of contributing factors of population growth has been provided (Batty 2008; Besussi et al. 2010; Kim and Law 2012; United Nations 2015). From that list, the contributing factors for Lilongwe include the city's central location and flat terrain; economies of scale brought by agglomeration in its establishment; and the country's institutional history of post-independence. While there is evidence to support a positive relationship between concentration of people and economic efficiency, city primacy also creates imbalances in the urban hierarchy and bias in the development processes (UN-Habitat 2008; Short and Pinet-Peralta 2009; United Nations 2015). The primacy of Lilongwe in Malawi's urban system is evident since 2015 (see Table 16.1; Fig. 16.3).

16.3 Urban Land Use/Cover Patterns and Changes (1990–2030)

16.3.1 Observed Changes (1990–2013)

Figures 16.4 and 16.5 indicate that the built-up areas grew sevenfold from 5.08 km^2 in 1990 to 34.73 km^2 in 2013 (Table 16.2). This represents an annual increase of 1.29 km^2/year over the study period. The built-up growth rate for the "1990–1999," "1999–2011," and "2011–2013" epochs was 0.90, 1.22, and 3.47 km^2/year, respectively (Table 16.3).

Figure 16.6 shows the observed land use/cover changes of 23 out of the 58 areas for the three epochs based and ranked on the observed land use/cover changes of the "1999–2011" epoch. These were obtained after performing a zonal analysis of the land use/cover changes of the three epochs on the administrative boundaries (areas). Between 1990 and 1999, out of the 58 city areas, significant change took place in ten areas of 25, 47, 57, 29, 50, 49, 56, 7, 8, and 44 (in descending order) which contributed 4.5 km^2 (59%) of the total change. Between 1999 and 2011, significant change occurred in the areas of 49, 47, 25, 57, 50, 18, 56, 44, 45, and 29 (in descending order) contributing 9.8 km^2 (71%) of the total change. Between 2011 and 2013, significant changes occurred in areas of 49, 25, 50, 47, 36, 56, 45, 18, 55, and 57 (in descending order) contributing 3.9 km^2 (63%) of the total change.

The high-density areas in Malawi are categorized into squatter, traditional, and permanent. A traditional house is one built of mud walls and grass thatch roof, while a permanent house is built of modern or durable materials such as iron sheets and burnt brick walls (Manda 2013). Out of the total number of houses in Malawi in 2008, 44% were permanent, 45.1% semipermanent, and 10.9% traditional (NSO 2010; Manda 2013). Majority of the houses in traditional and squatter residential areas are semipermanent built with mud walls but having iron sheets (see Fig. 16.2e). The traditional housing areas (THAs) are officially designated neighborhoods which provide serviced plots that allow traditional construction techniques while squatters are unserviced, unofficial plots in the city (Manda 2013; Ministry of Lands, Housing and Urban Development 2013).

Figure 16.7 shows the contribution of some of the land use categories against the total observed land use/cover change of the areas over the years. These were obtained after performing a zonal analysis of the land use/cover changes of the three epochs on the administrative boundaries (areas) with each area designated a land use category. The contributing land use categories include traditional high-density (areas 25, 7, and 8); squatter (areas 57 and 56); industrial (areas 29 and 50); permanent high-density (area 49), and medium-density (areas 47, 44) areas for interval 1990–1999. Between 1999 and 2011, the contribution includes traditional high-density (area 25); squatter (areas 57 and 56); industrial (areas 50 and 29); permanent high-density (areas 49 and 18); agricultural (area 45); and medium-density (areas 47 and 44) areas. And for 2011–2013, the changes include traditional high-density

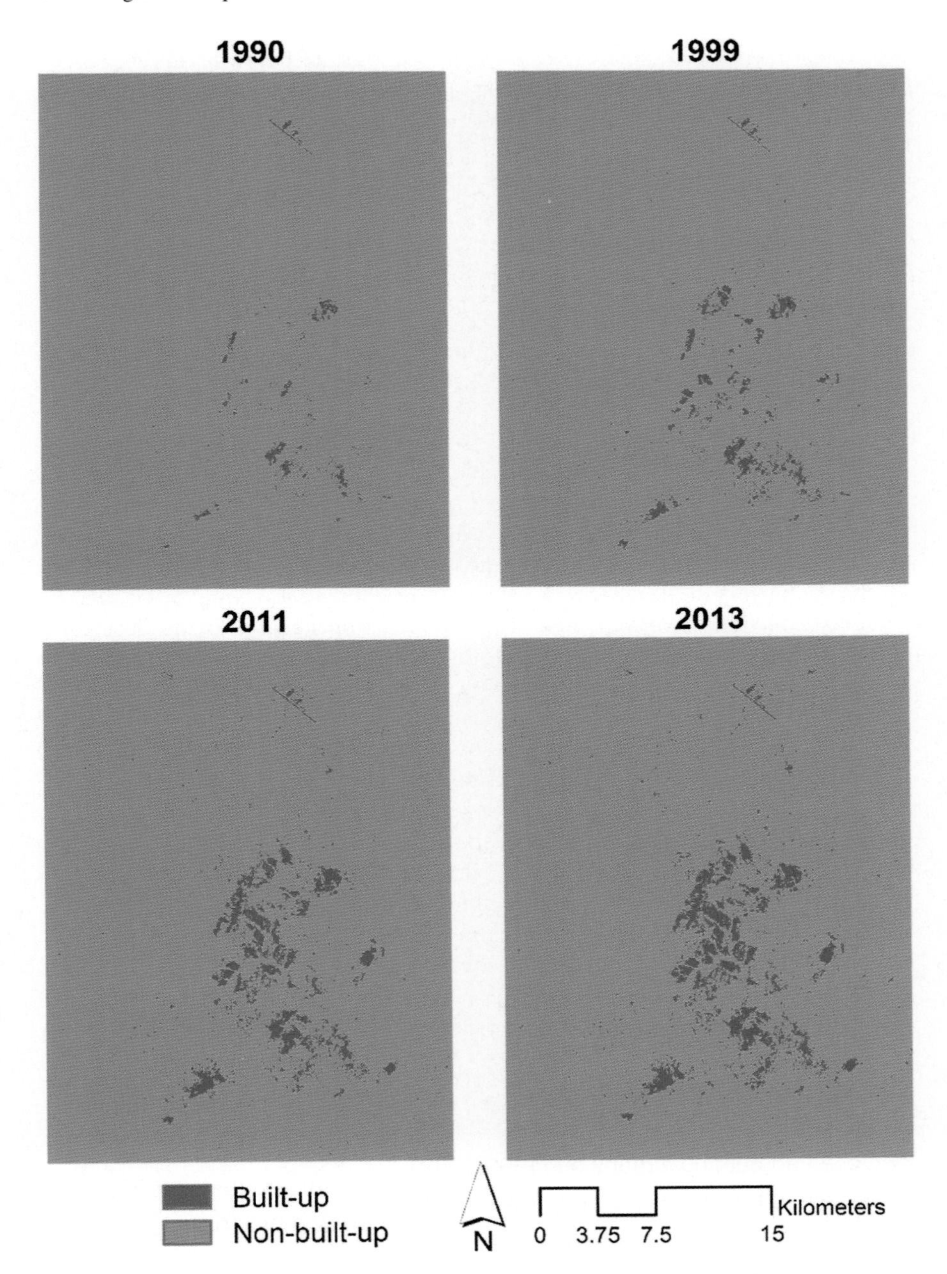

Fig. 16.4 Urban land use/cover maps of Lilongwe Metropolitan Area classified from Landsat imagery

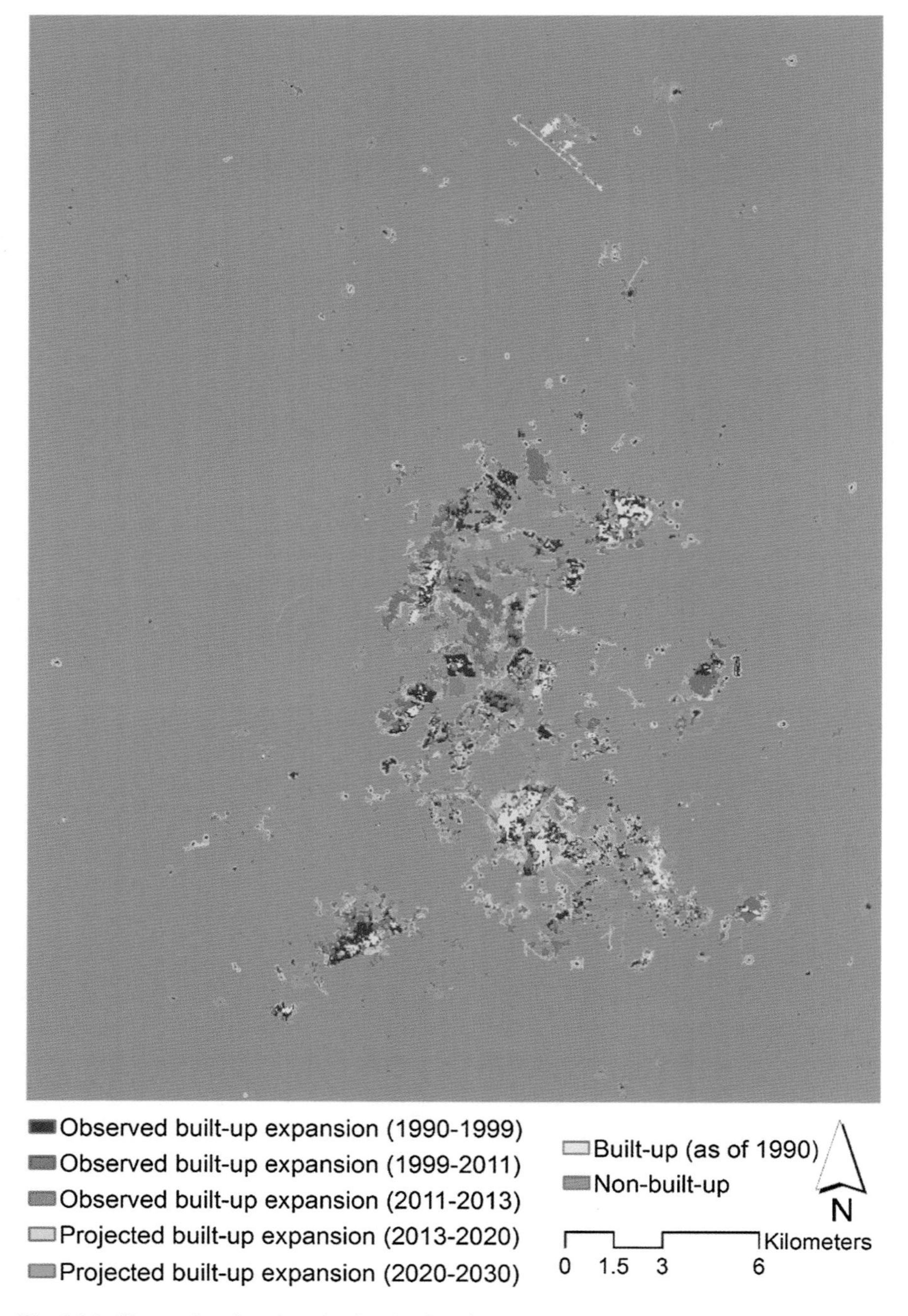

Fig. 16.5 Observed and projected urban land use/cover changes in Lilongwe Metropolitan Area

Table 16.2 Observed urban land use/cover of Lilongwe Metropolitan Area (km^2)

	1990	1999	2011	2013
Built-up	5.08	13.14	27.80	34.73
Non-built-up	911.27	903.20	888.55	881.61
Total	916.34	916.34	916.34	916.34

Table 16.3 Observed urban land use/cover changes in Lilongwe Metropolitan Area (km^2)

	1990–1999	1999–2011	2011–2013
Built-up	8.06	14.65	6.93
Annual rate of change (km^2/year)	*0.90*	*1.22*	*3.47*
Non-built-up	−8.06	−14.65	−6.93
Annual rate of change (km^2/year)	*−0.90*	*−1.22*	*−3.47*

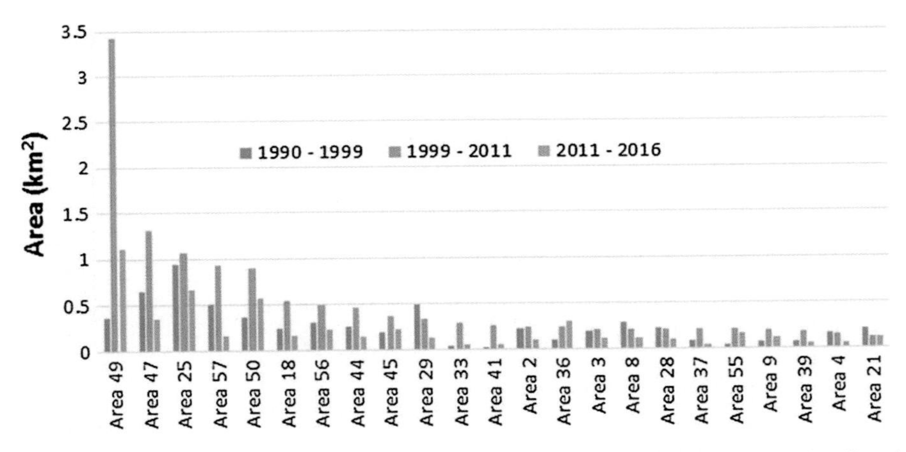

Fig. 16.6 Observed built-up expansion at the area level for "1990–1999," "1999–2011," and "2011–2013" epochs. *Note* See Fig. 16.11 for the spatial locations of the "areas"

(area 25); squatter (areas 57 and 56) with industrial (area 50), permanent high-density (areas 49, 36 and 18), agricultural (area 45), medium-density (area 47), and undetermined (area 55) areas. This indicates that traditional high-density areas and squatters contributed a combined land use/cover change of 21, 18, and 16% in 1990–1999, 1999–2011, and 2011–2013, respectively.

Table 16.4 shows the spatial metrics of the built-up areas. In terms of percentage of landscape (PLAND)—the proportion of the built-up class at a particular time to the whole landscape—the values for Lilongwe were 0.55, 3.03, 3.79, 4.50, and 5.79% for the years 1990, 1999, 2011, and 2013, respectively. The patch density (PD) metric measures the fragmentation based on the number of patches per unit area (in this case per km^2), in which a patch is based on an 8-cell neighbor rule. In 1990, Lilongwe had a patch density of 0.43, which increased to 0.84, 1.41, and 1.66

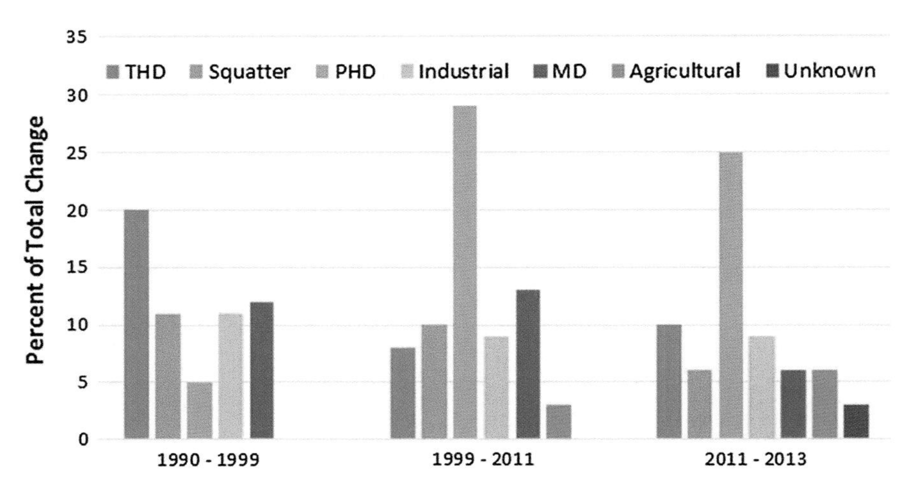

Fig. 16.7 Observed change per land use category (*THD* traditional high density, *PHD* permanent high density, *MD* medium density)

Table 16.4 Observed landscape pattern of Lilongwe Metropolitan Area

Class-level (built-up) spatial metrics	1990	1999	2011	2013
PLAND (%)	0.55	1.43	3.03	3.79
PD (number per km^2)	0.43	0.84	1.41	1.66
ENN (mean) (m)	239.89	236.17	191.51	177.72
CIRCLE (mean) ($0 \leq$ CIRCLE < 1)	0.39	0.39	0.37	0.36
SHAPE (mean) ($1 \leq$ SHAPE $\leq \infty$)	1.23	1.22	1.24	1.22

in 1999, 2011, and 2013, respectively. The Euclidean nearest-neighbor distance (ENN) metric is a measure of dispersion based on the average distance to the nearest neighboring patch of the same class. The mean ENN of the built-up land in Lilongwe decreased from 239.89 m in 1990 to 236.17, 191.51, and 177.72 m in 1999, 2011, and 2013, respectively. The decrease in mean ENN is largely due to growth of the old patches (increasing PLAND) and rise of new patches of built-up in between the old patches (increasing PD values). The CIRCLE metric measures the circularity of patches. The CIRCLE value ranges from 0 for circular or one cell patches to 1 for elongated, linear patches one cell wide. The mean CIRCLE value of built-up land for Lilongwe showed a general decreasing trend between 1990 and 2013. This signifies a more radial and/or circular pattern especially around the city's sectorial growth centers of Old Town, Capital Hill, Kanengo, and to a less extent Lumbadzi (the airport area). The SHAPE metric is a measure of complexity having a value of 1 when the patch is square and increases without limit as patch shape becomes more irregular. Lilongwe's mean SHAPE value fluctuated around 1.23 between 1990 and 2013, indicating a stable level of low complexity of the built-up land areas.

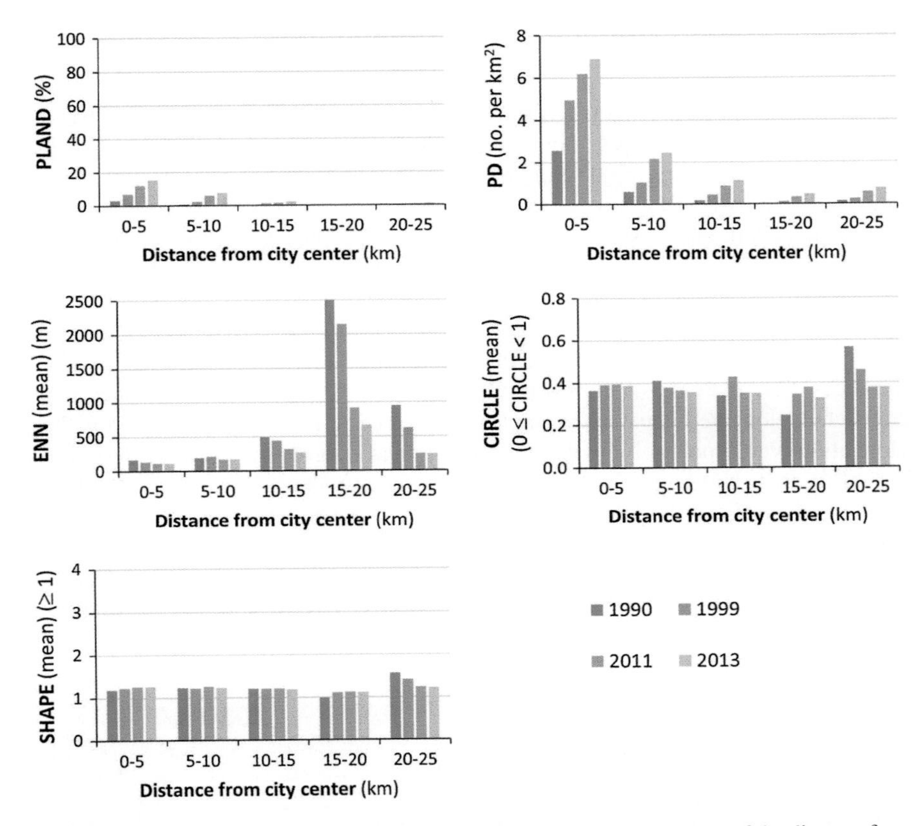

Fig. 16.8 Observed class-level spatial metrics for built-up along the gradient of the distance from Lilongwe zero city center of Lilongwe Metropolitan Area. *Note* The y-axis values are plotted in the same range as those in Fig. 16.10

Figure 16.8 shows all the spatial metrics for the built-up class of Lilongwe along the gradient of the distance from the city center across all time periods from 1990 to 2030. The figure shows that PLAND decreases as the distance from the city center increases, indicating that the proportion of built-up land near the city center is relatively higher. The trend is the same for PD except when the distances reach 25 km from the city center when it begins to increase. This indicates that there were more patches of built-up land near the city center and in the areas farther from the city center. The patches seem to have been more dispersed around the middle distances as indicated by the mean ENN values that increase up to a distance of about 15 km and then begins to decrease. This could be attributed to the same zoning scheme of the city planning. The result in Fig. 16.8 shows a stable trend of the mean CIRCLE and SHAPE values along the gradient of the distance from the city center. The former indicates that the patches of built-up land elongate or grow more linear in farther distances. However, the latter indicates that the complexity of the built-up land in Lilongwe City was uniform across the distances.

16.3.2 Projected Changes (2013–2030)

Based on the projected land use/cover maps (Fig. 16.9, see also Fig. 16.5), the built-up land for Lilongwe City would increase from 34.73 km^2 in 2013 to 41.24 km^2 by 2020 and 53.08 km^2 by 2030 (Table 16.5). This represents annual rates of 6.51 and 11.84 km^2/year, respectively (Table 16.6). The results of the landscape pattern analysis show that the predicted patches of built-up land in 2020 and 2030 would be more aggregated as indicated by the increase in PLAND and decrease in PD (Table 16.7). However, the increase in mean ENN indicates that average distance would increase between the newly formed bigger aggregated patches as PD decreases and the newly formed smaller patches as PLAND increases. The increase in the mean CIRCLE and SHAPE values indicates more connected, elongated/linear and complex patches of built-up land, respectively. The elongation would probably be due to built-up land outgrowing the four designated growth centers and beginning to follow road infrastructure (Figs. 16.4 and 16.5). This trend is already being observed at present despite the statistics failing to capture it significantly.

Along the gradient of the distance from the city center (Fig. 16.10), the PLAND of the predicted 2020 and 2030 patches of built-up land would also be higher at distances closer to the city center. The PD would decrease dramatically by 2020 and

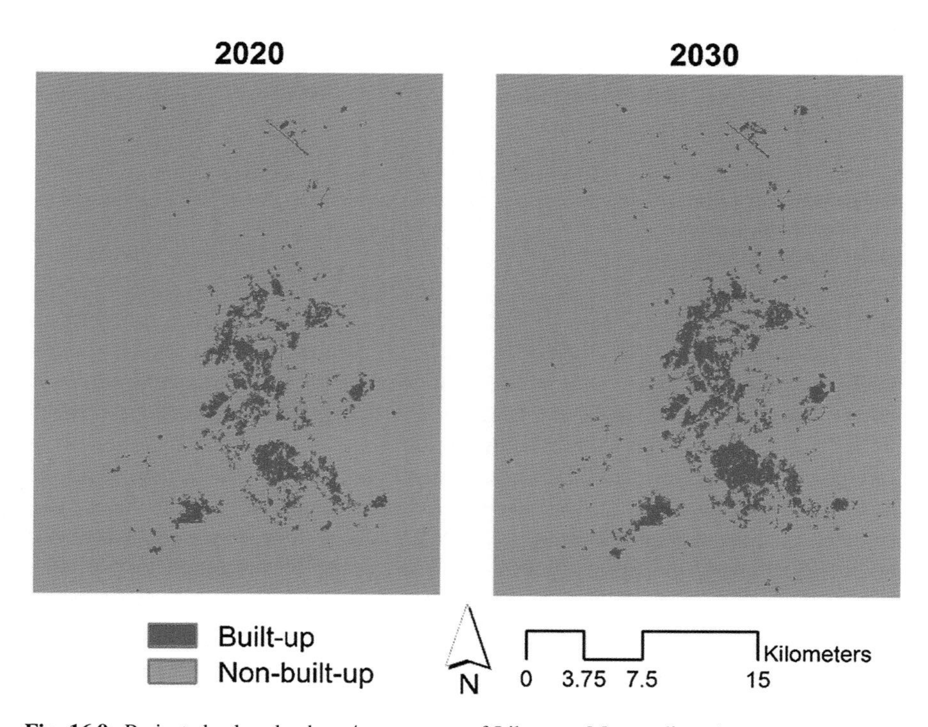

Fig. 16.9 Projected urban land use/cover maps of Lilongwe Metropolitan Area

Table 16.5 Projected urban land use/cover of Lilongwe Metropolitan Area (km^2)

	2020	2030
Built-up	41.24	53.08
Non-built-up	875.10	863.26
Total	916.34	916.34

Table 16.6 Observed urban land use/cover changes in Lilongwe Metropolitan Area (km^2)

	2013–2020	2020–2030
Built-up	6.51	11.84
Annual rate of change (km^2/year)	*0.93*	*1.18*
Non-built-up	−6.51	−11.84
Annual rate of change (km^2/year)	*−0.93*	*−1.18*

Table 16.7 Observed landscape pattern of Lilongwe Metropolitan Area

Class-level (built-up) spatial metrics	2020	2030
PLAND (%)	4.50	5.79
PD (number per km^2)	0.21	0.25
ENN (mean) (m)	460.74	428.90
CIRCLE (mean) ($0 \leq$ CIRCLE < 1)	0.59	0.61
SHAPE (mean) ($1 \leq$ SHAPE $\leq \infty$)	2.97	3.18

2030, but would still be relatively higher in the nearer and farther distances. The direct opposite would be true for the mean ENN which is predicted to increase drastically. However, the general pattern would be the same in that it will increase up to a distance of about 15 km and then begin to decrease. The mean CIRCLE value would also increase by 2020 and 2030 and would be stable/uniform across the distances. The mean SHAPE value would also increase by 2020 and 2030, but would gradually decrease across the gradient of the distance to the city center (Fig. 16.10).

16.4 Driving Forces of Urban Development

Factors affecting the urban development of Lilongwe include both physical and socioeconomic forces. Physically, Lilongwe is centrally located and easily accessible from all districts. Sitting on the Lilongwe plains, the terrain of Lilongwe area supports a great variety of activities that includes agriculture and forestry. The Lilongwe plain supports a substantial proportion of the tobacco industry that Malawi's economy relies on. However, much of the urbanization factors of Lilongwe City are socioeconomic in nature, which include population and economic growth. In examining these factors, there is a need to look into the existing legislations with regard to Malawi urban development.

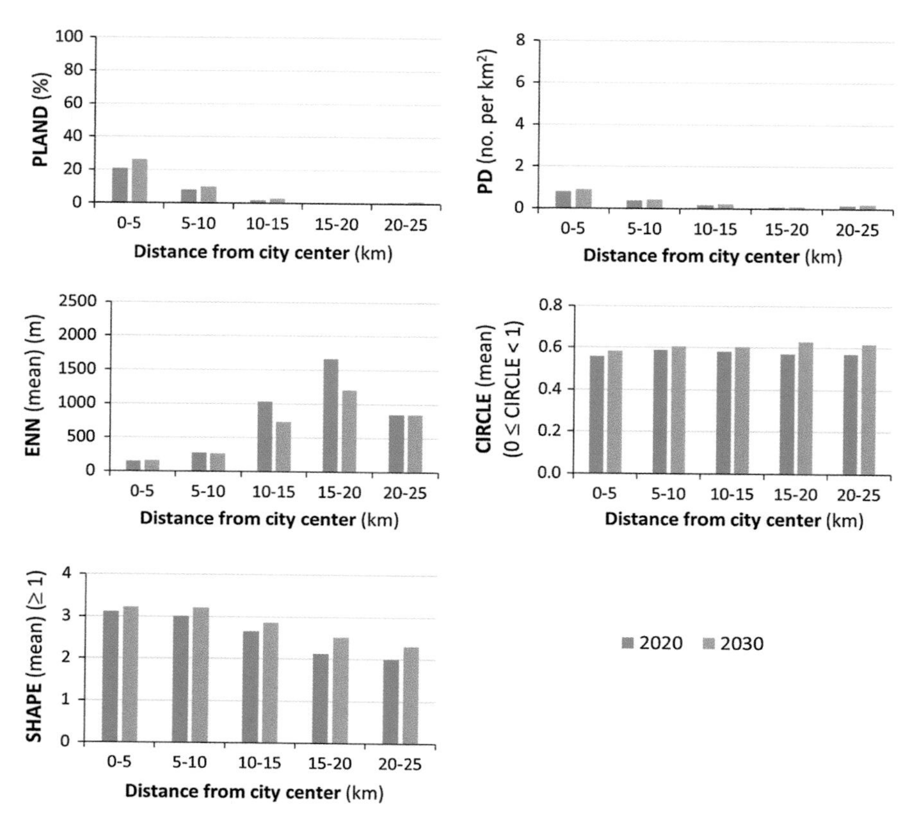

Fig. 16.10 Projected class-level spatial metrics for built-up along the gradient of the distance from city center of Lilongwe. *Note* The y-axis values are plotted in the same range as those in Fig. 16.8

16.4.1 Urban Development Legislation in Malawi

There are several legal and policy instruments that are utilized by an assortment of institutions involved in urban planning, management, and land governance. However, an analysis of some of the legislations and policies reveals several conflicts, contradictions, and implementation challenges (Manda 2013). Some of the critical legislations guiding urban development in Malawi are as follows:

- Town and Country Planning Act (TCPA). This is the principal law governing urban planning and urbanization in Malawi. The law mandates the Minister and through him the Commissioner and Planning Committees, to prepare and approve plans. The Minister has the power to approve, reject, and revoke plans. The law provides for definition of planning authorities, planning levels and types, and development control process.

- Land Act. This is the main law for land administration. Under section 26, local traditional leaders (chiefs) have power to allocate customary land for any use. Some of the customary land is within declared urban statutory planning areas. In Zomba City, for example, customary land constitutes nearly 90.5% of all the land (UN-Habitat 2011b).
- Chiefs Act. The law is generally meant to guide the operation of traditional leadership in the country. Under section 5, it is provided that chiefs have no mandate within city, municipal, and town councils unless they have written permission from the chief executive.
- Public Health Act. The law is meant to ensure adherence to environmental health. The Minister for Health has power to order demolition of buildings that fail to provide for adequate sanitary measures. The problem can emerge when such order contradicts approval by planning committees. Of interest is the placing of guidelines for the preparation and implementation of traditional housing areas (THAs). Detailed layout plans under this law with speculation rife that low-income housing is seen from the viewpoint of public health rather than as a development sector or right (Manda 2013).
- Malawi Housing Corporation (MHC) Act. The law was developed in 1964 to support the development, construction, and maintenance of housing estates and plots on land given by government on freehold basis.
- Environmental Management Act. The law provides the administrative framework for environmental management including impact assessment and audits for specific development projects outlined in sections 24–29. The law provides for the protection and sound management of the environment, conservation, and sustainable utilization of natural resources. The Environmental Affairs Department (EAD) is mandated to administer the Act. To this extent, it has deployed officers to both city and district councils.
- Local Government Act (LGA). The Act, No. 42 of 1998, was developed to promote devolution.
- Decentralization Policy. The policy was approved in 1998 to institutionalize democracy by allowing local communities to take leading roles in the development process. A significant provision was devolution of fiscal, political, and administrative functions including urban planning.
- National Land Policy. The policy was approved in 2002. There are two major provisions relevant here: (i) the entire country will be a planning area and (ii) the policy unequivocally states that compliance with planning does "not require the acquisition of and conversion of customary tenure to public ownership" (Malawi Government 2002)

While all these legal frameworks exist to ensure proper planning, the observed land use/cover changes for Lilongwe indicate that a significant portion of the urbanization in Malawi is unregulated and unplanned (16–21%, Fig. 16.7). For instance, the administration of the THAs under MHC Act required that the houses built fully comply with minimum building standards and regulations. However, since the devolvement of administration of THAs to the four major city governments and some town councils in the early 1990s, the city assemblies have failed to enforce the requirements. It is not surprising then to note that between 1990 and 1999, 21% of the urbanization in Lilongwe occurred in unregulated and unserviced traditional and squatter housing areas (Fig. 16.7). This to the most part has been due to political interference and emerging international anti-eviction campaigns in the housing sector (Manda 2013). However, the contradictory authorities the LGA and TCPA empower to manage the development plans compound the problem. The former empowers local governments (city and town assemblies) to prepare development plans including urban plans in their jurisdictions while the latter empowers Physical Planning Department at the central office to do the same.

The decentralization policy has also not been successful because several sectors including Physical Planning, which cited low capacity in the councils, have delayed or shown reluctance to devolve even town ranging positions that are more relevant to councils (Manda 2013). As a consequence, chiefs, who through the Chiefs Act have no mandate in town planning, take advantage to the extent that government institutions are almost powerless. Hence, the plan that led urban development is abandoned in favor of ad hoc locating of projects leading to unserviced and unregulated squatter and traditional housing areas. It is comforting, though, that over the years the urbanization in the traditional and squatter housing areas in reducing (see Fig. 16.7). Whether this is due to improvement in the legal structures or subsiding political influence remains to be seen.

16.4.2 Population Growth

At an average annual growth rate of 4.18%, the highest of all cities in Malawi, Lilongwe has seen its population grow from 440,471 in 1998 to 669,021 in 2008 (Table 16.1). This rate, though lower than national urban population growth (5.45% per annum), is higher than the national population growth rate (2.8% per annum) (NSO 2008a). It is also higher than Johannesburg's (3.18%, 2001–2011), Cape Town's (2.57%, 2001–2011) (Statistics South Africa 2015), and comparable to Lusaka's (4.9%, 2000–2010) (Zambia Government 2011) despite Lilongwe being a small economy comparatively. Between 1977 and 2008, the population of Lilongwe has grown sixfold with an annual population growth rate of 6.04%. With population growth being the most important common factor influencing urbanization, these statistics for Lilongwe are very crucial to its urban development. Much of this population growth has been due to rural–urban migration, natural increase, and reclassification of settlements (Manda 2013).

For instance, the number of children a woman in child bearing ages 15–49 can have before retiring (total fertility rate, TFR) is recorded to be 5.2 (NSO 2010) with an urban TFR of 4.5 (rural TFR = 6.3) which is very high for an urban population (Mpando 2000; Manda 2013). When it comes to rural–urban migration, the drivers can be classified as push and pull factors. The push factors are largely problems associated with rural life such as rural poverty, lack of alternative wage employment, declining productivity of the soils due to constant tilling and population pressure on limited land areas as well as "drudgery and boredom of rural areas" (Kalipeni 1997; NSO 2010; Manda 2013). The pull (attraction) factors of urban areas include better services (such as better health care and education services) (Manda 2013) and benefits of modernity like "good housing, a car, a stereo, and a television set" (Kalipeni 1997). The most important pull factor, however is, as outlined in the Harris–Todaro model, the anticipated potential of finding wage employment in urban areas (Todaro 1994).

16.4.3 Economic Growth

Economic and population growth tend to influence each other. As the population grew for Lilongwe, manpower for economic growth became readily available. The civil service employs about 27% of the city's work force with the private and self-employment sectors employing about 40 and 24%, respectively (UN-Habitat 2011a). Many of these are employed in finance, banking, retail trade, construction, transport, public administration, tourism, and tobacco manufacturing which are the main economic activities in Lilongwe. The informal sector in Lilongwe is also vibrant and employs an estimated 27% of the economy though it lacks adequate regulations and operates with minimal city council support (UN-Habitat 2011a). Most of the industrial activities are concentrated in Kanengo (Area 52, 53) and Mchesi (Area 8, 7, Old Town) (see Fig. 16.11). Lilongwe is also the center for tobacco sales and processing which is Malawi's main export earner (Manda 2013). Despite being an administrative city, it was not until 2005 when the central government decided to relocate its administrative head offices to the city following a presidential directive. While this did not really affect the population of the city that much, it heightened the concentration of the employment opportunities and economic growth.

16.5 Implications for Future Sustainable Urban Development

The role of urbanization in economic development is recognized globally usually with the expectation that the more urbanized a country is, the more developed it becomes (Manda 2013). The positive outcomes relate to economies of scale and concentration resulting from the increased size of the population and human density (Dretcher and Lanquinta 2002; Manda 2013). Many have then argued that the low level of urbanization in Malawi [at 15.3% with Zambia's at 30% (Manda 2013)] reflects the low level of economic development. It is argued that the urbanization

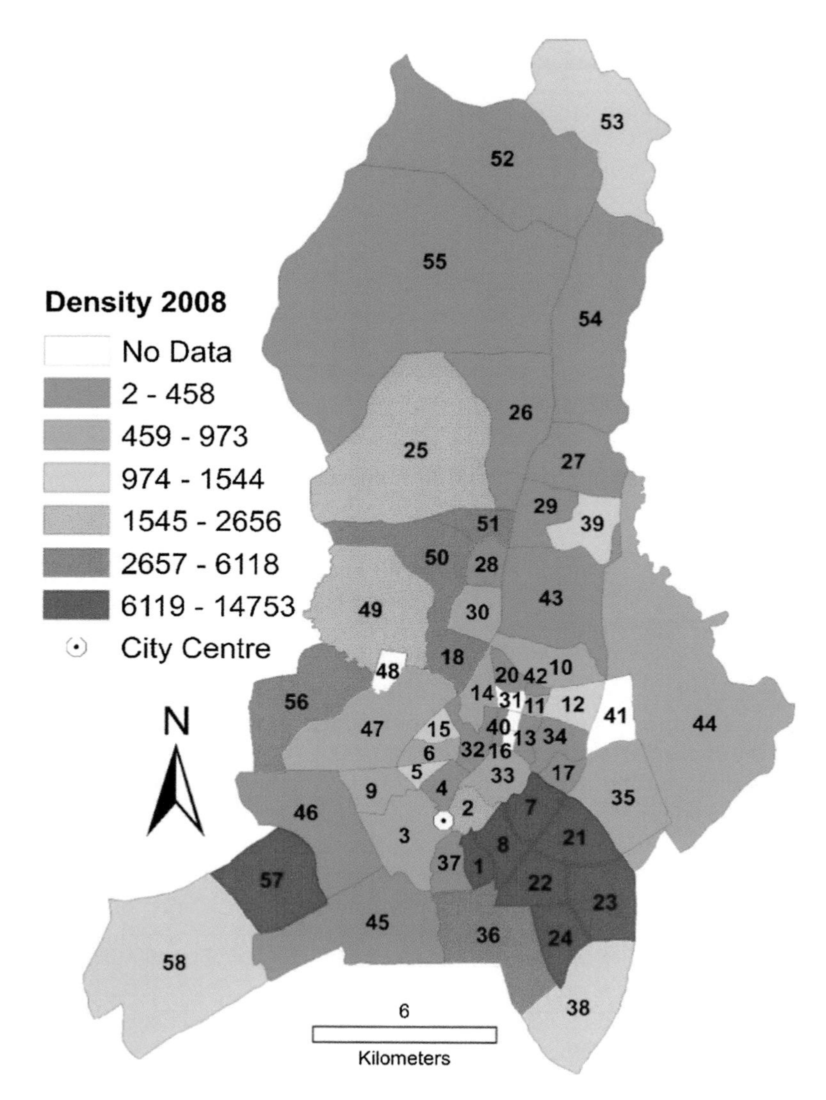

Fig. 16.11 Population density in 2008 of areas (*numbered*) in Lilongwe City (people per km^2)

which could have increased the population of urban centers diverted to other countries where nearly a quarter of the male labor force might have gone during the colonial era (Kalipeni 1997; Manda 2013). Again it also suggested that the low level of urbanization was a result of national investments that would attract migrants, such as Malawi Cargo Centers, being located outside the country (Potts 1985). The argument is further extended to say that urbanization started to grow only with the curtailment of these paid international labor exports (Kalipeni 1997).

Since then, cities like Lilongwe, have experienced rapid urbanization alongside improvements to social welfare and economic development. Lilongwe, taking over from Blantyre, has been the pillar of all socioeconomic improvement. However, this has come at a huge cost for the city.

16.5.1 Urbanization and Development

The State of the World Cities 2012/13 (UN-Habitat 2013, p. 22, 55) outlines the role of urban areas in development which can be realized if cities meet five conditions or are supported with these five aspects:

(i) Productivity through economic growth. The lack of infrastructure such as markets and good roads, lack of access to credit to expand and improve their businesses and lack of entrepreneurial skills are major obstacles that hamper economic growth in Lilongwe (UN-Habitat 2011a). The informal economy needs to be integrated with the formal economy and regulated to boost economic growth. Poverty in Lilongwe stands at about 25%, with 9% of the population being ultra-poor (Lilongwe City Assembly 2009). The poor do not have access to loans due to the high interest rates that financial institutions charge. This makes it difficult for the poor to improve their economic status and get out of poverty (UN-Habitat 2011a).

(ii) Consistent and targeted infrastructure development which is considered the "bedrock of urban prosperity" as it "provides the foundation on which any city will thrive" and attract investment (UN-Habitat 2013, p. 55). Growing at a rate of 4.3% and a population density of 1479 persons per km^2, as of 2008, the city had a total of 147,379 households against a population of 669, 021 (NSO 2008a). The housing typology includes low-density, medium-density, high-density permanent for intermediate-income earners, and THAs for low-income earners (UN-Habitat 2011a). Additionally, there are the informal settlements that are very high density. Formal housing is inadequate and has led to the increase in population in the informal settlements with over 70% of the city's population living in informal settlements (Lilongwe City Council 2010). The majority of these live in substandard housing conditions with inadequate social infrastructure and lack access to the basic urban services (UN-Habitat 2011a);

(iii) Promotion of quality of life through service delivery. Energy sources in Lilongwe include electricity, fuel wood, and paraffin. The Electricity Supply Commission of Malawi (ESCOM) supplies electricity with only 26% of the city residents connected to electricity. The electricity supply infrastructure is old and worn-out resulting in frequent power outages and high maintenance costs (UN-Habitat 2011a). Further, the cost of electricity supply is high and inaccessible to the poor. Similar statistics hold for water delivery in the city. Most of the formal city dwellers have access to piped water.

While communal water points (kiosks) are used in the informal settlements, the water is not enough to service all the informal settlers. As a result, some rely on unprotected water sources, such as rivers and boreholes, to supplement the water provided by the city. It is estimated that about 75% of the population was served by the Lilongwe water board in the 2009/10 up from 50% in 2008/09 (Lilongwe City Assembly 2009) and about 91% of the population was able to access piped water within a 30-min walking distance in 2009 (Malawi Government 2010). The condition of the public transport, land, and security is in similar state;

(iv) Promotion of equity and social inclusion. With almost equal proportions of females (49%) and males (51%), women do not participate as much as men in the city's management activities (UN-Habitat 2011a); and

(v) Environmental sustainability. The sewerage system in Lilongwe covers only 9% of the city forcing the majority (75%), especially in the informal settlements, to rely on pit latrines for human waste disposal (NSO 2008a). Waste management services are readily available in the high-income areas, but the low-income areas do not have access to these services (Lilongwe City Council 2010). Commercial areas such as markets have their waste collected, albeit irregularly. Most households, especially in the informal settlements, dispose their waste in open spaces, on riverbanks, and along roadsides (UN-Habitat 2011a). Such a poor sewerage infrastructure and high dependence on pit latrines have led to pollution of underground water (UN-Habitat 2011a) and air.

To this extent a city prosperity index (CPI) has been developed whereby a city is judged on how well it fares with regard to these five conditions or criteria (Manda 2013; UN-Habitat 2013). Lilongwe, though generally making progress, is failing to provide sustainable urban development.

16.5.2 Urbanization and Poverty

While there has been an increase in economic opportunities resulting from the rapid urbanization, the resultant population growth has tended to compromise its full potential. In general, poverty is seen in terms of income or consumption (Wratten 1995). The World Bank poverty line of $1 per day has become adopted internationally. The measure of $1 converts to K555 (October 2015 rates) which implies an earning of K16,650 per month (K199,800 annually). The Malawi National Statistical Office's (NSO) definition of poverty is a total annual per capita (food and non-food) consumption equivalent of a quarter of $1/day (NSO, Integrated Household Survey 2010–2011, 2012, p. 204) which translates to K51,340 (K31,851 for ultra-poor). In Malawi, very few can afford this including many in the urban population. However, urban poverty is also spatially expressed in the form of (lack of) access or availability of services and infrastructure (Manda 2013).

As such, though the general expectation would be that urban areas would have higher income (or that urban residents are non-poor), poverty levels have been exacerbated by lack of access to the basic urban services and a steady growth in the prices of basic goods (UN-Habitat 2011a). Acquiring loans for economic development is hard for the poor due to the high interest rates, and their participation in city development is minimal (Manda 2013). Generally, the poor spend about 43% of household income on food (UN-Habitat 2011a). It is not surprising then that Lilongwe has the largest (25%) [22% according to NSO (NSO Integrated Household Survey 2010–2011, 2012, p. 205)] incidences of poverty compared to the other cities in Malawi while 9% [4.1% according to NSO (NSO Integrated Household Survey 2010–2011, 2012, p. 209)] are estimated to be ultra-poor (UN-Habitat 2011a). From this discussion, and without delving into the process of urbanization of poverty, the 70% in Lilongwe living in low-income or informal settlements or under slum conditions fall under the category of poor (UN-Habitat 2011a) though living in an urban setup.

16.5.3 Built-Up Expansion Pattern

As shown in Table 16.4 and Fig. 16.8, the spatial metrics indicate that the proportion of built-up land near the city center is high and that there were more patches of built-up land near the city center and in areas farther from the city center. This could be explained by the Lilongwe Outline Zoning of 1986 plan that guided the early development of the Capital city. It adopted a linear, multicentered urban form in order to avoid congestion problems that can arise with a single center (Ministry of Lands, Housing and Urban Development 2013). The aim was to cluster residential, employment, and service areas around each center, so as to reduce the need to travel long distances. There are four such centers and each one was the focus of a sector of the city: Old Town Primary Commercial Center, comprising the twin established centers in Area 2 (Bwalonjobyu) and Area 3 (Kang'ombe), which is also the city center from which the distance statistics are being calculated from; City Center (Area 13, 16 and 19), serving the Capital Hill sector and just adjacent to Old Town; Kanengo Primary Commercial Center in Area 25/2 (Bvunguti); and Lumbadzi Primary Commercial Center, based on the established trading center in Area 52 and 53 (Kalimbakatha) (see Fig. 16.11) (Ministry of Lands, Housing and Urban Development 2013). The latter two are some >20 km away from the city center which could explain the increase in the number of patches at distances of 25 km and above. This is also being supported by the mean ENN values that increase up to a distance of about 15 km and then begins to decrease. Figure 16.8 shows a stable trend of the mean CIRCLE and SHAPE values along the gradient of the distance from the city center. The former indicates that the patches of built-up land elongate or grow any more linear in farther distances. And the latter indicates that the complexity of the built-up land in Lilongwe city was uniform across the distances.

The projected figures in Table 16.7 indicate that the landscape pattern analysis of the predicted patches of built-up land in 2020 and 2030 would be more aggregated and that more neighborhood (closer) patches would become connected. The elongation would probably be due to built-up land outgrowing the four designated growth centers and beginning to follow road infrastructure (Figs. 16.5 and 16.9). This trend is already being observed at present despite the spatial statistics failing to capture it significantly. For instance, despite the fact that the existing outline zoning scheme was planned to develop the four sectors, Old Town, City Center, Kanengo, and Lumbadzi, the actual urbanization has concentrated in the two big economic centers of Old Town and City Center (Ministry of Lands, Housing and Urban Development 2013). The current trends indicate urban expansion to the south (Area 1, 36), the southeast (Area 8), and the southwest (Area 57) evidenced by the high population densities in the areas concerned (see Figs. 16.6 and 16.11). Following from this development, the future urban structure for Lilongwe City adopted the cluster shape development with Old Town and City Center forming the inner ring; Kanengo being in the middle ring; and Lumbadzi in the outer ring (Ministry of Lands, Housing and Urban Development 2013).

An overarching implementation strategy has since been developed by the Ministry of Lands, Housing and Urban Development to see sustainable urbanization of Lilongwe involving the following three phases (Ministry of Lands, Housing and Urban Development 2013):

(a) Phase 1 (~2015): dubbed the significant period, this phase challenges the responsible organizations to arrange legal and institutional measures supporting the new zoning proposed in the cluster shape development of 2030 land use plan. It encompasses various aspects such as land registration, investment, promotion measures for foreign capital, industrial estates, housing bill, and regularization measures of unplanned settlements.
(b) Phase 2 (2015–2020): This is the active transformation phase of the urban structure in Lilongwe City such as the urban road network formation including the completion of the inner ring road, improved intersections and more effective north–south and east–west links. The aim is to propel change in the distribution of employment opportunities and to optimize the person-trip distribution throughout the area, thereby reducing economic opportunity cost. Urban redevelopment projects for densification as well as for establishment of high-rise commercial areas shall be mobilized on a full-scale dimension with deliberately accelerated attraction of foreign capital direct investment with the full use of investment promotion measures already introduced in Phase 1.
(c) Phase 3 (2020–2030): It is envisaged that the new urban structure will be completed in this phase. By 2030, Lilongwe City shall embody the four concepts; (i) abundant greenery within the city, (ii) efficient land use, (iii) strengthening of Lilongwe economy through economic diversification, and (iv) a compact urban land use. A compact urban land use shall be achieved through establishment of high-rise commercial and residential zones, and mixed-use redevelopments in the central areas. The twin-pole urban center

(City Center and Old Town) shall be developed to become attractive as well as efficient locations in order to promote Lilongwe City as an internationally competitive city among neighboring countries' capital cities. Simultaneously, comfortable living environment shall be maintained with the preservation of abundant greenery as well as creation of parks and recreational spaces.

16.6 Concluding Remarks

From its humble beginnings on the side lines of Zomba and Blantyre, Lilongwe City has grown rapidly to become the administrative and economic hub of Malawi and pushing its presence in the southern African region. Its growth had been slow until it became the capital in 1975 when built-up area grew sevenfold from 5.08 km^2 in 1990 to 34.73 km^2 in 2013. Furthermore, it is projected that its built-up area would have increased from 34.73 km^2 in 2013 to 41.24 km^2 by 2020, and 53.08 km^2 by 2030. While the proportion of built-up class at a particular time to the whole landscape is very low (only 5.79% in 2013), the built-up area for Lilongwe is less fragmented concentrating in two (Old Town and Capital Hill) of the four growth centers. The growth pattern of the built-up class indicates a radial pattern around Old Town, Capital Hill, and Kanengo with a stable low level of complexity evidenced by more regular patch shapes. Much of this growth has been due to rural–urban migration, natural increase, and reclassification of settlements and accessibility. The road has been long and drudgery, but with supporting legislation, the population and economic growth of Lilongwe have been central to Malawi's urban development. While, the amount of urbanization occurring in the informal settlements in Lilongwe is reducing over the years, the numbers are still significant to be ignored (16–21%). It is important that the legislation guiding the urbanization in Lilongwe be revisited to ensure that the growth of the city is according to plan. This growth has been at a cost; however, in that urban poverty is the highest in the capital than in its sister cities within Malawi largely due to lack of access to basic urban services for its dwellers. All is not lost, however, for the primate city, if only the strategic plans outlined for the city can be implemented.

References

Batty M (2008) The size, scale, and shape of cities. Science 319:769–771

Best AC (1970) Gaborone: problems and prospects of a new capital. Geogr Rev 60:1–14

Besussi E, Chin N, Batty M, Longley P (2010) The structure and form of urban settlements. In: Rashed T, Jurgens C (eds) The remote sensing of urban and suburban areas. Springer Science +Business Media B.V, Amsterdam, pp 13–31

Cole-King P (1971) Lilongwe: a historical study. Government Press, Zomba

Connell J (1972) Lilongwe: another new capital for Africa. East Afr Geogr Rev 1972:89–110

Dretcher A, Lanquinta D (2002) Urbanization—linking development across the changing landscape. SOFA, Rome
Englund H (2002) The village in the city, the city in the village: migrants in Lilongwe. J South Afr Stud 28:137–154
Friedman J (1973) Urbanization, planning and national development. Sage Publications, California
Kalipeni E (1997) Contained urban growth in post independence Malawi. East Afr Geogr Rev 19:49–66
Kalipeni E (1999) The spatial context of Lilongwe's growth and development. In: Kalipeni E, Zeleza P (eds) Sacred spaces and public quarrels: African cultural and economic landscapes. Africa World Press, Trenton
Kim S, Law MT (2012) History, institutions, and cities: a view form the Americas. J Reg Sci 52:10–39
Lilongwe City Assembly (2009) Lilongwe City development strategy for 2010–2015. Lilongwe City Assembly, Lilongwe
Lilongwe City Council (2010) The study on urban development master plan for Lilongwe in the Republic of Malawi. Final Report Summary. Lilongwe City Council, Lilongwe
Lindstrom N (2014) Survey of urban poor settlements in Lilongwe. ActionAid, Lilongwe
Malawi Government (2002) Malawi national land policy. Ministry of Lands Physical Planning and Surveys, Lilongwe
Malawi Government (2010) Welfare monitoring survey 2009. National Statistical Office, Zomba
Manda MA (2013) Situation of urbanisation in Malawi report. Ministry of lands, Housing and Urban Development, Lilongwe
Ministry of Lands, Housing and Urban Development (2013) The urban structure plan of Lilongwe City. Ministry of Lands, Housing and Urban Development, Lilongwe
Moore J (1982) Planning Abuja: the policies of new capital construction. Princeton University, Thesis
Mpando L (2000) Fertility levels and trends. National Statistical Office, Malawi Demographic and Health Survey 2000. National Statistical Office, Zomba
NSO (1978) Population census 1977, preliminary report: district population by traditional authorities (including urban areas). National Statistical Office, Zomba
NSO (2008a) 2008 population and housing census. National Statistical Office, Zomba
NSO (2008b) Population projection Malawi. National Statistical Office, Zomba
NSO (2010) Welfare monitoring survey 2009. National Statistical Office, Zomba
NSO (2012) Integrated household survey 2010–2011. National Statistical Office, Zomba
Potts D (1985) Capital relocation in Africa: the case of Lilongwe in Malawi. Geogr J 151:182–196
Potts D (1986) Urbanization in Malawi with special reference to the new capital city of Lilongwe. PhD thesis. University College London, London
Richards G (1974) From vision to reality: the story of Malawi's new capital. Lorton, Johannesburg
Salau AT (1977) A new capital for Nigeria: planning, problems and prospects. Afr Today 24:11–22
Short JR, Pinet-Peralta LM (2009) Urban primacy: reopening the debate. Geogr Compass 3:1245–1266
Statistics South Africa (2015) Statistics South Africa. Accessed on 12 Jan 2016 from http://www.statssa.gov.za/?page_id=964
Todaro M (1994) Economic development. Longman, New York
UN-Habitat (2008) State of the world's cities report 2008/2009. United Nations Human Settlements Programme, Nairobi
UN-Habitat (2011a) Malawi: Lilongwe urban profile. United Nations Human Settlements Programme, Nairobi
UN-Habitat (2011b) Malawi: urban profile for Zomba. United Nations Human Settlements Programme, Nairobi
UN-Habitat (2013) State of the world's cities 2012/2013: prosperity of cities. Earthscan/Routledge, London

United Nations (2015) World urbanization prospects: the 2014 revision. Department of Economic and Social Affairs, Population Division, New York

Vail L, White L (1989) Tribalism in the political history of Malawi. In: Vail L (ed) The creation of tribalism in Southern Africa. James Curry, London

World Bank (2015a) Surface area (sq. km). Accessed on 13 Jan 2016 from http://data.worldbank.org/indicator/AG.SRF.TOTL.K2

World Bank (2015b) Population ranking. Accessed on 22 Dec 2015 from http://data.worldbank.org/data-catalog/Population-ranking-table

World Bank (2015c) Population density (people per sq. km of land area). Accessed on 16 Jan 2016 from http://data.worldbank.org/indicator/EN.POP.DNST

Wratten E (1995) Conceptualising urban poverty. Environ Urban 7:11–38

Zambia Government (2011) 2010 census of population and housing. Central Statistical Office, Lusaka

Chapter 17
Harare Metropolitan Area

Courage Kamusoko and Enos Chikati

Abstract Sustainable urban planning and development require reliable and timely land use/cover change information. The objective of this chapter was to analyze observed and simulated land use/cover changes between 1990 and 2030. Based on land use/cover maps for 1990, 2002, 2009, and 2014, built-up areas increased substantially, while non-built-up areas decreased over the study period. The land use/cover change analysis revealed significant built-up expansion for the "1990–2002" epoch. However, built-up expansion slowed down during the "2002–2009" and "2009–2014" epochs. The built-up growth pattern and the spatial/landscape metrics revealed that infill, extension, and leapfrog developments were occurring in the study area. Future land use/cover simulations (up to 2030) indicated that the current land use/cover change trends such as the increase in built-up areas and decrease in non-built-up areas would continue to persist unless sustainable urban development policies are implemented. The observed and simulated land use/cover changes provide a synoptic view of built-up expansion as well as a plausible future urban growth scenario for Harare Metropolitan Area. This could potentially assist decision-makers with general built-up change information, which can be used to guide strategic sustainable urban land use planning and development for Harare Metropolitan Area.

C. Kamusoko (✉)
Asia Air Survey Co., Ltd., Kawasaki, Japan
e-mail: kamas72@gmail.com

E. Chikati
Department of Environmental Sciences,
University of South Africa, Pretoria, South Africa

Y. Murayama et al. (eds.), *Urban Development in Asia and Africa*,
The Urban Book Series, DOI 10.1007/978-981-10-3241-7_17

17.1 Introduction

17.1.1 Physical Geography

The Harare Metropolitan Area described in this chapter encompasses Harare Urban, Harare Rural, Chitungwiza, and Epworth (including Ruwa town) districts (Fig. 17.1). The metropolitan area extends between approximately 17° 40′ and 18° 00′S, and between 30° 55′ and 31° 15′E, encompassing an area of about 1937.9 km^2. The average altitude is approximately 1500 m above sea level. The metropolitan area is characterized by a warm, wet season from November to April; a cool, dry season from May to August; and a hot, dry season in October. Daily temperatures range from about 7 to 20 °C in July (coldest month), and from 13 to 28 °C in October (hottest month). The study area receives a mean annual rainfall ranging from 470 to 1350 mm between November and March. Vegetation varies from grasslands to open Miombo woodlands dominated by *Brachystegia spiciformis* trees as well as some introduced tree species such as *Jacaranda*. The metropolitan area is characterized by a complex of gabbro, dolerite, metagreywacke, phylite, and granites, while fersialitic and paraferrallitic soils are dominant (Nyamapfene 1991). Poorly drained areas occur in widespread vleis, which are mainly depressions with soils that are waterlogged during the rainy season.

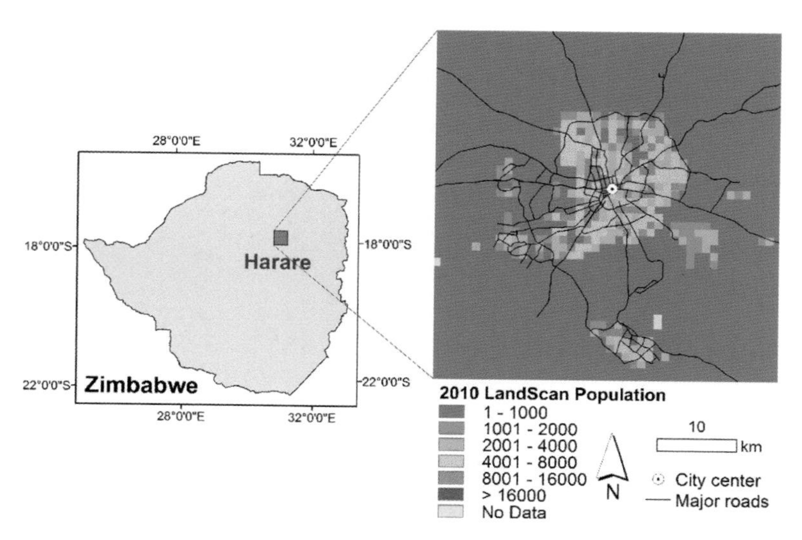

Fig. 17.1 Location and LandScan population of Harare Metropolitan Area, Zimbabwe

17.1.2 Origin and Brief History

Harare was founded in 1890 as a fort for the Pioneer Column, which was a volunteer army for the British South Africa Company under Cecil John Rhodes (Murray 2010). The Pioneer Column settled at the Kopje area, which was Chief Gutsa's territory (Weiss 1992). The fort was originally named Fort Salisbury after the British Prime Minister Robert Arthur Talbot Gascoyne-Cecil, the Third Marquess of Salisbury (Murray 2010). The settlers aimed to establish a city that could serve as a good manufacturing base as well as control African labor force (Rakodi 1995). A detailed cadastral plan was prepared with streets laid out on two intersecting grids (Zinyama et al. 1993; Murray 2010). The original site, which was located near the Kopje area (fort) and Mukuvisi River, was characterized by poor drainage (Weiss 1992). Therefore, new settlements and administrative offices for the British South Africa Company were developed a short distance from the Kopje area (Murray 2010). Salisbury was declared a municipality in 1897 and became a trading center when the railway line was constructed from Beira, Mozambique, in 1899 (Brown 2001). Since the colonial government was in need of African labor for industry, the first settlement for African workers was established in 1892. However, a legislation in 1906 designated separate residential areas for the native African workers. In 1907, the first African township was established at Harari (now Mbare), which is 3 km south of the city center (Zinyama et al. 1993). In 1923, Britain colonized Southern Rhodesia from the British South Africa Company (Rakodi 1995; Brown 2001). Salisbury became the capital of the British colony and was declared a city in 1935.

In a couple of decades, Salisbury had grown to become the industrial and commercial hub of the colony of Southern Rhodesia surpassing cities such as Fort Victoria (now Masvingo) and Bulawayo. Given the rising population of native Africans, the colonial government developed a second township for 2500 people on the state-owned Highfield Farm, about 8 km southwest of the city center in 1935. By the 1940s, Salisbury had developed into a metropolitan city, which attracted migrants from Nyasaland (now Malawi), Northern Rhodesia (Zambia), and Portuguese East Africa (Mozambique). The post-World War II witnessed a new wave of migrants from abroad, who introduced planning designs from western capitals (Brown 2001). During the post-World War II period, the economy of Southern Rhodesia grew significantly, which increased the demand for labor. Consequently, rural migration to Salisbury increased forcing the colonial regime to develop more policies for African urban areas such as the Natives Urban Areas Accommodation and Registration Act (AUAARA) of 1946 (Smout 1975). The city continued to develop along segregated lines, whereby townlands and large residential properties were developed for the white settler community (Zinyama et al. 1993).

The Central African Federation of Rhodesia and Nyasaland was created in 1953 (Lopes 1997), and Salisbury became the capital city, which cemented its position as one of the major cities in southern Africa. New residential areas were developed for the native Africans in the southern and western parts of the city including Dzivarasekwa and Kambuzuma in the 1960s. In 1963, the Central African Federation of Rhodesia and Nyasaland disintegrated (Lopes 1997). Salisbury became the capital of Southern Rhodesia again. However, the dominance of Salisbury continued since the collapse of Central African Federation of Rhodesia, and Nyasaland did not affect the social and migratory patterns it had generated (Lopes 1997). In November 1965, the Smith regime declared Unilateral Declaration of Independence (UDI), which resulted in Rhodesia gaining independence from Great Britain. Following UDI, the Smith regime strengthened minority rule by introducing drastic apartheid-style segregation laws (Ramsamy 2006). Nonetheless, despite the declaration of UDI and drastic legislation that intended to keep blacks out of the urban areas, the population of African citizens continued to increase (Ramsamy 2006). This led to the development of Glen Norah and Glen View to the southwest of the city in the 1970s. Meanwhile, the armed struggle for independence by the African nationalists had intensified during the 1970s, especially in the rural areas. Many people fled their rural homes to the city to escape the war. As a result, squatter areas, such as Epworth in Harare and Chirambahuyo in Chitungwiza, were created for the war refugees (Zinyama et al. 1993; UN-Habitat 2009).

Zimbabwe gained independence in 1980 after a protracted armed struggle by the African nationalists. The name of the capital city changed from Salisbury to Harare in 1982. The Government of Zimbabwe in collaboration with international donors initiated low-income housing projects, which led to the development of Warren Park (1981), Kuwadzana (1984), Hatcliffe (1984), and Budiriro (1988) municipalities. Despite these efforts, housing shortage continued to be a major problem, particularly in low-income, high-density areas. For example, during the 1980s, the country was producing only 18,000 houses annually against a backlog of 84,000 units (UN-Habitat 2008). While the City of Harare acknowledged the acute housing shortages, the city authorities were intolerant to any illegal and informal settlements. During the 1980s, most illegal settlements of the late 1970s were cleared (Rakodi 1995) and their populations moved to various holding camps such as Porta Farm (40 km west of the city center). However, Epworth (Fig. 17.1) remained the only regularized squatter area in Harare (Weiss 1992).

In 2000, the government embarked on the "Fast Track Land Reform Programme" (FTLRP), which focused on land redistribution in commercial farming areas. Taking advantage of the FTLRP, the "land-hungry" urban people formed housing cooperatives in order to acquire commercial farmland in the peri-urban areas (Marongwe 2003). Nonetheless, some of the housing cooperatives acquired and developed peri-urban land in violation of the regulations of the local authorities. Furthermore, some of the housing cooperatives were reported to have developed housing structures, which did not conform to the City of Harare building codes and standards. In April 2005, the Government of Zimbabwe embarked on "Operation Restore Order or Remove the Filth" (Operation Murambatsvina), which resulted in the destruction of approximately 700,000 illegally constructed homes or businesses countrywide (Chirisa 2009). Following criticism from the international community, the government initiated a reconstruction programme named "Garikai/Hlalani

Kuhle" (Live well) in June 2005 (Parliament of Zimbabwe 2006). This programme sought to promote large-scale delivery of low-cost housing as well as sites for small and medium businesses (Parliament of Zimbabwe 2006; UN-Habitat 2009). However, progress was slow due to lack of finance among other factors (UN-Habitat 2009). In addition, most of the housing structures failed to meet the required building codes and standards since "Garikai/Hlalani Kuhle" was a hastily prepared programme (UN-Habitat 2009). Given the slow implementation of "Garikai/Hlalani Kuhle" programme coupled with politics between the Central Government and local authorities, informal settlements with little or no basic services (e.g., clean water), such as Hopley in Harare South, have grown rapidly. Figures 17.2, 17.3, and 17.4 show selected sites in Harare Metropolitan Area.

17.2 Zimbabwe Urban System

17.2.1 Primacy in the National Urban System

In this section, we examine urban primacy of Harare Metropolitan Area based on population census data and household national income. As noted previously, the political and socioeconomic context during the colonial era profoundly influenced the urban system in Zimbabwe (Rakodi 1995; Brown 2001). The initial urban areas in pre-independence Zimbabwe were established as administrative centers, which were meant to serve the needs of the white colonial settlers. Therefore, these urban

Fig. 17.2 Central business district of the City of Harare. *Source* Author's fieldwork (2014)

Fig. 17.3 Aerial view of residential areas under development. *Source* Author's fieldwork (2014)

Fig. 17.4 Budiriro housing development scheme by the Central African Building Society (CABS) in the southwest of the City of Harare. *Source* Author's fieldwork (2014)

areas were located in the high veld areas with good climatic conditions and natural resources. These urban areas were linked by a railway line. For example, Harare was linked to Mutare to the east and Bulawayo to the southwest. The other smaller urban areas were later established for mining purposes or as service centers for white commercial farmers.

The definition of "urban" used in this chapter is based on the criteria established by the Zimbabwe National Statistical Agency (ZimStats 2012). According to ZimStats (2012), "urban" refers to a designated urban areas with a compact settlement patterns with more than 2500 inhabitants of which 50% are employed in non-agricultural sector. Figure 17.5a summarizes the details of the 11 most populous cities and urban centers in Zimbabwe. The urban population in Zimbabwe grew from 677,270 in 1962 to 3,409,848 inhabitants in 2012, representing 20%

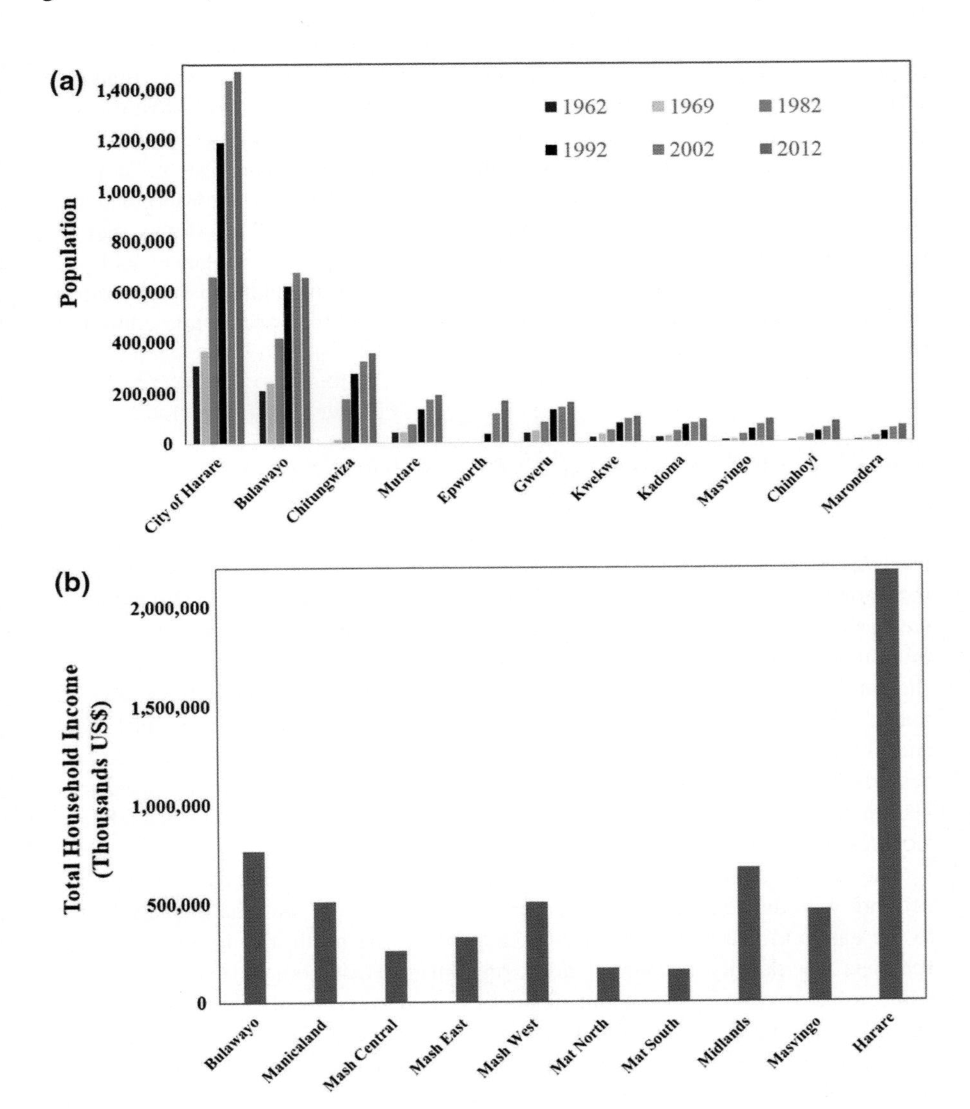

Fig. 17.5 **a** Population of the main cities in Zimbabwe. *Source* ZimStats (2012). **b** Total household income in Zimbabwe (2011–2012). *Source* ZimStats (2011)

urban population increase. According to ZimStats (2012), 33% of the population in Zimbabwe is urban. Currently, the total population of Harare and Bulawayo accounts for about 62% of the total urban population in Zimbabwe. Harare Metropolitan Area (including Chitungwiza, Epworth, and Ruwa) represents over 47% of total urban population in Zimbabwe, while Bulawayo, the second largest city, has 15% of the urban population. The rapid expansion of urban centers in Zimbabwe occurred after independence when restrictive controls were removed (Fig. 17.5a).

Figure 17.5a, b show that the City of Harare is conspicuously dominant in terms of population (ZimStats 2012) and income distribution (ZimStats 2011). Bulawayo is the second largest city, while other medium and small cities follow. Most notable, however, is the great upsurge of population in the City of Harare from 1992 to 2012, which shows a high population imbalance within the urban system. Based on population and household income, the urban primacy of Harare Metropolitan Area relative to the other urban centers in Zimbabwe is evident. This shows that the urban system in Zimbabwe is characterized by high spatial imbalance, with one dominant center. As mentioned before, the high spatial imbalance is partly attributed to colonial economic development policies and the socioeconomic conditions after independence.

17.2.2 Spatial Structure of Harare Metropolitan Area

Harare Metropolitan Area comprises four districts, namely Harare Urban (City of Harare), Harare Rural, Chitungwiza, and Epworth (Fig. 17.6). Harare Urban district incorporates the City of Harare, which is the capital and largest city in Zimbabwe as well as its administrative and commercial center. Given its good transport infrastructure (rail, road, and air transport), the city serves as a distribution point for the surrounding agricultural and gold-mining areas. Harare is a trading hub for tobacco, maize, cotton, and citrus fruits. The manufacturing industry includes textiles, steel, and chemicals. Before the economic crisis in Zimbabwe, factories produced processed food, beverages, clothing, cigarettes, building materials, and plastics in Harare. Currently, the deindustrialization of the economy in Zimbabwe is leading to closures of manufacturing industries (The Financial Gazette 2014).

The spatial structure of the City of Harare is characterized by a radial road network with the central business district (CBD) at its core, and the industrial areas to the east and south (Gamanya et al. 2009). To the north and northeast are the spacious low-density residential areas on plot sizes of about 1000 m^2 or more, while to the extreme east, south, southwest, and west are the high-density residential areas on plot sizes of about 300 m^2 (Gamanya et al. 2009). In addition, some medium-density residential areas measuring between 800 and 1000 m^2 are also found in the study area. Pre-independent City of Harare was divided along racial lines, whereas the post-independent city is divided along socioeconomic divisions. Services and amenities in low-income, high-density residential areas are poor and

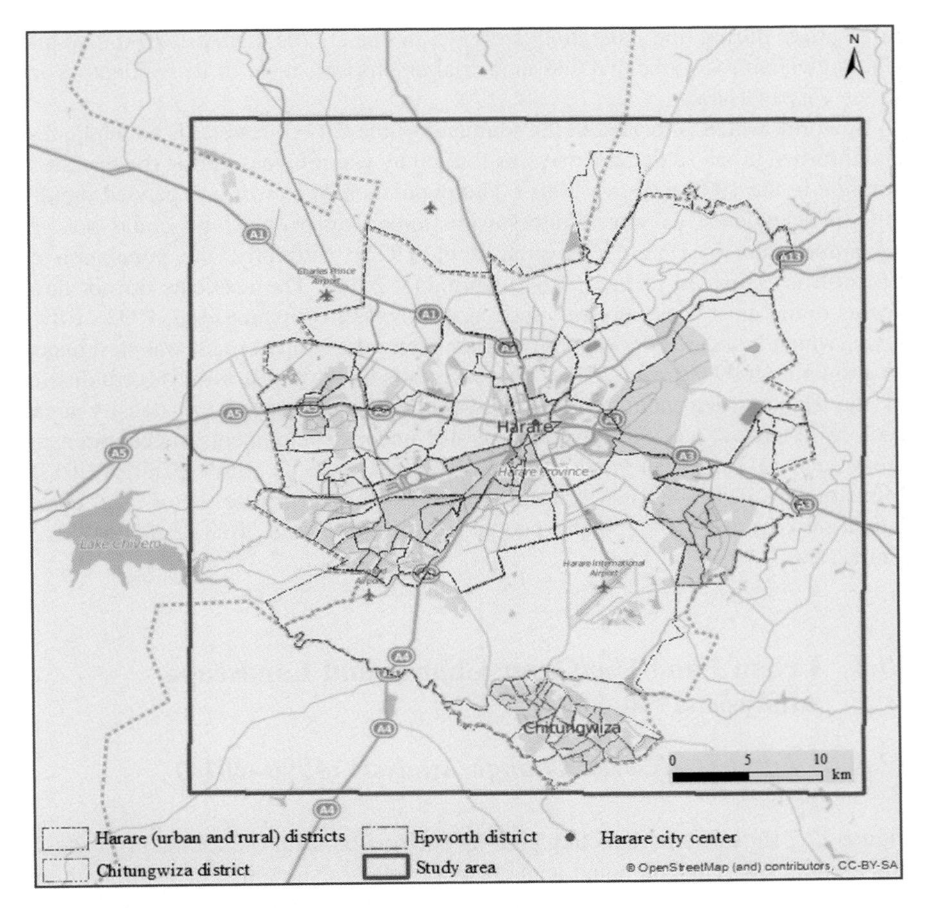

Fig. 17.6 Harare Metropolitan Province boundaries overlaid on OpenStreetMap

inadequate (Colquhoun 1993; Zinyama et al. 1993). The population in Harare Urban district has been increasing at a fast rate since independence in 1980, when migration controls were removed (Mutizwa-Mangiza 1986; Colquhoun 1993). For example, the population of the City of Harare increased from approximately 310,360 in 1962 to 1,435,784 in 2012 (Rakodi 1995; Central Statistical Office 2004; ZimStats 2012).

Chitungwiza city, which lies approximately 25 km south of the City of Harare, was developed out of St Mary's (formerly a settlement designated for missionary services and churches) and Seke townships in the early 1970s. The city was developed by the colonial government in order to locate residential areas for Africans. The population of Chitungwiza city expanded exponentially from approximately 15,000 in 1969 to 354,472 in 2012 (Zinyama et al. 1993; ZimStats 2012). Population expansion was mainly driven by people who migrated from the rural areas during the liberation struggle in the 1970s (Brown 2001). While Chitungwiza has commercial and industrial enterprises, most of its residents work in the City of Harare.

Epworth, which is located in the southeast of the City of Harare, is an unplanned and informal urban settlement that was formed by war refugees during the liberation struggle in the 1970s (Brown 2001). The population of Epworth expanded rapidly after independence as war refugees were joined by people who could not get accommodation in Harare (Zinyama et al. 1993). Currently, the population of Epworth is estimated to be 161,840 (ZimStats 2012). The residents do not have access to the most basic services, such as clean water (Zinyama et al. 1993). Ruwa town, which is located 25 km from Harare along the Mutare road, was designated as a town in 2009 (Chirisa 2013). The town falls under the Ruwa–Epworth district of the Harare Metropolitan Area. Ruwa town comprises of low-density areas such as Windsow and Sunway that are still under development, medium-density areas such as Damofalls, and high-density areas such as Ruwa (Chirisa 2013). Infrastructure in Ruwa town is mainly provided by the private sector. The population grew from 1477 to 56,678 people between 1992 and 2012 (ZimStats 2012; Chirisa 2013).

17.3 Urban Land Use/Cover Change and Landscape Analysis (1990–2014)

17.3.1 Land Use/Cover Change Analysis (1990–2014)

Figure 17.7 shows maps depicting built-up and non-built-up classes for 1990, 2002, 2009, and 2014. The computed land use/cover classes show that in 1990, built-up areas occupied 122 km^2, while non-built-up areas occupied 1815.9 km^2 of the study area. However, significant spatial expansion in built-up and subsequent decreases in non-built-up areas were observed in 2002. Built-up areas increased to 237.8 km^2, while non-built-up areas decreased to 1700.2 km^2. Visual analysis of the 2009 land use/cover map revealed further increases in built-up areas which occupied 312.8 km^2, while non-built-up areas decreased to 1625.1 km^2. For the 2014 land use/cover map, built-up and non-built-up areas occupied 358.6 and 1579.3 km^2, respectively. In general, built-up areas increased substantially by 236.6 km^2 between 1990 and 2014 (Fig. 17.7).

The rates of land use/cover changes varied during the "1990–2002," "2002–2009," and "2009–2014" epochs (Table 17.1; Fig. 17.8). Between 1990 and 2002, "non-built-up to built-up" change was approximately 115.8 km^2 at an annual rate of 9.6 km^2 (Table 17.1). The majority of changes occurred in the northeastern, southwestern, and western parts of the study area. However, during the "2002–2009" epoch, "non-built-up to built-up" change slowed down to approximately 75.1 km^2 at an annual rate of 6.3 km^2. The "2009–2014" epoch also revealed a

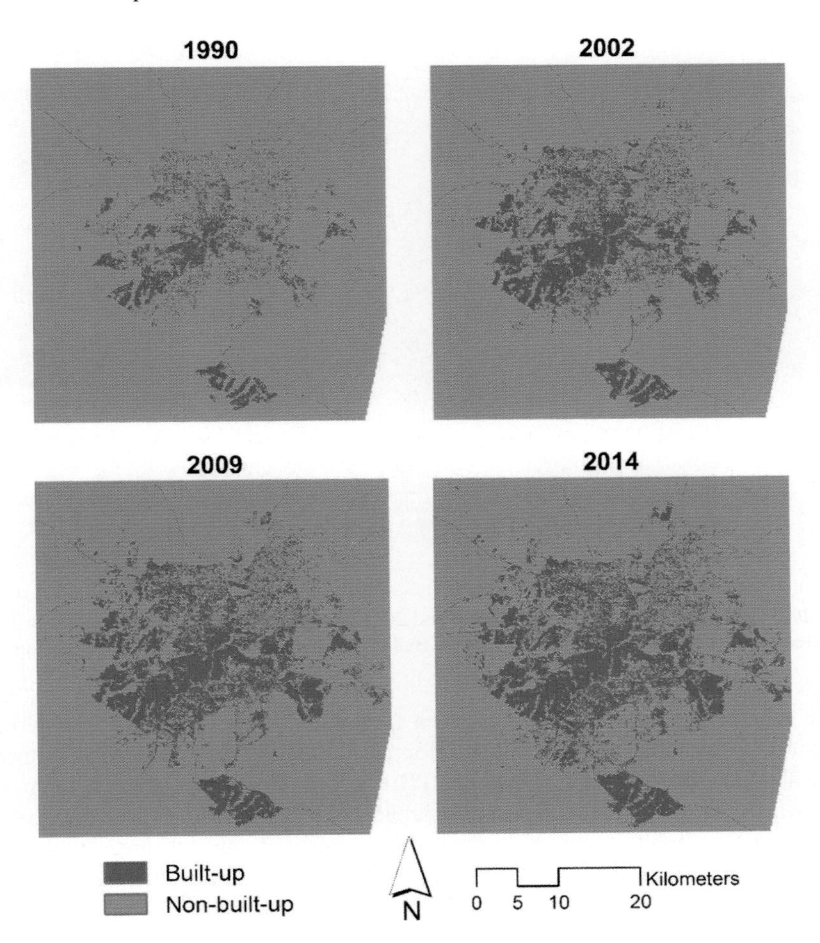

Fig. 17.7 Urban land use/cover maps of Harare Metropolitan Area classified from Landsat imagery

Table 17.1 Built-up changes for the "1990–2002," "2002–2009," and "2009–2014" epochs

	1990–2002	2002–2009	2009–2014
Built-up changes	115.8	75.1	45.7
Annual rate of change (km^2/year)	9.6	6.3	3.8

slight decrease in "non-built-up to built-up" changes (Table 17.1). The "non-built-up to built-up" change was approximately 45.7 km^2 at an annual rate of 3.8 km^2 (Table 17.1). In summary, the land use/cover change analyses revealed significant rate of built-up expansion for the "1990–2002" epoch, while built-up growth slowed down during the "2002–2009" and "2009–2014" epochs, respectively.

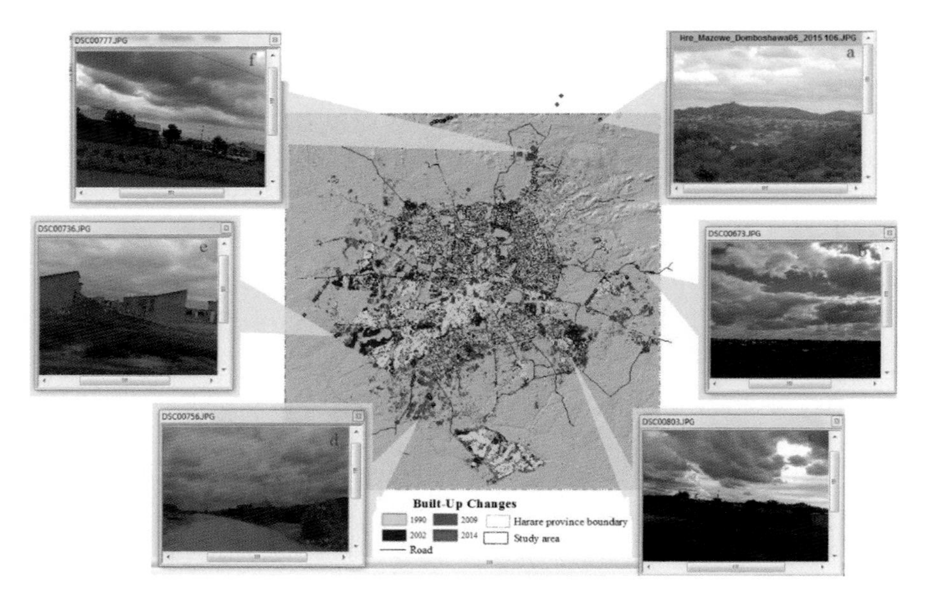

Fig. 17.8 Built-up expansion between 1990 and 2014 overlaid on hillshade: **a** Charlotte Brooke; **b** Caledonia; **c** Glenwood; **d** Southlea Park; **e** Budiriro Extension; and **f** Hatcliffe Extension

In general, built-up expansion between 1990 and 2014 was mainly characterized by infill, extension, and leapfrog developments (Fig. 17.8). These developments were also identified by Kamusoko et al. (2013). Infill development refers to growth of newly developed areas that are in the urbanized areas of the previous time period (that is, 1984), while extension refers to expansion of built-up areas within the urbanized areas (Yue et al. 2013). Leapfrog development is defined as newly developed areas that are converted from non-developed parcels outside of and unconnected with existing urban built-up areas (Yue et al. 2013).

Figure 17.8 shows that built-up expansion between 1990 and 2002 was largely dominated by extension and infill developments and to a smaller extent by leapfrog development. For example, the expansion of built-up areas in southwest and parts of Harare North (e.g., New Bluffhill) shows extension developments, while areas such as Crowborough, Tynwald, Westlea, and Nkwisi parks in the western parts of the city center show infill developments. However, patterns of leapfrog development such as Kuwadzana extension are observed in the western and southwestern portions of the study area. As observed during the "1990–2002" period, built-up expansion between 2002 and 2009 is also characterized by extension and infill developments. Conspicuous "non-built-up to built-up" change patterns suggest that the extension and infill developments observed in the metropolitan area were intensifying. Nonetheless, the location of new built-up areas, particularly in the northern and northwestern parts of the study area, indicates leapfrog development (Fig. 17.8). While extension (e.g., CABS Budiriro housing developments, see Fig. 17.8e) and infill development continues, the "2009–2014" period is also characterized by leapfrog developments. For example, Charlotte Brooke (Fig. 17.8a) in the northeast, informal settlement such as Caledonia in the east (Fig. 17.8b), Glenwood Park near Epworth (Fig. 17.8c) in the southeast, Southlea Park in the south (Fig. 17.8d), and Hatcliffe Extension (Fig. 17.8f) in the north show leapfrog developments.

17.3.2 Landscape Change Analysis (1990–2014)

We computed class-level metrics using FRAGSTATS 4.2, a spatial pattern analysis program designed for categorical maps (McGarigal et al. 2012). In this chapter, class-level spatial/landscape metrics represent the spatial patterns for the built-up class. The following class-level metrics were selected: (1) percentage of landscape (PLAND); (2) patch density (PD); (3) Euclidean nearest-neighbor distance (ENN); (4) related circumscribing circle (CIRCLE); and (5) shape index (SHAPE). The PLAND is a landscape composition metric, which measures the proportion of a particular class relative to the whole landscape (McGarigal et al. 2012). It is 0% when the corresponding class is rare in the landscape, while 100% when the entire landscape is dominated by a single class (McGarigal et al. 2012). The PD metric is a measure of fragmentation based on the number of patches per unit area. Low PD means patches are compact and less fragmented, while high PD implies high fragmentation. The ENN is a patch isolation metric, which measures the dispersion based on the average distance to the nearest neighboring patch of the same class (McGarigal et al. 2012). Generally, isolation measures the degree in which patches are spatially isolated. ENN approaches 0 as the distance to the nearest neighbor decreases. CIRCLE is a shape metric that focuses on geometric complexity (McGarigal et al. 2012). Note that CIRCLE provides a measure of overall patch elongation. For example, CIRCLE is equivalent to 0 for circular or one cell patches and approaches 1 for elongated, linear patches one cell wide. The SHAPE metric measures the complexity of patch shape compared to a standard shape (e.g., square or circle) of the same size (McGarigal et al. 2012). This metric has a value of 1 when the patch is square and increases without limit as patch shape becomes more irregular.

Table 17.2 shows the observed class-level spatial/landscape metrics for the built-up class between 1990 and 2014. The PLAND metric increased from 6.3 to 18.5% over the study period. While the increase in PLAND is observed, generally, the low PLAND metric shows that most of the study area is composed of non-built-up areas. The patch density (PD) decreased slightly from 5.8 to 4.9 per km^2 during the "1990–2002" epoch, suggesting that built-up patches became more spatially connected and compact. However, PD increased slightly from 4.9 to 6 per km^2 between 2002 and 2014, which implies additional new built-up patches.

Table 17.2 Observed landscape pattern of Harare Metropolitan Area

Class-level (built-up) spatial metrics	1990	2002	2009	2014
PLAND (%)	6.3	12.3	16.1	18.5
PD (number per km^2)	5.8	4.9	5.5	6.0
ENN (mean) (m)	80.8	79.5	88.9	88.3
CIRCLE (mean) ($0 \leq$ CIRCLE < 1)	0.3	0.3	0.3	0.3
SHAPE (mean) ($1 \leq$ SHAPE $\leq \infty$)	1.2	1.3	1.2	1.2

Although slight variations are observed over the study period, PD is relatively low which suggests that most of the built-up areas within the study area are less fragmented and more compact. The mean ENN decreased slightly from 80.8 to 79.5 m between 1990 and 2002. However, the mean ENN increased slightly between 2002 and 2009, and then decreased slightly between 2009 and 2014. While the mean ENN shows some variations, the values indicate low levels of dispersion between the built-up areas. The mean CIRCLE and SHAPE values remained almost constant over the study period, suggesting that built-up patches are less complex and regular in most parts of the study area.

Figure 17.9 shows spatial/landscape metrics for the built-up areas of Harare Metropolitan Area (1990–2014) along the gradient of the distance from the city center. Generally, PLAND increased between 1990 and 2014 within all the distance buffer zones. However, the magnitude of change is greater within the 0–15-km buffer zone, which shows that the proportion of the built-up class near the city center is relatively high. While the PD metric within the 0–5 km buffer zone is lower than the PD within the 5–10-km buffer zone, generally, PD decreased within the 0–10-km buffer zone over the study period. Therefore, the built-up class within the 0–10-km buffer zone became more compact and spatially connected to the city center. This is partly attributed to infill and extension development that occurred mainly within the 0–10-km buffer zones. However, the slight increase in PD within the 10–25-km distance buffer zone suggests that the built-up classes became less compact with distance. This is mainly attributed to outward and leapfrog developments. The mean ENN does not show a clear trend over the study period. Nonetheless, the mean ENN metric substantially increased within the 20–25-km buffer zones, which indicates high level of dispersion. This implies that outward and leapfrog developments were dominant over the 20-km buffer zones. Last but not least, the mean CIRCLE and SHAPE values do not show a clear trend over the study period. However, the relatively low mean CIRCLE and SHAPE values suggest that the built-up class is generally less complex and more regular. In summary, the analysis of the spatial/landscape with distance buffer zones indicates that built-up areas are continuous and highly dense near the city center, which is attributed to infill and extension developments. However, the built-up areas became discontinuous and scattered further away from the city center. This is attributed to outward and leapfrog developments.

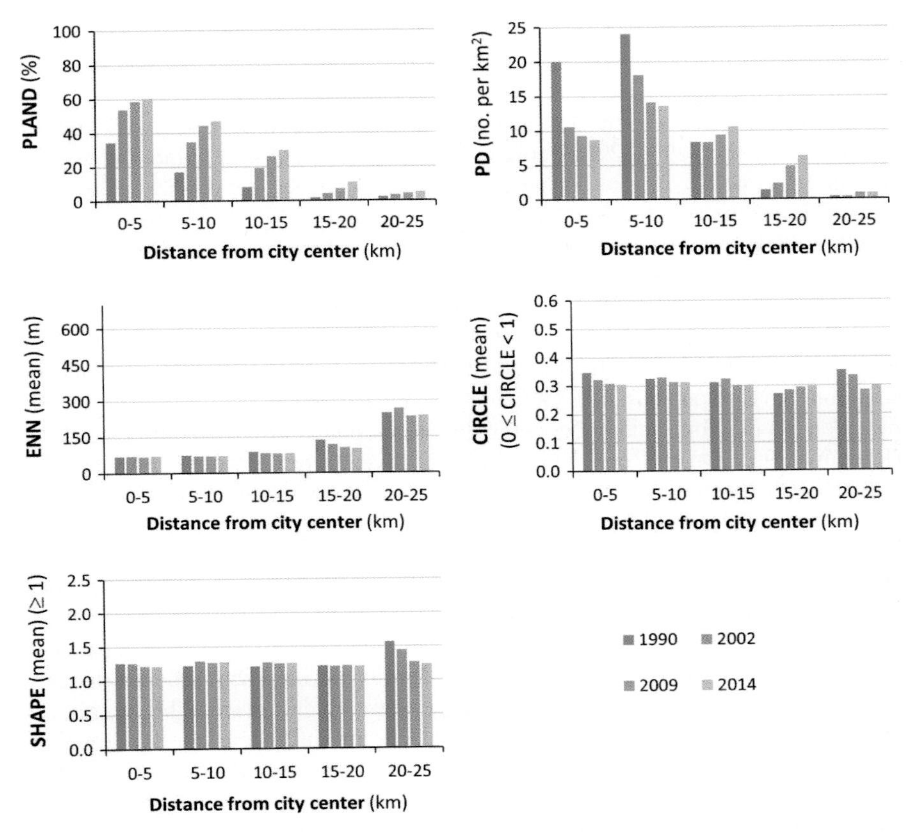

Fig. 17.9 Observed class-level spatial metrics for built-up along the gradient of the distance from city center of Harare Metropolitan Area. *Note* The y-axis values are plotted in the same range as those in Fig. 17.13

17.3.3 Driving Forces of Urban Development

Urban development is influenced by many complex underlying and proximate driving forces, which vary over space and time (Turner et al. 1995; Geist and Lambin 2001). Generally, the analysis of driving forces depends on the simplifications, theoretical, and behavioral assumptions hypothesized to influence land use/cover changes in a particular system. The analysis of the driving forces of urban development for Harare Metropolitan Area is challenging given the lack of disaggregate socioeconomic data at smaller administrative units (e.g., municipalities or ward level). In addition to the paucity of appropriate data, the available aggregate data do not generally correspond to the land use/cover change epochs considered in this chapter. Taking into account these limitations, we did not perform a quantitative analysis of the driving factors of urban growth for Harare Metropolitan Area.

In this chapter, we performed a qualitative analysis of the driving forces for the "1990–2002," "2002–2009," and "2009–2014" epochs based on the literature review and field observations.

The land use/cover change analyses revealed significant rate of built-up expansion for the "1990–2002" epoch, which is attributed to a number of socioeconomic and policy factors during the post-independence period (i.e., after 1980 when Zimbabwe got independent). According to the Central Statistical Office (CSO) (2004), population in Harare Metropolitan Area increased from approximately 1,478,810 in 1992 to 1,896,134 inhabitants in 2002. The urban population growth due to natural increase, rural–urban migration, and to some extent city–city migration increased the demand for houses (Rakodi 1995; Ministry of Economic Planning & Investment Promotion (MEPIP 2011)). In order to improve the housing situation, the Central and Local Government pursued housing development policies based on the previous colonial government master plans, which had strict standards (Rakodi 1995). On the one hand, housing development pursued infill housing development schemes given the growing concern about urban sprawl, high cost of service provisions, and long commuting distances (Zinyama et al. 1993). The infill housing development scheme focused more attention to the utilization of vacant land within existing high-income medium- to low-density residential areas (Zinyama et al. 1993). On the other hand, the government continued with the development of low-income housing schemes in high-density areas (Zinyama et al. 1993; Rakodi 1995). The low-income high-density development schemes focused on outward expansion because building costs were much lower in the outskirts of the city center (approximately 12–30 km) than for infill housing developments (Rakodi 1995; MEPIP 2011). However, following the introduction of the "Fast Track Land Reform Programme" (FTLRP) in 2000, many community and public service housing cooperatives acquired land on commercial farms or vacant land within peri-urban areas. This led to massive housing development and hence the expansion of built-up areas, particularly between 2000 and 2002.

Urban built-up expansion slowed down during the "2002–2009" epoch, which is attributed to a number of socioeconomic and policy factors. While the population in Harare Metropolitan Area increased from 1,896,134 in 2002 to 2,098,199 inhabitants in 2014 (ZimStats 2012), this period was marked by a rapid economic decline and political instability (UN-Habitat 2012). For example, hyperinflation coupled with the deindustrialization of the Zimbabwean economy led to company closures and thus high unemployment (The Financial Gazette 2015). Despite the serious economic decline, housing development continued within the vacant areas and peri-urban areas. However, many housing structures were destroyed during "Operation Restore Order or Remove the Filth" (Operation Murambatsvina) in April 2005 (UN-Habitat 2009). This slowed housing developments since people were afraid to start constructions. Following "Operation Restore Order or Remove the Filth," the government launched a reconstruction programme named "Garikai/Hlalani Kuhle" (Live well) in June 2005 (Parliament of Zimbabwe 2006; UN-Habitat 2009). However, the slow progress of the reconstruction programme combined with poor governance by local authorities (UN-Habitat 2012) led to the increase in informal settlements or slums again. According to the UN-Habitat (2012), the proportion of people living in informal settlements or slums in Zimbabwe increased from 4 in 1990 to 24 in 2009. Furthermore, quasi-legal land barons emerged to sell land to home seekers given the vacuum created by policy inconsistencies and politics between the central and local government authorities (The Financial Gazette 2015; UKAid 2015). This led to developments of informal settlements such as Caledonia in the east part of the City of Harare (Fig. 17.8b).

17.4 Projected Future Land Use/Cover Changes

17.4.1 Projected Land Use/Cover Changes

Future land use/cover changes for 2020 and 2030 (Fig. 17.10) were simulated based on the calibration scenarios between 1990 and 2010 (see Chap. 2). Based on the 1990 land use/cover base map, the multiple transition probabilities, and the "1990–2010" transition potential maps, the land change model projected that built-up areas would increase substantially from 358.6 km^2 in 2014 to 501.7 km^2 in 2030, while non-built-up areas would decrease significantly from 1579.3 km^2 to 1436.3 km^2 over the same study period (Fig. 17.11). The annual rate of built-up growth was projected to be 7.2 km^2.

The simulated built-up areas in 2020 and 2030, and the observed built-up areas in 1990 indicated that urban expansion will continue with infill and leapfrog developments toward the northwest, north, northeast, east, and southwest parts of the study area (Fig. 17.12). The simulation results show that vacant land will be

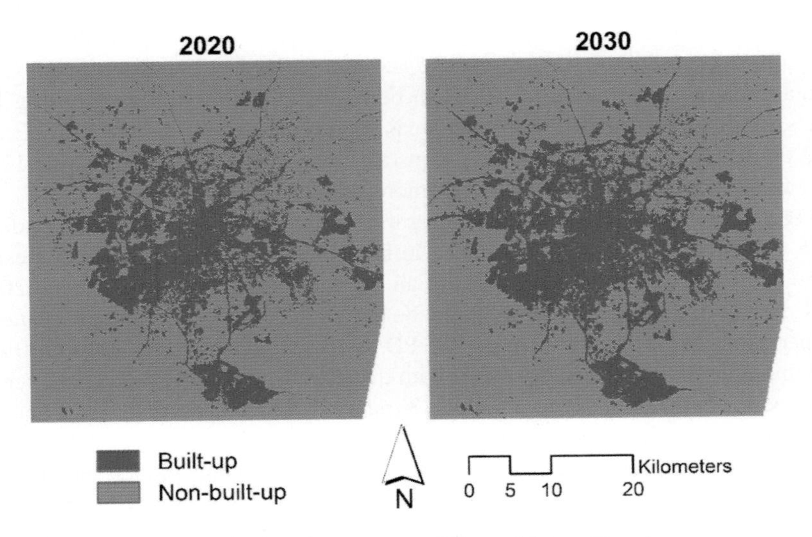

Fig. 17.10 Projected urban land use/cover maps of Harare Metropolitan Area

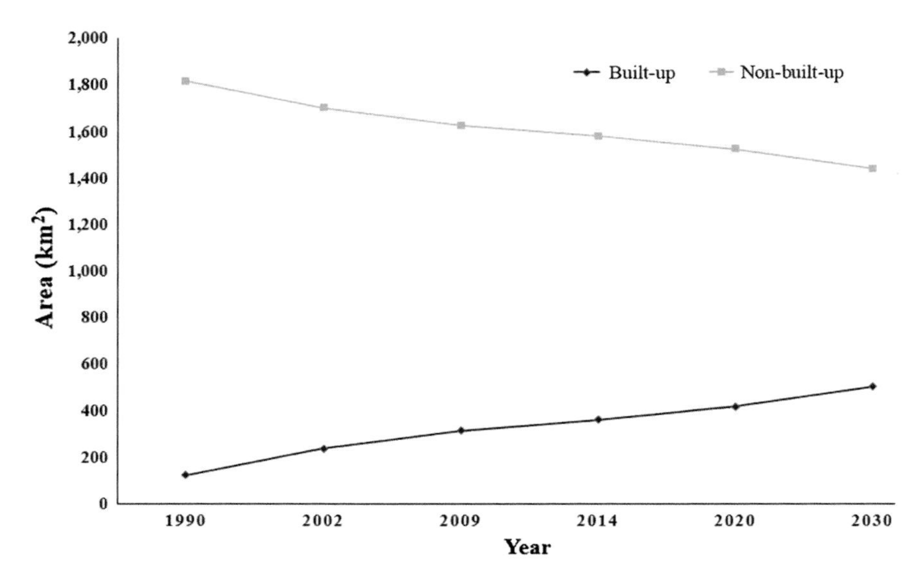

Fig. 17.11 Simulated future urban land use/cover trend under the current scenario

developed into built-up areas following the major roads. For example, built-up areas are expected to expand following the major roads to the northwest, north, and northeast of the study area. In addition, the expansion of built-up areas in the eastern and southwestern parts of the study area is also expected to follow the major roads. The spatial/landscape metric results (Table 17.3) for the built-up class indicate that PLAND would likely increase from 21.4 to 25.9% between 2020 and 2030. This implies an increase of built-up areas in the future. However, PD would decrease slightly from 2.4 to 1.7 per km^2 in the future, which implies that most of the built-up class will be less fragmented and more compact. The mean ENN would increase slightly from 111.4 to 118.4 m between 2020 and 2030, suggesting low levels of dispersion for the built-up areas in the class. The mean CIRCLE and SHAPE values would likely remain constant over the study period (Table 17.3), indicating that built-up would become more regular in the future.

Figure 17.13 shows spatial metrics for the built-up areas of Harare Metropolitan Area (2020 and 2030) within distance buffer zones from the city center. Results show that PLAND would decrease with an increase in distance for 2020 and 2030. This implies that built-up patches would be denser near the city center and scattered further away from the city center. The PD metric would increase between 0 and 10-km buffer zone and then decreases with distance from the city increases over the study period. This implies more built-up expansion, which is attributed to extension development, on one hand, and infill development on the other as built-up areas aggregate later. Note that cellular automata (CA)-urban change models tend to form compact and aggregate patterns since CA depends on local growth rules (Herold et al. 2003). Therefore, some of the compactness of the built-up areas might be

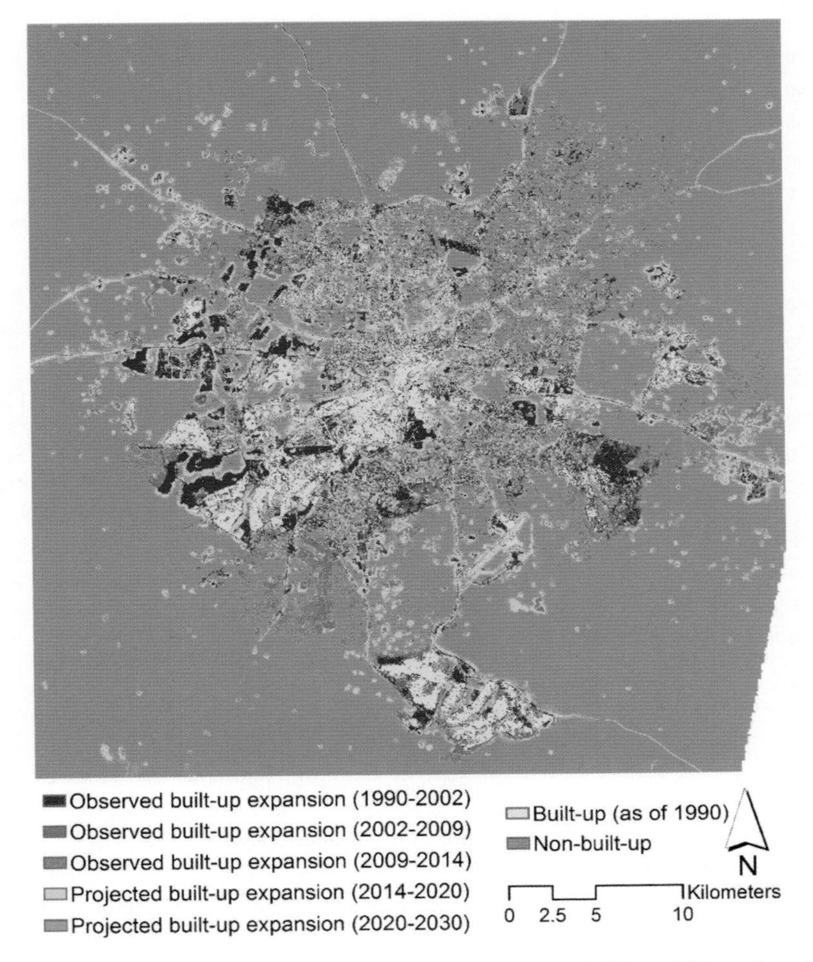

Fig. 17.12 Observed and projected urban land use/cover changes in Harare Metropolitan Area

Table 17.3 Projected landscape pattern of Harare Metropolitan Area

Class-level (built-up) spatial metrics	2020	2030
PLAND (%)	21.4	25.9
PD (number per km^2)	2.4	1.7
ENN (mean) (m)	111.4	118.4
CIRCLE (mean) ($0 \leq$ CIRCLE < 1)	0.3	0.3
SHAPE (mean) ($1 \leq$ SHAPE $\leq \infty$)	1.3	1.3

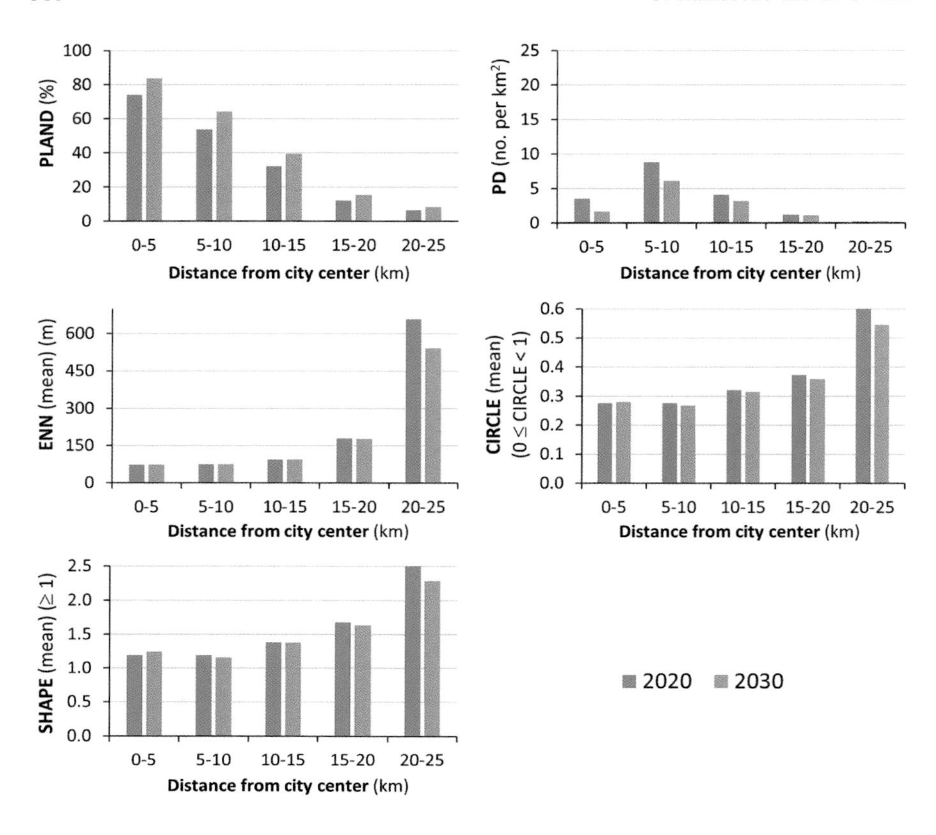

Fig. 17.13 Projected class-level spatial metrics for built-up along the gradient of the distance from city center of Harare Metropolitan Area. *Note* The y-axis values are plotted in the same range as those in Fig. 17.9

partly attributed to CA model framework and not the expected built-up changes in the future. However, results show that the mean ENN will increase with distance, which indicate high level of dispersion. This implies that built-up patches would likely become scattered from the 15-km and above distance buffer zone. In addition, the mean CIRCLE and SHAPE values show a slight increase with, which suggests that the built-up class would likely be more complex and irregular in the future. In summary, the results indicate that built-up areas would likely be continuous and dense near the city center in the future, which implies continuation of infill and extension developments. However, built-up areas would likely be discontinuous and scattered with further distance from the city center in 2020 and 2030. This implies outward and leapfrog developments in the future.

17.4.2 Implications for Future Sustainable Urban Development

Most of the problems, which the Harare Metropolitan Area is currently facing, are partly attributed to rapid urbanization coupled with lack of proper urban policies. This was evident during the pre-independence period when restrictive policies emphasized segregated urban development. For example, during the colonial era, white residents resided in low-density and high-income areas, while the native black residents resided in high-density and low-income areas. Despite removing the restrictive colonial era policies, the post-independence period witnessed the continuation of previous colonial trend, where the focus was on housing development based on socioeconomic distributions. To date, the Government of Zimbabwe, in general, and the City of Harare, in particular, have intensified efforts to improve sustainable urban development. Nevertheless, a multitude of challenges that include lack of political will, inadequate policies and dearth of clear scientific basis for evaluating different policy scenarios, have impeded the effective implementation of sustainable urban development.

Within this context, the simulated land use/cover changes can be used to explore the implications of future urban development. For example, our results indicated that simulated future built-up expansion under the current scenario would be continuous and dense near the city center, and discontinuous and scattered with further distance from the city center. Based on this analysis, the Harare metropolitan authorities should focus attention on reducing discontinuous and scattered built-up expansion. This can be accomplished by integrating different neighborhoods with mixed built-up density (high, medium, and low density) and mixed land use patterns. However, this approach requires a paradigm shift away from the colonial mono-functional city mentality (i.e., dominated by one city center) and segregated residential system (i.e., high-, medium-, and low-density residential areas) to a compact and connected city philosophy. The compact city would minimize transport and service delivery cost, and use open land in a cost-effective manner. This is expected to improve sustainable urban development as well as improve efficiency and service delivery.

While we are aware of the limitations of the observed and simulated land use/cover maps (e.g., limited driving factors and uncertainty of the future simulations), the simulated land use/covers provides a visual and quantitative representation of built-up expansion. Equally important is the observed and simulated land use/cover changes under the current scenario provide built-up change information, which can be used as an initial scientific or diagnostic tool for guiding sustainable urban development. The land change model applied in this chapter provides "what if" scenarios, which can assist researchers, policy makers, and other stakeholders in assessing the implications of future built-up expansion and urban development policy alternatives.

17.5 Summary and Conclusions

The objective of this chapter was to analyze observed and simulated land use/cover changes in Harare Metropolitan Area. The observed land use/cover changes indicated that significant built-up expansion occurred between 1990 and 2014. In addition, the results revealed that built-up expansion was characterized by a combination of infill, extension, and leapfrog developments. This was attributed to rapid urban population growth and government polices among other factors. The analysis of the spatial/landscape metrics with the distance buffer zones indicated that built-up areas were expanding in a continuous and high-density form near the city center due to infill and extension developments. In contrast, the built-up expansion was discontinuous and scattered as housing developments occurred further from the city center due to outward and leapfrog developments. Although a 15-year period is too short to completely understand the complex nature of urban growth dynamics, this chapter provided a bird's-eye view of built-up expansion in Harare Metropolitan Area.

Future land use/cover simulations (up to 2030) indicated that the current land use/cover change trends such as the increase in built-up areas and decrease in non-built-up areas would continue to persist unless holistic sustainable urban development policies involving all the stakeholders in the metropolitan areas are implemented. The spatial/landscape metric results indicated that future built-up areas would likely expand in a continuous and homogenous form near the city center. However, the built-up areas would likely be discontinuous and scattered with distance farther from the city center in 2020 and 2030. Because built-up expansion affects open vacant and agriculture land, the current land use/cover changes will have an impact on the future well-being of the ecosystem as landscape degradation is anticipated to increase.

The land use/cover change and simulation results in this chapter are important because: (1) rapid urbanization is a growing problem in most developing countries, particularly in sub-Saharan Africa and Asia, and (2) past and current attempts to implement sustainable urban development have been slow or not implemented due to the lack of timely geospatial information. This chapter has demonstrated that an integrated approach incorporating remote sensing, spatial/landscape metrics, and simulation tools improve our understanding of land use/cover changes in Harare metropolitan area. This could potentially assist decision-makers and other stakeholders with geospatial information on the extent of built-up areas, which can be used to guide sustainable urban land use planning and development in Harare Metropolitan Area. Lastly, the land use/cover change analysis and simulation approach facilitated a better understanding of built-up expansion in the southern African urban context. This approach can be applied to other metropolitan areas in sub-Saharan Africa, in particular, and other developing countries in general.

References

Brown A (2001) Cities for the urban poor in Zimbabwe: urban space as a resource for sustainable development. Dev Pract 11:263–281

Chirisa I (2009) Prospects for the asset-based community development approach in Epworth and Ruwa, Zimbabwe: a housing and environment perspective. Afr J Hist Cult (AJHC) 1:28–35

Chirisa I (2013) Housing and stewardship in peri-urban settlements in Zimbabwe: a case study of Ruwa and Epworth. University of Zimbabwe. Unpublished thesis

Colquhoun S (1993) Present problems facing the Harare City council. In: Zinyama L, Tevera D, Cumming S (eds) Harare: the growth and problems of the city. University of Zimbabwe Publications, Harare, pp 33–41

CSO (Central Statistical Office) (2004) Census 2002 population census: provincial profile Harare. Harare, Zimbabwe

Gamanya R, De Maeyer P, De Dapper M (2009) Object-oriented change detection for the city of Harare, Zimbabwe. Expert Syst Appl 36:571–588

Geist HJ, Lambin EF (2001) What drives tropical deforestation? A meta-analysis of proximate and underlying causes of deforestation based on subnational case study evidence. LUCC Report Series 4. Louvain-la-Neuve

Herold M, Goldstein NC, Clarke KC (2003) The spatiotemporal form of urban growth: measurement, analysis and modelling. Remote Sens Environ 86:286–302

Kamusoko C, Gamba J, Murakami H (2013) Monitoring urban spatial growth in Harare Metropolitan province, Zimbabwe. Adv Remote Sens 2:322–331

Lopes C (ed) (1997) Balancing rocks: environment and development in Zimbabwe. SAPES Books, Harare

Marongwe N (2003) The fast track resettlement and urban development nexus: the case for Harare. In: Symposium on delivering land and securing rural livelihoods: post independence land reform and resettlement in Zimbabwe, Mont Clair, Nyanga, 26–28 Mar 2003

McGarigal K, Cushman SA, Ene E (2012) FRAGSTATS v4: spatial pattern analysis program for categorical and continuous maps. Computer software program produced by the authors at the University of Massachusetts Amherst. Accessed 1 July 2015 from http://www.umass.edu/landeco/research/fragstats/fragstats.html

Ministry of Economic Planning & Investment Promotion (MEPIP) (2011) Zimbabwe—medium term plan 2011–2015. Republic of Zimbabwe

Murray P (2010) Zimbabwe, the bradt travel guide. Bradt Travel Guides Ltd, IDC House, The Vale, Chalfont St Peter, Bucks SL9 9RZ, England

Mutizwa-Mangiza ND (1986) Urban centres in Zimbabwe: inter-censal changes, 1962–1982. Geography 71:148–151

Nyamapfene K (1991) Soils of Zimbabwe. Nehanda Publishers, Harare

Parliament of Zimbabwe (2006) Second report of the portfolio committee on local government on progress made on the operation Garikai/Hlalani kuhle programme. Accessed on 12 Oct 2015 from http://archive.kubatana.net/docs/legisl/ppc_locgov_garikai_060606.pdf

Rakodi C (1995) Harare—inheriting a settler-colonial city: change or continuity?. Wiley, Chichester

Ramsamy E (2006) The World Bank and urban development: from projects to policy. Routledge, Taylor and Francis, New York

Smout MAH (1975–6) Urbanisation of the Rhodesian population. Zambezia 4:79–91

The Financial Gazette (2014) Zim industrial decline continues. Accessed on 20 Aug 2015 from http://www.financialgazette.co.zw/zim-industrial-decline-continues/

The Financial Gazette (2015) Politics of land baron-ship in Zim. Accessed on 27 Aug 2015 from http://www.financialgazette.co.zw/politics-of-land-baron-ship-in-zim/

Turner II BL, Skole D, Sanderson S, Fischer G, Fresco L, Leemans R (1995) Land-use and land-cover change. Science/Research Plan. IGBP Report No. 35, HDP Report No. 7. IGBP and HDP, Stockholm and Geneva

UKAid (2015) Urban infrastructure in Sub-Saharan Africa—harnessing land values, housing and transport: report on Harare Case Study. Report No. 1.9. African Centre for Cities

UN-HABITAT (2008) The state of African cities 2008—a framework for addressing urban challenges in Africa. Nairobi, Kenya

United Nations Human Settlements Programme (UN-HABITAT) (2009) Housing finance mechanisms in Zimbabwe. Nairobi, Kenya

Weiss P (1992) Focus on geography book 4: the human environment in Southern Africa. College Press Publishers, Harare

Yue W, Liu Y, Fan P (2013) Measuring urban sprawl and its drivers in large Chinese cities: the case of Hangzhou. Land Use Policy 31:358–370

ZimStats (Zimbabwe National Statistics Agency) (2011) Poverty income consumption and expenditure survey 2011/12 report. Harare, Zimbabwe

ZimStats (Zimbabwe National Statistics Agency) (2012) Census 2012: preliminary report. Harare, Zimbabwe

Zinyama L, Tevera D, Cumming S (eds) (1993) Harare: the growth and problems of the city. University of Zimbabwe Publications, Harare

Chapter 18
Johannesburg Metropolitan Area

Tabukeli M. Ruhiiga

Abstract The purpose of this chapter is to examine observed and projected land use/cover changes for Johannesburg based on remote sensing and GIS analysis. In addition, the driving forces that influence urbanization as well as the potential implications of urban land use/cover changes on future sustainable urban development in Johannesburg are discussed. The land use/cover change results indicated that built-up areas increased substantially between 1990 and 2014. The rapid increase in built-up areas was attributed to a number of driving factors such as the historical and recent developments in the mining industry, government urban policies, rural-urban migration, and urban land market changes. The observed and projected land use/cover changes provide valuable insights, which can be used to guide sustainable urban development in Johannesburg. This is important given the current distortions in land access and supply as well as poor provision of urban services in Johannesburg.

18.1 Origin and Brief History

18.1.1 The Origins of Johannesburg

Johannesburg (Fig. 18.1) developed after the discovery of gold in 1886 (Bremner 2000). The initial growth of the town—that later led to the growth of lateral hamlets to the east (East Rand), and to the south and west (West Rand)—followed the existence of rich gold veins that were mined using open-cast techniques. The gold mines required thousands of workers, which resulted in the setting up of all-male housing hostels. In addition, recruitment companies that brought in thousands of Africans from their homelands to the mining areas were founded.

T.M. Ruhiiga (✉)
Department of Geography and Environmental Sciences,
North West University, Potchefstroom, South Africa
e-mail: tabukeli.ruhiiga@nwu.ac.za

Y. Murayama et al. (eds.), *Urban Development in Asia and Africa*,
The Urban Book Series, DOI 10.1007/978-981-10-3241-7_18

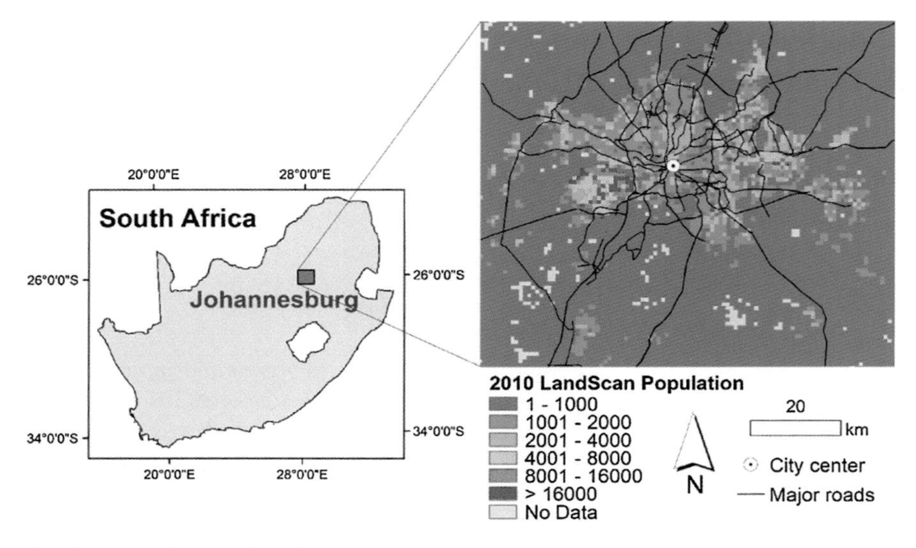

Fig. 18.1 Location and LandScan population of Johannesburg Metropolitan Area, South Africa

18.1.2 The Role of the Mining Industry

The mining industry has been critical in shaping the geography of Johannesburg and the politics of South Africa since 1870. The expansion of the mining industry led to the development of transportation infrastructure such as the harbors at Durban, Cape Town and Port Elizabeth. Furthermore, mining developments led to the construction of roads and railway lines that linked Johannesburg to Durban, Cape Town and Port Elizabeth, and later to Lourenco Marques (later renamed Maputo). Railway lines were constructed to bring in mining equipment, tools, machinery, and vehicles as well as imported products for the small trading stations, which developed into present-day towns along all the major lines.

The discovery of gold and later diamonds brought into the colony thousands of Europeans from the United Kingdom, France, and Belgium. A settler community of Europeans became the forefront of the mining industry. This was followed by the development of banking, insurance, and finance houses that controlled investments at that time (Bremner 2000).

The mining developments triggered the growth of coastal towns (Durban, Port Elizabeth, East London, and Cape Town), and the growth of inland towns linked to the mining industry (Johannesburg, Kimberley, Witbank, Secunda, Rustenburg, Carletonville, and Sasolburg). The increased demand for food by coastal and inland mining towns led to modernization in the agriculture sector. Johannesburg became the distribution point for agriculture produce, since it was surrounded by commercial farmlands in the Transvaal, Orange Free State, and parts of the then Cape Provinces.

18.1.3 Role of Colonialism

The earliest Europeans settled at the Cape as far back as 1652. The Great Trek saw the dispersion of the original European community of the Cape into the interior driven by the desire to be independent, get access to large tracts of land for live-stock, and carry out farming activities similar to what they had left behind in Europe. The Great Trek led to the setting up of several republics by Afrikaner nationalist leaders. After the 1899–1902 Anglo-Boer War, the republics were merged into the Cape, Natal, and Transvaal provinces. Johannesburg was already the most important commercial and industrial center in Transvaal province.

The second wave of European settlement was in the early 1830s. This was the result of ethnic and religious upheavals in Europe, which had dislodged significant numbers of migrants from Belgium and France, known as Huguenots to South Africa. Their arrival gave an added impetus to the "gold rush", which occurred at a time when the Afrikaner population group was already established, independent, and nationalistic. This was to be the cornerstone of the growth of colonial interests driven by the British. These developments led to conflict between Afrikaner nationalism and British imperialism. The setting up of a colonial administration by the British gave Afrikaners a free hand in the management of the colony. However, other population groups were excluded from sharing political power. As a result, discriminatory legislation against the nonwhite population was enacted. At the level of Johannesburg, these historical developments had a direct effect on the organization of the mining industry and on the nature of organized labor and its regulation. Another outcome was the setting up of separate residential districts for different population groups. These had a direct effect on the actual dynamics of the evolution of an urban economy.

18.1.4 Transport and the Urban Economy

The development of road and railway transport was soon followed by the building of power stations to supply electricity to the mines. Johannesburg was in close proximity to huge coal deposits to the east in Witbank and Secunda, and to the southeast near present-day Sasolburg. Electricity was also needed in Johannesburg, which had already developed into a large commercial and industrial hub by the early 1920s. The growth of Johannesburg therefore coincided with the global interest in gold. This, in turn brought a significant diversity of business investments. However, the growth of settlements next to each individual gold mine meant that from the beginning, Johannesburg developed as a dispersed settlement. These satellite towns eventually had their own central business districts (CBD). As more and more mines were opened, so did the emergence of satellite towns, each quite independent of the others before it. In addition, problems of surface drainage and the geology of the area influenced the choice of sites for settlement.

Following the end of World War I, the major developments in the mining industry required a regular and reliable supply of cheap labor to work in the mines. Therefore, labor recruitment agencies were set in South Africa and in all surrounding countries to guarantee supply. Pass laws were reinforced to control the movement of Africans from the rural areas to the cities. Furthermore, homelands were created through a policy of separate development and the enforcement of legislation tracing back to the 1913 Land Act. The Land Act dispossessed millions of black people of their lands, leading to large population relocations and a disruption in the original social landscape of Johannesburg. Therefore, racial discrimination referred to as apartheid ensured that the nonwhite population groups would never achieve similar levels of development as their European counterparts.

As a result of these discriminatory policies, residential districts for different ethnic groups and high-density African areas known as "townships" were developed. Radical changes in 1994 saw the official end of the apartheid government and its replacement by a democratic dispensation. However, the apartheid legacy of Johannesburg as a racially divided city (Tomlinson et al. 2003) has not been much altered. This is in spite of the process of "greying" in which different racial groups share the same residential district.

The mining industry spread to other parts of the country to include coal, iron ore, diamonds, and platinum, thereby widening the regional status of Johannesburg (Beavon 2005). The industrial base of manufacturing became closely linked to iron and steel, access to electricity, engineering, and chemicals. In line with the growth of the city and its manufacturing activities, the agricultural sector increasingly relied on locally manufactured inputs of tools, fertilizers, pesticides, equipment, and machinery. Johannesburg in the process became not just an industrial city but the country's largest market for manufactured goods and agricultural produce.

18.1.5 The Apartheid City

The apartheid city (Beavon 2005; Mabin 2006; Ruhiiga 2014) was built on a set of principles that made it a distortion of normal economic and demographic forces. A characterization of the key features will suffice here for it informs most of the development history of Johannesburg.

- Different races were allocated particular geographical areas as residential districts.
- Thousands of people were forcefully relocated to the demarcated geographical areas.
- Separate development meant that inter-racial mixing was legally prohibited and services were provided on scale with reference to ethnic identity.
- In addition, services were provided to these racially demarcated residential districts on a differentiated scale depending on household income patterns.

- Bulk infrastructure was also provided on the same racial basis which tended to disadvantage the non-European component of the population.
- The African population group was located farthest from the city center in what came to be known as "townships" and this forced such residents to rely either on mini buses or on train for their daily movement to and from places of work, shopping, and recreation.
- The Europeans were allocated residential areas close to services, work, and social amenities. The dominant means of movement became the use of a private car.
- These policies meant that different racial groups were subjected to socially engineered interventions that saw a differentiated dependence on public transport. This has remained in place up to the present.

Certain racial groups were disadvantaged by existing labor laws and a deliberate curtailment of opportunities for growth. There was a deliberate job reservation system in the labor market that ranked people from European, to Asian, mixed race, and African in an ascending order. For example, laws were enacted to prohibit the free movement of Asians, Indians, African, and mixed race people. This impacts negatively on the levels of urbanization as per race. While the Asian population group is fully urbanized at a level similar to that of the Europeans, the African population remains predominantly rural with a concentration in the former homelands. Laws were also enacted to make it difficult for men working in the mines to live with their families for it was illegal for wives and children to join and live with their husbands. This had a noticeable adverse effect on the family life of African people. For example, a high percentage of female household heads is a legacy of these policies. Furthermore, land use planning was deliberately used to ensure a clear separation between racially demarcated residential areas by the creation of buffer zones. These buffer zones (Todes 2006) created a "divided city" , and since 1994, the policy of spatial infilling (GPG 2011) is partly aimed at removing these zones.

18.2 Primacy in the National Urban System

18.2.1 Primacy Concept

Classical urban theory refers to the urban hierarchy in the form of ordering in which one city stands on top in terms of size, significance, role, influence, and status over others. Such a city has, over the years, dominated a country's urban system that it is often four times in size compared to the nearest second city. Where a network of towns, cities and conurbations (Pacione 2009), and ultimately urbanized regions develop in a country's spatial economic space, the concept of "primacy" acquires a particular meaning. While in most developed and developing countries, primacy remains a common feature of the urban system, the early history of urbanization in South Africa interfered with this process.

18.2.2 Johannesburg and Its Spatial Identity

Johannesburg shares borders to the east with Ekurhuleni, to the north with City of Tshwane, to the south with Sedibeng, and to the west with North West Province. The built-up areas of the city and those of the surrounding cities are continually extending. Consequently, the clear physical separation has become so blurred that it is difficult to identify on the ground. These developments are not unique to South Africa. This is part of the normal sub-processes that go with urbanization. While from a management perspective, Johannesburg is demarcated as a separate urbanizing region; in reality, what exists today is that Johannesburg is already part of a larger conurbation made up of cities to the east and north as initial processes toward eventually a typical “megalopolis”.

18.2.3 Johannesburg City Status

Following the 1998 Local Government Act (RSA 1998), new administration units were created after the restructuring of the local government. This had a direct impact on the place of cities in the county’s urban hierarchy. Note that the actual form and shape of Johannesburg is different from the land use/cover maps as it represents current administrative boundaries, while Fig. 18.1 depicts Landsat-derived land use/cover for Johannesburg and its environs.

For purposes of clarity, the term *Johannesburg* is consistently used throughout this chapter to refer to the city itself as a geographical area. The term “City of Johannesburg,” abbreviated to COJ is used in official parlance to mean Johannesburg together with surrounding satellite commercial, industrial, and residential districts that have been demarcated as a political and administrative unit. In terms of area, COJ is far larger than the original Johannesburg. However, in this chapter, COJ refers to the governance and management structure responsible for the city, while Johannesburg refers to the entire geographical area as currently demarcated.

18.3 Urban Land Use/Cover Patterns and Changes (1990–2030)

18.3.1 Observed Changes (1990–2014)

Figures 18.2 and 18.3 show the land use/cover maps of Johannesburg. Built-up areas increased from 482.2 km^2 in 1990 to 1400.3 km^2 in 2014 (Table 18.1). The built-up growth rate was 29.4 km^2/year during the “1990–2000” epoch

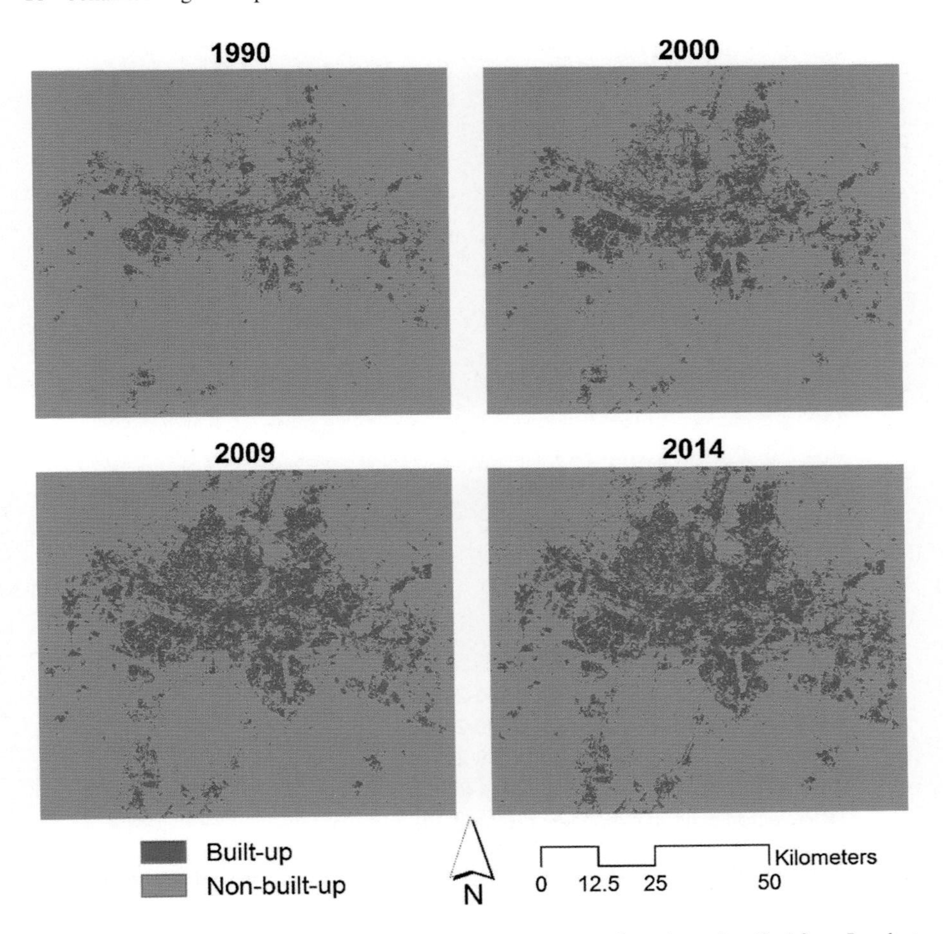

Fig. 18.2 Urban land use/cover maps of Johannesburg Metropolitan Area classified from Landsat imagery

(Table 18.2). However, built-up growth rate increased substantially to 42.3 and 48.7 km^2/year for the "2000–2009" and "2009–2014" epochs, respectively (Table 18.2). With reference to the direction and density of built-up areas in Fig. 18.2, for 1990, the built-up area shows an east–west trend as dominant but with a noticeable northward thrust from the CBD. For 2000, 2009, and 2014, there is a noticeable reinforcement of these trends except that the northward expansion of the city appears to have centered on three parallel corridors (Fig. 18.3). Within these areas, the existence of nonbuilt-up area is noted, but the relative significance declines from 1990 to 2014. To the south of the CBD, there is little expansion other than the appearance of three thin southward corridors.

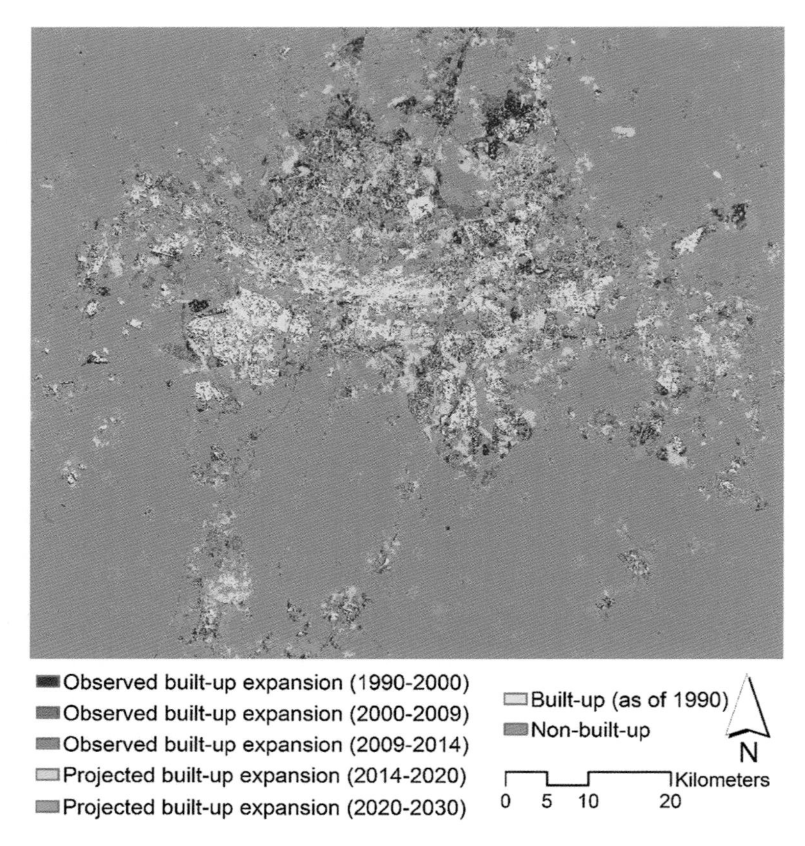

Fig. 18.3 Observed and projected urban land use/cover changes in Johannesburg Metropolitan Area

Table 18.1 Observed urban land use/cover of Johannesburg Metropolitan Area (km^2)

	1990	2000	2009	2014
Built-up	482.15	776.34	1156.83	1400.32
Non-built-up	6752.41	6458.22	6077.72	5834.24
Total	7234.55	7234.55	7234.55	7234.55

Table 18.2 Observed urban land use/cover changes in Johannesburg Metropolitan Area (km^2)

	1990–2000	2000–2009	2009–2014
Built-up	294.19	380.49	243.49
Annual rate of change (km^2/year)	29.42	42.28	48.70
Non-built-up	−294.19	−380.49	−243.49
Annual rate of change (km^2/year)	−29.42	−42.28	−48.70

Table 18.3 Observed landscape pattern of Johannesburg Metropolitan Area

Class-level (built-up) spatial metrics	1990	2000	2009	2014
PLAND (%)	6.66	10.73	15.99	19.36
PD (number per km^2)	1.96	2.10	1.86	2.09
ENN (mean) (m)	129.4	131.34	150.70	146.11
CIRCLE (mean) ($0 \leq \text{CIRCLE} < 1$)	0.43	0.41	0.44	0.43
SHAPE (mean) ($1 \leq \text{SHAPE} \leq \infty$)	1.37	1.35	1.36	1.34

Table 18.3 shows the observed class-level spatial/landscape metrics for the built-up class between 1990 and 2014. The percentage of landscape (PLAND) metric increased from 6.7 to 19.4% over the study period. Generally, the low PLAND metric shows that most of the study area is composed of nonbuilt-up areas. The patch density (PD) increased slightly from 2 to 2.1 per km^2 during the 1990–2000 epoch. However, PD decreased slightly from 2.1 to 1.7 per km^2 between 2000 and 2009, and then slightly increased between 2009 and 2014. Although slight variations are observed over the study period, PD is relatively low, which suggests that most of the built-up areas within the study area are less fragmented. The mean Euclidean nearest-neighbor distance (ENN) metric increased substantially from 129.4 to 146.1 m during the period 1990–2014, which suggests an increase in dispersion between the built-up areas. The mean related circumscribing circle (CIRCLE) and shape index (SHAPE) values did not change much over the study period suggesting that the temporal variation of the complexity of the built-up patches was low.

Figure 18.4 shows that PLAND decreased with increasing distance from the city center between 1990 and 2014. According to the urban growth theory, networks and building densities decrease with increasing distance from the center (Antrop 2000). Figure 18.4 indicates that PD varied between 1990 and 2014. Generally, PD increased between the 0–5 km and 5–10 distance buffer zones during the study period. For 1990 and 2000, PD decreased gradually from the 10–15 to 45–50 km distance buffer zones. However, for 2009 and 2014, PD increased from the 10–15 to 20–25 km distance buffer zones, then decreased from the 25–30 km distance buffer zones. Similar to PD, mean ENN also varied between 1990 and 2014. The variations in PD and mean ENN suggest that the Johannesburg is polycentric in form and function (i.e., comprised of two or more urban cores), and the city is influenced by both densification and sprawl processes. According to Burger and Meijers (2012), polycentric form refers to the location of cities of similar size close to each other. However, functionally polycentric regions are determined by the distribution of economic activities in a lot of hubs in the city or region. As mentioned before, Johannesburg developed as a dispersed settlement, with each satellite towns having its own central business districts (CBD).

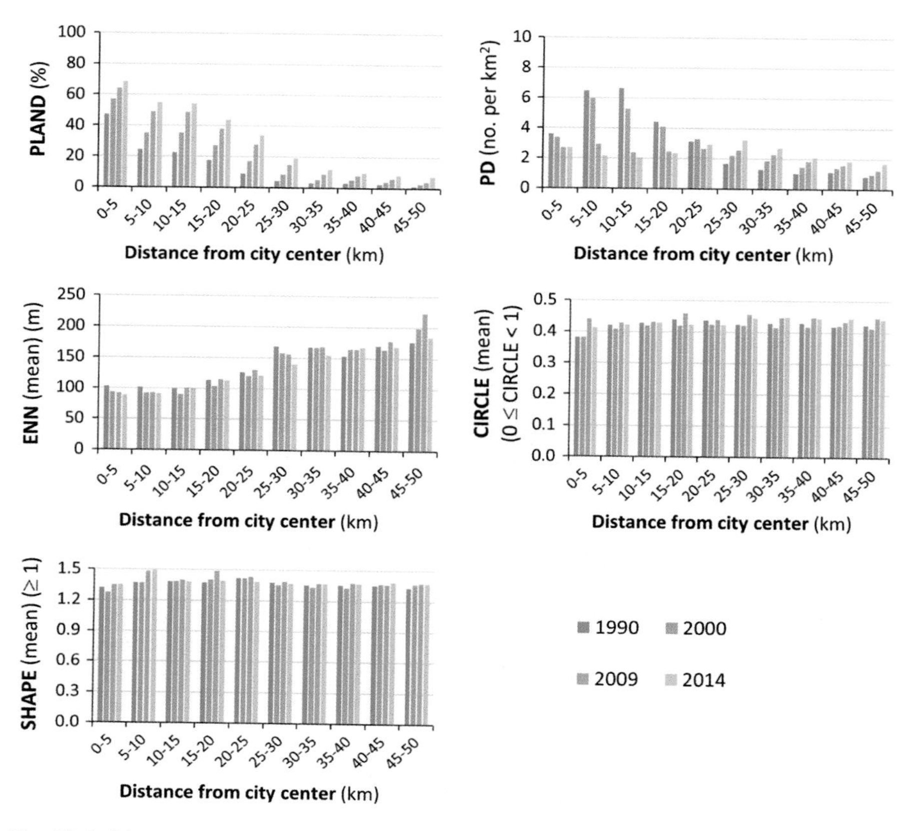

Fig. 18.4 Observed class-level spatial metrics for built-up along the gradient of the distance from city center of Johannesburg. *Note* The y-axis values are plotted in the same range as those in Fig. 18.6

18.3.2 Projected Changes (2014–2030)

Figure 18.5 shows the projected land use/cover maps. Built-up areas are projected to increase to 1500.1 and 1840.4 km^2 by 2020 and 2030, respectively (Table 18.4). The rate of built-up change for the 2014–2020 epoch is expected to be 16.6 km^2/year, and this will rise to 34 km^2/year between 2020 and 2030. Table 18.5 shows the projected class-level spatial/landscape metrics for the built-up class for 2020 and 2030. The PLAND metric is expected to increase from 20.7 to 25.4% over the study period. Generally, the moderate PLAND metric shows that most of the study area will be composed of nonbuilt-up areas. The PD will increase slightly from 5.8 to 7 per km^2 during the epoch, which is low suggesting that most of the built-up areas within the study area will be less fragmented. The mean ENN will hardly change, which points to an absence of change in dispersion between the built-up

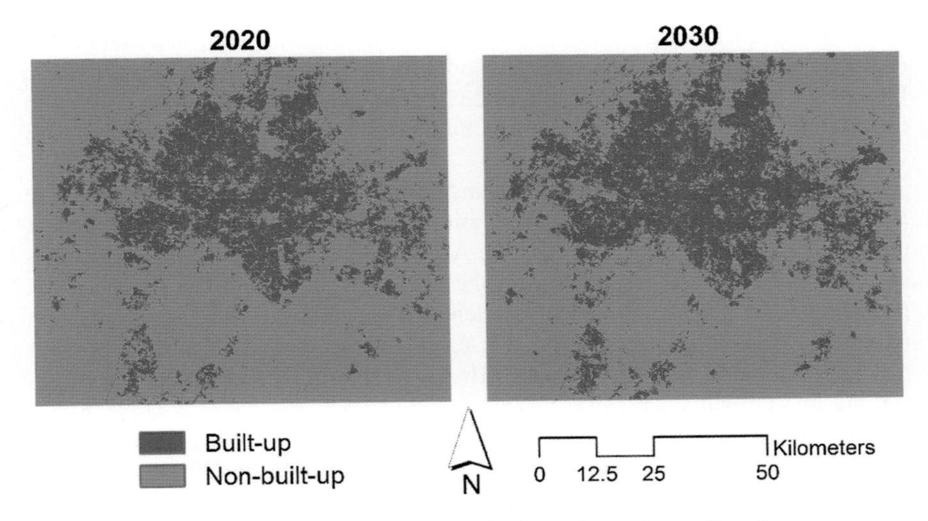

Fig. 18.5 Projected urban land use/cover maps of Johannesburg Metropolitan Area

Table 18.4 Projected urban land use/cover of Johannesburg Metropolitan Area (km^2)

	2020	2030
Built-up	1500.10	1840.44
Non-built-up	5734.46	5394.11
Total	7234.55	7234.55

Table 18.5 Projected urban land use/cover changes in Johannesburg Metropolitan Area (km^2)

	2014–2020	2020–2030
Built-up	99.78	340.35
Annual rate of change (km^2/year)	16.63	34.03
Non-built-up	−99.78	−340.35
Annual rate of change (km^2/year)	−16.63	−34.03

areas. The mean CIRCLE and SHAPE values are not expected to change over the study period.

18.4 Driving Forces of Urban Development

Urban development in Johannesburg is driven by many complex factors such as the historical and recent developments in the mining industry, pre- and post-apartheid rural-urban migration policies, urban development policies in particular, and government policies in general. The impact has been the growth of large low-cost housing settlements and related infrastructures in the east, west, and south. The apartheid imprint in the separate development of residential districts has left a legacy that will take a long time to replace. A combination of state urban policy and the control over significant land parcels by the private sector means that access to urban land remains constrained.

The most important push factor remains poverty in the rural areas. According to Ruhiiga (2011, 2012, 2014), much of this poverty can be traced to government policies that created labor reserves in the African homelands. Urban policy changes since 1988 have lifted restrictions on the inflow of predominantly African rural people into the city as permanent residents, which allowed orderly urbanization (Ruhiiga 2014). More importantly, the search for employment, especially among the youth, remains as a strong factor. This is in spite of the urban economy registering low growth since 1995.

Natural population growth from the already permanent resident population of Johannesburg remains a significant factor. It is noted from the 2011 census data (StatSA 2012) that in terms of annual growth rates, the African and mixed races still register over 2.0 rate compared to the Asian and Europeans (Pacione 2009; Ruhiiga 2013a, b, c). This shows that since the late 1980s, the African and mixed races in Johannesburg have witnessed explosive growth. However, this growth is from a low base following the long history of restrictions on the rural-urban migration (Beavon 2006). According to the LandScan population data for 2010, the highest population density is found in the large clusters of settlements to the west of the city's CBD (Fig. 18.1). This is a predominantly African residential district, called Soweto, set specifically during the apartheid to house Africans. To the immediate north, northeast of Johannesburg are predominantly high-income suburbs for the affluent working class (West Rand and North Rand). To the south of the CBD is a stretch of low-density areas predominantly for the mines and the old industrial areas. Finally, a high-density population cluster is developing in recent years to the southeast made of low-income resettlement schemes. This is largely comprised of people who were evicted from commercial farmland areas (Orange Farm and Sebokeng) since the mid-1980s. However, the population distribution patterns in Johannesburg have also been affected by government policy and the demand–supply forces operating in the urban land market (Ruhiiga 2011).

The road network shows a radial pattern in the form of an inner and outer set of freeways, while the rest tends toward a grid pattern with the CBD as the point of origin. The density of the road network reduces as distance increases from the inner circle earlier referred to. The need to maintain buffer zones contribute to the dispersion of mini satellites each with its own commercial center. To the south, the road network thins out because this region represents a newly urbanizing block which until recently was essentially large-scale commercial farmlands.

The actual location of the gold mines, existing, closed, or reopened recently due to new processing technologies, has had an impact on the location of industrial districts and the consequent positioning of the railway marshaling yards. They are concentrated in an east–west corridor south of the CBD. This happens to be the oldest part of Johannesburg and has not seen any significant developments either in population or in housing or in services for the last twenty plus years.

There is increasing settlement densification especially in predominantly African residential districts: Soweto, Orange Farm, and Diepsloot. These are essentially low-income planned settlements. Just to the north of the CBD, there exists a huge informal (slum) settlement where efforts at initiating urban renewal are constrained by the sheer high density of unplanned housing units. The increasing urban decay in the old CBD of Johannesburg following the typical processes of capital flight to the urban margins (Pacione 2009) is noted, especially around the oldest parts of the city.

The resulting relocation of the middle class and the affluent to the northern suburbs, the northeast, and northwest where housing densities are lower and where mixed land use planning is already entrenched (Ruhiiga 2014). However, the inner city renewal programs (Garner 2011) have been particularly noticeable in the case of Johannesburg. These have been implemented essentially through private public partnerships involving City of Johannesburg (COJ) and the private sector. In many cases, it has meant the passing of municipal by-laws through which COJ is able to take over abandoned or illegally occupied high-rise city blocks and renovate these into modern commercial and housing units. One of the unintended outcomes of these programs is that the resulting available housing units are expensive for the urban poor. So while these programs help in redeveloping the inner city, they also drive out the poor into the already overpopulated informal settlement areas.

The increasing growth in informal low-cost housing districts, partly as a response to the distortions of urban planning, is noticeable along the urban margins. This is partly due to housing market changes, the slow rate at which COJ releases land specifically for low-cost owner developed houses, and the slow rate at which COJ provides bulk services to these "new" settlement areas. In addition, the absence of strict housing regulations that operate in formally established residential development areas also results in informal settlements.

There are urban planning constraints (Mabin 2006; Todes 2006; GPG 2011; COJ 2012) where the long-term aim is the creation of flexible integrated settlements. Generally, integrated settlements are characterized by mixed land uses, which provide more sustainable urban future. However, these constraints cannot be totally blamed on governance systems. The constraints have been due to disagreements between the owners of urban land and COJ as to the true market value of such land, the pre-1994 urban form, the urban land market, and the preference of a significant high-end market consumer segment. Indeed, the requirement for negotiated settlement where radical restructuring of the urban form is the ultimate result tend to slow down the rate of urban change in the long run.

18.5 Implications for Sustainable Development

18.5.1 *The Post-Apartheid City*

According to the Johannesburg economic review (HSRC 2014), Johannesburg has the 27th largest city economy in the world. The city generates 17% of South Africa's wealth. The city registers an annual growth rate higher than the overall national figure for the country. The city accounts for 46.4% of the GDP of Gauteng, the province within which it is located. Gauteng contributes 35.6% of the country's GDP. The city has been growing its economy at a mean 4% per year and is estimated to register a GDP of R336 billion by end of 2015. In spite of this overall impressive performance, several characteristics that directly inform investment trends are worth noting.

Urban planning is now built on a spatial transformation framework (SDF) (COJ 2012, 2014a, 2014b, 2015) which ultimately envisages a city that is spatially and functionally efficient. The spatial development framework SDF (COJ 2003) as a point of departure sees and uses transportation networks to reorganise the urban space into a more efficient, compact, integrated system. The ultimate goal is to ensure fast movement through, to and from the city. In the process, the SDF aims at increase efficiency in production, shopping, services, and housing. Planning is driven by the desire to achieve mixed land uses and to counter the emergence of urban sprawl.

To this end, several principles underlie the framework: urban boundary, movement system, nodal development, environmental management, sustainable settlements, and corridor development (COJ 2003). A deliberate policy of urban infilling attempts to merge residential areas. This is meant to simultaneously make better and more efficient use of available land parcels (GPG 2011). There is evidence today that infilling is working to close open spaces between commercial, industrial, and residential land uses. This is shown in Fig. 18.2 for 2014 where observed infilling shows a clear densification in built-up areas in the central, northern, and southeastern sections of the city. However, significant parcels of land in the Johannesburg are not directly owned by COJ. This may explain, as in Fig. 18.3, why large tracts of undeveloped land still appear within the central districts of the city. Such privately held land parcels are traded on the urban land market (Ruhiiga 2013b, c).

18.5.2 *The Future of the City*

18.5.2.1 Densification

Densification is a natural process in urban areas (Chobokoane and Horn 2015) where a high demand for urban land drives a counter process of sub-divisions and developments within existing residential and commercial districts. Vacant land or land that is low in density is slowly taken over by new developments, thereby increasing land use density, population, and activities per unit of area. Densification is driven by essentially market forces: a recognition that certain areas are in such high demand that a strategic mix in land use zoning should attract investments into the area. But densification often sees the disappearance of recreation space and the building of concrete structures that undermine the need to sustain a liveable urban ecosystem (Ruhiiga 2014). In the case of Johannesburg, there is evidence that this process is occurring at varying rates in different parts of the city, especially the north, northeast, and northwest. Table 18.6 and Fig. 18.6 indicate that the rates of

Table 18.6 Projected landscape pattern of Johannesburg Metropolitan Area

Class-level (built-up) spatial metrics	2020	2030
PLAND (%)	20.74	25.44
PD (number per km^2)	5.75	7.04
ENN (mean) (m)	78.47	77.11
CIRCLE (mean) ($0 \leq$ CIRCLE < 1)	0.22	0.21
SHAPE (mean) ($1 \leq$ SHAPE $\leq \infty$)	1.07	1.06

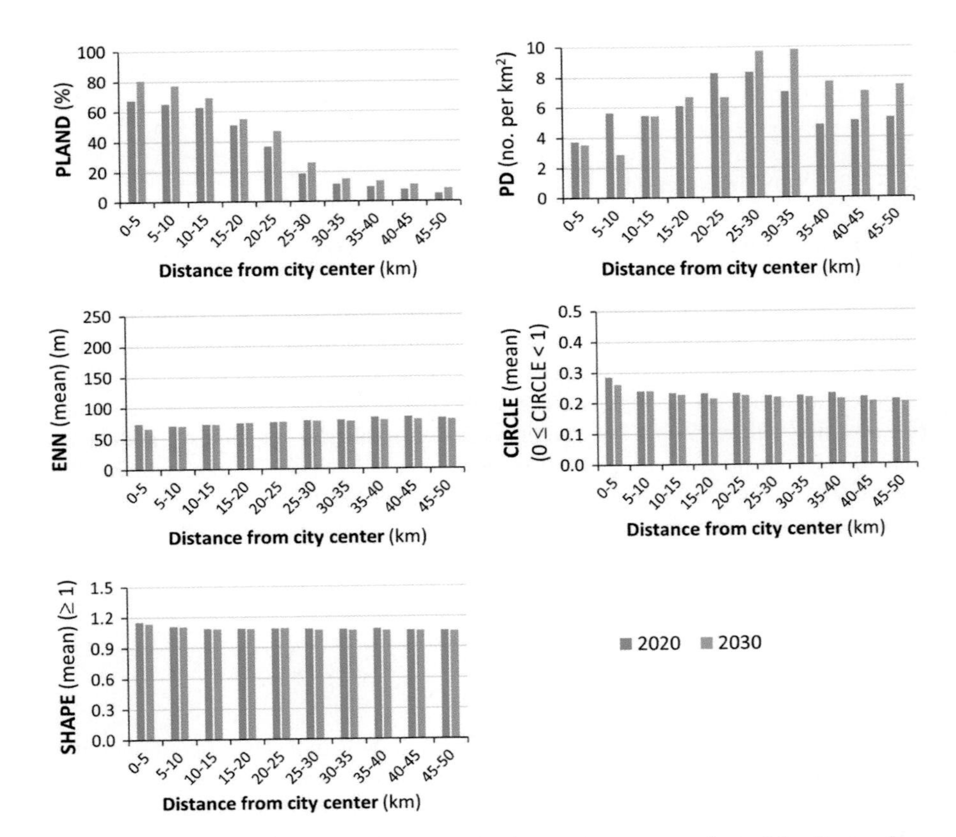

Fig. 18.6 Projected class-level spatial metrics for built-up along the gradient of the distance from city center of Johannesburg. *Note* The y-axis values are plotted in the same range as those in Fig. 18.4

densification have been accelerating since 2014, especially within the 20–25 km distance buffer zones.

18.5.2.2 Gentrification

The displacement of low-income residential units by middle- and high-income housing units is part of the process of urban change and urban decline, especially in the context of the traditional inner city. Significant literature exists on gentrification for advanced countries. In South Africa, this process is already underway (Rogerson 1995, 1996, 1997, 1998; Visser 2002). However, other studies report urban renewal programmes for Johannesburg, Durban, and Cape Town (Tomlinson et al. 2003). While the underlying theoretical debate is still stuck between "production" and "consumption" explanations, these forces ultimately do have a spatial imprint on the urban form. Figure 18.3 shows increasing built-up intensity, which is concentrated mainly in the northern part of Johannesburg during the 2000–2009 epoch. Similar built-up changes are observed during the 2009–2014 epoch. The projected results show that built-up changes are expected in the south and southeast areas of the city. However, there would be greater dispersion eastwards and in the northwest during the 2020–2030 epoch (Fig. 18.3).

18.5.2.3 Sub-Urbanization

Sub-urbanization as a process is reported in Visser (2003) for South African cities. Johannesburg city has been experiencing lateral sprawl at a scale unseen before 1995. There has been a tendency for corridor and arterial developments along the main roads into, through, and out of the city. Certain forms of business activity have shifted from the traditional industrial areas to the outskirts of the city partly as a response to skyrocketing rentals close to the CBD, and the need for greater space (for example show rooms for motor vehicles, industrial machinery, and heavy trucks). New residential districts also show this behavioral pattern, which often contradicts attempts at urban infilling—earlier referred to as the preferred urban planning policy of the state. The projected built-up changes imply that sub-urbanization would occur in the northern and eastern parts of Johannesburg (Fig. 18.3).

18.5.2.4 The Urban-Rural Interface

The traditional periphery (Pacione 2009) of every city is in a state of continual flux, and Johannesburg is no exception. Indeed, the encroachment of the city (Ruhiiga 2014) into farmlands that literally surround the city in all directions has been remarkable. The most notable change has occurred to the north, where as recent as

1990, large-scale commercial farms that created a buffer zone (Beavon 2006) between Johannesburg and Pretoria have literally disappeared. Today, they have been replaced by mixed land uses in the form of residential gated communities, warehouses, factories, and office parks.

18.5.2.5 Social Segregation in the Urban Space

Segregation in the urban space on the basis of ethnic, religious, language, political, cultural, and income and social status (class) is not unique to Johannesburg. It is a worldwide character of practically all urbanized regions. In the case of Johannesburg, contestation of urban spaces as a process through time has already occurred and will continue into the future. This is normal, but its spatial implications are often poorly understood by those tasked with urban planning. Nowhere are these tendencies are as explicit as in the residential housing industry (Beavon 2006) where a combination of the so-called municipal minimum housing regulations literally makes it impossible for members of certain social classes to acquire property and settle in particular parts of the city. Deliberate attempts at siting low-cost housing estates next to high-end gated communities have met stiff resistance on grounds that this increases crime in the area and lowers the market value of properties. The immediate implications of these conditions have been the mushrooming of informal settlements, which present service delivery and urban infrastructure development problems. Figure 18.3 shows evidence of outward built-up expansion in the future, which implies continued development of informal settlements under the current scenario. In addition, increased fragmentation as shown by high PD, particularly within the 25–30 and 30–35 km distance buffer zones in 2030 implies outward built-up expansion or continued development of informal settlements.

18.6 Concluding Remarks

This chapter has presented a geospatial analysis result of land use/cover changes, landscape metrics, and projected land use/cover changes for Johannesburg. Built-up areas expanded rapidly between 1990 and 2014, and the trend is projected to continue into the future. Rapid built-up changes and urbanization have been attributed several driving factors such as the historical and recent developments in the mining industry, government urban policies, rural-urban migration, and urban land market changes. The future of the city is discussed in the context of sustainable Johannesburg, which has been presented as a city in transition from an apartheid era toward a normal western city given its historical ties to Europe through colonialism, trade, and globalization. The city faces challenges around restructuring away from the apartheid past to create a modern, efficient urban form that is sustainable and responsive to the needs of its inhabitants. To this end, Johannesburg represents a complex interplay of often competing forces across time and place.

References

Antrop M (2000) Chagig patterns in the urbanized countryside of Western Europe. Landscape Ecol 15:257–270

Beavon KSO (2005) Johannesburg: the making and shaping of of the city. Pretoria, Unisa Press

Beavon KSO (2006) Johannesburg: a quest to regain world status. In: Amen MM, Archer A, Bosman MB (eds) Relocating global cities: from the center to the margins. Rowman and Littlefield Publishers Inc., New York, pp 45–75

Bremmer L (2000) Reinventing the johannesburg inner city. Cities, 17(3):185–193

Burger M, Meijers E (2012) Form follows function? Linking morphological and functional polycentricity. Urban Stud 49:1127–1149

Chobokoane N, Horn A (2015) Urban Compaction and densification in Bloemfontein, South Africa: measuring the current urban form against mangaung metropolitan municipality's spatial planning proposals for compaction. Urban Forum 26:77–93

COJ (2003) City of Johannesburg: spatial development framework—a growth management approach. Directorate of Development Planning and Facilitation, Johannesburg

COJ (2012) Johannesburg. IDP 2012//2014. City of Johannesburg, Johannesburg

COJ (2013) Johannesburg annual economic review. City of Johannesburg, Johannesburg

COJ (2014a) City of Johannesburg 2014/15 IDP. City of Johannesburg, Johannesburg

COJ (2014b) JOZI: a city @ work. 2014/15 IDP Review. City of Johannesburg, Johannesburg

COJ (2015) City of Johannesburg 2015/16 IDP. City of Johannesburg, Johannesburg

GPG (2011) Gauteng spatial development framework, 2011. Gauteng Provincial Government, Johannesburg

Garner G (2011) Johannesburg: ten ahead: a decade of inner-city regeneration. Double G-Media, Johannesburg

HSRC (2014) City of Johannesburg: a review of the state of the economy and other key indicators, HSRC, Pretoria

Mabin A (2006) Local government in South Africa's larger cities. In: Pillay U, Tomlinson R, Du Toit J (eds) Democracy and delivery: urban policy in South Africa. HSRC, Pretoria, pp 135–156

Nightingale CH (2012) A global history of divided cities. University of Chicago Press, Chicago

Pacione M (2009) Urban geography: a global perspective. Routledge, London

Rogerson CM (1995) South Africa's economic heartland: crisis, decline or restructuring? Africa Insight 25:241–247

Rogerson CM (1996) Dispersion within concentration: location of corporate head offices in South Africa. Development Southern Africa 13:567–579

Rogerson CM (1997) African immigrant entrepreneurs and Johannesburg's changing inner city Africa. Insight 27:265–273

Rogerson CM (1998) Restructuring of the apartheid space economy. Reg Stud 32:187–197

RSA (1998) Local government: municipal structures Act, 1998 (Act 117 of 1998), vol 42 No. 19614. Government Gazette, Cape Town

Ruhiiga TM (2014) Urbanization in South Africa: a critical review of policy, planning and practice. Afr Popul Stud 28:610–622

Ruhiiga TM (2012) Public transport and the decline of the traditional retail sector in South Africa. J Hum Ecol 39:49–60

Ruhiiga TM (2011) Land reform and rural poverty in South Africa. J Soc Sci 29:29–38

Ruhiiga TM (2013a) Growth of urban agglomeration nodes in Eastern Africa. J Hum Ecol 41:237–246

Ruhiiga TM (2013b) Managing explosive urbanization in Africa. J Hum Ecol 42:43–52

Ruhiiga TM (2013c) Reverse empowerment in post South Africa's Anti-Poverty Strategy. J Soc Sci 35:11–22

Ruhiiga TM (2014) Urbanisation in south africa: a critical review of policy, planning and practice. Afri. Popul Stud, 28(1): 610–622

StatSA (2012) Mid-year population estimates 2012. Statistical Release. StatSA, Pretoria

Todes A (2006) Urban spatial policy. In: Pillay U, Tomlinson R, Du Toit J (eds) Democracy and delivery: urban policy in South Africa. HSRC, Pretoria, pp 50–74

Tomlinson R, Beauregard R, Bremmer R, Mangcu X (eds) (2003) Emerging johannesburg: perspectives on postapatheid city. London, Routledge

Visser G (2002) Gentrification and South African cities: towards a research agenda. Cities 19:419–423

Visser G (2003) Spatialities of South African urban change perspectives on post-apartheid urban problems and the challenges at the beginning of the twenty-first century. Acta Acad Suppl 2003(1):79–104

Part IV
Urban Trend and Future

Chapter 19
Trends and Spatial Patterns of Urbanization in Asia and Africa: A Comparative Analysis

Ronald C. Estoque and Yuji Murayama

Abstract This chapter examines and compares the temporal and spatial patterns of urban land changes (ULCs), i.e., changes from non-built-up to built-up lands, in 15 major cities (metropolitan areas) in the developing Asia and Africa (*Asia*: Beijing, Manila, Jakarta, Hanoi, Bangkok, Yangon, Dhaka, Kathmandu, and Tehran; *Africa*: Dakar, Bamako, Nairobi, Lilongwe, Harare, and Johannesburg) based on remote sensing-derived urban land use/land cover (LULC) maps (c. 1990, 2000, 2010 and 2014) and GIS-simulated LULC maps (2020 and 2030). We used the land change intensity analysis technique to examine how the extent and rate of ULCs vary across time intervals. For the analysis of the spatial patterns of ULCs, we used spatial metrics. Between 1990 and 2014, Beijing, Bangkok, and Johannesburg had the highest gain of built-up in terms of area, whereas Bamako, Dhaka, and Lilongwe had the highest percentage increase. Five of the top seven cities in terms of total gain of built-up came from Asia, while four of the top seven cities in terms of total percentage came from Africa. Although some cities across Asia and Africa showed either stable or fluctuating intensities of ULCs, most of the cities had intensifying ULCs. During the same period, majority of the cities have become more fragmented. However, the simulated future urban LULC maps (2020 and 2030) indicated that their fragmented or diffused patches of built-up lands would eventually coalesce and result in more aggregated urban landscapes. In bigger cities such as Beijing, Manila, Bangkok, and Jakarta, ULCs are already occurring and moving away from the city center, whereas in smaller cities such as Bamako, Kathmandu, and Lilongwe, ULCs are still largely concentrated in closer proximities to the city center. We hypothesize that the bigger cities might have already undergone through the alternating process of diffusion–coalescence many times, whereas the smaller cities are still in the early stages of this cyclic process.

R.C. Estoque (✉) · Y. Murayama
Faculty of Life and Environmental Sciences,
University of Tsukuba, Tsukuba City, Japan
e-mail: estoque.ronald.ga@u.tsukuba.ac.jp; rons2k@yahoo.co.uk

Y. Murayama et al. (eds.), *Urban Development in Asia and Africa*,
The Urban Book Series, DOI 10.1007/978-981-10-3241-7_19

19.1 Introduction

19.1.1 Urbanization and Its Impacts

In today's Earth's geological epoch, the Anthropocene (Crutzen 2002), we are entering an urban era (Seto and Reenberg 2014). From approximately 13% (around 220 million) in 1900 (UN 2006), the world urban population has increased to about 30% (around 700 million) in 1950 and 54% (around 3.9 billion) in 2014 (UN 2015a). If the pace of urbanization and fertility rate were to remain constant at current levels, the world urban population would increase to 7.4 billion by 2050 (UN 2015a).

It has also been projected that rapid urbanization will take place in the developing countries, including those in the Asian and African regions (UN 2015a; Dunde 2015). Asia's urban population has increased from 17.5% (244.6 million) in 1950 to 44.8% (1864.8 million) in 2010 and has been projected to increase to 64.2% (3313.4 million) by 2050 (UN 2015a). Africa's urban population has increased from 14.0% (32.0 million) in 1950 to 38.3% (394.9 million) in 2010 and has been projected to increase to 55.9% (1338.6 million) by 2050 (UN 2015a). This shows that urban landscapes are continuously becoming more and more important for the everyday living of the Asian and African population in particular and the global population in general.

Urbanization has a dual nature. On the one hand, urbanization brings improvements to social and economic aspects of people's lives. Cities exemplify the creativity, imagination, and power of humanity. They are the cradles of innovation and knowledge creation, hearts of sociocultural transformations, and engines of economic growth (Wu 2010). Globally, cities are connected through political, economic, and technical systems, but more importantly through the Earth's biophysical life-support systems (Gómez-Baggethun et al. 2013; Jansson 2013). They are also part of the global sustainability agenda (Folke et al. 2011; Gómez-Baggethun et al. 2013; UN 2015b, c; Estoque and Murayama 2016).

On the other hand, urbanization also brings negative impacts on the natural environment, both locally and globally (Bloom et al. 2008; Grimm et al. 2008; Wu 2010; Seto et al. 2011; Dahiya 2012; Estoque and Murayama 2014a, 2015, 2016). This is especially so if urban 'development' is poorly planned, or worst, unplanned at all, resulting in various negative socioeconomic and environmental impacts, such as poor urban environment and poor quality of urban life. Urbanization itself is arguably 'the most drastic form of land transformation that results in irreversible landscape changes' (Estoque and Murayama 2014a, p. 943).

Various studies have shown that urbanization and land use/land cover (LULC) changes influence surface temperatures (e.g., Kalnay and Cai 2003; Chen et al. 2006; Bokaie et al. 2016; Estoque et al. 2017). Also, although urban areas occupy only a small portion of the Earth's land area, today's intensive burning of carbon fuels in the world's urban areas accounts for approximately 70% of global greenhouse gas emissions (Solecki et al. 2013; Bagan and Yamagata 2014). It has also been reported that a 10% increase in urban land cover in a country is associated with an increase of more than 11% in the country's total CO_2 emissions (Angel et al. 2011; Bagan and Yamagata 2014). In addition, unplanned (rapid) urban land changes (ULCs), i.e., changes from non-built-up to built-up lands, often result in the loss of urban green spaces. The loss of valuable urban green spaces can lead to the degradation of urban biosphere and the loss of valuable urban ecosystem services on which the environmental quality and liveability of urban areas depend.

There is therefore a need to examine the trend and spatial pattern of ULCs in urban areas because such analysis might help in the understanding of human–environment interactions and in the context of land use policy development and landscape and urban planning toward sustainable urbanization.

19.1.2 Capturing Trends and Spatial Patterns of ULCs: A Brief Overview

In the field of land change science, a technique called land change intensity analysis has been proposed for the purpose of gaining a better understanding of land change trends and patterns (Aldwaik and Pontius 2012). This technique is a mathematical framework that compares a uniform intensity against observed intensities of temporal changes among LULC categories. Intensity analysis at the interval level, for instance, examines how the size and speed of change vary across time intervals (Aldwaik and Pontius 2012). This technique has been applied in various cases (e.g., Villamor et al. 2014; Estoque and Murayama 2015).

Various studies have also shown the potential of spatial metrics for capturing and analyzing landscape connectivity and fragmentation and their relations to landscape patterns and processes (e.g., Dietzel et al. 2005a, b; Seto and Fragkias 2005; Kamusoko and Aniya 2007; Thapa and Murayama 2009; Wu et al. 2011; Estoque and Murayama 2013, 2016). Many of these metrics are documented in McGarigal et al. (2012). Some of which have been used to illustrate the diffusion–coalescence urban growth theory (Dietzel et al. 2005a, b; Wu et al. 2011; Estoque and Murayama 2016). The theory suggests that urbanization exhibits a cyclic pattern in time and space driven by two alternating processes: *diffusion*, in which new urban patches are dispersed from the origin point or seed location, and *coalescence* or the union of individual urban patches, or the growing together of the individual patches into one form or group (Dietzel et al. 2005a, b; Wu et al. 2011; Estoque and Murayama 2015, 2016).

This chapter examines and compares the temporal and spatial pattern of ULCs in 15 major cities (metropolitan areas) in the developing Asia and Africa (*Asia*: Beijing, Manila, Jakarta, Hanoi, Bangkok, Yangon, Dhaka, Kathmandu, and Tehran; *Africa:* Dakar, Bamako, Nairobi, Lilongwe, Harare, and Johannesburg) from c. 1990 to c. 2014 (observed period) and from c. 2014 to 2030 (projection or simulation period) by applying the concept of land change intensity and using various spatial metrics. The details of the LULC mapping, change analysis, and simulation modeling are given in Kamusoko (2017), while the resulting LULC maps and statistics, including the change maps and statistics, are presented in Murayama et al. (2017).

19.2 Extents and Rates of ULCs in Asia and Africa

Figure 19.1a shows that among the 15 cities examined, Beijing had the highest gain of built-up from c. 1990 to c. 2014 with 1574 km^2. It was followed by Bangkok, Johannesburg, and Tehran with 1401, 918, and 622 km^2 gain of built-up, respectively. Next in rank are the other two megacities in Southeast Asia, Manila and Jakarta, with 393 and 312 km^2 gain of built-up, respectively. The four cities with

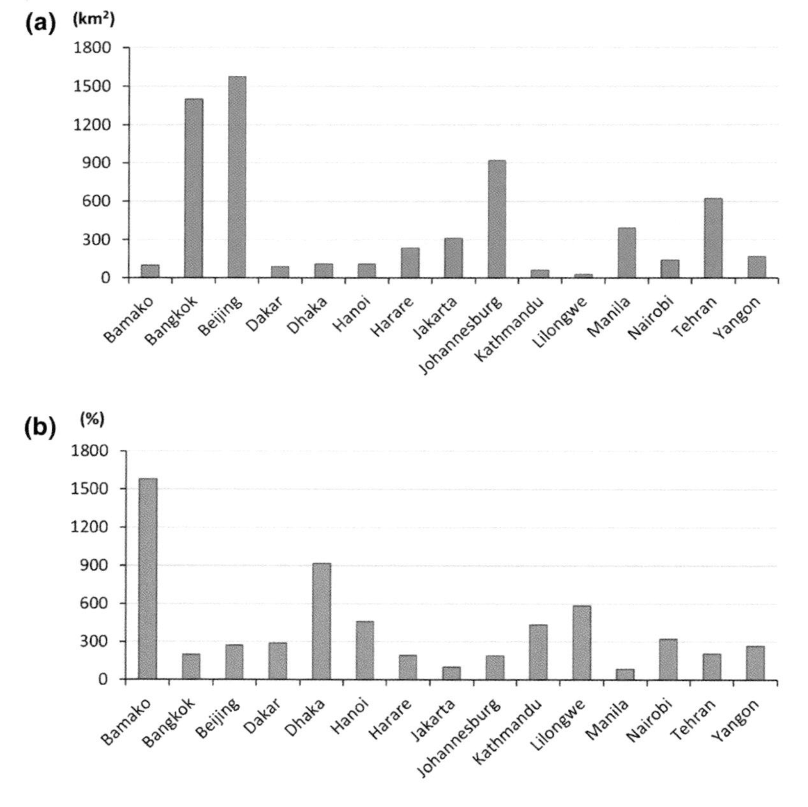

Fig. 19.1 Extents and rates of ULCs (c. 1990–2014) **a** gain of built-up; and **b** percentage increase of built-up

the lowest gain of built-up during the same period are Lilongwe (30 km^2), Kathmandu (62 km^2), Dakar (88 km^2), and Bamako (98 km^2).

In order to take into account the time difference between the LULC maps across some cities (Murayama et al. 2017), the gain of built-up was expressed per year. The difference was trivial, leaving the respective ranks of the cities the same as their respective ranks based on the total gain of built-up (Fig. 19.1a). Beijing had an annual gain of built-up of 60.6 km^2, while Bangkok, Johannesburg, and Tehran had 53.9, 38.3, and 23.9 km^2, respectively. Manila had 18.7 km^2 annual gain of built-up, while Jakarta had 13.0 km^2. Among the lowest ranked cities, Lilongwe had 1.3 km^2 annual gain of built-up, while Kathmandu, Dakar, and Bamako had 2.5, 3.5, and 4.1 km^2, respectively.

However, in terms of percentage increase of built-up during the same period, the results revealed a different pattern (Fig. 19.1b). Despite being one of the lowest ranked cities in terms of total gain of built-up, Bamako had the highest percentage increase of built-up (1579%) between c. 1990 and c. 2014, with an annual percentage increase of 65.8%. Dhaka (917%), Lilongwe (584%), Hanoi (458%), and Kathmandu (435%) complete the top five list. They had an annual percentage increase of 36.7, 25.4, 18.3, and 17.4%, respectively. Among the lowest ranked cities in terms of percentage increase of built-up during the same period are Manila (86%), Jakarta (100%), Johannesburg (190%), Harare (194%), and Bangkok (198%). They had an annual percentage increase of 4.1, 4.2, 7.9, 8.1, and 7.6%, respectively. Bangkok (see also Tehran) had lower annual percentage increase of built-up than Harare despite the former having a higher total percentage increase than the latter. This was due to the time extent used in the analysis. The time extent for Bangkok (and Tehran) was 26 years (1988–2014), while for Harare was 24 years (1990–2014). It should be noted, however, that the source LULC maps had different extent (size or area); they were not also based on administrative boundaries (see Murayama et al. 2017 for details). This aspect was not taken into account in this analysis.

19.3 Trends of Urbanization

19.3.1 Capturing Trends of ULCs

Land change intensity analysis at the time interval level involves examining how the extent and rate of change vary across time intervals (Aldwaik and Pontius 2012; Estoque and Murayama 2015). Thus, this analysis can help capture and compare the trends of ULCs in all the study sites. In this article, five time intervals (i.e., TI_1, TI_2, TI_3, TI_4, and TI_5), out of six time points (i.e., t1, t2, t3, t4, t5, and t6), were considered. Times t1, t2, t3, and t4 correspond to the years c. 1990, 2000, 2010, and 2014, while times t5 and t6 correspond to the years 2020 and 2030, respectively.

First, we determined the annual change intensity (ACI) of ULC for each time interval (Eq. 19.1) (Estoque and Murayama 2015).

$$\mathrm{ACI}\,(\%) = \frac{(\mathrm{ULC}/\mathrm{LA})}{\mathrm{TE}} \times 100, \tag{19.1}$$

where ACI is the annual change intensity for a given time interval (e.g., TI_1); ULC is the area of urban land change (from non-built-up to built-up) for the given time interval; LA is the area of the whole landscape; and TE is the time extent in years of the given time interval.

Second, we determined the uniform intensity (UI) of the ULCs. The UI is the rate of change, not for each time interval, but rather for the entire time extent of the land change analysis (Aldwaik and Pontius 2012; Estoque and Murayama 2015). The UI was determined based on the total ULC ($\mathrm{ULC}_{\mathrm{total}}$) and total time extent ($\mathrm{TE}_{\mathrm{total}}$) of the five time intervals (Eq. 19.2) (Estoque and Murayama 2015).

$$\mathrm{UI}\ (\%) = \frac{\mathrm{ULC}_{\mathrm{total}}/\mathrm{LA}}{\mathrm{TE}_{\mathrm{total}}} \times 100 \tag{19.2}$$

Finally, we compared the derived ACI for each time interval with the UI, and determined the ULC intensity category for each time interval. Aldwaik and Pontius (2012) proposed that if the ACI in a particular time interval (e.g., TI_1) is less than the UI, then the ULC intensity for that interval is considered slow, but if it is greater than the UI, it is considered fast.

A separate study attempted to expand the intensity scale by proposing six category levels (CL), namely very slow (<−60), slow (−60 to <−20), medium slow (−20–0), medium fast (>0–20), fast (>20–60), and very fast (>60), which can be determined using Eq. (19.3) (Estoque and Murayama 2015).

$$\mathrm{CL}\,(\%) = \frac{\mathrm{ACI} - \mathrm{UI}}{\mathrm{UI}} \times 100 \tag{19.3}$$

19.3.2 Trends of ULCs in Asia and Africa

Table 19.1 presents the results of the ULC intensity analysis for the 15 cities examined based on the two-category intensity scale: fast (F) and slow (S). During the first time interval (TI_1), only three out of the 15 cities had F ULC intensity. In TI_2 and TI_3, the number increased to eight and 14, respectively. This indicates that from c. 1990 to c. 2014, ULCs have intensified in most of the cities. However, the two-category intensity scale is broad. For instance, ACI 0.659 and ACI 2.927 are both considered F when compared to a UI value of 0.639, despite their large difference (see Jakarta, Table 19.1).

Table 19.1 Trends of ULCs across time intervals based on the two-category intensity scale: fast (F) and slow (S)

	ACI/time interval				
	TI_1	TI_2	TI_3	TI_4	TI_5
Bamako (UI = 0.886)	0.98	0.64	1.89	0.92	0.61
	F	S	F	F	S
Bangkok (UI = 0.334)	0.27	0.37	0.68	0.13	0.31
	S	F	F	S	S
Beijing (UI = 0.534)	0.189	0.651	1.258	0.576	0.457
	S	F	F	F	S
Dakar (UI = 0.177)	0.108	0.282	0.330	0.079	0.129
	S	F	F	S	S
Dhaka (UI = 0.536)	0.484	0.456	1.230	0.537	0.397
	S	S	F	F	S
Hanoi (UI = 0.288)	0.160	0.243	0.404	0.641	0.190
	S	S	F	F	S
Harare (UI = 0.490)	0.498	0.553	0.472	0.485	0.448
	F	F	S	S	S
Jakarta (UI = 0.639)	0.659	0.411	2.927	0.253	0.405
	F	S	F	S	S
Johannesburg (UI = 0.469)	0.407	0.584	0.673	0.230	0.470
	S	F	F	S	F
Kathmandu (UI = 0.315)	0.287	0.250	0.853	0.248	0.241
	S	S	F	S	S
Lilongwe (UI = 0.131)	0.098	0.133	0.378	0.102	0.129
	S	F	F	S	S
Manila (UI = 0.495)	0.484	0.428	0.735	0.603	0.374
	S	S	F	F	S
Nairobi (UI = 0.244)	0.178	0.254	0.958	0.026	0.157
	S	F	F	S	S
Tehran (UI = 0.174)	0.088	0.156	0.496	0.205	0.148
	S	S	F	F	S
Yangon (UI = 0.299)	0.292	0.327	0.353	0.245	0.284
	S	F	F	S	S

The previously proposed six-category intensity scale (Estoque and Murayama 2015) applied in this chapter presents a little more specific results and analysis (Fig. 19.2). For Bamako, its ULC intensity during TI_1 was medium fast (MF), but slowed down (S) during TI_2. During TI_3, it was very fast (VF). However, the simulation results revealed that its ULC intensity would decelerate to MF during TI_4 and finally to S during TI_5. By contrast, Bangkok's ULC intensity was medium slow (MS) during TI_1, but accelerated to MF and VF during TI_2 and TI_3, respectively. However, based on the simulation results, its ULC intensity during TI_4 and

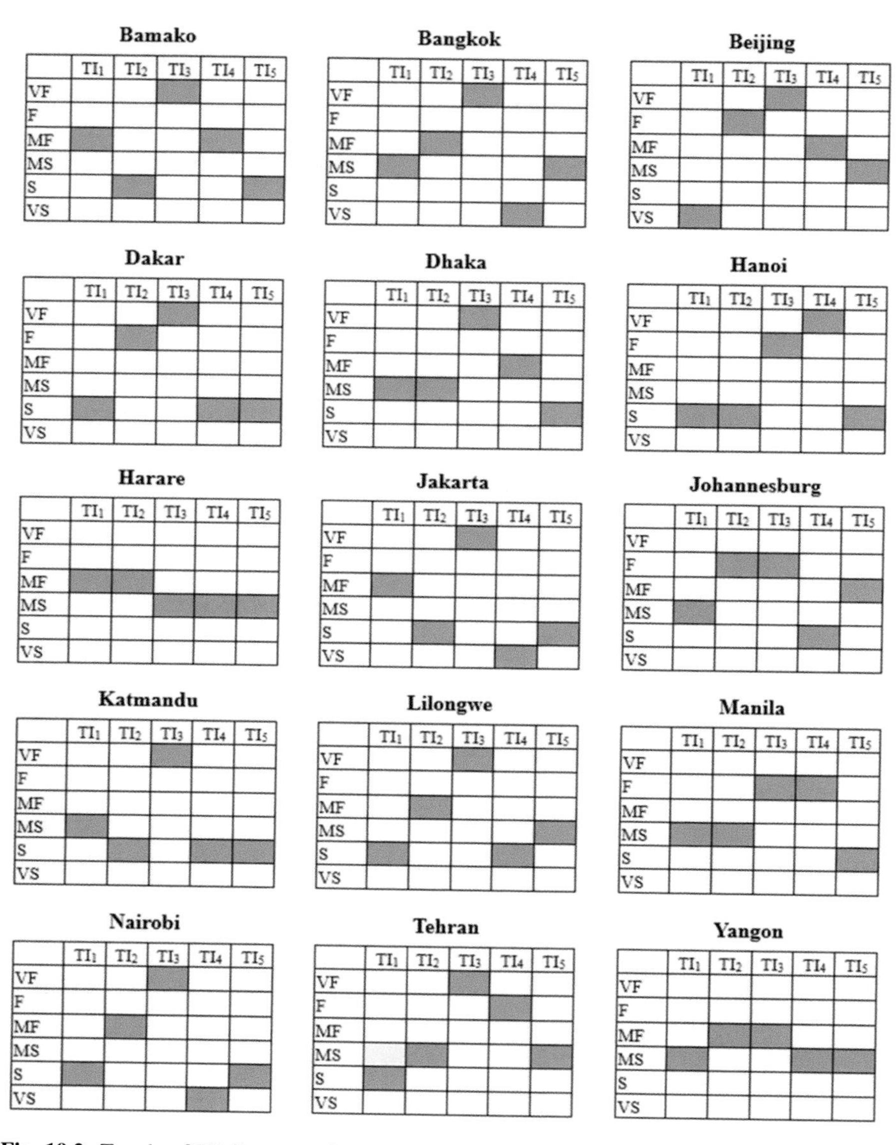

Fig. 19.2 Trends of ULCs across time intervals based on the six-category intensity scale: very slow (*VS*), slow (*S*), medium slow (*MS*), medium fast (*MF*), fast (*F*), and very fast (*VF*)

TI_5 would be very slow (VS) and MS, respectively. For Beijing, its ULC intensity during TI_1 was VS, but abruptly accelerated to fast (F) and VF during TI_2 and TI_3. The simulation results, however, showed that its ULC intensity would also decelerate to MF and MS during TI_4 and TI_5, respectively.

For Dakar, its ULC intensity during TI_1 was S, but accelerated to F and VF during TI_2 and TI_3, respectively. However, the simulation results revealed that its ULC intensity would decelerate to S during TI_4 and TI_5. On the other hand, Dhaka's ULC intensity was MS during TI_1 and TI_2, but abruptly accelerated to VF during TI_3. Based on the simulation results, however, its ULC intensity during TI_4 and TI_5 would be MF and S, respectively. For Hanoi, its ULC intensity during TI_1 and TI_2 was S, but accelerated to F during TI_3. The simulation results showed that its ULC intensity would continue to accelerate to VF during TI_4, but would later decelerate to S during TI_5.

For Harare, its ULC intensity during TI_1 and TI_2 was MF, but slowed down to MS during TI_3. The simulation results revealed that this intensity (MS) would continue across TI_4 and TI_5. By contrast, Jakarta's ULC intensity was MF during TI_1. It first slowed down to S during TI_2 before it accelerated to VF during TI_3. Based on the simulation results, its ULC intensity during TI_4 and TI_5 would be VS and S, respectively. For Johannesburg, its ULC intensity during TI_1 was MS, but accelerated to F during TI_2 and TI_3. The simulation results showed that its ULC intensity would decelerate to S during TI_4, but would later accelerate to MF during TI_5.

For Kathmandu, its ULC intensity during TI_1 and TI_2 was MS and S, respectively, but abruptly accelerated to VF during TI_3. However, the simulation results revealed that its ULC intensity would decelerate to S during TI_4 and TI_5. On the other hand, Lilongwe's ULC intensity was S, MF, and VF during TI_1, TI_2, and TI_3, respectively. However, based on the simulation results, its ULC intensity during TI_4 and TI_5 would be S and MS, respectively. For Manila, its ULC intensity during TI_1 and TI_2 was MS, but accelerated to F during TI_3. The simulation results showed that this intensity (F) would continue during TI_4, but would later decelerate to S during TI_5.

For Nairobi, like Lilongwe, its ULC intensity across TI_1, TI_2, and TI_3 was S, MF, and VF, respectively. However, the simulation results revealed that its ULC intensity would decelerate to VS and S during TI_4 and TI_5, respectively. Tehran's ULC intensity was S, MS, and VF during TI_1, TI_2 and TI_3, respectively. However, based on the simulation results, its ULC intensity during TI_4 and TI_5 would be F and MS, respectively. For Yangon, its ULC intensity during TI_1 was MS, but accelerated to MF during TI_2 and TI_3. The simulation results showed that its ULC intensity would be MS during TI_4 and TI_5.

19.4 Spatial Pattern of Urbanization

19.4.1 Capturing Spatial Patterns of ULCs

In order to capture, examine, and compare the spatial patterns of ULCs in all the cities, we used the results of the ULC spatial pattern analysis for each city. The details of the ULC spatial pattern analysis approach are described in the Methodology chapter (Kamusoko 2017). The approach included five class-level (built-up) spatial metrics (McGarigal et al. 2012): (1) the percentage of landscape (PLAND) metric, which measures the proportion of a particular class at a certain time point relative to the whole landscape; (2) the patch density (PD) metric, which is a measure of fragmentation based on the number of patches per unit area, in this case per 100 ha or 1 km^2, in which a patch is based on an 8-cell neighbor rule; (3) the Euclidean nearest neighbor distance (ENN) metric, which is a measure of dispersion based on the distance of a patch to the nearest neighboring patch of the same class; (4) the related circumscribing circle (CIRCLE) metric, which measures the circularity of patches. The value of CIRCLE is 0 for circular or one cell patches and approaches 1 for elongated, linear patches one cell wide; and (5) the shape index (SHAPE) metric, which is a measure of complexity. This metric has a value of 1 when the patch is square and increases without limit as patch shape becomes more irregular.

In addition, gradient analysis was also employed. Gradient analysis, developed in the context of vegetation analysis (Whittaker 1975), uses the concept of 'gradient,' i.e., the variation in the values of a given variable, for example, distance from the urban center. It has been used to investigate the effects of urbanization on species diversity, vegetation composition and structure, soil nutrients, water quality, and ecosystem properties (see Luck and Wu 2002; Weng 2007 for more details). More recently, the concept of gradient analysis has also been used in land change simulation modeling studies (Chen and Pontius 2010; Estoque and Murayama 2014b) and in the comparison of urban landscape patterns in various cities (Estoque et al. 2014; Estoque and Murayama 2016).

In this chapter, the five class-level spatial metrics were integrated with the gradient analysis approach, focusing on the gradient of the distance from the respective city centers of the 15 cities. The center of each city was identified on the basis of geographical and sociocultural (symbolism, historical) significance. For example, the landmarks used to represent the city centers of Bangkok, Jakarta, and Manila are the Grand Palace, Bangkok, the National Monument, Central Jakarta, and Kilometer Zero (KM 0) in Rizal (Luneta) Park, Manila, respectively. The gradient analysis approach also enabled a zone by zone analysis (e.g., 0–5; 5–10 km; and so on), which is important in this comparative analysis since the cities largely vary in terms of size or spatial extent.

19.4.2 *Spatial Patterns of ULCs in Asia and Africa*

19.4.2.1 Whole Landscape

Figure 19.3 presents the results of the spatial pattern analysis based on the five class-level (built-up) spatial metrics. The heights of the bars for the PLAND metric represent the proportion of built-up class relative to the whole landscape of each city, i.e., each city's extent or size based on the source LULC maps

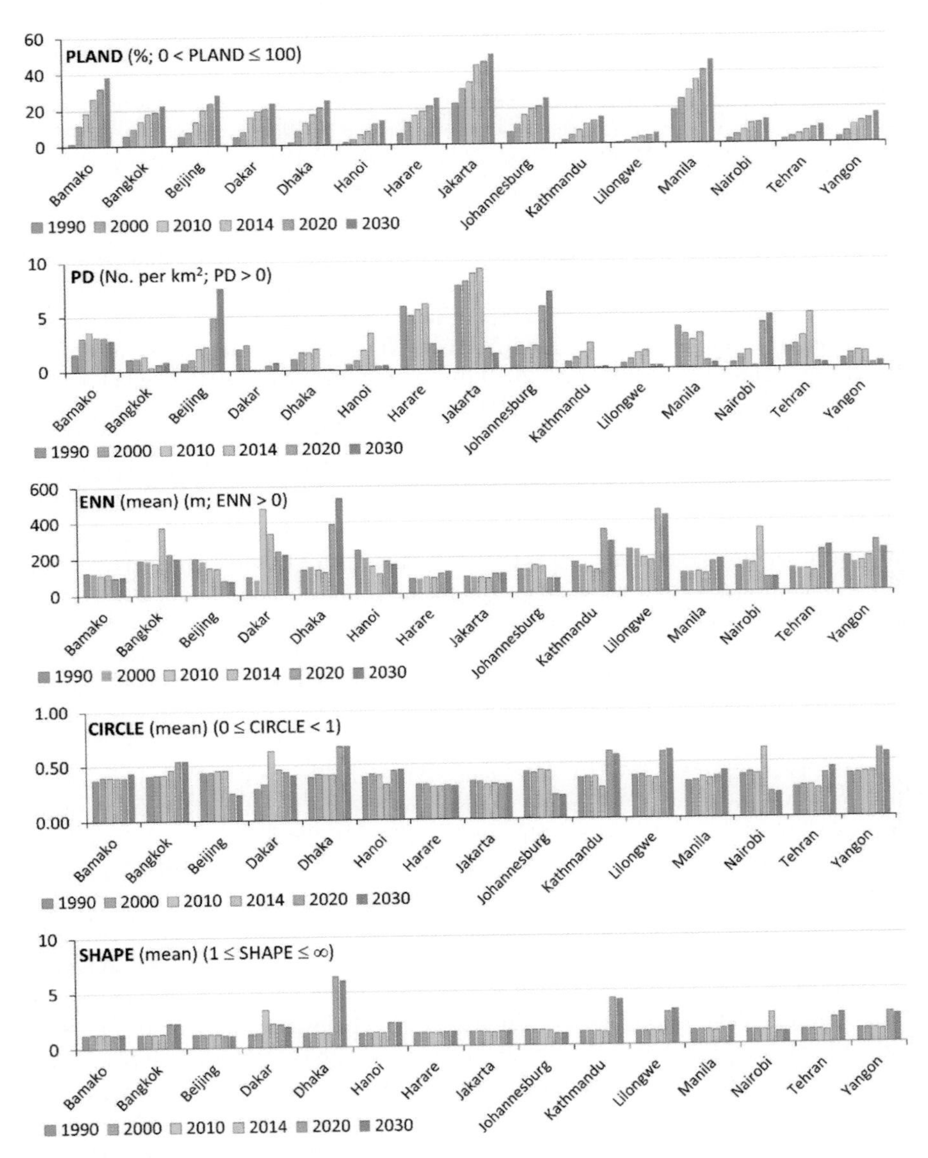

Fig. 19.3 Spatial patterns of ULCs based on the five class-level (built-up) spatial metrics

(Murayama et al. 2017), excluding the water class, across the six time points or five time intervals. However, since the extents or sizes of the source LULC maps were not based on the administrative boundaries of the 15 cities (Murayama et al. 2017), the heights of the bars of PLAND are not indicative of the relative sizes of the cities. The gradient analysis (discussed in Sect. 19.4.2.2) provides indications of the relative size of each city. From c. 1990 to c. 2014, the results revealed that the proportion of built-up lands in all the cities has been increasing and would continue to increase in 2020 and 2030. However, it can be observed that the rates of increase in PLAND also varied across the five time intervals in all the cities. The results of the ULC intensity analysis discussed above (Sect. 19.3.1) provide detailed characterizations of these varied magnitudes of ULCs.

The heights of the bars for the PD metric represent the number of built-up patches per km^2. An increasing PD indicates landscape fragmentation, while a decreasing PD indicates landscape aggregation and connectivity. Between c. 1990 and c. 2014, nine cities had an apparent increase in PD (i.e., Bamako, Beijing, Dhaka, Hanoi, Jakarta, Kathmandu, Lilongwe, Tehran, and Yangon) (Fig. 19.3). Four cities showed an increasing trend from c. 1990 to c. 2010 (becoming more fragmented) but showed a decreasing trend from c. 2010 to c. 2014 (becoming more aggregated or connected) (i.e., Bamako, Bangkok, Nairobi, and Yangon). Dakar's PD increased from c. 1990 to c. 2000 but decreased from c. 2000 to c. 2014, while Manila's PD decreased from c. 1990 to c. 2010 but increased from c. 2010 to c. 2014. Harare's PD decreased from c. 1990 to c. 2000 but increased from c. 2000 to c. 2014, while Johannesburg's PD had an 'up–down–up' trend from c. 1990 to c. 2014. During the observed period (c. 1990–2014), Jakarta and Harare had the most fragmented built-up lands among the 15 cities based on the PD metric (Fig. 19.3). Based on the simulation results, three cities would have a much higher PD in 2020, but more especially in 2030 (in comparison with c. 2014) (i.e., Beijing, Johannesburg, and Nairobi). The other cities would have a much lower PD in 2020, which would either increase (e.g., Lilongwe and Yangon) or decrease (e.g., Harare and Manila) in 2030.

In the same figure (Fig. 19.3), the heights of the bars for the ENN metric represent the mean distance between the nearest neighboring built-up patches. ENN is a measure of dispersion. It can be noted that the aggregation of built-up patches can lead to a much higher mean ENN between the newly aggregated built-up patches and other isolated built-up patches. By contrast, the expansion of the old patches and the occurrence of new patches of built-up lands in between old patches (but not necessarily connected patches) can lead to a much lower mean ENN. Within the observed period (c. 1990–2014), six cities had an evident decreasing mean ENN (i.e., Beijing, Hanoi, Jakarta, Kathmandu, Lilongwe, and Tehran). It can be observed that these cities also had an increasing PD during the observed period (c. 1990–2014). These results are consistent with the above-mentioned interpretations for the ENN. The cities with fluctuating PDs during the observed period also had fluctuating mean ENN values during the same period. The trends of the simulated mean ENN values for 2020 and 2030 for all the cities are also consistent with the trends of the simulated PD values.

The heights of the bars for the CIRCLE metric represent the elongation or circularity of built-up patches. An increase in the value CIRCLE is indicative of the development of more elongated built-up patches exhibiting a linear pattern. By contrast, a decrease is indicative of the development of more circular built-up patches. Within the observed period (c. 1990–2014), many cities showed an almost stable mean CIRCLE values, i.e., with only a slight increase or decrease between time points. However, there are also some cities that showed remarkable changes in their respective mean CIRCLE values. For example, Dakar's mean CIRCLE values increased abruptly from c. 2000 to c. 2010 then decreased in c. 2014. The mean CIRCLE values of Bangkok and Nairobi increased substantially from c. 2010 to c. 2014, while those of Hanoi and Kathmandu decreased substantially. Based on the simulation results, Dhaka, Kathmandu, Lilongwe, and Yangon are among those cities that would have much higher mean CIRCLE values in 2020 and 2030, relative to their respective CIRCLE values in c. 2014. By contrast, Beijing, Johannesburg, and Nairobi are among those cities that would have much lower mean CIRCLE values in 2020 and 2030, in comparison with their respective mean CIRCLE values in c. 2014.

Lastly, the heights of the bars for the SHAPE metric represent the irregularity or complexity of built-up patch shape compared to a standard shape (square) of the same size. An increase in the SHAPE values indicates that built-up patches became more irregular and complex, and vice versa. Within the observed period (c. 1990–2014), the complexity of the built-up patches of all the cities, except Dakar and Nairobi, was more or less stable (Fig. 19.3). The simulation results revealed that Bangkok, Hanoi, Manila, Tehran, and Yangon, but more especially Dhaka, Kathmandu, and Lilongwe, would have much higher mean SHAPE values in 2020 and 2030, relative to their respective mean SHAPE values in c. 2014.

19.4.2.2 Along the Gradient of the Distance from City Center

Figures 19.4, 19.5, 19.6, 19.7, and 19.8 present the results of the gradient analysis using the five class-level (built-up) spatial metrics along the gradient of the distance from city center. The maximum distance considered in the preparation of the bar graphs for all the metrics in all the cities was set to 50 km. However, it should be noted that the maximum distance from city center for some of the cities, such as Bangkok, Beijing, and Tehran, based on the extent of their respective LULC maps goes beyond 50 km (Murayama et al. 2017). Nevertheless, the graphs can still provide indications of the relative size of the cities.

In all the cities, the proportion of built-up land (PLAND) was higher in areas closer to the city center (Fig. 19.4). PLAND gradually decreases as the distance goes farther from the city center. However, it can be observed that in bigger cities such as Beijing, Manila, Bangkok, and Jakarta, ULC rate was higher in middle distances, e.g., 10–30 km. In smaller cities such as Bamako, Kathmandu, and Lilongwe, it was still high in closer proximities to the city center, e.g., 0–10 km.

Within the observed period (c. 1990–2014), in some of the bigger cities such as Manila, Jakarta, and Bangkok, PD was higher in middle distances, while for the some of the smaller cities such as Harare and Yangon, it was higher in closer proximities to the city center (Fig. 19.5). However, in most of the cities, PD has been decreasing in areas closer to the city center across time intervals, whereas in farther distances, it has been increasing. The decreasing trend might have been due to the effect of the coalescence of built-up lands, while the increasing trend might have been due to the diffusion process. The simulation results revealed that PD

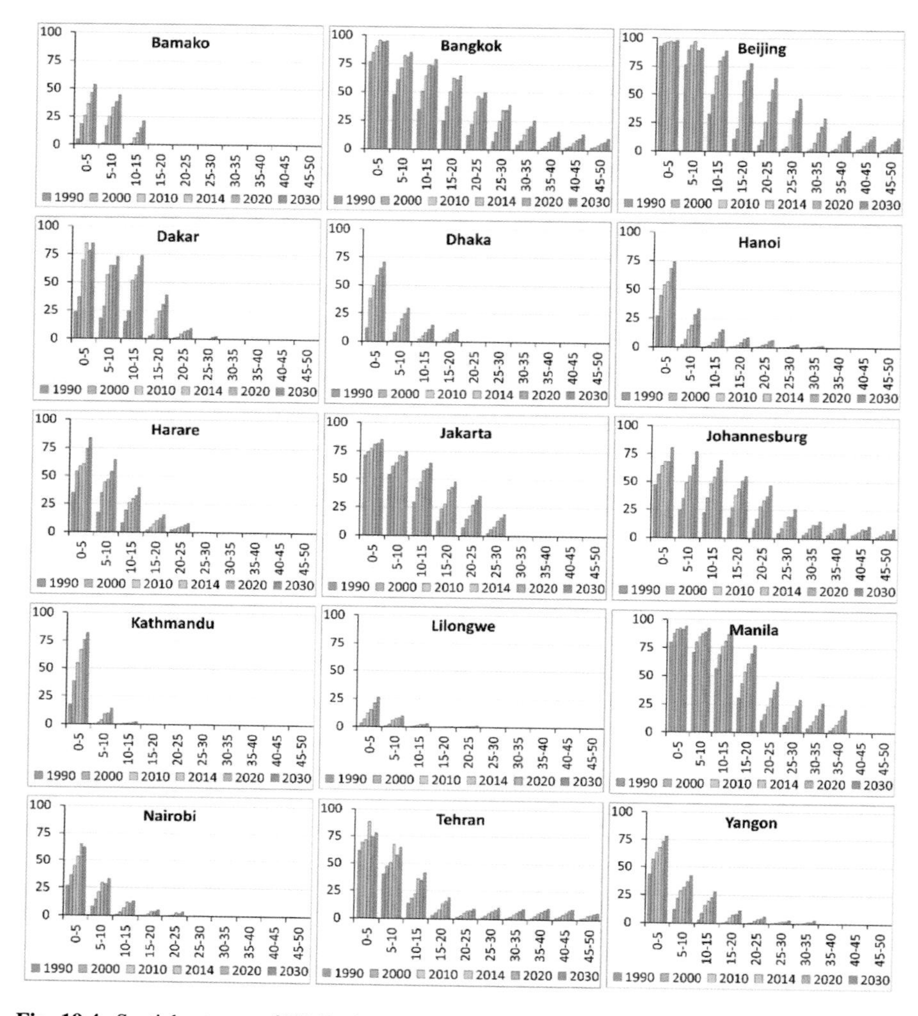

Fig. 19.4 Spatial patterns of ULCs along the gradient of the distance from city center (km) based on the class-level (built-up) **PLAND** metric (%; $0 < \text{PLAND} \leq 100$)

would either increase, e.g., Beijing, Johannesburg, and Nairobi, or decrease, e.g., Harare, Manila, and Tehran, from c. 2014 to 2020 and 2030 especially in farther distances.

In general, mean ENN increases as the distance from the city center increases (Fig. 19.6). Within the observed period (c. 1990–2014), some cities showed a decreasing mean ENN especially in farther distances, e.g., Beijing and Tehran, while some cities showed a fluctuating mean ENN, e.g., Bangkok and Dakar. The simulation results showed that mean ENN would either increase, e.g., Hanoi, Lilongwe, and Manila, or decrease, e.g., Beijing, Johannesburg, and Nairobi, from c. 2014 to 2020 and 2030.

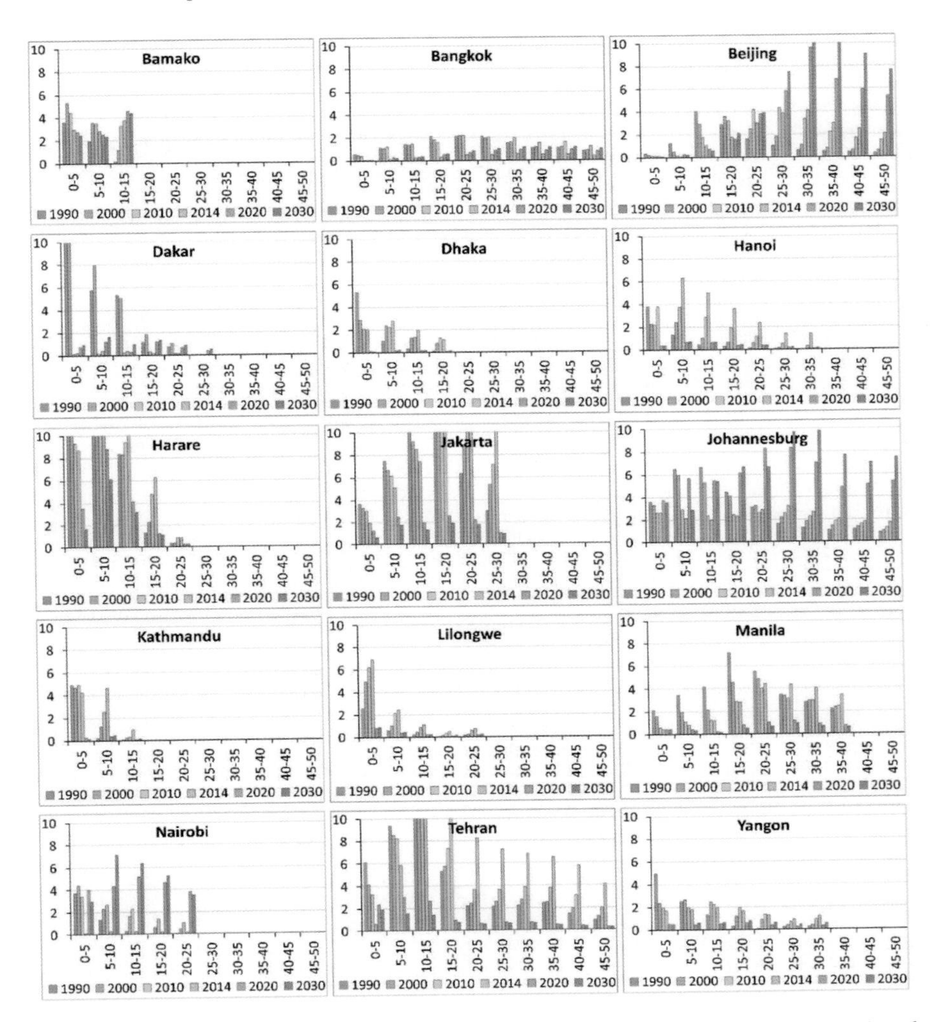

Fig. 19.5 Spatial patterns of ULCs along the gradient of the distance from city center (km) based on the class-level (built-up) **PD** metric (No. per km^2; PD > 0)

Also, in general, the mean CIRCLE values do not vary very much along the gradient of the distance from city center in all the cities (Fig. 19.7). Within the observed period (c. 1990–2014), many of the cities had fluctuating mean CIRCLE values, e.g., Hanoi and Nairobi, while some cities were more or less stable, with only slight fluctuations, e.g., Johannesburg and Yangon. The simulation results revealed that mean CIRCLE values would either increase, e.g., Lilongwe, Manila, Tehran, and Yangon, or decrease, e.g., Beijing and Johannesburg, from c. 2014 to 2020 and 2030.

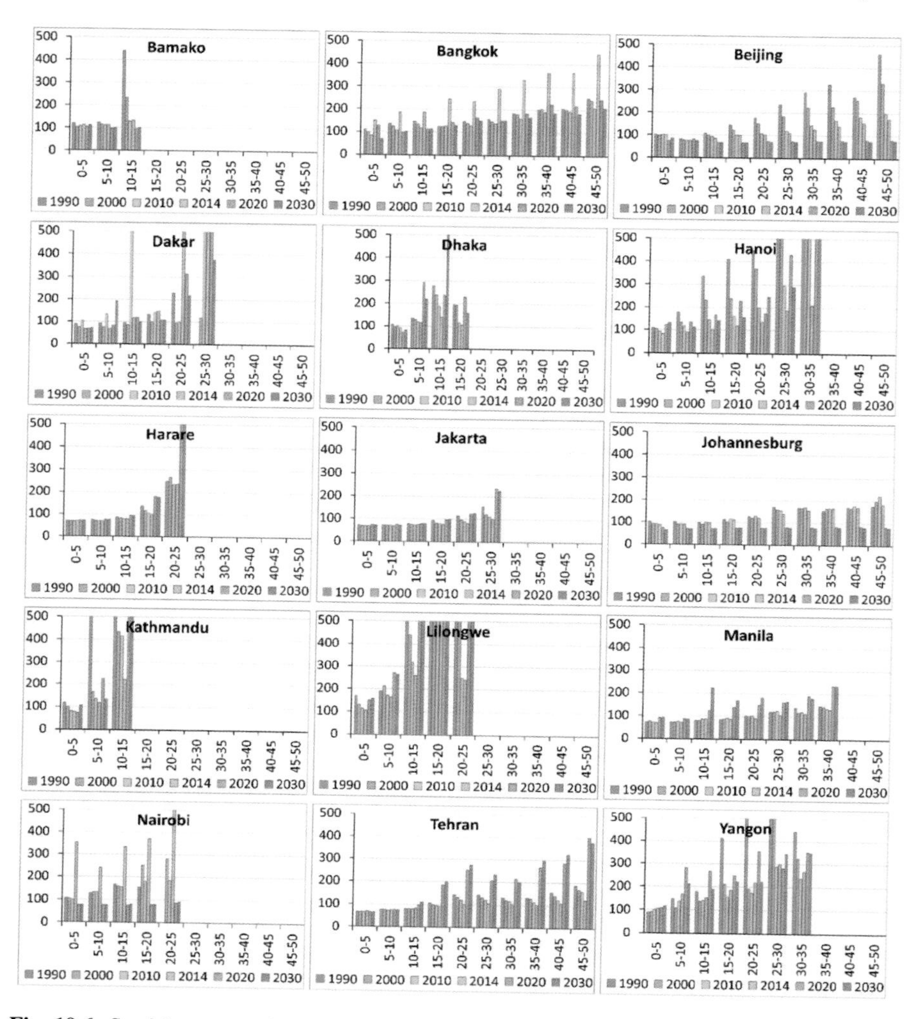

Fig. 19.6 Spatial patterns of ULCs along the gradient of the distance from city center (km) based on the class-level (built-up) **ENN** (mean) metric (m; ENN > 0)

Lastly, like the CIRCLE metric, the mean SHAPE values in general do not vary very much along the gradient of the distance from city center in all the cities (Fig. 19.8). Within the observed period (c. 1990–2014), while some cities had fluctuating mean SHAPE values, e.g., Dakar and Nairobi, most of the cities had more or less stable, with only slight fluctuations. Also, like the CIRCLE metric, the simulation results showed that mean SHAPE values would either increase, e.g., Dhaka, Hanoi, Kathmandu, Lilongwe, Tehran, and Yangon, or decrease, e.g., Beijing, Johannesburg, and Nairobi, from c. 2014 to 2020 and 2030 especially in farther distances.

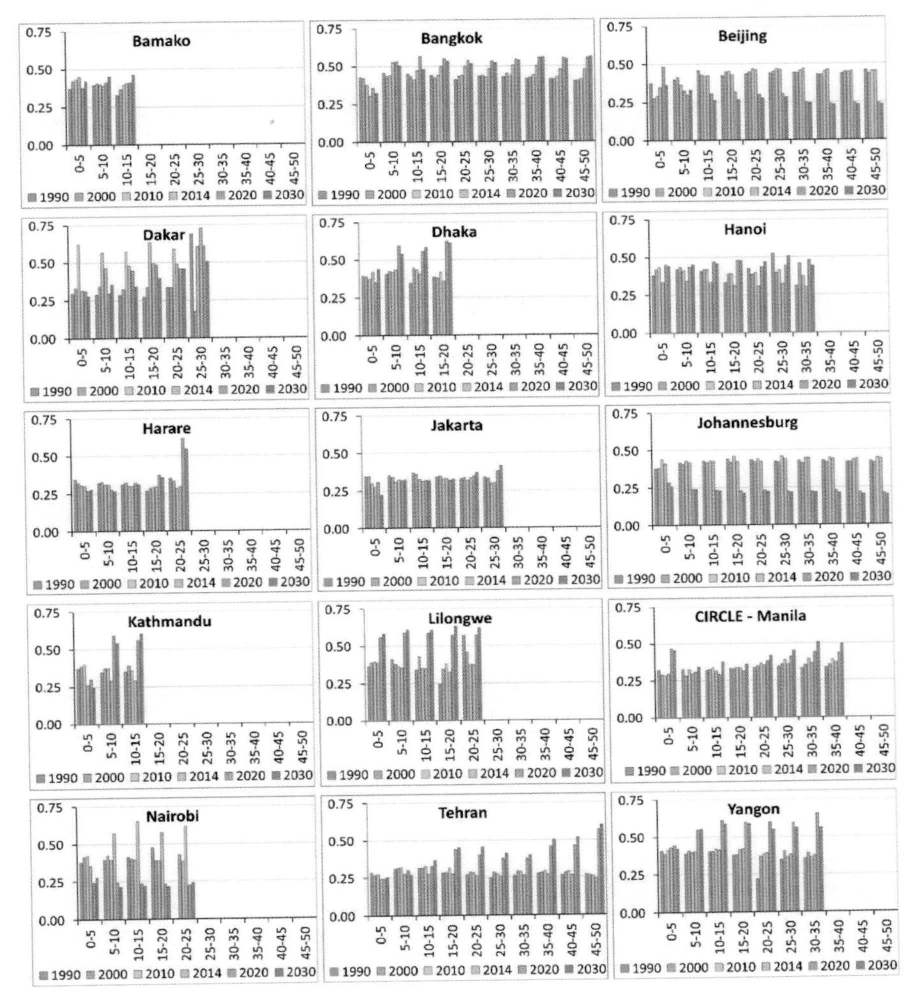

Fig. 19.7 Spatial patterns of ULCs along the gradient of the distance from city center (km) based on the class-level (built-up) **CIRCLE** (mean) metric ($0 \leq$ CIRCLE < 1)

19.5 Discussion

While some bigger cities like Beijing had higher gains of built-up from c. 1990 to c. 2014, they did not have the highest percentage increase over the same period (Fig. 19.1). By contrast, although some smaller cities like Bamako had lower gains of built-up, they had a high percentage increase over the same period (Fig. 19.1). Interestingly, five of the top seven cities in terms of total gain of built-up came from

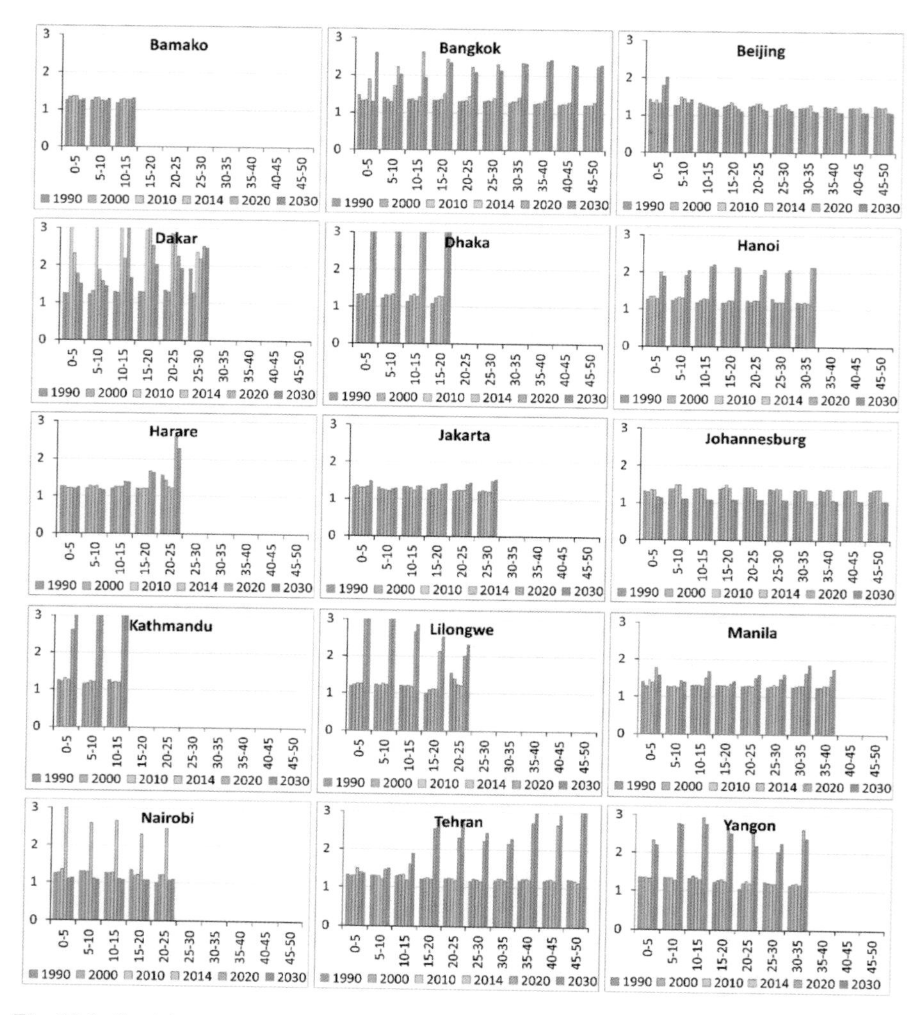

Fig. 19.8 Spatial patterns of ULCs along the gradient of the distance from city center (km) based on the class-level (built-up) **SHAPE** (mean) metric ($1 \leq \text{SHAPE} \leq \infty$)

Asia (Fig. 19.1a). However, in terms of total percentage increase of built-up, four of the top seven cities came from Africa (Fig. 19.1b).

Among the 15 cities examined, none have shown a steady rate or intensity of ULC across the five time intervals considered in the analysis, i.e., from TI_1 to TI_5. Harare and Yangon had ULC intensities that were almost stable, which fluctuate between medium fast (MF) and medium slow (MS) (Fig. 19.2). Bangkok, Johannesburg, Lilongwe, and Nairobi showed a similar trend in their respective ULC intensities, characterized by an 'up–down–up' trend (Fig. 19.2). The 'down–up–down' trend is also another general trend that characterizes the respective ULC intensities of the other cities such as Bamako, Jakarta, and Kathmandu. By contrast, Beijing, Dakar, Dhaka, Hanoi, Manila, Tehran, and Yangon showed an 'up–down' trend.

However, because of the uncertainties associated with the simulated ULCs (c. 2014–2020; 2020–2030) (Murayama et al. 2017), more attention can be given to the observed trends over the c. 1990–2014 period (Fig. 19.2). From c. 1990 to 2014, 11 cities had an increasing trend in their respective ULC intensities, i.e., Bangkok, Beijing, Dakar, Dhaka, Hanoi, Johannesburg, Lilongwe, Manila, Nairobi, Tehran, and Yangon. Three cities, i.e., Bamako, Jakarta, and Kathmandu, had a fluctuating trend, and one city, i.e., Harare, had a decreasing trend.

The ULC intensities for the Asian cities vary across time intervals and thus cannot be characterized and generalized by a single 'trend' only. The same can be said for the African cities. In this analysis, the results did not reveal any distinct regional characteristics of the Asian and African cities in terms of ULC intensity. In fact, cities from the two regions can have a similar characteristic based on ULC intensity (Fig. 19.2).

For the spatial patterns of ULCs between c. 1990 and c. 2014, many of the cities have become more fragmented as indicated by the increase in their respective PLANDs and PDs, and the decrease in their respective mean ENNs (Fig. 19.3). By contrast, some cities have become more aggregated. For example, Bangkok, Dakar, Manila, and Nairobi showed an overall increase in their respective PLANDs and decrease in their respective PDs between c. 1990 and c. 2014. Bamako and Yangon also showed an increase in their respective PLANDs and a decrease in their respective PDs from c. 2010 to c. 2014 (Fig. 19.3).

The fragmentation and aggregation patterns can help illustrate and provide evidence for the diffusion–coalescence urban growth theory mentioned in Sect. 19.1.2 of this chapter. Based on the diffusion–coalescence urban growth theory, the results seem to indicate that the more fragmented urban landscapes of many of the cities in c. 2014 were due to the diffusion of built-up patches since c. 1990. By contrast, the more aggregated urban landscapes of some of the cities were due to the coalescence of their built-up patches. The process of continuous diffusion and expansion can result in a sprawl development pattern, whereas the process of coalescence can result in an infilling growth pattern (Estoque and Murayama 2015).

The results also seem to indicate that the diffusion process that caused many of the cities to have a decreasing mean ENN also involved a type of an infilling growth pattern, a type where new 'infill' built-up patches are 'not yet connected.' The simulation results revealed that many cities would have a much higher PLAND and mean ENN, but much lower PD in 2020 and 2030 relative to c. 2014. This suggests that the 'not yet connected' new 'infill' built-up patches in c. 2014 would eventually coalesce and redefine the average distance between neighboring built-up patches. This process would eventually result in a more aggregated urban landscapes in the near future. For the various potential environmental implications of diffusion and coalescence, as well as sprawl and infilling patterns, the reader is referred to the literature (e.g., Dietzel et al. 2005a, b; Houck 2010; Brooks et al. 2011; McConnell and Wiley 2011; Wu et al. 2011; Estoque and Murayama 2015, 2016).

The intensity analysis technique provides understanding that cannot be directly grasped from the results of the spatial metrics alone, and vice versa. Thus, these geospatial techniques complement each other. Also, the integration of gradient analysis provides a platform, in which the ULCs in a particular city can be better compared with those of the other cities, either zone by zone or within a specific distance from the city center. For instance, within 20 km distance, it can be observed that Beijing, Manila, Bangkok, Johannesburg, Jakarta, and Dakar are among the most densely urbanized cities based on their respective proportions of built-up lands.

19.6 Concluding Remarks

In this chapter, we examined and compared the temporal and spatial patterns of ULCs in 15 major cities (metropolitan areas) in the developing Asia and Africa. Beijing, Bangkok, and Johannesburg had the highest gains of built-up in terms of area from c. 1990 to c. 2014, whereas Bamako, Dhaka, and Lilongwe had the highest percentage increase during the same time period. Five of the top seven cities in terms of total gain of built-up came from Asia, while four of the top seven cities in terms of total percentage came from Africa. Although some cities across Asia and Africa showed either stable or fluctuating intensities of ULCs, most of the cities had intensifying ULCs. During the same period, majority of the cities have become more fragmented. However, the simulated future urban LULC maps (2020 and 2030) indicated that their fragmented or diffused patches of built-up lands would eventually coalesce and result in more aggregated urban landscapes.

In bigger cities, ULCs are already occurring and moving away from the city center, whereas in smaller cities, ULCs are still largely concentrated in closer proximities to the city center. The bigger cities might have already undergone through the alternating process of diffusion–coalescence many times, whereas the smaller cities are still in the early stages of this cyclic process.

References

Aldwaik SZ, Pontius RG Jr (2012) Intensity analysis to unify measurements of size and stationarity of land changes by interval, category, and transition. Landscape Urban Plann 106:103–114

Angel S, Parent J, Civco DL, Blei A, Potere D (2011) The dimensions of global urban expansion: estimates and projections for all countries, 2000–2050. Progress Plann 75:53–107

Bagan H, Yamagata Y (2014) Land-cover change analysis in 50 global cities by using a combination of Landsat data and analysis of grid cells. Environ Res Lett 9:064015

Bloom DE, Canning D, Fink G (2008) Urbanization and the wealth of nations. Science 319: 772–775

Bokaie M, Zarkesh MK, Arasteh PD, Hosseini (2016) Assessment of urban heat island based on the relationship between land surface temperature and land use/land cover in Tehran. Sustain Cities Soc 23:94–104
Brooks N, Donaghy K, Knaap G (eds) (2011) The Oxford handbook of urban economics and planning. Oxford University Press, New York
Chen H, Pontius RG Jr (2010) Diagnostic tools to evaluate a spatial land change projection along a gradient of an explanatory variable. Landscape Ecol 25:1319–1331
Chen XL, Zhao HM, Li PX, Yin ZY (2006) Remote sensing image analysis of the relationship between urban heat island and land use/cover changes. Remote Sens Environ 104:133–146
Crutzen PJ (2002) Geology of mankind: the Anthropocene. Nature 415:23
Dahiya B (2012) Cities in Asia 2012: demographics, economics, poverty, environment and governance. Cities 29:S44–S61
Dietzel C, Herold M, Hemphill JJ, Clarke KC (2005a) Spatio-temporal dynamics in California's Central Valley: empirical links to urban theory. Int J Geogr Inf Sci 19:175–195
Dietzel C, Oguz H, Hemphill JJ, Clarke KC, Gazulis N (2005b) Diffusion and coalescence of the Houston metropolitan area: evidence supporting a new urban theory. Environ Plann B 32: 231–236
Dunde R (2015) Urbanization will change the (developing) world. http://www.forbes.com/sites/danielrunde/2015/02/24/urbanization-development-opportunity/. Accessed 3 Dec 2015
Estoque RC, Murayama Y (2013) Landscape pattern and ecosystem service value changes: implications for environmental sustainability planning for the rapidly urbanizing summer capital of the Philippines. Landscape Urban Plann 116:60–72
Estoque RC, Murayama Y (2014a) Measuring sustainability based upon various perspectives: a case study of a hill station in Southeast Asia. AMBIO: J Human Environ 43:943–956
Estoque RC, Murayama Y (2014b) A geospatial approach for detecting and characterizing non-stationarity of land change patterns and its potential effect on modeling accuracy. GISci Remote Sens 51:239–252
Estoque RC, Murayama Y (2015) Intensity and spatial pattern of urban land changes in the megacities of Southeast Asia. Land Use Policy 48:213–222
Estoque RC, Murayama Y (2016) Quantifying landscape pattern and ecosystem service value changes in four rapidly urbanizing hill stations of Southeast Asia. Landscape Ecol 31:1481–1507
Estoque RC, Murayama Y, Kamusoko C, Yamashita A (2014) Geospatial analysis of urban landscape patterns in three major cities of Southeast Asia. Tsukuba Geoenviron Sci 10:3–10
Estoque RC, Murayama Y, Myint SW (2017) Effects of landscape composition and pattern on land surface temperature: An urban heat island study in the megacities of Southeast Asia. Sci Total Environ 577:349-359
Folke C, Jansson A, Rockström J, Olsson P, Carpenter SR, Chapin FS III et al (2011) Reconnecting to the biosphere. AMBIO: J Human Environ 40:719–738
Gómez-Baggethun E, Gren A, Barton DN, Langemeyer J, McPhearson T, O'Farrell P et al (2013) Urban ecosystem services. In: Elmqvist T, Fragkias M, Goodness J, Güneralp B, Marcotullio PJ, McDonald RI et al (eds) Urbanization, biodiversity and ecosystem services: challenges and opportunities: a global assessment. Springer, Dordrecht, pp 175–251
Grimm NB, Faeth SH, Golubiewski NE, Redman CL, Wu J, Bai XM, Briggs JM (2008) Global change and the ecology of cities. Science 319:756–760
Houck MC (2010) In livable cities is preservation of the wild. The politics of providing for nature in cities. In: Douglas I, Goode D, Houck M, Wang R (eds) The Routledge handbook of urban ecology. Routledge, Abingdon, pp 48–62
Jansson A (2013) Reaching for a sustainable, resilient urban future using the lens of ecosystemservices. Ecol Econ 86:285–291
Kalnay E, Cai M (2003) Impact of urbanization and land-use on climate change. Nature 423:528–531

Kamusoko C (2017) Methodology. In: Murayama Y, Kamusoko C, Yamashita A, Estoque RC (eds) Urban development in Asia and Africa—geospatial analysis of metropolises. Springer Nature, Singapore, pp 11–46

Kamusoko C, Aniya M (2007) Land use/cover change and landscape fragmentation analysis in the Bindura District, Zimbabwe. Land Degrad Dev 18:221–233

Luck M, Wu J (2002) A gradient analysis of urban landscape pattern: a case study from the Phoenix metropolitan region of USA. Landscape Ecol 17:327–329

McConnell V, Wiley K (2011) Infill development: perspectives and evidence from economics and planning. In: Brooks N, Donaghy K, Knaap G (eds) The Oxford handbook of urban economics and planning. Oxford University Press, New York, pp 473–502

McGarigal K, Cushman SA, Ene E (2012) FRAGSTATS v4: Spatial pattern analysis program for categorical and continuous maps. Computer software program produced by the authors at the University of Massachusetts Amherst. http://www.umass.edu/landeco/research/fragstats/fragstats.html. Accessed 1 July 2015

Murayama Y, Kamusoko C, Yamashita A, Estoque RC (eds) (2017) Urban development in Asia and Africa— geospatial analysis of metropolises. Springer Nature, Singapore

Seto KC, Fragkias M (2005) Quantifying spatiotemporal patterns of urban land-use change in four cities of China with time series landscape metrics. Landscape Ecol 20:871–888

Seto K, Reenberg A (eds) (2014) Rethinking global land use in an urban era. Strungmann Forum Reports, vol 14, Lupp J, series editor. MIT Press, Cambridge

Seto KC, Fragkias M, Guneralp B, Reilly MK (2011) A meta-analysis of global urban land expansion. PLoS ONE 6:e23777

Solecki W, Seto KC, Marcotullio PJ (2013) It's time for an urbanization science. Environ Sci Policy Sustain Dev 55:12–17

Thapa RB, Murayama Y (2009) Examining spatiotemporal urbanization patterns in Kathmandu Valley, Nepal: remote sensing and spatial metrics approaches. Remote Sens 1:534–556

UN (United Nations) (2006) World urbanization prospects: the 2005 revision. United Nations, New York

UN (United Nations) (2015a) World urbanization prospects: the 2014 revision. United Nations, New York

UN (United Nations) (2015b) Transforming our world: the 2030 agenda for sustainable development. United Nations, New York

UN (United Nations) (2015c) The millennium development goals report 2015. United Nations, New York

Villamor GB, Pontius RG Jr, van Noordwijk M (2014) Agroforest's growing role in reducing carbon losses from Jambi (Sumatra), Indonesia. Reg Environ Change 12:825–834

Weng Y (2007) Spatiotemporal changes of landscape pattern in response to urbanization. Landscape Urban Plann 81:341–353

Whittaker RH (1975) Communities and ecosystems. MacMillan, New York

Wu J (2010) Urban sustainability: an inevitable goal of landscape research. Landscape Ecol 25:1–4

Wu J, Jenerette GD, Buyantuyev A, Redman CL (2011) Quantifying spatiotemporal patterns of urbanization: the case of the two fastest growing metropolitan regions in the United States. Ecol Complex 8:1–8

Chapter 20
Future of Metropolises in Developing Asia and Africa

Yuji Murayama and Ronald C. Estoque

Abstract This chapter discusses the future of metropolises in the developing Asia and Africa in the context of five themes: (1) urbanization in developing countries; (2) metropolitanization; (3) nodes in the national urban system; (4) nodes in the world urban system; and (5) developing methodology and future research challenges. First, the metropolises of developing Asia and Africa are experiencing rapid urbanization, giving rise to an urban population explosion. Second, as the metropolitan area expands, the spatial pattern gradually changes from a unipolar structure, where the city center is the urban core, toward a multipolar structure, which incorporates peripheral areas. Third, when considering the country's economic growth as a whole, concentrating economic functions in the capital appears to be the most efficient strategy in the short term. However, excessive concentration in the capital can widen the gap between the capital and local areas, which in turn affects regional cities through a straw effect. Fourth, the top cities around the world today that serve as the nodes in the world urban system are now referred to as world cities. And fifth, remote sensing and GIS are among the emerging geospatial tools and techniques today for advancing urban studies in the geographical context.

20.1 Urbanization in Developing Countries

The metropolises of developed countries have started losing population, and the counter-urbanization phenomenon is becoming widespread, particularly in Western countries. The results of the 1970 United States Census presaged this phenomenon. Today, the metropolises of the West are in an age of population decline.

On the other hand, the population explosion in developing countries shows no sign of abating, and the concentration of the population in metropolises, particularly

Y. Murayama (✉) · R.C. Estoque
Faculty of Life and Environmental Sciences,
University of Tsukuba, Tsukuba, Japan
e-mail: mura@geoenv.tsukuba.ac.jp

Y. Murayama et al. (eds.), *Urban Development in Asia and Africa*,
The Urban Book Series, DOI 10.1007/978-981-10-3241-7_20

capital cities, continues. Since the turn of the twenty-first century, the developing countries of Asia and Africa have exhibited remarkable economic growth, and migration from rural areas to metropolises has been gaining momentum. While there continues to be a labor glut in rural areas, metropolises offer growing employment opportunities due to the march of industrialization and the service economy. The economic gap between rural areas and metropolises is widening year by year. This gap is strengthening the push-pull effect and generating a one-way flow toward metropolises. The reverse of this flow (migration from metropolises to regional cities or urban areas) is scarcely observed. Young people make up a large portion of the population that moves from rural areas to metropolises. When these young people find a job in the city, they marry and have children. In many cases, when they secure a stable livelihood, they invite their friends and family, and such chain migration is a driving force behind population concentration. Thus, the metropolises of developing countries are experiencing social and natural increases simultaneously, giving rise to an urban population explosion. It is anticipated that this trend will continue until the late twenty-first century.

The rapid population concentration in metropolises has been causing deterioration in the residential/living environment. Chronic gridlock has led to economic inefficiencies, including a decline in mobility and delay in distribution. Atmospheric pollution and water pollution are on the increase. There is also an increasing risk of contracting infectious diseases. It is anticipated that health-related problems will increase. Thus, it is apparent that urban problems will become more serious.

The vibrant economic activities in cities that are experiencing rapid growth are contributing to the centralization of energy consumption, leading in turn to the urban heat island phenomenon. Localized torrential downpours, flooding, landslides, and the like have become a frequent occurrence in Asia. Urbanization in monsoon Asia is removing abundant green spaces, prompting fears of a decline in urban ecosystem services.

With increasing population size and density due to urban growth, cities are becoming more vulnerable to natural disasters, increasing the risk of major damage. Prompt action is necessary for cities to achieve sustainable development. Land use regulations and other legislation must be swiftly developed as appropriate with a view to conserving green areas, ensuring a safe and secure society, and achieving orderly urban growth.

20.2 Metropolitanization

In regions with undeveloped transport and information networks, low-income groups tend to live near the places of employment. Minimizing commuting time allows extra time to be spent in work and also saves transport costs. Those in the poorest population are forced to live next to their place of work; as a result, there are many who take up residence in the slums in the vicinity of city center. As a dense mesh of factories and workers' housing spreads out into the suburbs, the suburbs too become marked by poor-quality housing and, in some cases, slums. Slums tend to appear in areas that are poorly suited for human habitation such as riverbanks and wet zones.

In the urbanized areas of developing countries, residents live together in a small crowded area, leading to the formation of densely populated urban spaces. However, in recent years, migration from city center to the suburbs has become the prevailing trend, and an increasing number of cities are experiencing suburban dispersal of population. In a number of metropolises in Southeast Asia, one can observe the linking of the city center and suburbs through highways, and commuter rail, underground rail, and monorail services, leading to the formation of expansive metropolitan areas. Metropolises such as Manila, Bangkok, Hanoi, and Jakarta have systematically constructed suburban spaces for offices, large commercial facilities, and new residential areas, and as a result, the suburbs of these cities now form the core of urban development. In the peripheries of suburban cores, labor-intensive factory complexes have been built, and the number of houses for low-income workers is continuously increasing. In the urban frontier zones, there has been an emergence of un-tilled land and a forest of warehouses and storage sites. Many farmers are abandoning farming. The rise of factories and service industries has increased employment opportunities in the suburban cores and their peripheries, and so employment in such industries becomes a better source of high income than farming. Thus, in order to avoid chaotic development and slum formation in the urban frontier zones, it is necessary to develop land use legislation during the initial stage of metropolitan expansion.

As the metropolitan area expands, the spatial pattern gradually changes from a unipolar structure, where the city center is the urban core, toward a multipolar structure, which incorporates peripheral areas. Developed countries already experienced this development in the late twentieth century. The cities of developed countries have a plethora of transport routes that extend from the city center in a radial pattern. This prompts residential areas to emerge in all directions from the center to the periphery and star-shaped urban spaces to form along transport routes. Developing countries, however, do not have this myriad of transport routes linking the center and periphery. In many cases, the urban area expands along transport routes in a fragmented fashion, resulting in irregular urban spaces. The process and pace of metropolitan expansion and decentralization vary depending on the stage of economic development. Metropolitan expansion is already well underway in Southeast Asia, but it is still in its early days in Africa, particularly sub-Saharan Africa.

20.3 Nodes in the National Urban System

The primate cities (largest city in country) among developing countries function as nodes in the national urban systems. In many cases, this role is played by the country's capital, the political, economic, and cultural center. Since the turn of the twenty-first century, the capital cities of the developing countries have increased their economic influence over other cities within the nation year by year, resulting in increased urban primacy (ratio of primate city to next largest city). For example, when multinational companies start new business operations in a host country, they tend to locate the regional headquarters in the capital, where there is a concentration of financial and informational functions. For factory production activities, they tend to position a factory in the capital's suburbs. When launching a new manufacturing operation, it is important to consider land, labor, electricity, transport, and water resources. Generally speaking, these conditions are better in the capital's suburbs than in provincial areas. Transport facilities and services are important considerations when erecting a factory. With government backing, traffic routes in the capital suburbs are usually developed to facilitate the transportation of resources and products.

When considering the country's economic growth as a whole, concentrating economic functions in the capital appears to be the most efficient strategy in the short term. However, excessive concentration in the capital can widen the gap between the capital and local areas, which in turn affects regional cities through a straw effect. From a medium- to long-term standpoint, countries run the risk of ending up with an unbalanced national urban system.

The rapid economic development of capital cities can also lead to the kind of social polarization already witnessed in developed countries (increasing gap between rich and poor, and weakening of the middle class) (Murayama 2000). There is concern that the economic benefits will go only to a handful of local capitalists, entrepreneurs, and foreign elite business people, prompting social friction between them and the poor, who support their economic activities and lifestyles. What the authors wish to emphasize here is that the central governments of the developing countries must devise thoroughgoing national land policy that can rectify over-concentration in the capital, prevent excessive urbanization, and promote regional development.

20.4 Nodes in the World Urban System

Economic globalization and the march of information technology have increased the interdependence of metropolises around the world, such that people are now witnessing the formation of a world urban system interlinking cities throughout the world. The socioeconomic connections of regions of the world and the growth or decline of metropolises are governed less by the power of cities themselves and more by the strength of their relationships with other urban centers. The trend is occurring against the backdrop of expanding trade, increasing migration, global movements of workers, flows of capital, increasing political interdependence, increasing foreign tourists, and deepening ecological crises that transcend national borders (UN 2015). We can expect this trend to grow in intensity in the future.

The top cities around the world that serve as the nodes in the world urban system are now referred to as world cities. At the end of the twentieth century, three cities claimed this title: London, New York, and Tokyo. However, "semi-world cities" have now emerged and started playing a role in the world urban system. These cities are starting to serve as core nodes in a world urban system that is based on the interconnections between and among countries in a continent or region. In Europe, for instance, Paris, Berlin, and Amsterdam are examples of semi-world cities, and as such, they play the role of core cities in the EU zone. In Asia, one could cite Singapore or Hong Kong; these cities play key roles in the ASEAN bloc.

Today, the cities of developing countries are being incorporated into the world urban system, in which the largest ones now serve as important nodes. The cities of Asia and Africa in particular are becoming increasingly economically dependent on the cities of developed countries and mutually dependent on the cities in neighboring countries. The incorporation of developing countries' metropolises into the world urban system as key nodes denotes that these cities have acquired an opportunity to achieve economic development; however, at the same time, it denotes that they have been thrust into a fierce arena where they must hold their own against competing cities around the world. If they get their economic strategy wrong, they will suddenly plunge into decline. To ensure sustainable development, cities must pursue strategic urban management. To this end, city mayors and municipal authorities have an important role to play.

20.5 Developing Methodology and Future Research Challenges

Analyzing the plethora of metropolises in developing countries by quantitatively comparing their metropolitan expansion processes and making future predictions using standard criteria constitutes a considerable challenge. It is challenging because there is much less geospatial information of cities in the developing countries than in the developed countries. Census reports are a valuable source of data that enables scientific analysis. However, census years and the details of information usually vary across countries. There are also many countries that have not conducted a census for more than 10 years. The statistical units also vary by country, and only a few countries have available local statistics and boundary maps that can be used for spatial and temporal analysis.

Under these circumstances, satellite images have become a prevailing data source in urbanization studies among developing countries. Recently, medium- to high-resolution satellite imagery from Landsat, SPOT, ALOS, IKONOS, and QuickBird have become available for free or at minimal cost, and technologies such as geographic information system (GIS) and remote sensing (RS) can now be used to facilitate highly precise temporal and spatial analysis. There is a formidable accumulation of satellite images, particularly in the case of Landsat, which has global image data stretching back to the 1970s, and this offers considerable utility because it enables time series analysis of many metropolises based on standard criteria.

Researchers are developing and testing GIS and RS methods to categorize land use/cover and deduce spatial patterns (Murayama and Thapa 2011; Murayama 2012). The use of elaborate techniques such as random forest classification and support vector machine classification has enabled researchers to examine urban land use patterns and changes more accurately. Recently, researchers have produced valuable findings regarding urban land use/cover change prediction by applying compound science, including cellular automata and agent-based modeling. Such research is expected to make a considerable contribution toward the formulation of a master plan for urban development.

However, the urbanization outlook for the cities of developing countries includes many more uncertainties compared to that in developed countries; therefore, it is important to point out that the inductive reasoning used in this book, whereby the process of urbanization to date is used as evidence to make predictions about the future of urbanization, might not be watertight. Many of the countries in Asia and Africa are politically unstable. Political instability has a direct impact on the economy and residents' lives, and it imperils sustainable urban development. Furthermore, the occurrence of sudden events can wreak economic and social havoc in cities, owing to the vulnerability of the urban infrastructure. The larger a city is, the more vulnerable it is to droughts, floods, and other natural disasters, and the more time it will take to recover. The framework used in this book cannot adequately account for all these uncertainties; therefore, it is necessary to develop new models for spatial scenarios by applying the science of complexity.

As a final note about the methodological issues in research going forward, the reader is enjoined to focus his/her attention to the necessity of promoting research on the vertical development of urbanization. A notable development in recent years is the construction of high-rise apartment buildings and office blocks in the centers of Asian and African metropolises. This book was unable to analyze or discuss such vertical expansion of urban land use. LiDAR technology is now able to measure the heights of buildings and trees, and it is now possible to obtain elevation data automatically using satellite imagery from ALOS and other satellites. The application of such technology will make it easier for researchers to obtain height data by calculating the difference between the elevation values obtained from a digital surface model (DSM) and a digital terrain model (DTM). This approach will enable an all-round assessment of urbanization process, taking into consideration the horizontal and vertical aspects of urban development.

References

Murayama Y (ed) (2000) Japanese urban system. Kluwer Academic Publishers, Dordrecht

Murayama Y (ed) (2012) Progress in geospatial analysis. Springer, Tokyo

Murayama Y, Thapa RB (eds) (2011) Spatial analysis and modeling in geographical transformation process: GIS-based applications. Springer, Dordrecht

UN (United Nations) (2015) World urbanization prospects: the 2014 revision. United Nations, New York

Index

Y. Murayama et al. (eds.), *Urban Development in Asia and Africa*,
The Urban Book Series, DOI 10.1007/978-981-10-3241-7